D1383683

FEDERAL WORKS AGENCY
LIBRARY

SURPLUS - 1
LIBRARY OF CONGRESS
DUPLICATE

GEOLOGY AND ENGINEERING

FEDERAL WORKS AGENCY
LIBRARY

"Nature, to be commanded, must be obeyed."

—FRANCIS BACON.

Wm Smith
Civil Engineer

15 Buckingham Street
York Buildings
London

June 1818

This portrait of "The Father of British Geology"—who was also an eminent civil engineer—is reproduced as a tribute to his memory, and as a fitting introduction to this study of what was his life's work, more than a century ago, work which is still an inspiration to geologists and to civil engineers in many countries of the world.

(The portrait is reproduced by permission of the Council of the Geological Society of London; the signature, by permission of Mr. T. Sheppard, author of "William Smith; His Maps and Memoirs.")

QE
33
.L34
c.2

GEOLOGY
AND
ENGINEERING

BY

ROBERT F. LEGGET, M.Eng. (L'pool)

Assistant Professor of Civil Engineering, The University of Toronto; Associate Member, Institution of Civil Engineers; American Society of Civil Engineers; Engineering Institute of Canada

WITH A FOREWORD BY

P. G. H. BOSWELL, D.Sc., F.R.S.

Imperial College of Science and Technology, London

First Edition
Second Impression

FEDERAL WORKS AGENCY
LIBRARY

McGRAW-HILL BOOK COMPANY, Inc.
NEW YORK AND LONDON
1939

COPYRIGHT, 1939, BY THE
McGRAW-HILL BOOK COMPANY, INC.

PRINTED IN THE UNITED STATES OF AMERICA

All rights reserved. This book, or parts thereof, may not be reproduced in any form without permission of the publishers.

35624

THE MAPLE PRESS COMPANY, YORK, PA.

To

MY WIFE

whose continued encouragement and expert assistance almost warrant the dispelling of this anonymity by the appearance of her name, jointly, on the title page

GEOLOGY

may be defined as that branch of natural science devoted to the study of the physical features of the earth, the composition and structure of the rocks composing it, the forces at work in altering it, and the record of the animals and plants that have lived on its lands and inhabited its seas.

CIVIL ENGINEERING

was originally the counterpart of military engineering, its origins thus rooted in the depths of antiquity. One of the first attempts to define its scope was embodied in that splendid phrase still found in the charter of the Institution of Civil Engineers—"the art of directing the great sources of power in Nature for the use and convenience of man." A more modern description of the art is that it includes the design and construction of all structures other than simple buildings, and the investigation, design, and construction of all systems of transportation, of natural power development, of water supply and sewage disposal, as well as the direction of natural forces for the use and convenience of man.

GEOLOGY AND CIVIL ENGINEERING

It is with the application of the science of Geology to the art of Civil Engineering that this volume is concerned. To this end, a brief outline of the science is presented as Part I, preceding a review of the application of geological studies to the several main branches of civil engineering involving geological problems. Part III brings the volume to a close by recording in convenient form reference data which may be useful to the civil engineer engaged in any geological study. Part I may be omitted by all to whom the fundamentals of Geology are known, for it is no more than an introduction to the science. Part III is purely a reference section. Part II constitutes the principal part of this work.

It is hoped that, in addition to being of service to practicing engineers, this book will prove useful as a text for students. Instruction in Geology, in the usual college course, generally precedes the student's introduction to civil engineering practice. The consequent inadequate appreciation of the relevant value of Geology is unfortunate. It can be obviated only by enabling students, while they are studying pure Geology, to see evidence of its applications. This book is an attempt to provide such evidence in convenient form. If the book is thus used for engineering students, Part I will serve merely as the framework for more detailed notes, or as an incentive to the use of texts on Geology, study of which will help to provide the background essential to the full appreciation of Part II.

FOREWORD

By Professor P. G. H. Boswell, D.Sc., F.R.S.
(*Imperial College of Science and Technology, London*)

In the realm of science, as in other spheres, background is an important factor. Its influence is especially apparent in borderline subjects, among which engineering geology may be properly placed. There are already several textbooks on this branch of geology—works which present conscientious expositions of geological principles, with appropriate expansions in particular directions; but they have been written by geologists and their background is obvious. Where engineering applications are discussed, in particular, they indicate what the *geologist*, from his rather different viewpoint, imagines that the engineer should know.

My own approach to the wide domain of engineering geology has been along geological paths, but such prospects of that interesting territory as my experience has afforded me have made me wish that it could be viewed through the eyes of an engineer.

When, therefore, I had the pleasure of reading the manuscript of Professor Legget's book, I recognized at once that this was the picture that I had long envisaged. This book is the work of an engineer with the additional training of a geologist, and I am convinced that it is planned on the right lines because it has the proper background. The necessary viewpoint can be attained only with great difficulty by a geologist, even when experienced as a consultant on engineering problems. The present volume will therefore fill a long-felt want, and may well serve as a foundation for other works on similar lines.

At present it is not unusual for a geologist to be asked to give advice after the engineer has begun his work. He has then to assist, or even to defend, a course of procedure that is already in progress—a task that is unnecessarily difficult because it involves the solution of problems that need not have arisen if he had been consulted in the first instance. Geological problems arise in most of the large-scale engineering undertakings, but their full significance is not always apparent in the early stages. In such stages, careful records of natural conditions, such as overground and underground water, made during the progress of the work, would prove of inestimable value. Once man's handiwork has begun, the natural balance is disturbed, and it is then by no means easy to gain an accurate knowledge of original conditions. In many instances within my own experience I would have given much to know them, for only a knowledge of such conditions can enable a geologist to apprehend the problems that are likely to arise and to gauge the ultimate effects of the completed works.

An engineer trained in geology should be able to meet most of the difficulties that spring from earth structure, rock constitution, and water supply, and should also (which is really important) be able to decide when it is necessary to consult an expert. It is well known that adequate geological investigation of natural conditions in engineering undertakings in the past would have resulted in the saving of human life by the prevention of disasters; it has frequently been the means of effecting a considerable reduction in the cost of engineering works and of avoiding the trouble and expense of litigation.

ACKNOWLEDGMENTS

It is difficult to express adequately my appreciation of all the assistance that I have received in the preparation of this book. In the early stages of the work, I received great encouragement from Dr. J. J. O'Neill of McGill University; and as it neared completion, Dr. C. P. Berkey favored me with his advice and criticism; to both these distinguished geologists I express my thanks. From Mr. F. J. Fraser, of the Geological Survey of Canada, I have received continuous and invaluable assistance, mention of which in this place—with grateful appreciation—cannot be restrained even by the ties of close personal friendship.

It would, however, be an impossible task for me to list all the engineers, geologists, librarians, and public officials who have so generously answered my queries and allowed me to use the valuable information that they have provided. In the case of photographs, acknowledgment of their source is made in the text. In every case of assistance so kindly given, I have attempted to express my thanks by personal communication, but I hope that this general and public acknowledgment of my gratitude will be accepted by all who have so helped me as if it applied individually to each one of them. Despite the care with which this information and assistance have been used and with which all references have been checked, it is possible that some errors still remain in the finished work. If those who note mistakes and omissions in the text will advise me of them, I shall be grateful, as I shall be also for any suggestions for amendments or improvements to what is here presented. Should any information have been used without due acknowledgment or without application for permission to the proper authorities, I ask those concerned to accept my apologies and my assurance that any such error has been made unwittingly.

Only by reason of this generous cooperation has it been possible to provide in this book practical illustrations of all the main applications of Geology to civil engineering work. By singular good fortune, I have been able to see many such applications during my daily contact with civil engineering throughout more than a decade prior to taking up academic duties, but my own experience is indeed limited compared with that necessary for the provision of a full range of illustrative examples for this book.

I have been further assisted by the valuable comments and criticisms of personal friends and others who have examined parts of the manuscript during its preparation. In this connection, I should like to record my particular indebtedness, for the time and attention so generously given in this way, to Professor M. B. Baker, Mr. S. F. Kelly, Dr. George E. Ladd, Mr. J. W. Lucas, Mr. G. H. Matthes, Dr. O. E. Meinzer, Dr. H. H. Read, Mr. E. Viens and Dr. John Wyllie. Finally, it is a special pleasure to acknowledge my indebtedness to Mr. W. T. Halcrow of Westminster, to whom I shall always be grateful for the first years of my engineering training and for the use that he has allowed me to make since then of much of the valuable experience gained while associated with him.

My interest in Geology was naturally aroused by my first instruction in it, instruction that enabled even an impatient engineering student to realize that this branch of pure science was intimately associated with the practical work of his chosen profession. For that initial inspiration, for continued encouragement in all my ensuing studies of the subject, and for the suggestion that this book should be written, I am indebted to Professor P. G. H. Boswell, F.R.S., who has done me the honor of examining the whole of the book in manuscript form. To Professor Boswell I owe a debt that I cannot easily repay, save only by work in the field described in this book, a field in which I hope long to be an inquiring student.

ROBERT F. LEGGET.

TORONTO, ONT.,
June, 1939.

CONTENTS

PAGE

EXPLANATORY NOTE ix

FOREWORD BY P. G. H. BOSWELL xi

ACKNOWLEDGMENTS xiii

PART I

AN INTRODUCTION TO GEOLOGY

CHAPTER I

GEOLOGY: AN OUTLINE OF THE SCIENCE 3

Introduction—Early History—Some Pioneers—William Smith—The Science Today—Methods of Investigation—Past Achievements and the Future—Applied Geology—Conclusion.

CHAPTER II

THE COMPOSITION OF THE EARTH'S CRUST 11

Introduction—Rocks—Minerals—Main Rock Groups—Igneous Rocks—Sedimentary Rocks—Metamorphic Rocks—Distinguishing Rock Types.

CHAPTER III

THE STRUCTURE OF THE EARTH'S CRUST 25

Introduction—The Geological Cycle—Special Structural Features—Sedimentary Rock Bedding—Igneous Rock Features—Jointing—Folding—Faulting—Denudation—Unconformity—The Geological Succession—Palaeontology—Recent Changes.

CHAPTER IV

GEOLOGICAL FIELDWORK AND MAPPING 41

Introduction—Geological Maps—Published Geological Maps—Topographic Maps—Field Equipment—Field Observations—Drift and Solid Geological Maps—Special Aids to Fieldwork—Method of Recording—Geological Sections—Interpretation of Geological Maps—Fossils—Photography.

PART II

GEOLOGY AS APPLIED IN CIVIL ENGINEERING

CHAPTER V

THE CIVIL ENGINEER AND GEOLOGY 61

Introduction—The Science and the Art—Training in Geology—Practical Experience—Employment of Specialists—Geologists and Civil Engineer-

PAGE

ing Work—Cooperation of Geologists and Civil Engineers—Geology and Topography—Planning and Design—Contract Plans and Specifications—Earth and Rock Excavation—Construction Operations—Inspection and Maintenance—Conclusion.

CHAPTER VI

PRELIMINARY AND EXPLORATORY WORK 85
Introduction—Economics of Preliminary Work—Geological Surveys—Underground Exploratory Work—Exploratory Test Borings—Test Pits—Test Drilling—Special Exploratory Methods—Conducting Exploratory Work—Recording Exploratory Work—Interpretation of Exploratory Results—Utilization of Exploratory Results—Conclusion.

CHAPTER VII

APPLIED GEOPHYSICS AND CIVIL ENGINEERING 115
Introduction—Magnetic Methods—Seismic Methods—Gravitational Methods—Electrical Methods—Applications in Civil Engineering—Geophysical Methods and Groundwater—Conclusion.

CHAPTER VIII

TUNNELS 139
Introduction—Historical Notes—A General Note—Preliminary Work—Tunnels for New York—Tunnel Shapes and Linings—Pressure Tunnels—Construction Methods—Overbreak—Groundwater—Grouting—Construction Records—Conclusion.

CHAPTER IX

OPEN EXCAVATION, EMBANKMENT FILLS AND RETAINING WALLS 182
Introduction—Open Excavation—Economics of Open Excavation—Open Excavation of Rock—Open Excavation of Unconsolidated Material—Open Excavation in Water-bearing Strata—Embankment Fill and Dredging—Quarrying in Civil Engineering Practice—Earth Pressure and Retaining Walls.

CHAPTER X

EARTH MOVEMENT AND LANDSLIDES 211
Introduction—Stability of the Earth's Crust—Earthquakes—Ground Subsidence—Landslides—Preventive and Remedial Work—The Stability of Earth Slopes—Rockfalls.

CHAPTER XI

TRANSPORTATION ROUTES 247
Introduction—Route Location—Road-building Materials—Road Construction—Railways—Rivers—Canals—Docks and Harbors—Airports.

CHAPTER XII

THE FOUNDATION OF BRIDGES AND COFFERDAM PROBLEMS 270
Introductory Historical Note—The Importance of Bridge Foundations—Special Preliminary Work—The Design of Bridge Piers—The Design of Bridge Abutments—Some Special Features in Design—Some Construction Requirements—Cofferdam Construction—Inspection and Maintenance.

CHAPTER XIII

PAGE

THE FOUNDATION OF DAMS 304

Introduction—Historical Notes—Dam Failures—Review of Dam Construction—General Preliminary Work—Exploratory Work during Construction—Soundness of Bedrock—Possibility of Earth Movement—Permeability of Bedrock—Grouting of Foundation Beds—Dams on Permeable Foundation Beds—Materials of Construction—Construction Problems—Scouring below Spillways—Inspection and Maintenance.

CHAPTER XIV

RESERVOIRS AND CATCHMENT AREAS 353

Introduction—Leakage from Reservoirs—Detection of Leakage—Prevention and Elimination of Leakage—Secondary Effects of Reservoir Flooding—Catchment Areas—Relation of Geology to Run-off.

CHAPTER XV

EROSION AND SILTING 369

Introduction—Coastal Erosion—Littoral Drift—Erosion by Stream Flow—Transportation of Solid Material by Stream Flow—Silting up of Reservoirs—Erosion below Dams—River-training Problems—Silting of Deltas and Maintenance of Estuaries—Soil Erosion and Wind Action.

CHAPTER XVI

WATER SUPPLY 414

Introduction—Historical Note—Sources of Water Supply—Water Supply from River Flow—Water Supply from Impounding Reservoirs—Unusual Sources of Supply—Quality of Surface Waters.

CHAPTER XVII

GROUNDWATER 428

Introduction—Historical Note—Groundwater Characteristics—Influence of the Nature of Rocks—Quality of Groundwater—Influence of Geological Structure—Springs—Groundwater Surveys—Underground Dams—Wells and Boreholes—Salt Water in Wells—Artesian Wells—Some Outstanding Groundwater Supply Systems—Underground and Surface Drainage—Replenishment of Groundwater—Conservation of Groundwater.

CHAPTER XVIII

BUILDING FOUNDATIONS 474

Introduction—The Influence of Geological Conditions on Foundation Design—Subsidence of Buildings—The Influence of Groundwater on Foundation Design—Preliminary and Exploratory Work—Cooperative Subsurface Investigations—Building Regulations—Some Unusual Foundation Bed Conditions.

CHAPTER XIX

MATERIALS OF CONSTRUCTION 505

Introduction—Earth Dams—Puddle Clay—Bricks—Tile and Other Structural Clay Products—Building Stone—Limes and Plasters—Cements—Sands—Crushed Rock—Concrete.

CHAPTER XX

PAGE

SOILS AND SOIL MECHANICS 538

Introduction—Historical Note—Soil Mechanics—The Origin of Soils: Rock Weathering—Residual Soils—Transported Soils—Glacial Action and Glacial Deposits—Soil Characteristics—The Future of Soil Studies.

PART III

REFERENCE SECTION

APPENDIX A

GLOSSARY OF GEOLOGICAL TERMS COMMONLY ENCOUNTERED 587

APPENDIX B

GEOLOGICAL SURVEYS OF THE ENGLISH-SPEAKING WORLD 596

APPENDIX C

GEOLOGICAL SOCIETIES AND PERIODICALS 614

APPENDIX D

REFERENCES CITED IN THE TEXT 616

INDEX 633

PART I

AN INTRODUCTION TO GEOLOGY

CHAPTER I

GEOLOGY: AN OUTLINE OF THE SCIENCE

Nature will answer faithfully if we interrogate her—but not if we interrogate ourselves. CAPTAIN E. K. CALVER, R.N.[1.1]

The geologist is concerned with every aspect of the composition and structure of the earth's crust. His sphere of work is therefore world-wide; his main laboratory is the great out-of-doors wherein he examines rocks as they actually occur in nature. His considerations range to the beginning of time and into the future, even to that time when man will no more be known on the earth. He studies all that composes the crust of the earth sphere and especially those materials of use to his fellow men. Yet geologists were for long generally regarded as unduly academic persons, engaged in an unceasing hunt for fossils.

The paradox is due perhaps to the extent of the field covered by the science of Geology and to its relative remoteness from the things of every day, that is, when considered from the purely scientific standpoint. In a more general way, and quite unrecognized as such, geological appreciation is widespread, the mason, the quarryman, and the gardener being but a few of those whose very tasks bring them into daily touch with the earth and its constituent materials. Even in common speech, such phrases as "firm as a rock" are reminders of the basis of all geological study.

1.2. Early History.—This strange paradox, the contrast between the wide extent and importance of all that is now studied in the field of Geology and the generally restricted appreciation of its significance, is found to exist from the earliest days of scientific endeavor. The idea of devoting a special branch of science to the study of the earth itself was not then contemplated, and even the references in early writings to "Earth" as one of the four elements bear little relation to the modern concept of Geology. Nicolaus Steno (1631–1686), at one time Bishop of Hamburg and Vicar Apostolic of Denmark, is generally regarded as the founder of Geology as an independent branch of science. Of special interest to engineers is the fact that Robert

[1.1] All references thus noted will be found listed in Appendix D, under chapter numbers.

Hooke (1635–1703), who was Professor of Geometry at Gresham College when he was thirty, and whose name is now associated with the law relating stress and strain, was the first man to suggest that fossils could be used to construct a chronology of the earth. The name *Geology* does not appear to have been used until the end of the eighteenth century. It is derived from the Greek *ge*, the earth, and *logos*, a speech or discourse, the name thus being truly descriptive. When once it was recognized as a branch of scientific inquiry, it developed rapidly, and it was not long before the true extent of its scope was appreciated. The wide field thus made available for study encouraged, rather than discouraged, interested investigators, with the result that Geology today, despite its late start, ranks as a leading branch of natural science.

1.3. Some Pioneers.—The human aspect of the development of the science is full of interest, the pioneers of Geology including many famous characters. The international nature of any reference to these early workers is in itself significant and a just reminder of the true internationalism of science. The first workers were mainly English and Italian, but other European nations soon made their special contributions to the advance. J. G. Lehmann, a German who died in 1767, was the first, for example, to record the possibility of *order* in the arrangement of the rocks of the earth's crust. De Saussure (1740–1799), a Swiss who studied the Alps carefully, was the first to give the generic name to the science; previously it had been regarded as a part of mineralogy, a position that today is reversed.

James Hutton (1726–1797), a Scotsman of Edinburgh, was perhaps the first great name in the annals of the science; his "Theory of the Earth," published in 1785, republished in 1795, and finally popularized by Playfair (1748–1819) in 1802, provided the basis for much of the great advance made during the nineteenth century. Hutton first distinguished the three main types of rock which provide the basis for modern theories of earth sculpture. A. G. Werner (1749–1817), a German, was also a most potent influence in Geology through these same years, drawing pupils to his classes at Freiberg Mining Academy from all over the world. W. Maclure (1763–1840) is known today as the "Father of American Geology," and Sir William Edmund Logan (1798–1875) will long be remembered for his pioneer work in Canada. Agassiz (1807–1873) is yet another name associated with the North American Continent; he was a Swiss who emigrated to the United States of America when he was forty-two years old and there established his fame with his work on fossil fishes. He is generally regarded as the founder of the modern school of Glacial Geology. William

Nicol, a lecturer at the University of Edinburgh, by his invention of the Nicol prism and of the method of preparing thin rock sections for examination through a microscope (later to be developed by Sorby), placed the whole science in his debt and laid the basis for much of the laboratory work of today. The names of Sir Charles Lyell (1797–1875) and Charles Darwin (1809–1882) cannot be omitted from this review, brief as it is; both were distinguished geologists, although Darwin is perhaps better known as a biologist. Their publications, notably "The Antiquity of Man" and the "Origin of Species," achieved fame far beyond the confines of the scientific world.[1,2]

1.4. William Smith.—One distinguished name has been omitted from the foregoing list so that it may be given special mention, that of the remarkable man whose portrait forms a frontispiece to this volume. William Smith is now generally acknowledged as the "Father of British Geology." Born on March 23, 1769, of humble origin, he spent an appreciable part of his life of fifty years as a canal engineer in Somerset, England, and became later a land agent near Scarborough. The science was well founded when he first became interested in the structure of the earth, but in his earlier days he seems to have worked independently of other students of Geology and developed the basis of his wide knowledge by original observation, especially on his journeys along the post roads of England. He was one of the first to introduce the concept of quantitative study and regular stratigraphical succession and so helped to establish the science on its foundation of today. The work for which he will ever be remembered was the preparation of the first real geological map of England and Wales.

The "Map of the Strata," as it was first called, was constantly in his mind for a period of over twenty years, but it was not until the year 1813 that the work was definitely put in hand, William Cary being the engraver. The finished map, colored not unlike a modern geological map, was presented to the Society of Arts in April, 1815, being awarded a premium of £50 ($250), designated for the first such map to be produced. Throughout his life, William Smith practiced extensively as a consultant on geological matters, often with special reference to their application to engineering work. Before his death, he was honored by the Geological Society of London by being awarded in 1831 its first Wollaston Medal. The Society had been founded in 1807, proof indeed that the comparatively new science was already active and virile. It is of significance to note that the Society was thus founded some years before the Institution of Civil Engineers and almost fifty years before the American Society of Civil Engineers.

1.5. The Science Today.—Despite its tardy recognition, Geology has made great headway in the last two centuries. It is now primarily a quantitative rather than a descriptive science; it is studied alike in the laboratory and in the field, and modern instruments and methods are available to aid the geologist in his work. Subdivision of the subjects of special study has been inevitable; the present status of these subdivisions is shown diagrammatically in Fig. 1.1. In this diagram, the close relation of geologic study with other branches of scientific work is indicated, although it will be realized that the main contacts only are noted, the various subdivisions being intimately associated in many minor ways.

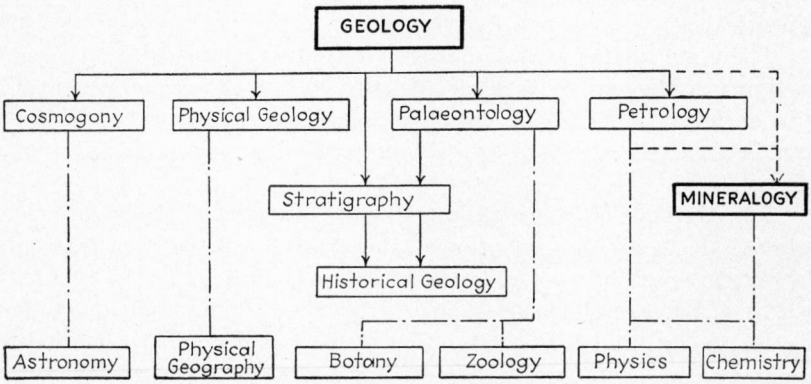

FIG. 1.1.—A diagram illustrating the main branches of geology, and their correlation with other branches of science.

It will be convenient to consider briefly the nature of the work covered by each of the subdivisions that are suggested.

1. *Cosmogony.*—This is the study of the relation of the earth to the solar system and to the universe, with special reference to its material composition; it considers also the origin and early stages of the terrestrial sphere.

2. *Physical Geology.*—This comprises the study of the present form of the earth's surface, its structure, manner of origin, and the nature of the modifying processes at work upon it being, in effect, an extension of the study more generally known as *Physical Geography.*

3. *Palaeontology.*—The study of ancient life as recorded by fossils, with special reference to the time relationships of the genera whose remains are frequently preserved in rocks.

4. *Stratigraphical Geology* (or *Stratigraphy*).—This division deals with the history of the earth and its geography in past ages by considerations of present structural relationships of the component materials

in the earth's crust and of their fossil contents. The division is by some considered to consist of two sections, historical geology and structural geology, the latter being the study of existing structural relations.

5. *Mineralogy and Petrology.*—Mineralogy deals with the crystal form, chemical composition, physical properties, and occurrence of minerals, which are the native elements and relatively pure chemical compounds constituting the inert mass of the earth and the meteoric bodies which come in contact with it. Petrology, in the widest sense, is concerned with the mineral constitution, texture, and origin of rock bodies. The descriptive part of the subject is called Petrography, the name Petrology being often restricted to the theory of origin of igneous rocks.

There are further some minor subdivisions which should, at least, be noted. They are *Seismology* (the study of earthquake movements), *Vulcanology* (the study of volcanoes and volcanic action), and *Glacial Geology* (the study of the action of glaciers and their effect on topography).

1.6. Methods of Investigation.—Field Geology covers a part of the work of all the divisions mentioned, since the science has been built up by patient observation of and consequent speculation upon the detailed nature of the accessible part of the earth's crust. The results of study in the several branches are recorded by means of standard methods, photographic prints now supplementing paintings and drawings as a means of visual record. Results of investigations in Stratigraphy are recorded by means of geological maps, sections, and tables of succession, features to be explained later in this volume. Detailed study of rock and other specimens by means of chemical tests, the microscope, the use of polarized light, the comparative study of fossil forms, and more recently the adoption of mechanical methods of analysis have enabled many aspects of the science to be studied in the laboratory as well as in the field. Within comparatively recent years, the development of geophysical methods of prospecting have put into the hands of the geologist a new and invaluable tool the significance of which is rapidly being recognized.

Chemical and physical methods of investigation and analysis are generally used in geological laboratory work, and in this direct way are these sister sciences associated. Mechanics and even mathematics are now being applied to the study of soils, and the detailed interpretation of geological maps often calls for the use of plane and solid geometry. Photography, as in so many other branches of scientific work, is of constant service in every branch of Geology.

1.7. Past Achievements and the Future.—Much has been achieved in Geology; much still remains to be done. Indeed, it is not too much to say that in many respects the science is at but the start of its development; new avenues of endeavor are still being opened up. The main rock groups to be found in the world have now been distinguished and generally correlated in time; the physical structure of rocks is fairly well understood; and the manner in which they reached their present form can be reasonably well determined. The preparation and publication of geological maps is an index (but no more) to the work already done, and so it may be mentioned that the whole of the British Isles has already been included in published geological maps, but only to a scale of one inch to one mile; a considerable part of the United States of America has been covered in a similar manner. The general geological mapping of some other countries has been barely started, so that it can be seen that in the field of mapping alone, the future contains great possibilities for work.

Stratigraphical study of a general nature presents work to be done that is liable to take many years for its completion. Correlation of rock strata between continent and continent is a line of study that has been but started, the consummation of which can hardly be conceived. The study of underground conditions is constantly proceeding as fresh mine workings and subterranean civil engineering works are opened up. Prediction of volcanic and earthquake disturbances may yet prove to be a profitable line of inquiry. The study of glacial-drift deposits and the detailed investigation of the composition of sands and clays are further examples of other branches of study still providing fertile fields for investigation.

1.8. Applied Geology.—The findings and methods of the pure science have naturally been applied to utilitarian purposes in view of the use that is made of the earth's surface and of the many valuable materials obtained from that part of the earth's crust that is at present accessible from the surface. Such applications are sometimes known as *Applied Geology* or *Economic Geology*. The general subject can conveniently be divided into four main groups:

1. Geology applied to the discovery and working of metallic ores.
2. Geology applied to the discovery and working of useful non-metallic substances.
3. Geology applied to the discovery and working of natural fuels such as coal and oil.
4. Geology as applied in civil engineering.

With the first three sections this volume is not directly concerned. The work achieved in each case through the application of Geology has been of inestimable benefit to mankind. The work done by geologists in developing and extending Canada's mining fields, to mention but one case, is sufficient warrant for this statement so far as the first section is concerned, and many examples could be cited of the direct aid given by careful geological study to the discovery and working of deposits of useful nonmetallic substances, particularly clay deposits. Even more striking have been the services of Geology in the discovery of natural fuels. It has been stated that from an extensive series of American oil-well records 85 per cent of the wells sunk on geological advice were successful, whereas only 5 per cent of those sunk at random were productive. The Kent coal field of England, to consider just one case relating to solid fuel, was foretold by a geologist, Godwen Austen, more than 30 years before borings revealed its actual existence at a depth of over 1,000 ft. below the surface.[1,3]

The fourth section is that to which this volume is devoted.

1.9. Conclusion.—This introductory chapter has been no more than an outline of the science. It is hoped that despite this, a sufficiently accurate summary has been given to enable a real appreciation of the relative significance of the various branches and their correlation to be obtained. Such an appreciation is essential to any study of the application of the science to the art of civil engineering, for although each individual contact may be limited to a certain aspect of one branch of Geology, the proper application of that branch can be made only if the particular problem can be considered in relation to the science as a whole. This truism is stated without apology, since all too often it is neglected.

A further object of this brief review has been to present at the outset a definite and clear-cut picture of what Geology actually is, the extent of its scope, and what it attempts to do. This is held to be necessary since misunderstanding on these points and consequent preconceived ideas as to its application must have led many to misjudge Geology as a whole. Distrust of the science, especially among engineers, has also been fostered by the unfortunate existence of pseudo geologists, of whom there are not a few in the mining field. Yet another misconception has been that of regarding Geology as something closely and inevitably associated only with mining or, again, as typifying an unceasing hunt for fossils by elderly gentlemen with little better to do. To some, such references will sound facetious, but not to those who have studied in any way the ideas of the layman with regard to Geology. Of no other branch of natural science do such indefinite and even quite erroneous ideas exist as is the case with

Geology, and it is therefore the more essential that all civil engineers should, on the contrary, know clearly the basic facts about this vital servant of their profession.

As this chapter presents but an outline of Geology, so will succeeding chapters do little more than introduce the reader to the special subjects discussed in them. There exists already a considerable bibliography covering generally the application of Geology to civil engineering, and, as may be appreciated, the textbooks dealing with the pure science will prove valuable aids to further study. An important part of each division of this volume will therefore be a guide to continued and wider reading. To this end, the following list is first suggested for the benefit of those who wish to review in more detail the main branches of the science:

Suggestions for Further Reading

General Introductions to Geology

Chamberlin, T. C., and R. D. Salisbury: "College Textbook of Geology," rev. ed., Henry Holt & Company, New York, 1930.

Coleman, A. E., and W. Parks: "Elementary Geology with Reference to Canada," J. M. Dent & Sons, Ltd., Toronto and London, 1922.

Emmons, W. H., G. A. Thiel, C. R. Stauffer, and I. S. Allison: "Geology," McGraw-Hill Book Company, Inc., New York, 1932.

Lake, P., and R. H. Rastall: "Textbook of Geology," Edward Arnold & Company, London, 1910.

Longwell, C. R., A. Knopf, and R. F. Flint: "Textbook of Geology," Part I, Physical Geology, John Wiley & Sons, Inc., New York, 1932.

Miller, W. J.: "Elements of Geology," D. Van Nostrand Company, Inc., New York, 1931.

Trueman, A. E.: "An Introduction to Geology," Thomas Murby & Co., London, 1937.

Note: Many so-called *Prospectors' handbooks* will be found to give excellent introductions to Geology. These are available from several sources, being frequently produced in conformity with special local demands. An excellent and typical example is Prospecting in Canada, *Canada Geol. Survey Econ. Geol. Bull.* 7, written by officers of the Survey.

On William Smith

Phillips, John: "Memoir of William Smith, Ll.D.," John Murray, London, 1844.

Sheppard, T.: "William Smith, His Maps and Memoirs," A. Brown & Sons, Hull, England, 1920, from the *Proc. Yorkshire Geol. Soc.*

CHAPTER II

THE COMPOSITION OF THE EARTH'S CRUST

> . . . and some rin up hill and down dale, knapping the chucky stanes to pieces wi' hammers, like sae mony road-makers run daft—they say it is to see how the warld was made!
>
> SIR WALTER SCOTT, "St. Ronan's Well."[2.1]

Detailed knowledge of the nature and composition of the earth sphere is confined to an exceedingly thin crust—much thinner than is generally realized. The diameter of the earth is slightly less than 8,000 miles. Geological investigation by means of boreholes has at present extended to about 10,000 ft. below the surface, representing no more than 0.02 per cent of the total thickness. By inference, utilizing stratigraphical relationships, investigations may be held to extend in some places possibly to a depth of 200,000 ft., but even this is relatively little below the surface. Even the irregularities on the earth's surface—mountain ranges and ocean depths—are of little moment when compared with the whole globe. For the highest mountain, Mount Everest, is 29,141 ft. above sea level; the greatest depth to the ocean bed so far discovered is in the Pacific Ocean, a distance of about 35,400 ft. below sea level. The total of these extremes amounts only to about 64,541 ft., or 0.15 per cent of the total diameter; and if average land height and ocean depth be taken (1,400 and 12,000 ft., respectively), the percentage is only 0.03 per cent.

2.2. Rocks.—About three quarters of the exposed surface of the terrestrial globe is covered with water, leaving only one quarter as exposed ground, which may therefore be studied by geologists. Some studies and investigations are pursued into the rocks forming the bed of the sea, but normally (*e.g.*, except in the case of subaqueous mine workings) they cannot be checked up experimentally. The great land masses are made up of rocks of many different types and conditions, structurally arranged in an often surprising but always systematic manner. The structural relationships of the various rock strata are considered in that branch of geological study called *Stratigraphy* and will be dealt with in the next chapter. The study of rocks as materials is classed as *Petrology* (from the Greek *petra*, a rock, and *logos*, speech). Used in this, its widest, sense, the term *rock* includes all the solid constituents of the earth's surface, whether solid (such as granite), granular (such as sand and gravel), or earthy (such as clay). To the civil engineer, on the other hand, rock is a term used to signify solid rock masses that cannot normally be excavated by manual methods alone. In Part I of this book, the geological sense of the word will be retained; in the second part, the engineering term will be used, as will be therein explained.

2.3. Minerals.—All rocks are composed of minerals. Minerals therefore are the unit constituents of the earth's crust, and as such they are of great interest and importance to the geologist. Their relative importance is changed somewhat in the case of the civil engineer, who is normally interested mainly in rocks not as assemblages of minerals but as solid materials, to be utilized or moved. In the case of rocks used for building materials, however, mineral composition is of importance to the engineer, and it enters to some degree into all applications of the science to engineering work. The study of minerals—*Mineralogy*—cannot therefore be entirely neglected by the civil engineer.

Minerals are of many kinds, about a thousand separate and distinct kinds being known. Each has its own chemical composition and atomic structure and tends to form crystals whose forms are determined by this structure. The study of these crystal forms is now in itself a distinct branch of science, termed *Crystallography*. Each mineral has a certain fixed combination of specific physical and optical properties, and it is generally by the investigation of these characteristics that minerals are distinguished. Although chemical and X-ray analyses are the final determining factors, the analysis of individual mineral crystals is a long and tedious operation outside the limits of ordinary routine laboratory work.

This detailed study and investigation of minerals in the laboratory is a fascinating occupation, but the civil engineer, unless engaged in some special task, will not have recourse to such a refinement in geological work. It may, however, be mentioned that the laboratory work involves the use of sections of rocks and minerals, mounted on glass slides, and ground so thin that they can be examined through a microscope. This is done using ordinary light and also polarized light, and optical and physical properties are thus determined. Other laboratory tests are also used in the the study of minerals, such as the determination of specific gravity. Those who are interested will find details of such investigations in some of the volumes listed at the end of the chapter. Special attention may be directed to H. A. Smith's "Minerals and the Microscope," a small book that provides a lucid introduction to the use of the petrological microscope, the special instrument used in such investigations.

Minerals have long been classified on the basis of chemical composition and crystal form. The larger divisions are determined by chemical similarities; the smaller groups and series within these divisions rest on similarities of crystal form. The more important chemical divisions are given in Table A.

TABLE A.—SOME CHEMICAL DIVISIONS IN MINERALOGY[1]

Division	Typical example	Chemical composition
Elements	Graphite	Carbon
Sulphides	Pyrite	Iron sulphide
Chlorides	Rock salt	Sodium chloride
Oxides	Quartz	Silicon dioxide
Sulphates	Gypsum	Hydrous calcium sulphate
Phosphates	Apatite	Calcium phosphate
Carbonates	Calcite	Calcium carbonate
Silicates	Feldspar:	
	Orthoclase	Potassium-aluminum silicate
	Plagioclase	Sodium-calcium-aluminum silicate
	Hornblende	Complex metallic silicates, containing potassium, magnesium, iron, etc.
	Augite	
	Mica	
	Muscovite	
	Biotite	
	Olivine	Magnesium-iron silicate
	Chlorite	Complex hydrous magnesium-aluminum silicate
	Serpentine	Hydrous magnesium silicate

[1] Distinguishing features cannot usefully be summarized in tabular form; reference may be made to standard books on mineralogy, listed at end of chapter.

2.4. Main Rock Groups.—All rocks can be divided into three main groups:

1. Igneous.
2. Sedimentary.
3. Metamorphic.

The third group is a derivative of either one of the other two. Although all rocks are composed of a combination of one or more minerals, all exposures of minerals in the earth's crust cannot be classed as rocks. The exceptions are, however, negligible in comparison with the extent of rocks, consisting of veins and certain unusual mineral aggregations. The names of these three great classes are descriptive. *Igneous* rocks (from the Latin *igne*, of fire or fiery) have been formed by the cooling of parts of the bodies of molten material, which is called, in general, *magma*, erupted from or trapped beneath the earth's crust; *sedimentary*, or *derivative*, rocks have been deposited in some geological age mechanically (through the agency of water, wind, or ice action), chemically, or organically; and *metamorphic* rocks (from the Greek *meta*, between, as denoting change, and *morphe*, form or shape) are rocks changed in some way from either an original igneous or a sedimentary form.

When it is recalled that all sands, gravels, clays, and shales are classed as sedimentary rocks, the fact that such rocks cover about three quarters of the earth's land surface will readily be appreciated. This relation in area covered cannot be applied indiscriminately as indicating relative importance. Nor is it a guide to the relative significance of the main rock groups to the geologist, for many of the most complicated geological structures requiring investigation are those presented by igneous formations. To the civil engineer, however, the significance of the main rock groups is to some degree comparable with their relative occurrence, in view of the general nature of civil engineering work. Construction operations are generally confined to the proximity of the surface of the earth, and consequently sedimentary rocks are more frequently encountered than igneous and metamorphic rocks.

2.5. Igneous Rocks.—These are of two main classes:

1. Extrusive (poured out at surface).
2. Intrusive (large rock masses which have not formed in contact with the atmosphere).

Initially, both classes were rock in a molten state. Their present state results directly from the way in which they solidified. If a violent volcanic eruption took place, some material would be emitted into the atmosphere with gaseous extrusions, there to cool quickly and eventually fall to the earth's surface as volcanic ash and dust. This type of action continues even today, dust on the dome of St. Paul's Cathedral in London having been found to be of volcanic and possibly cosmic origin. The main product of volcanic action is of the type of a lava flow, emitted from the earth as a molten stream which flows over the surface of the existing ground until it solidifies. Extrusive rocks are distinguished, in general, by their usual glasslike texture and by the "baking" of whatever rock stratum happens to underlie them.

Intrusive rocks, cooling and so solidifying at great depths and under pressure, containing entrapped gases, are wholly crystalline (holocrystalline) in texture, these conditions of cooling being conducive to crystal formation. Such rocks occur in masses of great extent, often going to unknown depths. Although always originally formed deep underground, intrusive rocks are now widely exposed as the result of earth movement and erosion processes explained in the next chapter.

Hypabyssal rocks are intermediate in position between extrusive and major intrusive rocks and so are, in general, partially crystalline in texture. They occur in many forms, the main types of which are indicated in Fig. 2.1. *Dykes* are large wall-like fillings in the earth's crust cutting across normal bedding planes. *Sills* are large horizontal sheets intruded into other formations. Common to all these occurrences is the baking of adjacent rock strata on *both* sides of the intrusion, as distinct from baking on the underside only as in the case of extrusive rocks.

Chemical analyses of igneous rocks show that they are essentially composed of the following nine elements: Silicon, aluminum, iron,

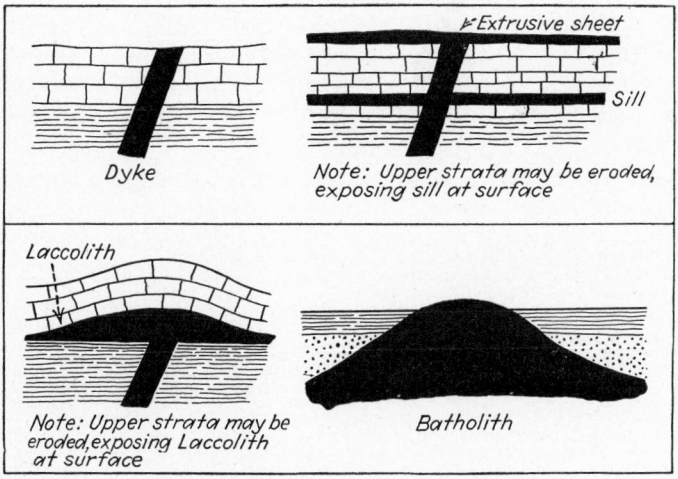

FIG. 2.1.—Some forms taken by igneous rock.

calcium, magnesium, sodium, potassium, hydrogen, and oxygen. These are, of course, in combination, generally as silicates, oxides, and hydroxides. Although the proportion of the oxides varies considerably, it is found that the chemical composition of rocks, considered as a whole, varies within quite narrow limits. Within these limits, despite the steady gradation from one to the other and the varying and quite distinct mineral constitution of different rocks, certain groups can be distinguished. Chemical and mineral composition have therefore been adopted as a basis for a general classification of crystalline igneous rocks, in association with the modes of origin. Silicon dioxide (silica), often crystallized as quartz, is one of the main mineral constituents of igneous rocks, and by the silica content a broad dividing line is fixed.

The following table gives a list of the main types of igneous rocks and suggests broad lines of classification.

TABLE B.—CLASSIFICATION OF IGNEOUS ROCKS[1]

	Acid	Intermediate		Basic
	Quartz	Little or no quartz		No quartz
Commonest minerals	Orthoclase, Oligoclase, Mica, Hornblende, Augite	Orthoclase, Biotite, Hornblende, Augite	Plagioclase, Biotite, Hornblende, Augite	Plagioclase, Olivine, Hypersthene, Augite
Plutonic	Granite	Syenite	Diorite	Gabbro
Hypabyssal	Quartz porphyry	Orthoclase porphyry	Porphyrite	Dolerite
Volcanic	Rhyolite	Trachyte	Andesite	Basalt

[1] Reference should be made to the text for comment on the indefinite nature of the divisions noted.

2.6. Sedimentary Rocks.—This great group of rocks may fitly be regarded as secondary rocks, since they consist generally of the result of weathering and disintegration of existing rock masses. These rocks are somewhat loosely named, *sedimentary* being truly descriptive of but one section; *aqueous rocks* is another name sometimes used, but it is not strictly correct, since water is not a depositing agent in the case of all these secondary rocks, nor are they always found *stratified*, another title that has been suggested. In view of the common use of the word sedimentary, it has been retained in this section and elsewhere in this book to denote the whole group of these secondary rocks.

The wide area over which sedimentary rocks are found throughout the world calls at once for the observation that this is only as a result of the great land movements that have taken place in past geological eras. This movement is often vividly demonstrated by the existence of marine deposits including fossil sea-shell remains, now elevated considerably above the nearest lake or seacoast. Marine deposits are to be found in the upper regions of the Himalayas and in many similar parts of the world now thousands of feet above sea level. The top of Mount Everest, the highest point of land in the world, is of limestone, a sedimentary rock. The movements that have caused these strange rock arrangements will be described in the following chapter.

Stratification has been mentioned. The sedimentary rocks generally are found in quite definitely arranged beds, or strata, which can be seen to have been horizontal at one time although today sometimes displaced through angles up to 90 deg. This bedding, or stratification, is a direct result of the method of formation, material having been deposited evenly over some lake or sea bottom or in some tropical jungle swamp. Sedimentary rocks are being formed similarly today by the silt and mud being washed down by rivers into lakes and seas and likewise by such marine organisms as the coral of the tropical seas. Sand and mud are constantly being brought into the Great Lakes of the North American continent, for example, and yet the water at the exit of Lake Ontario is quite clear, thus demonstrating that the sediments have been deposited on the lake bottoms. In a similar way, the building up of great river deltas is further evidence of sedimentary action. Measured by human standards of time, the building-up process is almost infinitesimally slow, geological time periods being a requisite basis for any comparison. Yet Swedish geologists have found that the clays laid down in lake basins during the northern retreat of the last ice sheet in Sweden were deposited as fine laminae, each sandy band representing the deposit of one summer; and the overlying clay band, that of one winter. By counting and plotting the thickness of these laminae, it has been possible to estimate what period of time has elapsed since the retreating edge of the ice sheet crossed Scania and to estimate the time since it crossed the Baltic and entered Sweden.

Three general groups form the natural divisions into which sedimentary rocks can conveniently be classified—briefly, those rocks which are

1. Mechanically formed.
2. Chemically formed.
3. Organically formed.

The so-called *mechanical processes* leading to rock formation are the action of wind, frost, rain and snow, and daily temperature changes, all of which can be classed as weathering influences leading to the formation of ordinary soil, rock scree (or scree breccia), and fine deposits of rain-washed material and blown dust such as brick earth and loess, as well as some special types of clay. The action of running water is a second type of mechanical action, leading to the formation of rocks that are true sediments such as conglomerate, grit and sandstone, and some types of clay. Finally, glacial action has been and still is a potent factor in rock formation, having given rise to the extensive glacial deposits which are met with over wide areas of the world.

These deposits are of great importance in civil engineering; they are discussed, in this connection, in Chap. XX.

Limestone is the best known of the organically formed rocks, being generally the accumulation of the remains of marine organisms; it occurs in many different forms. Coal, coprolites, and guano are other rocks that are obviously of organic origin.

2.7. Metamorphic Rocks.—Agencies that have led to the changing of sedimentary and igneous rocks into metamorphic rock types are many. The principal ones are the intense stresses and strains set up in rocks by severe earth movements and by excessive heat. This may be either from the cooling of intrusive rocks or from permeating vapors and liquids. The action of permeating liquids appears to have been particularly important. Another possible cause has been an interchange in the solid state. The results of these actions are varied, and the metamorphosed rocks so produced may display features varying from complete and distinct foliation of a crystalline structure to a fine fragmentary partially crystalline state caused by direct compressive stress, including also the cementation of sediment particles by siliceous matter. It is the foliation mentioned that is the characteristic of the main group of metamorphic rocks, the word meaning that the minerals of which the rock is formed are arranged in felted fashion, each layer being lenticular and composed of one or more minerals, the various layers not always being readily separated from one another. It will be appreciated that these characteristics are different from the flow structure of lava and also from the deposition bedding occurring in unaltered sedimentary rocks. A schist is the name commonly applied to such a foliated rock, and the various types of schistose rock are among the best known metamorphic rocks.

The problem as to the original rocks from which metamorphic rocks have been formed has been and still is a matter of keen discussion and the subject of much inquiry. Briefly, it may be said that some are definitely sedimentary in origin, some were originally igneous rocks, and some are of indeterminate origin. The presence of fossil remains in certain crystalline metamorphic rocks is proof enough of their sedimentary origin; and, on the other hand, uninterrupted gradation from granite and other igneous rock masses to well-defined schistose rocks is equal proof of the igneous origin of other metamorphic rock types. Of the former types, sedimentary in origin, marble (altered limestone) is a notable example, its appearance frequently demonstrating the organic remains of which it was originally formed. Schistose conglomerate is another well-defined sedimentary type. Quartzite can often be seen to have been formed from sand grains, and the great variety of slates are all obviously of clay or mudstone origin. The rocks classed generally as schists are of varying compositions, mica schist being a crystalline aggregate of mica and quartz with occasionally additional minor minerals. *Gneiss* is a term somewhat loosely used but generally to distinguish a group of rocks similar to the schists but coarsely grained and with alternate bands of minerals of different composition.

Certain sedimentary deposits, the geological history of which can readily be traced, give a clue to the metamorphism to which they have been subjected, in that structural relationships show that for some long time they have been located at great depths below the surface of the earth and thus subjected to great pressure at high temperatures. Metamorphism is clearly demonstrated adjacent to certain igneous intrusions where *contact metamorphism* is displayed in the gradual alteration of sedimentary rocks which may be pierced by a sill, say, of igneous rock, the most marked effects being naturally near the surface of contact.

2.8. Distinguishing Rock Types.—The civil engineer has a natural interest in the study of rocks in view of the use to which he puts them, not only as materials of construction but also as foundation beds for many of the structures that he designs and erects. He should therefore be able to distinguish at least the main rock types that he encounters in the field; he should know something of the origin and structural relationships of the rocks so encountered; and he will naturally want to find out if a rock will be suitable for the work for which it has been selected. It will be convenient to treat the testing of the suitability of various rock types under the separate divisions that constitute the main part of this book. Notes are given in the following chapter on the structural aspects of geological formations. Some guide to the main distinctions between different rock types is also necessary, and an outline is therefore here presented.

When a student of rocks first casts his eyes deliberately over any group of rock exposures or a selection of rock specimens, he may be dubious as to his ability ever to become acquainted with the main types thus displayed. Such doubts will quickly be dispelled by even a limited application of some simple practical rules. These are outlined below; and it can be suggested that for all the practical purposes of the civil engineer, these field tests together with the simple microscopic examinations mentioned as necessary in occasional instances will suffice. More detailed petrographical investigation will be called for only in exceptional cases; it will clearly be the work of an expert.

Simple equipment is all that is needed for the field investigations mentioned—a geologist's hammer, a tool that will accompany the civil engineer interested in geology on all his out-of-door excursions; a pocket magnifying lens for examination of the smaller crystal structures; a steel pocket knife for hardness tests; and a small phial of hydrochloric acid for the determination of mineral carbonates. A small magnet is often useful, as the mineral magnetite can be separated from other associated minerals (when crushed) by running a magnet through the mixture. The hardness test mentioned is based on a hardness table for minerals, a selected scale based on relative hardnesses, of great use in preliminary testing. The selected order is:

1. Talc 2. Gypsum	Can be scratched with a fingernail
3. Calcite	Can very easily be cut with a penknife
4. Fluorspar 5. Apatite	Can easily be scratched with a penknife
6. Feldspar	Can be scratched with a penknife but with difficulty
7. Quartz 8. Topaz 9. Corundum 10. Diamond	Cannot be scratched with any ordinary implement. Quartz will scratch glass; topaz will scratch quartz; corundum, topaz; and a diamond, corundum

The numbers given are used as relative hardness numbers, relative only since the actual hardness value of talc is about 0.02, whereas that for a diamond runs into the thousands.

Silt, clays, sand, and gravels can readily be distinguished by superficial examination; they are generally described from their appearance, color and coarseness of particle sizes being the two properties most usually selected for special note. These sedimentary deposits are of much importance in civil engineering work and are becoming so increasingly a subject for detailed study and research that they are considered separately in Chap. XX. *Boulder clay,* or glacial drift, is a special type of deposit characterised by its gritty nature and the presence in it of worn pieces of rock varying in size from that of gravel to large boulders; it is also discussed more fully in Chap. XX.

Investigation of the consolidated rocks necessitates closer examination of a hand specimen obtained from the outcrop by means of a sharp blow from the hammer. It must be emphasized that for all examinations a good clean and fresh rock surface must be used. The word *fresh* is used to indicate that the rock surface must have been freshly broken, since most rocks weather on their exposed surfaces to some extent and in this way do not expose their true character unless newly broken. From an examination of such a surface, it can usually be seen whether the rock is crystalline in structure or not.

Of the noncrystalline rocks, *shales* may be mentioned first, these being consolidated fine sediments, usually hardened clay or mud, having a characteristic fracture. Generally dull in appearance, all can be scratched with a fingernail. If it will break into irregular laminae, the shale will be argillaceous; if gritty, arenaceous; if black in color, it may be bituminous; if gray in color and effervescent on the application of acids, calcareous. *Slate* can easily be recognized from its characteristic fracture or cleavage and by its fine uniform grain composition; in color it may vary from black to purple or even green. All these rocks demonstrate their argillaceous nature by emitting a peculiar earthy smell when breathed upon.

Limestone is one of the best known rocks in a general way; it can be distinguished often by its obvious organic origin, but a surer mark of distinction is that it effervesces briskly when dilute hydrochloric acid is applied to it. Marble is a crystalline (metamorphic) form of limestone, generally distinguished by its crystalline texture but always effervescing when treated with dilute acid. *Dolomitic limestone* is generally darker in color and effervesces more slowly with cold hydrochloric acid, although more quickly when the acid is warm. *Flint* and *chert*, compact siliceous rocks of uncertain chemical or organic sedimentary origin, occur often as nodules in limestone beds.

Conglomerates and *sandstones* are another interesting group of sedimentary rocks, their granular structure suggesting their sedimentary origin. Conglomerates are, as their name implies, masses of waterborne gravel and sand, as denoted by rounded shapes, cemented together in one of several ways into a hard and compact mass. Sandstone is the general term used to describe such sedimentary consolidation of sand alone. *Quartzite* is a metamorphosed type of sandstone in which the grains of rock have been cemented together with silica so strongly that fracture will take place through the grains and not merely around them. *Grit* is a term sometimes used to denote a coarse-grained hard sandstone containing angular fragments.

The investigation of igneous and some of the metamorphic rocks not yet mentioned is not quite so straightforward as the determinations so far described. They are usually crystalline, but the crystals may vary in size from those of coarse-grained granite to a state of such minute crystallization that field distinction is almost impossible, and examination under a microscope becomes essential. Of the remaining metamorphic rocks, *serpentine*—a rock composed wholly of the mineral of the same name—is generally green to black in color, fairly soft and greasy or talclike to the touch; the color may not be uniform. It is important to the civil engineer as being a potent cause of instability in rock excavation. *Gneiss* may be recognized by its rough cleavage and typical banded structure, showing quartz, feldspar, and mica with a coarse structure. *Schists* may be distinguished from gneiss by their essentially fissile character—in all of them, there is at least one mineral that crystallizes in platy forms (mica, talc, or chlorite) or in long oblong blades or fibers, thus giving the rock a cleavage parallel to the flat surface. Of the paraschists, those characterized by mica or chlorite have probably been derived from shales or clay rocks; talc schists are from serpentine, and quartz schists are a metamorphosed sandstone. Orthoschists, on the other hand, are derivatives of igneous rocks; if white mica predominates as a constituent, it indicates an acid igneous rock as origin; and if hornblende or black mica, a basic igneous rock as origin.

Granite may be taken as a typical example of igneous rocks, since it is widely distributed and as it constitutes an important igneous rock type. It is composed mainly of quartz (clear) and orthoclase feldspar (white or pink) with some mica and possibly hornblende. All the crystals are about the same size, or even-granular, and the quartz (as the last mineral to separate out) occupies the angular spaces between the other crystals, this characteristic being known as a *granitic* structure. Color may vary from pale gray to deep red, depending on the mineral content; an average composition, however, is 60 per cent of feldspar, 30 per cent of quartz, and 10 per cent of dark minor minerals.

The rock described, granite, is but a starting point for considering igneous rocks. It will, however, be realized that although tabular statements may, and do, represent various distinct classes of igneous rocks, there is actually an indeterminate gradation from granite along the several directions of igneous rock classification with no specific boundary lines demarking the various named rock types. In a similar way, the texture of igneous rocks varies from a coarse even-granular structure to an aphanitic structure in which crystallization cannot be seen with the unaided eye. A porphyritic structure is characterized by one of the constituent minerals occurring as much larger crystals than the remainder (the large crystals being called *phenocrysts*); these other minerals may appear as a crystalline groundmass, or alternatively they may be aphanitic. Finally, there are a few rare igneous rocks found in glasslike form which have therefore not crystallized at all.

It may be mentioned that, as a general rule, the acid igneous rocks tend to be lighter in color than the basic rocks; granite, therefore, is pale in color because of the predominance of feldspar, a light-colored mineral. *Diorite* has a similar texture to granite, but it contains no free quartz. In *gabbro* (the corresponding basic rock), feldspar is subordinated but is still an important constituent, hornblende, pyroxene, and olivine being dark minerals which make the rock dark in color and give it a high specific gravity. *Dolerite* is a similar basic rock with a smaller grain size; *basalt* is the corresponding aphanitic rock, although occasionally phenocrysts will be found in it.

Granite porphyry and *diorite porphyry* are similar to granite and diorite in composition, but they have a porphyritic structure, feldspar being the most usual phenocryst. The porphyries are a common group of rocks, hypabyssal in nature, thus occurring as sills, dykes, laccoliths, and also in lava flows. *Rhyolite* is the corresponding extrusive rock to granite (an intrusive rock); it contains phenocrysts to varying extent. *Andesite* bears a similar relationship to diorite, no quartz being present. The aphanitic type of both rhyolites and andesites is known as *felsite*, a name that includes a large group of rocks—most of the light-colored aphanitic igneous rocks—the dark corresponding rocks being classed as *basalts*. Glasslike rocks will be found only in the vicinity of cooled lava flows, *obsidian* being the commonest variety, a lustrous, dark glasslike rock. *Pumice* is simply a frothed type of glassy rock.

The foregoing list includes the main common varieties of igneous rock; it must again be emphasized that as between the one variety and the next to it in scale, there is no hard and fast dividing line. This fact will explain the prevalence of names for varieties of igneous rocks, many of them local, but some in general use. Although these are frequently found in geological literature, there is little need for the engineer to attempt to become familiar with many more than those already mentioned. In special circumstances, close study may have to be made of a rock formation, but this will generally be the work of an expert. It cannot be too strongly urged that study of rock structures *in situ* is the only truly reliable method of becoming familiar with the rock types described.

As a preliminary to outside work, much can be gained by studying hand specimens of rock such as are available in the geological departments of universities and in natural history museums. Color, feel, surface appearance, and the nature of fracture are all so important in distinguishing rock types that only through contact with actual specimens can a start be made at accumulating that practical experience which is the essence of true geological study. For this reason, no photographs of rock specimens are here reproduced; for those who are familiar with common rock types, they would be superfluous; whereas for those who are not, they might be misleading. Better by far than a volume of photographs is a rock specimen in the hand, particularly if it has been collected by the student himself in the open.

Suggestions for Further Reading

The books listed at the end of Chap. I in their presentation of the fundamentals of Geology naturally include discussions of rocks and rock-forming minerals. The following books deal with these subjects in more detail:

General

MERRILL, G. P.: "A Treatise on Rocks, Rock Weathering, and Soils," The Macmillan Company, New York, 1913.

PIRSSON, L., and A. KNOPF: "Rocks and Rock Minerals," John Wiley & Sons, Inc., New York, 2d ed., 1926.

SHAND, S. J.: "The Study of Rocks," Thomas Murby & Co., London, 1931.

TYRELL, G. W.: "The Principles of Petrography," Methuen & Company, Ltd., London 1929.

Igneous and Sedimentary Rocks

DALY, R. A.: "Igneous Rocks and the Depths of the Earth," McGraw-Hill Book Company, Inc., New York, 1933.

NORTH, F. J.: "Limestones; Their Origins, Distribution, and Uses," Thomas Murby & Co., London, 1930.

TWENHOFEL, W. H.: "Treatise on Sedimentation," Williams & Wilkins Company, Baltimore, Maryland, 2d ed., 1932.

Mineralogy

FORD, W. E.: "Dana's Manual of Mineralogy," John Wiley & Sons, Inc., New York, 14th ed., 1929.

READ, H. H.: "Rutley's Elements of Mineralogy," Thomas Murby & Co., London, 23d ed., 1936. NOTE: In the immediately preceding editions of this well-known book will be found a preface of great interest to civil engineers, by G. T. Holloway. It had to be omitted from the 23d ed.

SMITH, H. G.: "Minerals and the Microscope," Thomas Murby & Co., London, 3d ed., 1933.

CHAPTER III

THE STRUCTURE OF THE EARTH'S CRUST

> The relation of land to sea changes, and a place does not always remain land or sea throughout all time.
>
> ARISTOTLE.[3.1]

Many of the types of rock that have been considered in the foregoing chapter were deposited originally in approximately horizontal beds. Sedimentary rocks clearly were so formed when, for example, sediments were deposited from suspension in water; volcanic igneous rocks, whether formed as lava flows or from the deposition of fragmentary volcanic material, originated similarly. Today, however, few rock beds are found to be truly horizontal, nor are strata of rock always of uniform thickness. The nature of the irregularities visible at many an outcrop of rock, and the causes of them, are varied and to some extent complicated; the study of them has been classed as Structural Geology. Associated branches of study are Physical Geology and Physical Geography (subjects blending the one into the other) which are concerned generally with present-day and recent modifying influences on the structural relationship of rocks. It may usefully be emphasized that despite the severe movements and contortions to which all rock beds have been subjected to some degree, their correlation yet remains quite definite, and it can still be determined with a reasonable degree of accuracy in all but exceptional cases.

In considering the present position of beds of rock with reference to the existing ground surface, two terms are in general use, *dip* and *strike*. As these will be encountered throughout this work, they may usefully be explained at this point. The strike of any rock stratum is the term given to the direction of a line considered to be drawn along an exposed bedding plane of the rock so that it is horizontal; obviously, there will only be one such direction for any particular rock layer. The dip of the bed is the angle made by a line considered as drawn on the exposed bedding plane at right angles to the line of strike, thus being a measure of the inclination of the bed to the horizontal plane. These two particulars, together with the usual topographical data, definitely locate a rock surface in space; as such, they are invaluable to the civil engineer, as they are also to the geologist.

3.2. The Geological Cycle.—The study of the processes that have led to the existing structural arrangement of the earth's crust may be approached by considering the similar processes at work in the world today. These are described generally as the *geological cycle* and may be represented as in Fig. 3.1. The cycle is naturally not so straightforward as the diagram might suggest, but the main lines of change are those indicated.

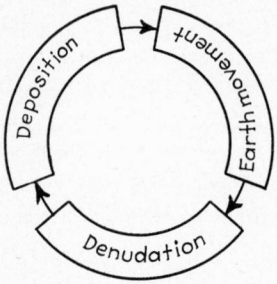

FIG. 3.1.—The geological cycle.

Denudation of existing rocks, or earth sculpture, is carried out by agencies such as severe temperature changes and especially by the freezing and thawing of ice in cracks, wind action (especially in desert regions), the action of rain water on exposed soluble minerals and on rocks decomposed by water, the action of running water in eroding rock surfaces, the disintegrating and transporting power of the slow-moving ice of glaciers, and finally the erosive action of the sea on almost all coast lines. All these processes are fairly generally appreciated, and the change wrought by some of them even during the course of a few years can sometimes actually be seen, whereas an examination of old maps and photographs will often reveal considerable alteration of topographical detail in the space of a few centuries.

Sedimentation is due to either wind or water action, principally the latter, the action of wind being generally confined to desert regions. Deltas, at river mouths, are an obvious result of sedimentation, and similarly the deposits that accumulate behind artificially constructed dams. Other material is carried by rivers into the beds of oceans or lakes, although not all the material derived from earth sculpture reaches this resting place. But in this way a slow change is wrought in the configuration of the earth, and eventually modifications lead to uneven pressures on the subcrustal portions of the earth, and some movement is bound to result as these pressures disturb the equilibrium of the local crustal structure.

These two parts of the general cycle can readily be visualized, as they are being carried on to a limited extent today. To obtain any true appreciation of the third part of the cycle requires rather more imagination and an excursion into the realm of geological time. It is true that there are some evidences of minor earth movement occurring now. As an example may be mentioned the appreciable sinking of large areas of the Mississippi flood plain near New Madrid, Missouri, in 1811; and later in this book reference will be made to subsidence of the shores of Lake Maracaibo in Venezuela. Coal- and salt-mining subsidences will be generally familiar.

Within recent years, there have also been some violent manifestations of volcanic action, notably the complete disintegration of the volcano Katmai in Alaska, last seen in 1912 and supposed to have been completely blown up by a severe volcanic explosion. Such effects are small as compared with the great earth movements of the past, which have given rise to the complicated structures known today. They are significant, however, as indicating that the solid earth is not a conglomeration of unusual structural materials but rather a body that is just as susceptible to stresses and their strain effects as any other structural material. The definite daily recordings of load and relief obtained on seismographs that are situated near tidal water is another telling reminder of the elastic properties of the earth's crust. Even the activities of the civil engineer can lead to such earth movement, as will be shown later by examples such as the troubles encountered during major excavation work on the Panama Canal.

With the foregoing in mind, it may not be quite so difficult to imagine great sections of the rock strata at surface level being distorted and crushed when subjected to unusual stresses and so strained beyond their capacity to resist. The movement can be divided roughly into general raising or lowering of large sections of the earth's crust and general bending and crumpling of sections of the crust. After such disturbing action, natural physical conditions readjust themselves to the changed configuration of the earth's surface, and denudation will again start. This will affect first the most exposed strata, and they may quickly be worn away. An underlying stratum will then be disclosed at the surface and brought into view. So the cycle will continue. By imagining the results of several such cycles, it may not be difficult to appreciate the main reasons for the present complicated nature of the structure of the earth's crust.

3.3. Special Structural Features.—It will be useful to describe briefly the main structural features likely to be encountered in normal civil engineering work and which, by their variation from what may be called *standard stratification,* warrant being described as special. It must be emphasized that the unusual features to be described are no more than a few of the more general examples of special structural arrangements; each type described can be subdivided into many detailed classifications, and other more unusual features may be encountered. Reference books listed at the end of the chapter provide a guide to further study of this subject.

3.4. Sedimentary Rock Bedding.—The method of formation of sedimentary rocks inevitably leads in many cases to an uneven disposition of material and to an uneven distribution of pressure on deposits. This can result in a distinct variation in the physical qualities of a

FIG. 3.2.—Well-stratified limestone near Port St. Mary, Isle of Man.

FIG. 3.3.—Crossbedded Wingate Sandstone, Johnson Canyon, Kane County, Utah. (*Photograph by H. E. Gregory. Reproduced by permission of the Director, U. S. Geological Survey.*)

sedimentary bed at different levels and also in changes in the thickness of the same bed. Sedimentary strata will often be found to thin out completely, and this can often cause confusion in geological mapping. Unusual surface markings constitute another special feature of sedimentary rocks, ripple marking (similar to that seen on a flat sandy beach after the recession of the tide) being one example. These variations are of minor importance, but another subsidiary feature is worthy of special note. This is what is called *apparent,* or *false, bedding,* and it is caused by a special process of deposition.

3.5. Igneous Rock Features.—These have already been explained, in so far as the principle types of formation are concerned. They are mentioned here again to emphasize the fact that these several types of igneous formation may be encountered, in structural geological

FIG. 3.4.—Jointing in white granite south of Tule Well, Pima County, Arizona. (*Photograph by J. Gilluly. Reproduced by permission of the Director, U. S. Geological Survey.*)

investigations, either perfect in form (as shown in the Fig. 2.1) or imperfect owing to subsequent distortion or denudation.

3.6. Jointing.—If any typical rock mass, igneous or sedimentary, is generally studied, it will soon be seen that in addition to any bedding planes that may be visible (in a sedimentary rock) fractures occur in other planes also, roughly at right angles to bedding planes, giving rise to a blocklike structure, even though the blocks may not be separated one from another. Such fractures are known generally as *joints,* or *joint planes;* they are caused either during cooling of the rock concerned or by structural displacement; they are not unconnected with the cleavage planes of the constituent minerals. Joints are sometimes filled with newer igneous rock which has penetrated into the cracks while still liquid; some remarkably intricate formations of this type may be found, with filled joints so small as only just to be visible. In sedimentary rocks, jointing is generally regular; in granite, it is often irregular; in basalt, it leads to the peculiar hexagonal column formation that is so striking a feature of locations such as the island of Staffa

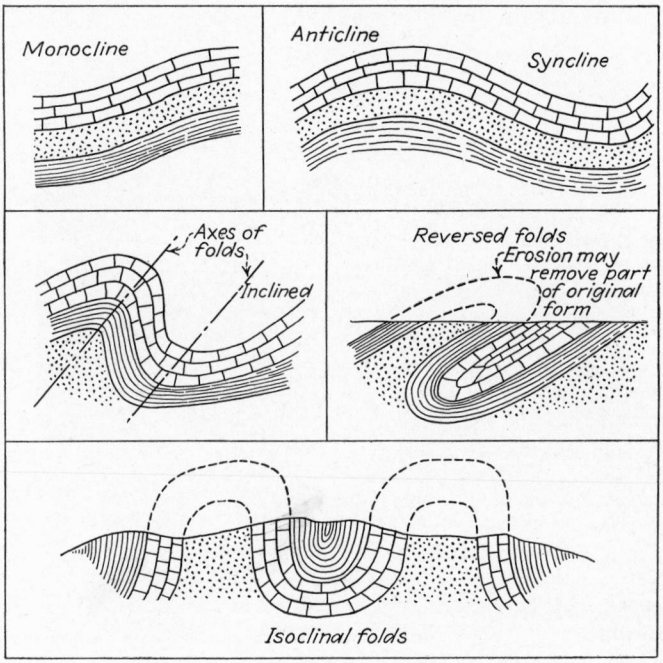

FIG. 3.5.—Some types of folds (diagrammatic).

off the west of Scotland. Joints are of great importance to the civil engineer, and numerous references to them will be found in the main part of this book.

3.7. Folding.—Considering now the distortion of rock strata that can take place, folding is perhaps the simplest structural feature apart from a general raising or lowering of the whole of one part of the earth's crust. In its basic form, it consists of the formation of simple regular folds, as shown in Fig. 3.5, the portions being termed *anticlines* and *synclines*, respectively, according to the type of the bend. An anticline and a syncline constitute a fold. This essentially simple type is not always found in practice. The first, and most general, variation is the inclination of the axis of the fold with the result shown. In certain special cases, the folding may not be confined to the one direction, a rock mass being subjected to bending stresses all around, causing a domelike structure. If the simple basic type is extended, many variations can be obtained such as double folds, reversed folds, and even fanlike structures such as are indicated. At first sight, the latter might appear to be a purely hypothetical case, but several outstanding examples of this type of structure could be given, including the European mountain group of which Mont Blanc is a leading member.

FIG. 3.6.—Folded limestone, much contorted, on foreshore at Scremerston, Northumberland, England.

(Geological Survey of Great Britain photograph; British Crown Copyright; reproduced by permission of the Controller of H. M. Stationery Office, London.)

Photographs show something of the appearance of these folds in actual practice; references at the end of this chapter lead to full and detailed descriptions of all types of folding in geological strata. It will be realized that, if the angle through which rock is folded is at all appreciable, unequal stresses will be set up in the rock mass. Normally, this would be of no consequence, but to the civil engineer it can mean a great deal. If such distorted material is underground, there being no way in which the unequal stresses can normally be released, underground civil engineering operations may interfere with this condition, and most unexpected happenings result when the previously restricted stresses are released.

3.8. Faulting.—When subjected to great pressure, the earth's crust may have to withstand shear forces in addition to direct compression. If the shear stresses so induced become excessive, failure will result; movement will take place along the plane of weakness at which failure occurs until the unbalanced forces are equalized; and a *fault* will be the result. Figure 3.7 shows the most general types of fault encountered, in simple diagrammatic form, the relative displacement of the various strata being clearly shown. In the case of the simple fault, the terms most generally used are indicated, the first fault shown being a normal fault, in which the *hade* (or inclination with

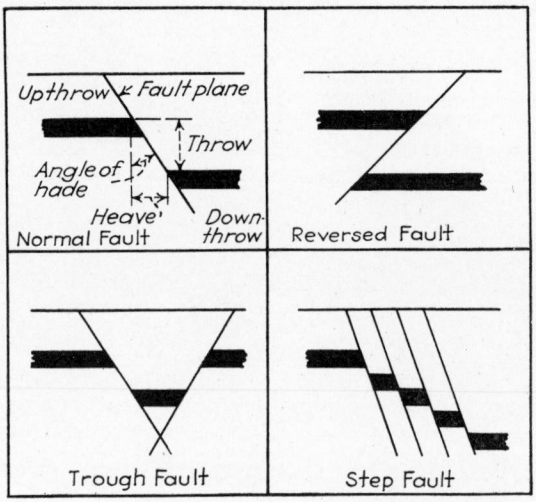

FIG. 3.7.—Some types of faults (diagrammatic).

the vertical) is always in the direction of the *downthrow*, the *throw* of the fault being the vertical displacement, and the *heave*, the lateral displacement. In the diagram, the fault is shown as a plane surface; in practice, this is not always found, the rock on one (or both) sides of the fault frequently being badly shattered into what is termed *fault breccia*, finely fractured fragments of rock, often retained in such close contact that they appear at first sight to be a solid mass. The direction of the main line of a fault is dependent on so many factors that it may bear no relation to the dip or strike. Thus it is that, for convenience, faults are characterized by their general direction as *dip faults* and *strike faults*. *Step faults* and *trough faults* are terms that will readily be understood by reference to the diagram. *Reversed faults* are the opposite of normal faults, having the hade away from the direction of the downthrow.

A cursory consideration of these simple diagrams will suggest the complicated structures that can and do result from faulting, especially where the fault plane cuts across several relatively thin strata. All these will be displaced, relatively to one another, at the fault; after denudation has taken place, their surface outcrops will be confusing. It is in the study of such structural results that the basically simple nature of faults must be kept in mind, coupled with the three-dimensional nature of the faulted section. The civil engineer's usually clear

FIG. 3.8.—A normal fault in beds of sandstone and fakes exposed in excavation at Clydesdale Iron and Steel Works, Mossend, Lanarkshire, Scotland.

(Geological Survey of Great Britain photograph; British Crown Copyright; reproduced by permission of the Controller of H. M. Stationery Office, London.)

appreciation of three-dimensional drawings, based on solid geometry, will here be of great service. The study of solid geometry in particular leads to considerable simplification in the quantitative (as distinct from the qualitative) investigation of faulting, together with all allied calculations.

Faults may vary greatly in size; in length, they may range from a few hundred feet to almost as many miles; whereas in throw, they include movements of a few inches or less (even pebbles being found with perfect fault sections in them) to thousands of feet, as in the case of those in the Plateau region of Arizona and Utah, some of which cross the Grand Canyon. Such dislocations may cause no unusual disturbance of topographical detail, whereas in other cases they may be the main factor affecting the physical character of large areas. In Canada, there are many notable examples of such conspicuous faults, Lake Temiskaming, which separates Ontario from Quebec in the North, lying in a fault depression.

3.9. Denudation.—Denudation listed under special structural features will appear to be almost a contradiction in terms. Yet the direct effects of the erosive action of wind, water, and ice are so markedly a characteristic of almost all geological structures that the use of the term to include these several effects may properly be permitted. On all approximately horizontal beds, the effect of such action will not, as a rule, be unusual, being generally uniform except where watercourses have cut their way somewhat deeply into the upper strata. Inclined strata, however, may be worn down unevenly, resulting in exposures of varying thicknesses. Erosive action will vary in the intensity of its results according to the hardness of the strata encountered. The resulting differential action may give rise to unusual structural features such as that indicated in Fig. 15.10 (page 384). Other notable results of erosive action are to be found in the history of river systems, this study being of especial interest to civil engineers in view of the difficulties always caused by buried river valleys, a feature that will be repeatedly mentioned later in this book. The deposition of material so eroded need only be mentioned, delta formation and other sedimentary effects having already been considered. Apart from the erosive action thus generally described, glaciation produces special topographical effects which may almost be classed as unusual structural features. These are considered in some detail in Chap. XX in view of their importance to civil engineers.

3.10. Unconformity.—There is an important effect of the deposition of sedimentary strata on previously deposited sedimentary beds, which is most clearly demonstrated by a diagram (Fig. 3.9). *Unconformity* is the term used to designate the unusual juxtaposition of these several newer beds on the older, and it will be seen that an exposure of this contact would not show up a regular stratigraphical succession. Another term used in this connection is *disconformity*, which describes a juxtaposition of two series of beds with parallel bedding planes, the upper surface of the lower group having been eroded prior to the deposition of the upper group.

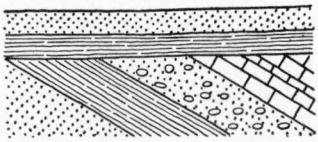

FIG. 3.9.—Unconformity of strata.

3.11. The Geological Succession.—Despite all the earth movements briefly indicated in the foregoing notes, the original relation of the various rock strata and groups of strata, not only in one particular area but throughout the world, can usually be determined. The completion of the correlation is still an unfinished task, but much has been done in the solution of the many problems involved, work toward this end being perhaps the supreme task of all geologists. In countries such as Great Britain, the general correlation can be regarded as complete, work now proceeding generally on detailed local problems. In other parts of the world, however, no final results can yet be reported, and there is still much to do in the linking up of the respective local systems, now studied as a whole. In the working out of all problems associated with stratigraphical succession, the study of the several special geological structures already described plays its part as

TABLE C.—THE GEOLOGICAL SUCCESSION

Group	Per cent of total	System	Character	Derivation of system name
Recent	..	Quaternary: Present day	Age of reason	Fourth advance in animal life
Cainozoic		Quaternary: Pleistocene	Glacial; man developed	
	4	Tertiary: Pliocene, Miocene, Oligocene, Eocene	Continental period; age of birds and mammals	Third advance in animal life
Mesozoic	4	Cretaceous	Continental and marine periods; age of reptiles and ammonites	The chalk, of Europe, is of this age
	2	Jurassic		From the Jura Mountains, where well-developed.
	2	Triassic		System has three divisions in Europe
Palaeozoic	3	Permian	Continental and marine periods; age of amphibians	From Perm, U. S. S. R., where well developed
	5	Carboniferous		From presence of coal, in Europe
	3	Devonian	Marine period; age of fishes and invertebrates	From Devonshire, England, where first studied
	2	Silurian		From name of race in Wales, where first studied
	4	Ordovician		From name of race in Wales, where first studied
	7	Cambrian		From Cambria or Wales
	64	Pre-Cambrian	Beginning of life	Before the Cambrian

one of the methods of the geologist. When once the local succession of beds has been elucidated, it must be compared with similar arrangements of rock strata elsewhere in order to determine its place in the general geological timetable. This correlation is assisted by comparison of the detailed nature of the various rock types, their specific mineral composition, their general characteristics, and their fossil contents. For as early an investigator as William Smith found that similar geological horizons had similar fossilized remains, a discovery that has withstood well the test of time and that forms one of the foundations for stratigraphical work of today.

Long years of intense study led gradually to the building up of a general table of rock groups, called the *geological succession;* it may usefully be likened also to a geological timetable. A summary of this table is given on page 35, one that is fairly generally applicable. Naturally, each one of the main geological measures is subdivided into many different rock groups, each of which is capable of further local subdivision; the table, however, shows what are now regarded as the standard divisions. The explanatory notes with the table will assist in making clear the special significance of some of the better known measures.

3.12. Palaeontology.—The importance of palaeontology, the study of fossils, will now be evident, its vital contact with and use in stratigraphy being alone enough to warrant this. Even a slight knowledge of the nature of common fossil forms will demonstrate the variety of types of life now entombed in rock strata, so that it will readily be appreciated that the zoological aspects of fossil study constitute a further wide and complete field of investigation. Palaeontologists themselves in many cases now concentrate on certain sections of their general subject. To these specialists, the civil engineer will always have recourse in connection with inquiries relating to fossils. These are by no means so remote from normal civil engineering practice as might be supposed, albeit they may bear only indirectly on the immediate work in hand. As affecting civil engineering practice, fossils may on occasion have to be used to determine exactly the nature of certain strata with special reference to other beds. It is in the other direction that connection will more generally be made, the works of the engineer often revealing sections of fossiliferous strata that would otherwise ever remain hidden to the geologist. Although not strictly a fossil remain, the skull of the now famous *Rhodesian Man* may be mentioned as an example of the services of engineers to science. This was discovered, quite by chance, during extensive excavation work of the Broken Hill Mining Company in South Africa, in searching for

Anthozoa (corals); *Halysites catenularia*, Linnaeus; from Silurian rocks, Ontario. One-half natural size.

Trilobita; *Isotelus gigas*, DeKay; from Ordovician rocks, Ontario. Three-fourths natural size.

Pelecypoda; *Trigonia subtriangularis*, Parkinson; from Jurassic rocks, Germany. Three-fourths natural size.

Echinoidea (sea-urchins); *Cidaris coronata*, Goldfuss; from Jurassic rocks, Germany. Natural size.

Ammonoidea; *Amaltheus gibbosus*, Schlotheim; from Liassic rocks, Germany. Natural size.

Brachiopoda; *Spirifer pennatus thedfordensis*, Grabau and Shimer, from Devonian rocks, Ontario. Natural size.

FIG. 3.10.—Typical invertebrate fossils.

(*Photographs by Dr. L. S. Russell. Specimens kindly loaned by the Royal Ontario Museum of Palaeontology, Toronto, Canada; Dr. M. A. Fritz, Asst. Director.*)

lead, zinc, and vanadium, being noticed by the engineer in charge of the work and saved by him from destruction.[3.2] The discovery was of the greatest importance to all anthropological study. More recently, during excavation work for the new graving dock at Southampton, England, a heavily fossiliferous section of the Eocene Lutetian beds was encountered. Apparently, the unusual fossils present were not noticed on the job; but through the cooperation of the engineers in charge of the work, geologists were enabled to examine the records of the strata, and scientific knowledge was thereby enriched. Other

FIG. 3.11.—Typical fossil plants: Pecopteris sp., from the Coal Measures, Nova Scotia, Canada. One-half natural size.

(Photograph by Dr. L. S. Russell. Specimen kindly loaned by the Royal Ontario Museum of Palaeontology, Toronto, Canada; Dr. M. A. Fritz, Asst. Director.)

examples could be quoted, but all would have the same intent—emphasis on the importance, in the interests of science, of civil engineers realizing something of the nature of and not a little of the importance of fossiliferous remains so as to be able to bring to the attention of geological authorities any such evidence that they may encounter in the course of their work.

Fossil remains may be of either animal or vegetable origin, and they may vary in size from minute organisms to be examined microscopically to the great skeletons sometimes seen in museums. They consist of a mold, the petrified remains or the chemically transformed body of a once-living creature of some sort—animal, insect, fish, bird, flower, or tree—preserved down the long annals of time in the surrounding rock stratum. Their preservation can generally be understood; the details of the processes involved need not here be discussed. Naturally, only certain types of life have been preserved, at certain geological eras and under certain favorable conditions. All rock strata are not therefore fossiliferous, and it will be appreciated that fossils are found only in sedimentary and in some metamorphic rocks. The span of geological time being commensurate with the span of life on the earth, it will be realized that some demonstration of evolution of living types might be expected in rocks of different ages, and this is found to be the case.

No attempt can here be made to describe even the leading typical fossils: two plates must suffice to demonstrate a few types as an introduction to fossil forms and possibly as an incentive to further study in standard reference books. It should perhaps be emphasized that most fossil remains found are imperfect to a degree, complete and perfect animal fossils being the exception; thus is the task of the palaeontologist made the more difficult. It may also be pointed out that the nature and type of fossils found in rock provide an accurate guide to both the climatic and geographical conditions obtaining on the earth sphere at the time that the bodies now fossilized were alive. Arctic and tropical climates have alternated (broadly speaking) throughout certain stretches of geological time; such conditions naturally affected the type of life extant. The fossil types are likewise a sure guide to the nature of the deposition of the rock in which they are found, coal beds (the most generally appreciated repository of fossil life), beds containing fossil snails, and amber being three clear indications of the existence of some type of land surface. Other fossils are just as typical of lacustrine conditions, whereas others equally well denote marine conditions. Earth movements that have taken place since the original deposition during which the fossils were first formed have resulted in wide and varied displacements in space of even marine deposits, fish beds being found far inland and often at great heights above sea level. A word may be added finally to explain, if not to defend, the nomenclature adopted for the naming of fossils. Although so strange-looking to the newcomer, this is nothing more or less than a simple type of scientific shorthand, adapted to give a flexible system of classification, invoking as an aid to this task the classical languages. If this is remembered, *Strongylocentrotus drobachiensis* will become a term of some meaning.

3.13. Recent Changes.—All the subsidiary modifying actions that have in past ages affected the structure of the earth's crust are still visibly at work in the world today, as has inevitably been indicated in describing them. Glacial action is one example, although now restricted to a few mountainous areas and the polar regions whereas in former periods much of the earth's surface was covered with ice. The erosive action of wind is still a potent force, not only in desert areas but also in semiarid lands, where cultivation has not been controlled. Water action is perhaps the most generally appreciated agent of denudation at work at present and equally well a means of sedimentation. The silting up of reservoirs described in Chap. XV is one direct result of the erosive action of the rivers that are dammed farther up their courses. And the effect of ice, rain, frost, and wind in disintegrating exposed rock faces gives rise to ever increasing deposits of rock talus. All these features will repeatedly call for mention in the main part of this book, in relation to engineering work, but they should always be visualized as a part of the incessant action of natural forces on the face of the earth.

Suggestions for Further Reading

The books listed at the end of Chap. I in their presentation of the fundamentals of Geology naturally deal, at least briefly, with geological structures. The following books treat of this one branch of Geology alone:

LEITH, C. K.: "Structural Geology," Henry Holt & Company, New York, rev. ed., 1923.

NEVIN, C. M.: "Principles of Structural Geology," John Wiley & Sons, Inc., 2d ed., 1936.

STOCES, B., and C. H. WHITE: "Structural Geology," Macmillan & Company, Ltd., London, 1935.

WILLIS, B., and R. WILLIS: "Geologic Structures," McGraw-Hill Book Company, Inc., New York, 3d ed., 1934.

CHAPTER IV

GEOLOGICAL FIELDWORK AND MAPPING

> Now, if it were noted how far these [soils] extended, and the limits of each soil appeared upon a map, something more might be comprehended from the whole and from every part than I can possibly foresee, which would make such a labour well worth the pains. For, I am of the opinion, such upper soils, if natural, infallibly produce such under minerals, and for the most part, in such order. But I leave this to the industry of future times.
>
> MARTIN LISTER, M.D., F.R.S., in a paper delivered to the Royal Society, March, 1683–1684.[4.1]

That Geology is essentially an out-of-door scientific study is a truism and yet one that cannot be repeated too often, especially as the subject has usually to be introduced to the student within the confines of the classroom. All inside geological work, whether with pen, microscope, or chemical reagents, is subsidiary and supplementary to the basic study of the earth's crust in its normal natural setting. It is only in the open that Geology can properly be studied, a close examination of even a small area of ground made under proper guidance being of more avail to the majority of students than a host of classroom lectures. Fieldwork is therefore the basis of the science of Geology. To the civil engineer, this outside activity is perhaps of even greater importance than to the geologist, since the bulk of inside geological work is concerned with the study of detailed matters of no special significance in civil engineering work.

4.2. Geological Maps.—The main, and often the only, aim of fieldwork is the determination of the present structure of the section of the earth's crust being examined and its relation to neighboring and similar structures. Thus it is that the means of record adopted in these studies is through the agency of maps. By a logical extension of the principles of topographic mapping, shortly to be explained, it is possible to indicate clearly and unmistakably the geological structure determined by observation. The production of geological maps in this way is the ultimate goal of every geological surveyor. It is this geological mapping that is of vital importance in civil engineering. To both geologist and civil engineer, accuracy in the maps produced is of paramount interest; and although the results generally obtained are not comparable in accuracy with that of topographic maps, remarkable results can be, and often are, attained by proficient geological fieldworkers.

An important distinction must be carefully noted at the outset of even this brief description of fieldwork. To the geologist, the geological map—when suitably correlated—is an end in itself. To the civil engineer, it is merely a means to the end represented by his major work. There is thus a distinct difference in emphasis according to the point from which the mapping work is viewed, and it is a difference that must be recognized and cheerfully tolerated in any cooperative work between geologists and civil engineers. If independent geological investigations of use can be made, without undue delay, during any fieldwork undertaken specifically to assist some construction operations, the civil engineer should appreciate its value and encourage the saving of effort thus achieved. Similarly, it behooves geologists engaged on engineering work to realize that what the civil engineer wants is not a complete geological survey of an entire area but sufficient data to assure him of the conditions to be expected in those parts of the earth's crust liable to be affected by his operations. Mutual regard for these necessarily divergent viewpoints will lead to finer results from all such joint effort. To this may be added the important suggestion that throughout many civil engineering construction operations, sections will be uncovered that may provide data of inestimable value to geologists, in some cases data that can be obtained only at that particular time. The courtesy of bringing these to the attention of geologists (the director of the local geological survey and the geological staff of the nearest university) cannot be too strongly commended.

Finally, in connection with all geological mapping work to be related to civil engineering practice, the civil engineer in charge should be able correctly and readily to interpret the geological maps supplied to him, either by expert geological advisers or by geologists on his own staff. It is of the greatest importance, therefore, that the basic principles of geological mapping should be appreciated widely by civil engineers who, even if they never have to undertake such geological fieldwork themselves, will thus at least be able to make full use of information supplied to them.

4.3. Published Geological Maps.—In most leading countries of the world, a geological survey is now an established scientific branch of central governmental activity. The publication of national geological maps, and complementary memoirs, is one of the main functions of official geological survey departments. Somewhat naturally, it is only in such highly developed small countries as Great Britain that national geological mapping has achieved any marked degree of completion.

In all leading countries, however, some official geological map series are available, and these will often prove useful in civil engineering work although necessarily in a general way only, since even the most widespread construction projects will call for some detailed geological investigation. Appendix B of this book is a guide to the publications of geological surveys of the English-speaking world; this will be of assistance in leading those interested to the best source of such published maps as are available.

In addition, the following summary will serve as a general guide to the present state of official geological mapping, the results of which have been published. It is based (by permission) on data given on page 119 of "Methods in Geological Surveying," by Drs. Greenley and Williams, a book that can be highly commended to all interested in geological fieldwork. Surveying has been undertaken in about 65 countries. Taking the British 6-in. and 1-in. maps as 1:10,000 and 1:60,000 respectively, the following summary indicates progress up to 1930:

1:2,500	Some small parts of the British Isles, mainly privately, a liberal estimate being less than 500 sq. miles
1:10,000	British Isles, not yet completed; some small parts of China and one or two other countries
1:20,000	Central Europe; part of South Africa; Victoria
1:60,000	The rest of Europe to the west of U.S.S.R., and the east of Spain; parts of North America
1:100,000 and smaller, often less than 1:1,000,000	Nearly all the rest of the surveyed countries, including parts of U.S.S.R., Spain, most of North America, India, Australia, Africa, and Japan
Untouched	Much of U.S.S.R., North America, Africa, and Australia; practically the whole of continental Asia (extra India), Central and South America, the East Indian Archipelego, the Pacific Islands, the Arctic and Antarctic lands

It thus appears that about two thirds of the earth's land surface has still to be mapped geologically.

It will thus be doubly clear that in the application of Geology to civil engineering work, geological mapping will always be necessary, to a more or less detailed degree. The following notes therefore indicate the general lines adopted in such work and illustrate basically the use of the resulting maps. They are but an indication; references, as usual, suggest sources of fuller information, but it must again be emphasized that nothing can take the place of actual experience in the field.

4.4. Topographic Maps.—As the only satisfactory means of providing conveniently for the accurate recording of the locations of all geological field observations, and as the necessary plan diagram on which to indicate the three-dimensional geological structure deduced from such observations, topographic maps of standard form and usual scale may truly be regarded as the basis of all geological map work. In this book, it is not necessary to discuss such maps or their preparation in any detail. Where official topographic maps are available to a suitable scale, as is the case in developed country, they can be used directly for geological work. The activities of the civil engineer are frequently well in advance of all such development and so often in advance of any general detailed topographic mapping. It is then necessary to combine the special topographic and geological surveying required for the project in view. It is doubly advantageous if the two branches of work can be carried out simultaneously and by the same surveying parties, both for preliminary reconnaissance surveys and for final detailed mapping. In all this work, standard methods of topographic surveying will be followed for the preparation of the base map. Whenever it is necessary for geological mapping to be undertaken separately in unsurveyed country (as is the case with much of the splendid pioneer work of the United States and Canadian geological surveys in their respective countries), the plane table provides what is undoubtedly the most convenient and satisfactory of all surveying methods; it will be familiar to all civil engineers, and its special value in locating the sites of individual geological observations will be evident.

4.5. Field Equipment.—Geological fieldwork normally involves an amount of traveling at least equal to that involved in ordinary civil engineering reconnaissance work; if the area to be covered is large, with widely scattered exposures, the traveling necessary may become considerable. The greatest attention must therefore be paid to the equipment carried by the investigator. General rules with regard to the rugged nature of hand instruments, adequate bindings for notebooks, and, above all, correct personal attire (minor matters all, but of importance sufficient to warrant brief mention) are the same as those familiar to civil engineers in topographical work.

Ample plain ruled notebooks are the first essential of the geological surveyor. Equally important is the topographic map to be used as a base (considering only surveying in country already mapped topographically, there being no need to detail the surveying equipment otherwise required). The map of the area available should be cut up into small sheets of convenient size, each of which can be fitted into a strong leather map holder. Field observations are then marked directly on to these sheets, first in pencil, to be inked immediately on return from the field. For locating exposures, a prismatic or Brunton compass is advisable, pacing from some determined points giving the necessary distances. A pocket barometer is often useful, where contours are not frequent on the topographic map to be used; such an instrument must be used with the care usually called for in topographic work. Finally, a clinometer is a pocket instrument almost indispensible for the determination of the dip of exposed rock strata. The remaining equipment is composed of items that can be carried in the pocket. Chief of these is the geologist's hammer; this is an essential for the examination of all rock exposures, to ensure that a fresh face be studied and not weathered rock. A pocket magnifying glass is yet another essential. On some excursions, a pocket phial of hydrochloric acid is useful for testing limestone and other carbonates. There will be few excursions from which it is not advisable to bring back rock specimens; cloth sample bags and labels (one for each specimen) should be in another pocket. Finally, a hand camera can often prove invaluable in geological fieldwork if only to assist the surveyor's memory. It is difficult to appreciate fully the service rendered to all fieldwork by photography unless old geological records are examined and the freehand sketches, often so laboriously prepared under trying conditions, are carefully studied. Alike in civil engineering and geological fieldwork, a camera should be a constant companion of the careful fieldworker.

4.6. Field Observations.—The initial simplicity of geological field observations is almost dangerous, certainly to a beginner. Although so simple, involving but little more than accurate observations with the unaided eye, they call for an unusual concentration of attention and involve a background of experience. Following a general reconnaissance of the area to be surveyed, a start must be made at examining all exposures of solid rock. At each of these, the nature of the rock, its dip and strike, must carefully be noted and recorded. These direct observations are, however, but the start of investigation. For although in topographic mapping, record is made only of what can be seen, in geological mapping the contacts of adjacent beds are the main goal of the survey, and these may more than often be hidden from sight by surface deposits of drift material. Thus it is that the direct evidence offered by actual rock exposures must be followed up by detailed examination of all topographical features that may give a clue to the hidden geological structure beneath. The type of soil found in an area is generally a good guide to the rock strata beneath: even vegetation is noticeably affected in some districts by a change in the underlying strata. In Norway, Vogt has found even the health of grazing cattle affected by the changes in the chemical composition of soils derived definitely from different rock beds. Topographical features are naturally a most important guide to hidden geological structure.

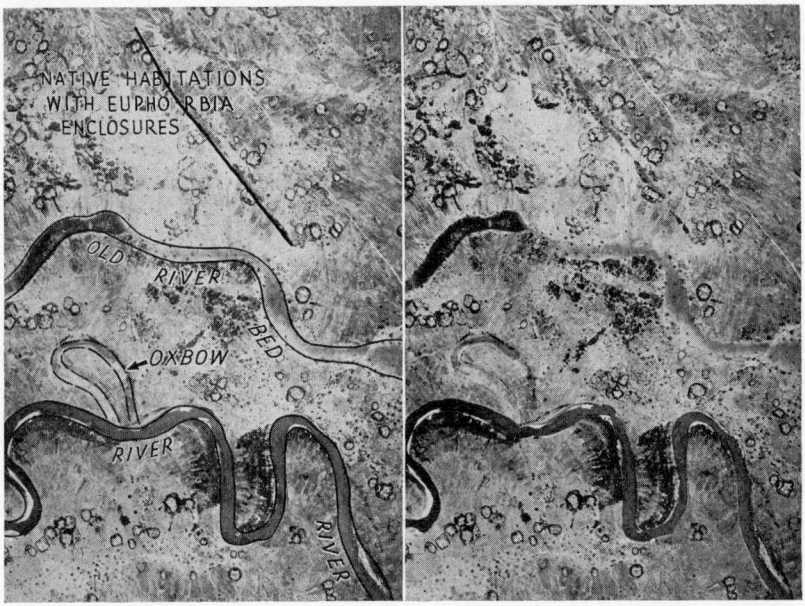

FIG. 4.1.—An example of aerial photography as applied in geological surveying—showing old and present-day river beds, an old oxbow, and a dyke near the shores of Lake Victoria, Africa.

(Reproduced, by permission, from "Geological Interpretation of Aerial Photographs" (Fig. 7), by J. J. Van Nouhuys, Trans. A.I.M.E., 1937, 126, *Metal Mining and Mining Geology, pp.* 622–623.)

Escarpments, indicating a relatively hard stratum; unusual deviations of watercourses and waterfalls, also often indicative of a hard rock layer; the elevation and location of springs—these and many other topographical features come within the critical review of the geologist in the field. The natural slope of ground being examined must always be remembered, especially when consideration is being given to such surface features as are demonstrated by means of the soil, in which case the *upper* limit of all special effects is clearly the only one that can be used with certainty.

It is naturally only rarely that a contact of two adjacent beds can be traced, or even seen, in other than isolated locations: its general location must be obtained by inference from the accumulated evidence of other observations. In a similar way, the various structural features noted in the last chapter will not often be revealed in anything but minor and infrequent locations. Occasionally complete sections of rock folds are exposed; but in all general cases, folding will be revealed only after the collection of numerous dip observations. Faults are shown up sometimes a little more clearly, especially if they are of comparatively recent origin so that their effect on topographic detail is still unobscured. Glacial structures and effects are sometimes clearly indicated, as will be suggested later by the description of these features. It cannot be too strongly emphasized that these notes are no more than a general review of geological field methods, mention being made only of the principal points to be noted. Experience in the field is the only sure guide, although much help can be obtained from the experience of others through the medium of the printed word.

A note may be added with regard to the use in recent years of the airplane for geological reconnaissance work. This adaptation of air travel will be familiar to civil engineers in connection with topographical reconnaissance work, so that no descriptive note is necessary, observation of general geological features being comparable to topographical airplane work. Aerial photography, interpreted by a skilled geologist, will often provide at least a general idea of the essential geological structure of the country shown. Figure 4.1 is a good illustration of the possibilities that aerial photography presents in geological survey work.

4.7. Drift and Solid Geological Maps.—The impression may have been given by the preceding notes that geological surveying can be carried out only in country exhibiting noticeable topographical variations and in which outcrops of solid rock occur with reasonable frequency. This is not the case, although it must be admitted that the mapping of the structure of exposed solid sections of the earth's crust as compared with mapping in a drift-covered area is comparatively easy. In addition to solid maps, however, there can exist for most districts a second complete set of geological maps which show in map form the nature and structural arrangement of the superficial deposits above the solid ground formation, generally characterized as *drift* (although including all recent deposits and not only glacial drift), the maps being described as *drift maps*. The importance of this branch of geological surveying to the civil engineer will immediately be obvious. Attention was first given to the mapping of drift deposits by the Geological Survey of England in 1863, and since that time they have gradually come to be regarded as on an equal footing with solid maps.

Special developments have arisen from the original simple mapping, notably the detailed surveys of soils with special reference to agricultural requirements.

Mapping methods for drift deposits are essentially similar to those for solid rock strata, with added emphasis, of course, on the close examination of soil constituents. The surveyor may be unwittingly aided by the excavation work of burrowing animals, not only because of the spoil dumps which they accumulate outside their burrows but also because of the well-known habits of the main species, *e.g.*, rabbits choosing dry sand deposits for preference. In addition, the hand auger provides a simple and ready means of obtaining actual sections through deposits with the expenditure of little energy or cost, this being in marked distinction to the use of test drilling in rock strata.

Drift maps are naturally complementary to solid maps, covering only those areas of ground over which solid rock is not exposed. The actual structure of any particular area covered by both solid and drift maps may therefore be visualized by imagining the latter superimposed on the former. Difficulties will often arise as to the exact line of demarcation between solid and drift deposits, but no general statement can be made on this point, each case having to be considered on its merits. Similarly, there are areas in which it will be difficult, if not impossible, to attempt to map the underlying rock strata, if situated at a great depth below the surface, in which case the drift map will constitute the only geological map for the area.

4.8. Special Aids to Fieldwork.—In normal geological survey work, it is generally economically impossible to obtain any evidence additional to that recorded by field observations, through the agency of special mechanical or electrical methods. The only check information that can usually be obtained is that provided by the strata records obtained in oil- or water-well boring operations and civil engineering operations of sufficient extent and magnitude to open up an appreciable part of the undisturbed crust of the earth. As the location of all these operations will be determined (naturally) from utilitarian rather than from abstract geological reasons, the data so provided will as often as not prove to be of little special value. Well records should perhaps be excluded from this rather wide assertion, but unfortunately it is not universally obligatory for well records to be made publicly available, so that a good deal of useful information is not accessible.

When geological fieldwork is carried out specifically for application to civil engineering construction, the situation may, and probably will be different. Funds should be available for adequate preliminary survey work, so that special work can be done if necessary; and the accuracy demanded of the work and the degree of certainty necessary will almost always render essential the checking of even the most definite results from ordinary field observations. In the case of superficial deposits, borings may be taken as far as rock level; test pits may also be dug. In solid rock strata, diamond-drill borings may be taken to determine the soundness of the rock and possibly the location and nature of underlying strata. In addition, there are now available what are called geophysical methods of underground surveying. These are a relatively recent development. Chapter VII gives a general description of the main methods and a bibliography.

It will readily be realized that localized information is of little use unless it can be considered in relation to general geological structure. Mechanical and geophysical methods, therefore, must definitely be regarded as complementary to general fieldwork and of major importance only in confined areas such as building foundations where ordinary geological survey methods cannot be followed. It is in these cases that published written records of previous or adjacent geological survey work can also be of assistance, a matter that will be duly emphasized later in this book. The matter is mentioned here, since a necessary preliminary to all geological fieldwork is a close study of all publications that have even a slight bearing on the area to be examined. These special aids to fieldwork, as is indicated, are generally associated with special survey work, as for civil engineering; they will therefore be discussed in detail when the application of Geology to preliminary civil engineering work is considered.

4.9. Method of Recording.—Geological maps, as distinct from the fieldwork on which they are based, must now be considered in some detail. The main types of information to be recorded on field maps have already been indicated. In the field, this is all noted carefully on the topographic map in use, for which purpose it is naturally advantageous to adopt a species of special shorthand in extending the signs used on topographic maps to cover the geological features now recorded. Most important of these special signs is that indicating the direction of strike and dip, one sign being generally used for this purpose, as shown in Fig. 4.2, the cross line being drawn in the direction of the strike, and the arrow in the direction of the dip. The angle of dip may be indicated by the use of the number of degrees made with the horizontal, at the foot of the arrow, or approximately by a variation in the length of the arrow from some standard length set by the observer. The limits of rock outcrops must also be indicated clearly, one method being to draw a line around the boundary of the outcrop,

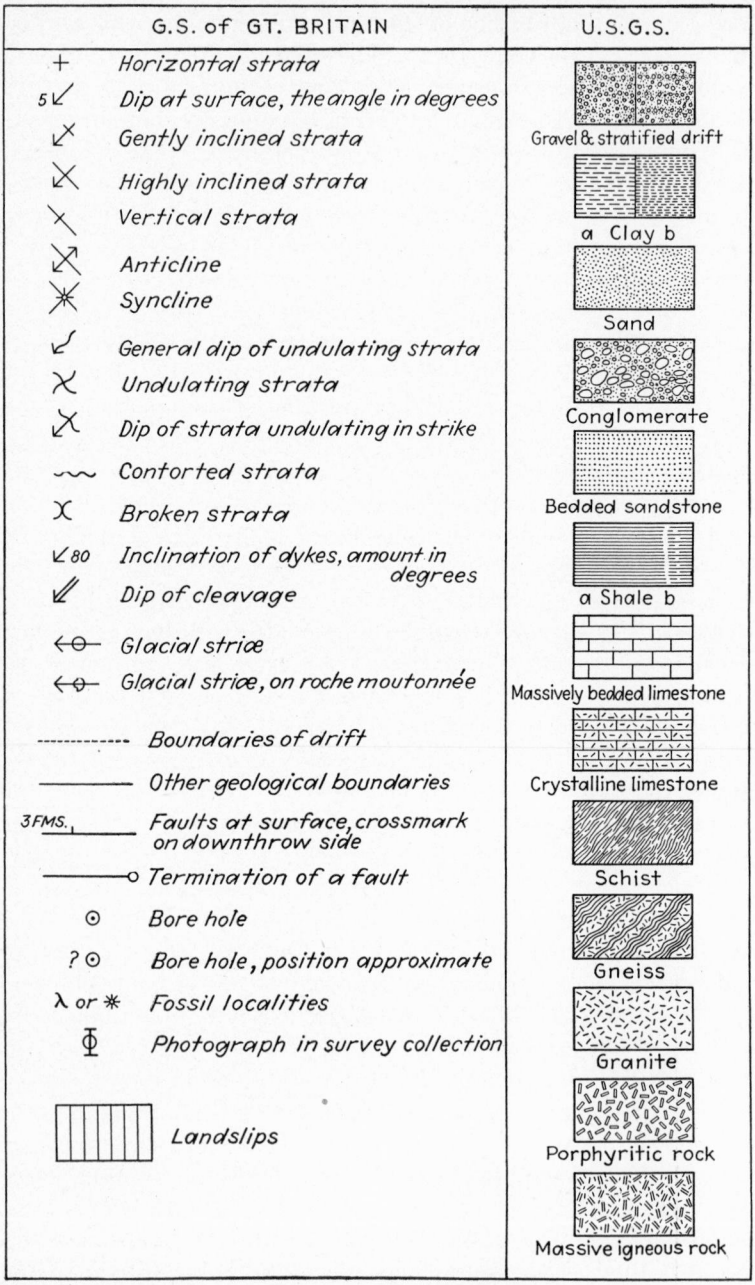

FIG. 4.2.—Typical signs and shading used for geological maps. (NOTE: The legends upon local maps should always be carefully studied, as conventional practice varies slightly.)

coloring or shading the side of this line corresponding *to the start of the drift coverage* of the solid rock. The diagram in Fig. 42 illustrates the more usual signs in general use; local variations will naturally be introduced by every observer into his own work, although it is essential that a rigid system be adhered to in all fieldwork to avoid possibility of error. Different types of deposits may be indicated either by lettering or by characteristic shading or coloring for each exposure.

At the conclusion of each day's work, all observations must be clearly and accurately inked in, either on the map used in the field or on a special copy of the base map retained for office use only, preferably on both of these. This daily checking of fieldwork cannot be overemphasized, since if notes are left over beyond the day on which they are made, all too often they may be misunderstood or found lacking in some minor detail when it is too late to return to the location for further observation.

As such notes are gradually accumulated on a map, geological structure will slowly reveal itself. The limits of the surface exposure will first appear in a general way; later, the boundary contact lines will become traceable with some degree of accuracy; structural features such as faults will appear in correlation with the rock strata; and finally the cause of certain topographical features may be made plain. Naturally, the progress of this elucidation of geological structure will vary from one locality to another, but usually the surveyor will reach a point at which he can finish off his map with some degree of confidence.

In finished form, the map will have the various strata exposed at the surface differentiated by several wash colors or by special shading, the former being preferable, the latter often essential for published maps where color printing cannot be adopted. Different surveys have different standard color forms; and wherever possible, the local standard arrangement should be followed. Observed contacts can be marked as full black lines; inferred contacts, as broken black lines. Faults are often marked as heavy (full or broken) white lines superimposed on the base colors. Other details noted on the original map will be retained, in a general way, indicated by pen and ink markings. Such finished maps should always be accompanied by an index table or legend of the colors used and possibly of the special signs incorporated in the map.

4.10. Geological Sections.—Remembering that geological maps are a two-dimensional representation of an essentially three-dimensional structure, it can readily be appreciated that prior to the completion of many a geological map it is necessary to plot a cross section along some special line across the area under investigation in order to study the inferred structure from another angle. Similarly, when a geological map has been completed, one or more sections will almost

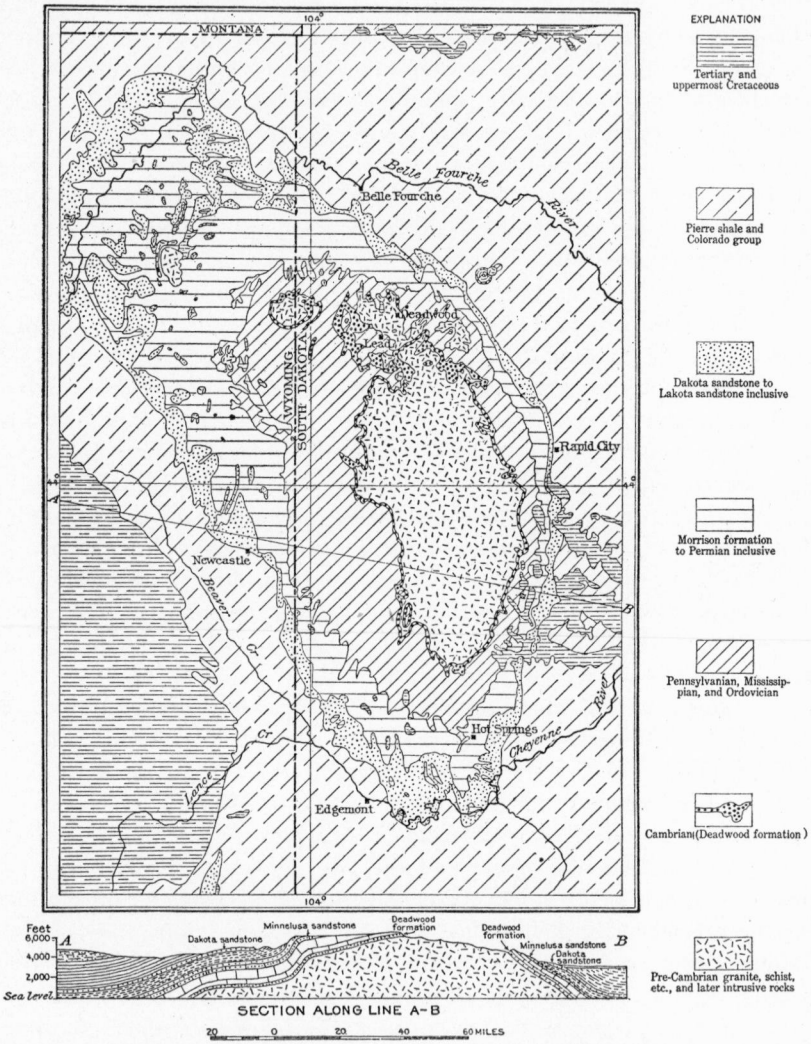

FIG. 4.3.—A typical geological map and section—of Black Hills in South Dakota and Wyoming, U.S.A., illustrating a structural dome.

(*After N. H. Darton. Reproduced by permission of the Director, U. S. Geological Survey, from U. S. G. S. Water Supply Paper No.* 489, *by Dr. O. E. Meinzer, Washington,* 1923.)

always be advisable in order to present a more vivid idea of the structure studied than can ever be obtained from a study of the surface map alone. It is not unusual to find one or more such sections printed in the borders of officially published geological maps. Naturally, the vertical scale used is distorted, but to engineers this feature of geological sections will not be strange.

The preparation of these sections from the base map obtained is a quite straightforward matter, needing no description here. The dip of the rock strata at the several contact planes plotted will indicate the general direction of the beds; and when these are drawn out, the fundamental structure of the rocks considered will often become clear straight away. In the use of such sections, an important point to be noted is that the section taken may not be exactly aligned with the direction of the dip of the beds, in which case a projected dip will show in the section. The relation of the direction of the section and the dip of the main strata should always be stated.

4.11. Interpretation of Geological Maps.—A completed geological map for country of normal structure will be a fairly simple looking specially colored and marked topographical map. It must be remembered, however, that it represents far more than the usual topographical map ever indicates and as such is worthy of careful and detailed study. An indication of this is to be found in the number of publications dealing only with the quantitative interpretation of such maps, a matter that can be but touched upon here. The basis of all interpretative study is a realization of the fact that the topographic data and, in particular, the contour lines represent the surface of the ground only, whereas the geological markings indicate subsurface structure as well.

The relation of the two leads to certain definite results which will gradually be realized after some study of geological maps. An elementary observation is that the projected distance between contacts shown on the map is no indication whatsoever of the thickness of the stratum in between, since this depends on the inclination of the beds and on the difference in elevation of the ground surface between the two contacts. An appreciation of this is fundamental. Certain characteristics will become evident after study of different types of maps, the most common of which can readily be memorized as four general rules:

1. Lines on the map representing boundaries of strata that are horizontal will be parallel with contour lines.

2. Lines on the map representing boundaries of strata that dip *into* a hillside will wind less than do the corresponding contours.

3. Lines on the map representing boundaries of strata that dip *away from* a hillside will wind more than the contours of the hill except when the dip exceeds the average inclination of the hill.

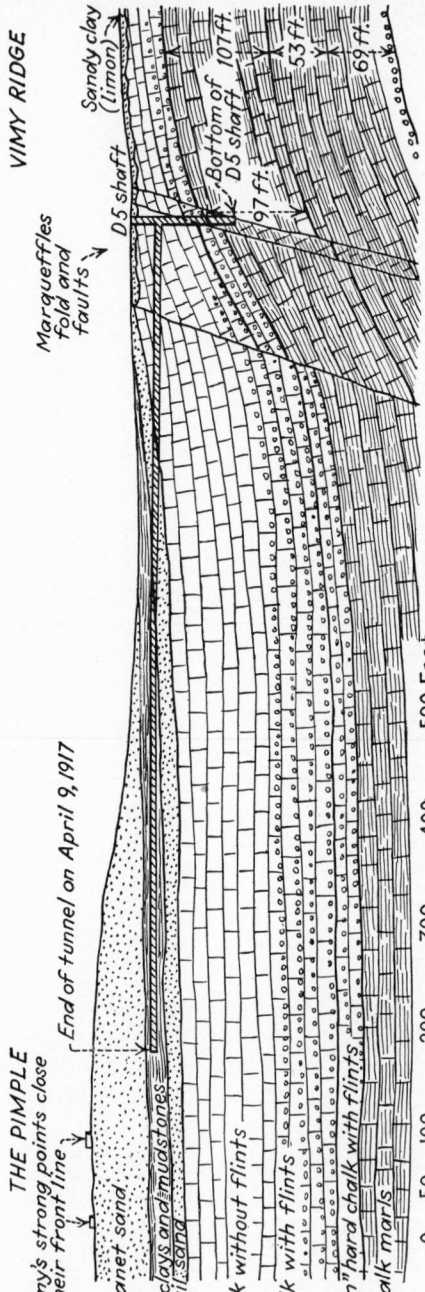

FIG. 4.4.—A typical geological section, as used for an unusual utilitarian purpose. (The section shows mining operations carried out during the World War of 1914–1919 between Vimy Ridge and "The Pimple," the tunnelling being kept in clay as far as possible to ensure silent working.)

(*Reproduced by permission of the Institution of Royal Engineers, England, from "Geological Work on the Western Front"—Fig.* 6.)

4. In all valleys, strata will show up on maps with a V shape leading *upstream* in all cases except those in which the dip is downstream at a greater angle than the inclination of the valley floor.

When once the reasons for these rules have been understood from a study of geological maps of different and differing regions, the interpretation of new maps will become much easier, and it will be found possible to obtain some idea of the structural relationship of the various rocks in any one area from a study of the geological map alone, without the aid of drawn-out sections. There are naturally other features that will become familiar, in addition to those suggested by the rules, and similar features will often be found displayed in the maps of any one group of rocks for one special area. Geological sections provide a general guide to the history of the region being studied and so enable a table to be prepared showing the geological succession for the area.

The foregoing notes have dealt with the interpretation of maps in a general or qualitative manner only. Quantitative study is also possible, and often desirable, and so has arisen a small subsidiary branch of specialized study which may almost be called *geological geometry* or *geological graphics*. By means of measurements taken directly from maps, of boundary locations and elevations, accurate calculations can be made of the actual (not projected) thicknesses of the beds being studied. As such geometrical work will not be strange to civil engineers, it need not be further described here, references at the end of this chapter including treatises dealing solely with this subject.

4.12. Fossils.—The relation of palaeontology to the other branches of geological study has already been outlined; its scientific significance and importance will probably now be apparent. It is therefore desirable that in all geological field investigations, including those undertaken for strictly civil engineering purposes, careful watch should be kept for any fossiliferous exposures. When any such strata are found, note should be made of the location and nature of the rock in which the fossils occur, and the attention of the nearest palaeontologist drawn to the discovery in case it is of any special significance. It may often happen that no palaeontological expert is available, in which case it will fall to the surveyor to obtain such specimens as he can conveniently transport to his headquarters. The selection of specimens must be a matter of judgment; their collection will be made possible by the judicious use of a geological hammer, no attempt being normally made in the field to obtain specimens free from the rock body in which they are found. The extraction of large fossils is, of course, necessarily the work of an expert. It is of vital importance that the exact location, and horizon in the particular strata, of each specimen be carefully recorded and that the specimens be well marked and carefully packed prior to transportation. It is not unlikely that such attention will lead many a civil engineer to a study of palaeontology, which will soon be found to have a fascination all its own.

4.13. Photography.—This brief survey cannot be properly closed without a few words of appreciation of the importance of photography in geological fieldwork. To the individual observer, even the results of quick work with a hand camera can prove of inestimable value when fieldwork is being followed up and checked up in the office, often serving as an aid to memory of indisputable veracity. As a means of recording details of temporary exposures such as are often encountered in civil engineering work, prior to these being hidden again from view, photography is unrivaled, although its use in this connection is now so usual that its value is apt to be underestimated. To the student, good photographs form a splendid complement to actual field experience (although in no way supplanting this), by enabling special structural features to be examined in comparison with those observed in fieldwork, and in many similar ways which need no enumeration. Finally, in all geological publications, photographic illustrations are an indispensable counterpart of the printed matter, and it is hoped that the examples used in this volume will themselves bespeak the value of this particular application.

No special technique is involved, the ordinary standards of exposure, developing, and printing sufficing. A small but important detail is that in all field photographs some means should always be used to provide a scale to the view depicted, by the inclusion of a human figure in general views or the geologist's hammer in close-up scenes. The true value of many photographs is lost if this precaution is not taken. The fullest possible notes of exposure should always be made; and in addition to the usual particulars, a special note of the exact spot from which the photograph was taken and of the direction in which the photographer faced as the exposure was made. In this way, the resulting prints can be correlated with other observed data. The technique of obtaining record photographs of microscopic views of thin rock sections is a highly specialized matter; and as it will but rarely come within the province of the civil engineer, it need be no more than mentioned here.

Finally, note must be made of the extensive collections of geological photographs now available from some of the national geological surveys, a few of which are reproduced by permission in this volume. Comprehensive indices of the many views thus available are issued, and these will often prove invaluable not only to the student whose sphere of fieldwork is for some reason limited but to all interested in the science in even a general way.

Suggestions for Further Reading

Although the books listed at the ends of the preceding three chapters deal generally with matters that have to be studied out-of-doors, they do not specifically treat of the methods to be followed in making geological observations, nor do they describe geological maps in any detail. The following books may usefully be consulted with regard to these two aspects of geological work:

Geology in the Field

Brown, C. B., and F. Debenham: "Structure and Surface," Edward Arnold & Company, London, 1929.

Geikie, J.: "Structural and Field Geology," Oliver & Boyd, Ltd., Edinburgh, 1905.

Greenly, E., and H. Williams: "Methods in Geological Surveying," Thomas Murby & Co., London 1930.

Lahee, F. H.: "Field Geology," McGraw-Hill Book Company, Inc., New York, 3d ed., 1931.

Geological Maps

Dake, C. L., and J. S. Brown: "Interpretation of Topographic and Geologic Maps," McGraw-Hill Book Company, Inc., New York, 1925.

Dwerryhouse, A. R.: "Geological and Topographical Maps," Edward Arnold & Company, London, rev. ed., 1919.

Earle, K. W.: "Dip and Strike Problems, Mathematically Surveyed," Thomas Murby & Co., London, 1934.

Elles, G. L.: "The Study of Geological Maps," Cambridge University Press, Cambridge, England, 1921.

PART II

GEOLOGY AS APPLIED IN CIVIL ENGINEERING

CHAPTER V

THE CIVIL ENGINEER AND GEOLOGY

> Geology stands to [civil] engineering in the same relation as faith to works. . . . The success or failure of an undertaking depends largely upon the physical conditions which fall within the province of geology, and the "works" of the engineer should be based on the "faith" of the geologist.
>
> BOYD DAWKINS, F.R.S.[5.1]

Every branch of civil engineering has some contact with the surface of the earth, the works designed by the civil engineer being supported by or located in some part of the earth's crust. The practice of civil engineering includes the design of these works and the control and direction of their construction. Geology is the name given to that wide sphere of scientific inquiry which studies the composition and arrangement of the earth's crust. This book is concerned with the application of the results of this scientific study to the practice of the civil engineer.

The relation of the science to the art is at once so obvious and so intimate that general comment upon it might appear to be uncalled for; unfortunately, this is not the case. It may be that the very intimacy of the relationship has been the main reason for its frequent neglect. Whatever the cause, the fact remains that, for a considerable period, civil engineering work was carried on, in all countries, with but little conscious reference to Geology or to geologists. In the early years of modern engineering, at the start of the nineteenth century, when engineering had not become the highly specialized practice known today, many civil engineers were active geologists also, William Smith being the outstanding example of these pioneers. Robert Stevenson combined geological study with his early work in railway construction, as is related in Chap. VIII; and other well-known figures in the annals of engineering history carried out distinguished work in both spheres of action.

There is today a growing appreciation of the vital importance of the science to all who practice the art. Geology is now a usual feature in courses of training for civil engineers; current civil engineering papers will be found to contain frequent references to the geological features of the sites of works described; and the newly founded scientific approach to soil studies provides a common meeting ground for civil engineer and geologist which has already done much to foster their cooperation. The divorce of the science from the art persisted for so long, however, that their complete correlation may not yet be realized for some time; some thought may therefore be given to general aspects of the contacts between them before the subject is considered in more detail, so that the reader may have at the outset a clear idea of their fundamental relationship.

5.2. The Science and the Art.—Frequently in the past, geological considerations have featured prominently in the study and discussion of failures of civil engineering works; in fact, to some engineers Geology may still be thought of as merely a scientific aid to the correct determination of the reasons for some of the major troubles that develop during or subsequent to construction operations. Valuable as is the assistance rendered by geologists and by the study of geological features, in such "post-mortem" considerations, the very fact that geological features may have had something to do with the failures being considered would seem to suggest, with abundant clarity, that the best time to consult a geologist or to study geological features is before construction begins. In this way, the science could serve the art in a constructive rather than merely an analytical manner. It will later be seen, as such applications of Geology are considered in some detail, that not only can this constructive service of the science be in the nature of preventing possible future troubles but it can also suggest new solutions to engineering problems and often reveals information of utility and economic value even in preliminary work.

As an example of the more obvious effects of geological features on major civil engineering works, consider the underground railway services in two such cities as London and New York. In London, owing to the fact that the city is built on a great basin of unconsolidated material including the stratum of well-known London clay, tube railways—located well below ground level, easily and cheaply excavated in the clay—have provided an admirable solution to one part of the city's transportation problems. In New York, on the other hand, as the surface of Manhattan Island on which the city is located is underlain to a considerable extent by Manhattan schist, underground railways have had to be constructed in expensively excavated rock cuttings just below surface level, as the innocent visitor to that great city is so loudly reminded if he happens to be standing on a ventilation grating when a train passes in the subway below!

Many similar instances of the profound effect of local geological characteristics upon major civil engineering works could be cited, but all would serve to emphasize the same point—how closely the science and the art are related and how dependent civil engineering work, in general, must always be upon Geology. It is, indeed, no mere figure of speech to say that the science of Geology stands in relation to the art of the civil engineer in just the same way as do physics, chemistry, and mathematics. The importance of these sciences to the civil engineer is never questioned; they are always thought of as the necessary and inevitable background to a civil engineering training. It would be inconceivable to imagine any engineer worthy of the name who was not familiar with the chemistry of simple materials. So should it be in the case of the civil engineer in respect to the nature of the materials on which or in which he is to construct his works.

There is a minor point of difference to be noted, when thinking of the relation of Geology and of parallel sciences to civil engineering. The latter are utilized directly by the engineer himself, mathematical and physical methods being important in many branches of design work. In the case of Geology, the service rendered to civil engineering is by the *application* of the findings of the pure science to the specific problems of the engineer. In the case of construction problems, for example, it is the task of the geologist to state the probable difficulties, and that of the engineer to overcome them; in the case of materials of construction, it is for the geologist to say where they may be found, and for the engineer to obtain them and put them to use. It is a minor point of difference, but is thus mentioned since it leads to another thought of some importance. As has already been pointed out, the geologist has the whole of the earth's surface as his laboratory, and in every locality upon it he will encounter purely local problems, having characteristics that may not be duplicated elsewhere. Thus it is that every application of the results of the science, or its methods, to engineering work will likewise be in some respect unique. In this sense, too, there is some difference between the relationships with civil engineering of Geology and of associated sciences; but although local characteristics may vary, the fundamental geological principles applying to them do not. And it is these guiding principles that constitute the basis of geological study—study that, it is hoped, will be seen to be so essential a part of the mental equipment of every civil engineer.

5.3. Training in Geology.—As most civil engineers who enter upon their professional careers do so by way of a university or technical college education or the equivalent course of study leading to the examinations of a leading professional society, considerations of the training of civil engineering students in Geology may for convenience be confined to university courses. It will be noted that reference is made to training in *Geology* and not to training in Civil Engineering Geology or some such suggested course. There is no special brand of Geology applicable to civil engineering. There is, however, a special course of study possible and desirable in considering the application of fundamental geological principles and methods to civil engineering problems, and this must be complementary to the study of Geology, as such. This is an important distinction which is emphasized in the title of this book, and a study of the curriculums of many universities and colleges will show that it is generally appreciated in university work.

The training in Geology necessary for civil engineers must obviously be general in nature, enabling students to obtain a good grasp of the nature of the subject and of the character and interrelation of its various branches. Thereafter, in the usual college course, time will not permit of detailed attention being paid to all branches of the subject, nor does this seem desirable. Attention will have to be concentrated on those branches of Geology the applications of which are of special importance in civil engineering practice. These include Physical Geology, Structural Geology, and Petrology. Study of geological maps and sections can be a valuable laboratory aid to the first two of these branches, and examination of rock slides in the petrological microscope is an eminently satisfactory complement to lectures on Petrology. The actual making of rock sections is something that the student will never have to do in engineering practice and so can be neglected in laboratory work; similarly, it does not seem advisable that civil engineering students should be called upon to spend much time on the examination of fossils, although a little time spent in introducing students to practical palaeontological work will be a valuable stimulus to their interest in this branch of study.

Of fundamental importance in all training is experience in the field. It may be said with propriety that no course in Geology for civil engineers can be regarded as in any sense complete without a reasonable period of time spent on actual geological survey work. Local conditions will dictate how fieldwork must be arranged; but if it is at all possible, a continuous period (of one or two weeks) spent in a suitable locality will usually be found more effective than any number of short periods fitted into a regular schedule. It is to be regretted that local circumstances usually conspire to make it impossible to combine a geological field camp with the usual topographical survey camp, combination of the two giving an ideal arrangement.

Turning, now, to the second part of geological training for civil engineers, it has been suggested that this should be an introduction to the study of the applications of what has been learned of the science to the problems encountered in actual engineering practice. The word *introduction* is used advisedly, as this note refers specifically to college work. The average student will have had no extensive experience on outside civil engineering work and will therefore have much to learn with regard to civil engineering construction. If, however, he can start to do this in association with an introductory study of the applications of Geology to such work, a double purpose will be served, to great advantage. The usual college course does not permit of much time being devoted to practical subjects, but courses on foundations and construction methods present possible opportunities for instruction of the nature indicated.

All instruction in the classroom is but a cursory introduction to what is a lifelong study for the majority of civil engineers, considerations of the applications of Geology to their work being, either consciously or unconsciously, an essential part of that basic experience which is the most prized possession of all members of the profession. How much the better this experience can be gained if the engineer possesses at the outset a fundamental knowledge of Geology will be evident from what has already been written. Without a correct attitude of mind toward Geology, no civil engineer will benefit from instruction in the science no matter how well this may be presented. It is, indeed, the development of this mental attitude that should be, and generally is, the principal aim of those charged with instructing engineering students in Geology. The task is no easy one, especially in view of the fact that students will not have had any preliminary opportunity of appreciating the vital importance of geological features in actual construction work.

Fortunately, appreciation of the science as an important aid to the civil engineer is steadily becoming more widespread. Perhaps the best indication of this change is to be found in the nature of the papers describing civil engineering work presented to the various engineering societies. If the transactions of these societies of twenty or thirty years ago are examined, mention of the geology of the site of the works described will be found only infrequently; if similar transactions of recent years are studied, few such papers will be found that do not include at least a brief reference to the local geology affecting the works described. Frequently, entire sections of engineering papers will be found to be devoted to geological considerations; many examples of this kind will call for mention in later chapters of this book. This development is of real significance and is encouraging, for the presentation of any description of an engineering structure without some reference to foundation-bed conditions or other corresponding geological data is equivalent to presenting a paper on a bridge without mentioning the loading used in design or omitting all reference to the nature of the materials used in construction.

5.4. Practical Experience.—The appreciation of geological features and particularly of the characteristics of the materials that make up the earth's crust is not acquired solely by training in Geology. Many engineers, and particularly those who have spent much time on construction, possess this appreciation unconsciously as a product of their wide practical experience. Many engineers there have been, and some there probably still are, who have never troubled to find out what Geology is and yet who know instinctively many of the matters herein discussed solely in geological terms. This point is emphasized, since the dividing line between what may truly be called *practical experience* and a cultivated appreciation of geological features is quite indeterminate. Convenience will be served if references are confined, throughout this book, to this trained attitude of mind toward geological features, but it is always to be understood that the practical experience of engineers unversed in Geology is included, although not stated, in such references.

This knowledge derived from wide experience is not the common thing that some people suggest. It is almost intuitive in nature and certainly much more than merely factual knowledge derived from long observation. Some men could not acquire such intuitive judgment even after a lifetime of outside experience; to others it comes easily and early in their lives. The suggestion may therefore be advanced that all so-called practical experience may not prove to be this sound intuitive judgment. All who are to be in contact with men engaged on construction work will do well to remember that sometimes the so-called "practical man is the one who practices the theories of thirty years ago."

Whether appreciation of Geology comes by training or by intuition, it is in the field that it can so well serve the civil engineer. Thus equipped, he will be able to appreciate the beauties of scenery all the more vividly and in his topographical work realize the reason for the many special features that he is called upon to survey. He will be able to direct his exploratory work all the more accurately appreciating, at least to some degree, what may lie hidden beneath the surface of the ground. And on construction, every step that is taken in connection with excavation and foundation work will take on a new and added significance to the resident engineer who has had his interest awakened to all that Geology can mean in his supervision of such work.

5.5. Employment of Specialists.—One of the most important results of a desirable attitude toward Geology on the part of civil engineers should be that they will then know when to call for the services of an expert geologist. As a general rule, engineers cannot hope to be more than the equivalent of amateur geologists, familiar with the science and its methods, appreciative of its value, but not qualified to carry out any detailed investigations either in the field or in the laboratory. On small jobs and for routine civil engineering work, this general familiarity with the science will enable the engineer to tackle practically every problem that he encounters. On large jobs, however, and for special work, specialists must be consulted, and a realization of the necessity for consultation can be taken to be one of the hallmarks of a true engineer, rather than the reverse, as is sometimes erroneously believed to be the case.

On occasion, the engineer may meet with opposition to such a course, the leading argument met with in this connection being that similar works have been carried out successfully in the past, without the special aid of a geologist; to employ a geologist means the expenditure of more money; so why not carry on the work without this extra assistance? Superficially, such an argument may appear to be difficult to refute unless thought is given to the parallel case of fire insurance premiums! There was a time when building owners never thought of paying out this extra money, and owners then continued to live and prosper. But how many today would neglect to take advantage of modern fire coverage, despite the adequacy of modern fire-fighting equipment?

Thus may the argument be answered in commercial terms. Of far more importance, however, particularly for the engineer's own satisfaction, is the impressive record of the great services already rendered to the art by geologists, known and unknown, a record to which the examples cited in this book pay some tribute. How many disasters have been prevented through the use of geological advice can never be estimated, but the record of failures that have occurred when such advice has not been taken is at least an indication of what this constructive contribution has meant to civil engineering achievement. The important work of Dr. C. P. Berkey and his coworkers in connection with the Catskill Aqueduct for the water supply of the city of New York is a telling example. The geological investigations associated with the tunnels that now exist under the River Mersey at Liverpool, England, and particularly the work of Professor P. G. H. Boswell in connection with the latest vehicular tunnel, offer similar evidence. The great Boulder Dam on the Colorado River in the

United States, the new Madden Dam in the Panama Canal Zone—these and many other successful dams testify silently to the value of the assistance that civil engineers may obtain from specialist geological advisers.

These few names could be extended to fill many pages, but the examples of the cooperative work quoted throughout this book will probably be found to be an equally impressive record. Usually, geologists will be called upon to assist on engineering work in the capacity of consultants, the practice of having a geologist as a member of the official board of consultants so often followed in North American work being a most satisfactory arrangement. Only occasionally will it be possible to have geologists permanently associated with an engineering staff, but it may be of interest to note that the engineering staff of the Tennessee Valley Authority (T.V.A.) of the United States Government, the great federal agency set up to develop the natural resources of the valley of the Tennesse River, has (in 1938) the benefit of cooperation with the special Division of Geology of the Authority, headed by Major E. C. Eckel and consisting of a staff numbering almost thirty geologists, the work of which is essentially that of applying geological studies and methods to the engineering and mineral resources problems of the Authority.

5.6. Geologists and Civil Engineering Work.—This book is not intended for the use of geologists, and it would be invidious for the author to comment upon the attitude of geologists towards civil engineering work. It may, however, be useful for the benefit of engineers to note that geologists appear to welcome gladly any opportunity afforded to them of cooperating on civil engineering work, their only regret being that these opportunities do not occur more frequently. This suggestion is confirmed by the contributions made by geologists to the discussion of engineering papers, such as the following statement by Dr. T. Robinson of the Geological Survey of Great Britain, during the discussion of a paper at the Institution of Civil Engineers, London:[5.2]

> The records of the Geological Survey showed conclusively that closer co-operation between the geologist and the engineer would be greatly to the advantage of both, and it was a pity that there was no very direct way in which geologists could be kept informed of the progress of important excavations.

Practically the only way in which geologists can be advised of the new exposures shown up by civil engineering operations is by the engineer in charge of the work. May it be urged again, therefore, that, whenever possible, civil engineers should advise the director of the appropriate geological survey and the head of the geological department of the nearest university of all excavation work of interest under their charge, so that geologists may at least have the opportunity of seeing the exposures before they are covered up. This courtesy demands little of the engineer, but it may often lead to scientific data of great value.

5.7. Cooperation of Geologists and Civil Engineers.—When geologists are called in to advise upon civil engineering work, they will have to act in conjunction with the engineers responsible for the work. Thus arises the need for cooperation between the civil engineer and the geologist, the practical builder and the man of science, which can so often prove to be a partnership of great value and one often leading to considerable personal pleasure. It is, in some ways, a union of opposites, for even the approach of the two to the same problem is psychologically different! The geologist analyzes conditions as he finds them; the engineer considers how he can change existing conditions so that they will suit his plans. From his analysis, the geologist cites problems that exist and suggests troubles that may arise; the engineer's main task is to solve the problems and overcome the troubles. The final responsibility for the decisions involved must rest, always, with the engineer; but in coming to his conclusions, the latter will be guided by and will probably rely upon the factual data given to him by the geologist.

This joint work, therefore, calls for a fine degree of real cooperation. As was suggested in Chap. IV, the geologist has to remember that what the engineer wants is a clear picture of the geological conditions which will be related to his work, presented to him as concisely as possible and with their practical utilization in view. On the other hand, the engineer must remember that the geologist *is* a geologist and not an engineer and need not expect to receive from him a report of exactly the same nature as he would receive from another engineer. In many cases, the most effective results can be achieved if the engineer is able to give to the geologist, at the outset of his work, a list of specific questions to be answered if possible. These may relate to geological conditions, to the necessity for and location of further exploratory works such as boreholes and test pits, and to similar matters. He should also be willing to cooperate with the geologist in allowing the latter to take advantage of his special engineering work in order to pursue, within reasonable limits, any aspects of purely scientific interest that may develop in the course of his main task.

5.8. Geology and Topography.—Before beginning to study applications of the science to special branches of civil engineering work, some thought may be given to the relation of Geology to certain basic aspects of engineering procedure. When a new project comes up for consideration, the civil engineer will first wish to know something of the locality in which the work is to be carried out; preliminary investigations and studies will then be made; and when these are approved, complete contract drawings and specifications will be prepared. When financial arrangements are made, tenders will be called for, a contract awarded, and construction begun; when the project is complete, it must be periodically inspected and maintained in good order. Brief comment on the application of Geology to each of these main divisions of civil engineering procedure will be made.

Information with regard to an area in which work is to be carried out will generally be obtained by a civil engineer by visiting the area, even if only for a hurried tour of inspection, and by studying any literature generally descriptive of the district. If geological reports are included in the literature selected, and if the topography of the area is studied with due regard to the significance of the local geology, then a more vivid and more accurate picture of the district will be obtained than would be the case if the geology were neglected. Elementary study of the geology will demonstrate its relation to local scenery, so that it should not be difficult for a civil engineer trained in Geology to visualize, after the necessary inspection and investigation, the general structure of the ground with which he is concerned and the origin of leading features of the local topography that will be of importance in his work.

As an example of broad concepts of local geology, the following extract may be quoted from a Special Lecture delivered in 1935 at the Institution of Civil Engineers, in London, by Dr. H. E. Gruner of Switzerland, on "Hydro-Electric Power Development on the Rhine":[5.3]

In the stretch which concerns us, the Rhine cuts through the jurassic system, the tertiary system, the lower layers of trias, the gneiss massive of the Black Forest and some layers of the permian system. The power-plants are founded on each of the different layers, and the results have proved favourable in every case. Of still greater importance in their effect on the longitudinal profile of the Rhine, and thus upon its character as a source of energy, are the historical occurrences during its origin. Even in pre-glacial times the Rhine flowed through the same valley as it does today, but, owing to the enormous quantities of gravel that the glaciers deposited during their retreat, the Rhine lost its old course and was partially forced into a new bed. Whilst scouring out its new bed it eroded the underlying rock, which it took more time to carry off than the gravel. This produced the different falls and rapids as we find them at Neuhausen, Reckingen, Schwaderloch, Laufenburg, Ryburg-Schworstadt, Rheinfelden and on the Kembsersill. . . . The soundings and geological studies that were needed for the different power-plants, enabled us to determine the old river-bed for almost its entire length. These pre-glacial river-beds of the Rhine are technically important in many ways; they always carry water and are thus well suited for water-supply works. During the construction of power-plants they may also become the source of disagreeable surprises if they are not thoroughly examined at first.

This descriptive note gives a general picture of the Geology of the sites of the plants described by Dr. Gruner; it could not, of course, have been written without study of relevant geological reports. Although so general, it indicates some of the geological problems that had to be faced during the construction of the Rhine water-power plants and suggests what preliminary exploratory investigations would have had to be undertaken. Many similar notes could be quoted, but all would serve to illustrate the value of general conceptions of local geology and, incidentally, of such notes as introductions to descriptive papers on civil engineering work.

Preliminary studies may be undertaken in more detail, especially if civil engineering work is to be carried out over a large area. General reconnaissance surveys will probably have to be made, and general topographic maps either prepared or checked. Simultaneously with this work, geological reconnaissance can always be carried out with advantage, the local geology being studied in more detail and correlated (although still in a general way) with engineering requirements. It is not often that civil engineers will be called upon to undertake extensive work of this nature; and when the need does arise, special organizations will usually be recruited to undertake it; the subject will not, therefore, be discussed in detail. It is of interest, however, to glance at a concise statement of the results of one important geological reconnaissance of this nature, covering an area of 40,000 sq. miles—that under the control of the T.V.A., in the United States. Table D, which is reproduced (by permission) from a geologic bulletin of the T.V.A., shows clearly the preliminary results of the work of Major Eckel and his staff, and its value as a guide to the engineering staff of the Authority can easily be appreciated. The table illustrates also the relation of topography to the local geology. Attention may be specially directed to the information given under the headings Stone, Clay, and Cement Supplies, the value of which in the preliminary stages of project work will be evident.

5.9. Planning and Design.—Having obtained a general idea of the district in which his work is to be carried out, the civil engineer will next proceed with his preliminary plans and estimates. Gradually these will be evaluated and discussed until finally an accepted scheme

TABLE D.—ENGINEERING GEOLOGY OF THE FOUR T.V.A. SUBREGIONS[1]

Age of rocks	Pre-Cambrian and metamorphic (the oldest rocks)	Cambrian, Ordovician, Silurian, Devonian, and Mississippian	Pennsylvanian (Coal Measures)	Cretaceous, Tertiary, Recent (the youngest formations)
Prevailing rocks	Gneisses, schists, granites, and other igneous rocks, chiefly slate, quartzite, marble in outliers and along west flanks of mountains	Shales, limestones, sandstones	Shales, sandstones, conglomerates	Sand, gravel, clay, chalk
Geologic structure	Usually steep dips to southeast. Faulting common but not noticeable	East of Coal Measures close folds, chief dip southeast. Faults numerous and often large. West of Coal Measures dips flat, normally southwest; faults rare and small throw	Dips flat except along east edge of coal plateaus where steep dips and faulting are common	Dips low to west or southwest. Faults rare and unimportant from engineering standpoint
Stone supplies	Ample supplies of granite, diabase, gneiss, and slate; local supplies of marble. Sand and gravel in local stream deposits only	Extensive supplies of limestone; widespread though less abundant supplies of sandstone. Chert gravels extensive west of coal fields	Sandstones abundant and locally furnish attractive building stone	No real stone in whole area. Very extensive sand and gravel beds in several formations
Clay supplies	Common clays rather scarce; kaolins common in certain areas	Brick and tile clays common in river flats; good shales available for similar products	Brick clays rare. Fire clays common in all coal fields	Extensive clay deposits for all types of clay products
Cement supplies	No cement material. No lime material except in the marble regions. No gypsum or slag	Extensive supplies of limestone, slag, etc., for portland and slag cement. Lime abundant. Gypsum extensive in southwest Virginia	No lime, cement, or plaster materials anywhere	Lime supplies scarce except clayey chalks in Cretaceous. No gypsum or slag
Surface waters	Runoff rapid. Waters soft, usually fairly pure	Runoff is moderately rapid. More so than the slopes indicate, because the surface is tight. Waters usually very hard and rarely pure	Runoff rapid. Waters soft, usually moderately pure	Waters soft, rarely pure. Runoff usually slow

[1] Reproduced, by permission, from E. C. ECKEL, Engineering Geology and Mineral Resources of the T.V.A. Region, *Knoxville Tennessee, T.V.A. Geol. Bull.* 1, 1934.

TABLE D.—ENGINEERING GEOLOGY OF THE FOUR T.V.A. SUBREGIONS.[1]—(*Continued*)

Age of rocks	Pre-Cambrian and metamorphic (the oldest rocks)	Cambrian, Ordovician, Silurian, Devonian, and Mississippian	Pennsylvanian (Coal Measures)	Cretaceous, Tertiary, Recent (the youngest formations)
River topography	Narrow gorges with frequent falls and rapids. Occasional small flats with relatively narrow flood plains. Gradient high	Streams mostly bordered by low flood plains, increasing in area from north to south and west. Gradient moderate at north; low at southwest	Narrow gorges with frequent falls and rapids. Flood plains and flats very rare and narrow. Gradient high	Streams bordered by low and very broad flood plains. Gradient very low
Caves and sinks	No sinks and no real caves	Caves and sinks common and extensive in several of the purer limestone formations	None originating in Coal Measures, but solution of underlying limestone causes rare sinks	No caves or sinks
Underground waters	Underground supply limited; localized along cleavage planes or other fracture surfaces	Underground supply extensive but often contaminated. Occurs in solution channels in the limestones and in sandstones.	Underground supply limited by smallness of gathering area; in sandstones and usually pure	Extensive underground supplies in several sand formations
Soil erosion	Erosion slight except in the deeply weathered granite areas and on the Piedmont	Erosion extensive in most pure limestone areas. Slight in shales and nonexistent in sandstones	Erosion slight	Extensive gullies form rapidly in many clay and sand areas

[1] Reproduced, by permission, from E. C. ECKEL, Engineering Geology and Mineral Resources of the T.V.A. Region, *Knoxville, Tennessee, T.V.A. Geol. Bull.* 1, 1934.

or design is evolved which can then be prepared in detail. All this work can properly be carried out only if the engineer possesses an adequate and detailed knowledge of the ground in which his work is to be located and of the natural materials available at or near the site. This essential information will be obtained by means of detailed geological fieldwork and by exploratory investigations such as boreholes and test pits. This preliminary exploratory work is so important that it is considered in detail in the following chapter; it is mentioned here in order to show the logical association of Geology with this leading phase of civil engineering work.

5.10. Contract Plans and Specifications.—The final design of the civil engineer will usually be incorporated in a set of contract plans and specifications, on the basis of which tenders for performing the work involved will be called for from contractors. In the case of those projects which are carried out by direct administration, instead of by contract, a complete set of drawings equivalent to a set of contract plans will be necessary, and the equivalent of a contract specification for the guidance of the engineers in actual charge of construction operations.

The preparation of these documents marks a definite change in the nature of the engineer's work and in the extent of his responsibilities; for when once issued to a successful tenderer and made the basis of a formal contract, they become legal documents and entitle the contractor to certain rights, in this way taking absolute control of construction operations to some extent out of the hands of the owner and of the engineer as his representative. If, therefore, the application of Geology can in any way assist in making the preparation of contract documents more effective and less open to question, the science will be rendering particularly valuable service. It would appear probable, from considerations already advanced, that there are several ways in which assistance can thus be rendered.

It has already been mentioned that the chief results of preliminary exploratory work, and associated geological studies, will be to provide accurate knowledge of subsurface conditions at the site of the proposed work and with regard to the availability of suitable construction materials in the vicinity. The first of these results will affect the design of the engineer and also the construction methods to be adopted by the contractor; the second may have some bearing on the design adopted, especially from the economic angle, and will have an appreciable effect on construction planning. It is clear, therefore, that every effort should be made by the engineer to include in his contract documents as much as possible of the information with regard to the site and available materials that has been obtained through the preliminary work.

On the contract plans, this can be done by showing full details of the records of boreholes, test pits, and other subsurface explorations. These should be shown not only in section but also in a general plan showing their correlation with the location of the work to be constructed. It is difficult to conceive of a set of civil engineering contract drawings (apart from those showing only superstructures) on which the presence of borehole or similar records is not essential, so that the contractor may take advantage, if he is so minded, of all the information relative to his work that the engineer has available. In many cases, and when possible, it will be advisable to show accurately the nature of the geological structure adjacent to foundations, instead of the usual pictorial representation of rock or unconsolidated material. In certain special cases, such as tunnel work, it will be desirable to show on the profile drawings full details of the geological formations that are anticipated along the line of the work to be built. Reference to, and illustrations of, special data on drawings will be given in subsequent chapters.

The same criterion should be aimed at in the preparation of civil engineering specifications. As a general rule, the opportunity for use of geological data in specifications will arise in one or more of four ways:

1. In the provision for possible alterations in design due to variations encountered in subsurface conditions.
2. In the provision of data relating to available materials of construction.
3. In the clauses relating to methods of construction to be adopted.
4. In reference to the measurement and payment for excavation.

The first of these divisions is closely related to preliminary exploratory work and so may conveniently be left over for consideration in Chap. VI. The second calls for the clearest and fullest explanation of all the facts known relative to the materials available; it will also be referred to in the next chapter. The third touches upon a most important matter but one that is generally and advisedly left to the selection of the contractor, usually with such a qualification as this:

> . . . The contractors are to submit to the engineers a statement with drawings showing how they propose to carry out the works, but any approval of the engineers is not to relieve the contractors of any liability that devolves upon them under this contract.

This provision, as indeed the whole essence of a civil engineering contract, places with the engineer the moral responsibility for giving the contractor all available information which may assist him in his construction planning and methods. Data included in the specification in addition to that shown on the contract drawings will assist in fulfilling this obligation, and the fullest use should therefore be made of the opportunity thus provided.

The fourth division, relating to excavation, is of such importance that it will be considered in some detail.

5.11. Earth and Rock Excavation.—Contracts for civil engineering projects are designed to result in the work being carried out to the satisfaction of the owner, to the requirements of the engineer, and with due safeguards for the contractor. General conditions define the scope of the contract; specification and contract drawings detail the design of the engineer; and quantities and unit prices (in the usual type of contract) define the extent of the contractor's operations and his remuneration. Great care is always exercised in preparing all these contract documents with a view to avoiding disputes, but success is not always achieved, as the record of court cases regarding engineering contracts makes clear. It is probably safe to say that no one feature has been responsible for more disputes arising out of civil engineering contracts than classification of material to be excavated, as "earth" or "rock," and consequent payment for this part of the work performed. As unit prices for rock excavation may be ten or twelve times as much as the corresponding unit prices for earth excavation, the possibility of disputes arising from questionable classification of material removed will be obvious.

The excavation of solid rock which has to be drilled and blasted is rarely in question, at least in this connection; similarly, the removal of loose sand and gravel or soft clay is an operation about which there will rarely be discussion as to whether it is earth or rock under the terms of the contract. But in between these two extremes there may be encountered materials that cannot easily be classified, if no reference basis has been adopted before the contract operations began. Such doubtful material is often termed *hardpan*. The use of this name should be avoided by engineers, if they wish to obviate trouble. It is essentially a popular term, being sometimes applied to special local gravel deposits of unusual hardness due to partial cementing of the rock fragments. More frequently, it is used to describe soil strata that prove more difficult to excavate than a contractor has estimated, and in consequence it has been featured as the subject of many lawsuits. It is a name not generally recognized in geological nomenclature, and there are few satisfactory definitions of it in engineering literature. The best of those available is given in Appendix A, but this should be used only when absolutely necessary. Boulder clay is often described as *hardpan*, but in this case—as in practically all others—the material can be more accurately described as compact gravel, sand, and clay with boulders (or a modification of this). An indication may accompany such a description of the methods necessary for the excavation of the material, if it has to be removed during construction operations.

It might be thought that this is a situation in which a direct application of geological terminology would be of assistance. Unfortunately, this is not the case, although an appreciation of the geological character of the materials involved will be of great assistance to the engineer. In Geology, the term *rock* is used to describe all the main constituent materials of the earth's crust (as explained in Chap. II), so that before going any further into geological terminology it can be seen to be inapplicable. It may also be noted that, as a general rule in excavation work, the engineer is not interested in the type of rock that has to be excavated—and the contractor less so—but only in its character in relation to excavation methods. Engineers have on occasion attempted to utilize geological data available to them in describing materials to be excavated, sometimes with unfortunate results. As an example of this may be mentioned the use of the term *cemented Triassic formation* in connection with an important contract for the construction of 6,000 ft. of the outfall of the Passaic Valley Sewer in New York Harbor. This term was said to be a good description of glacial drift, but it was held by the courts to be misleading, as the contractor encountered varieties of sand and clay that were not cemented, leading to construction troubles which finally became the subject of legal action.[5,4] It will be clear, therefore, that the use of geological terms in connection with excavation work may not be helpful, and the detailed divisions of the main geological rock types sometimes suggested in this connection need not therefore be considered.

How, then, may the problem be resolved? The following suggestions are offered as possibly assisting toward a satisfactory solution:

1. A careful study should be made, prior to the completion of contract plans, of all material to be excavated, by means of boreholes, etc.

2. If any material intermediate between loose soils and compact rocks is discovered, special tests should be conducted to discover its character in so far as excavation is concerned. (This can be done in connection with test boring by an experienced crew.)

3. In the specification, the terms to be used to define different classes of excavation should be rigidly defined *in terms of the methods to be used to excavate the material.*

4. These engineering definitions should, if possible, be correlated with the geological descriptions of the materials to be encountered, using correct terminology which has been used in connection with the test-boring results, and no indefinite and questionable terms like *hardpan* should be included.

5. The designation used for describing the material to be excavated should be as few and as simple as possible, *e.g.:*

Hard-rock excavation (excavation of all granite and limestone which has to be carried out with the aid of drilling and blasting).

Loose-rock excavation (excavation of blocky limestone which does not require blasting but which can not be carried out by means of hand shovels or picks).

Soft-rock and earth excavation (excavation of disintegrated granite, clay, sand and gravel which can be carried out economically by means of hand shovels and similar tools.[5.5]

It may finally be noted that throughout the rest of this book, the term *soil* will be used generally to describe unconsolidated natural materials, *rock* being used in its engineering sense as referring only to solid bedrock, either in place or as excavated.

5.12. Construction Operations.—Every cubic yard of excavation that is removed during construction, every unusual loading that is applied to a natural foundation bed, every pile driven into the ground—in fact, every operation in construction in which the existing condition of the earth's crust is affected—is associated with geological features of some kind. Preliminary investigations of the relevant geology should therefore be of considerable value not only to the resident engineer on construction work but also to the contractor who is undertaking the work. Throughout this book, most of the examples mentioned will be found to confirm in one way or another the importance to contractors of advance geological information.

The geological data that may be available at the beginning of a job can be fully effective, however, only if they are constantly checked up with actual geological conditions as they are revealed during the progress of construction. It is essential, therefore, that on all civil engineering construction work, a regular and constant watch be kept on all geological formations as they are revealed and that an adequate and complete record of all leading geological features be kept on the job in addition to the usual construction progress records. This can the more easily be done if at least one of the engineers on the job has been trained in geological fieldwork; in some cases, geological training has been made a prerequisite for appointment to a resident engineer's staff.

Geological data thus obtained as construction progresses have a threefold value. They will act as a check—in the first place—on the assumptions made with regard to geological conditions in the preparations of the final designs for the works being constructed, any variation noted from these conditions usually enabling alterations to be made to the design before it is too late. Secondly, the revelation of the actual geology of the working site will enable the contractor to keep check on the suitability and efficacy of his construction plans and plant. Finally, if the geological progress record is kept in a satisfactory manner, it may prove of inestimable value at some future time if further work has ever to be carried out at the same location.

Brief reference may here be made to the last point, and mention made of a most interesting confirmation in connection with some cement grouting work carried out in parts of the famous Severn Tunnel in the southwest of England. This tunnel, which carries the Great Western Railway under the River Severn, was started in 1873 and completed only after a fight against tremendous difficulties by the contractor, Mr. T. A. Walker. Mr. Walker kept a complete geological record throughout the construction of the tunnel, which pierces the Trias and Coal Measures of England, rocks encountered varying from marl and shale to limestone and sandstone.[5.6] The existence of this valuable and complete record assisted the engineering authorities of the railway company almost sixty years later in deciding to undertake the extensive cementation program which was described by Mr. R. Carpmael in a paper presented to the Institution of Civil Engineers.[5.7] On the other hand, many examples could be given of the trouble and expense caused by the lack of comparable records for other works. A well-known tunnel in the United States, for example, is reported to have had to be resurveyed, within seven years of its completion, in connection with the installation of a concrete lining, no records being then available of the timber sets used or of the final tunnel cross sections.

5.13. Inspection and Maintenance.—All civil engineering works have to be regularly inspected and maintained in good condition by such measures as may be called for as a result of the inspection. This routine work must always be most carefully carried out, and it is necessary, therefore, only to mention that not only man-made structures must be regularly inspected but also the ground adjacent to that on which they rest. Inspection of bridge piers, to check against scouring, and of dams, to check against erosion of foundation-bed material by means of leaks, are two of the more important branches of such work in which geological features may be of special significance. They will be referred to in the respective chapters dealing with these branches of work, and other references will be found to the regular checking over and maintenance of other structures and works in which careful attention to geological features can add appreciably to the value of the inspection.

5.14. Conclusion.—In this chapter, an attempt has been made to review generally the work of the civil engineer with special reference to the possible applications of Geology to its many branches. The review has been made, as far as has been possible, by considering in a general way the attitude of the civil engineer to the science, an attitude of appreciation which, if properly developed, can be of great assistance in connection with all the main divisions of the engineer's work. The discussion has had to be conducted along general lines, reference being made to only a few specific works to illustrate salient points of interest. This has been necessary not only in order that the whole field might be covered (if only in a cursory manner) but also because the applications of Geology to each of the main branches of civil engineering work follow naturally the characteristics peculiar to each of the branches. To discuss the applications usefully, therefore, necessitates their consideration in groups corresponding to the various leading sections of engineering work.

This course is adopted in the chapters that follow. Preliminary and exploratory work is first considered, and then geophysical methods of investigation. The applications of Geology in tunnel construction are next dealt with; and thereafter, in a logical sequence, other branches of civil engineering work are discussed with special reference to the service that can be rendered in their prosecution through the application of Geology. Although convenient, this treatment of the subject suffers from the disadvantage that the civil engineering works dealt with have to be considered from one point of view only, and this inevitably tends to present a distorted picture of them. Similarly, in view of the concentration of attention on the applications of Geology to civil engineering, the examples given have necessarily been selected because of their special geological interest. They are not therefore typical of all civil engineering work in view of the many projects in which local geological features are such that they do not call for any unusual precautions. The examples will serve, however, to direct attention to those features of the works considered which must always be watched as construction proceeds until it is certain that no unusual conditions are to be encountered. If these two qualifications are kept in mind, the reader should be able to assess the true significance of the citations to be made from world-wide engineering practice.

All the work to be described is carried out by men for the benefit of their fellows. It may well be remembered, therefore, that behind all impersonal discussions of scientific applications and engineering endeavor lie the relations of man to man and the cooperation of many men working toward a common goal in rendering useful service to their community. This book would not be complete without at least a passing reference to this human background. It is in keeping with this thought that this chapter was headed The Civil Engineer and Geology and that as a frontispiece a portrait has been reproduced—a portrait of a man who epitomized all that has been suggested as the attitude that a civil engineer should entertain toward Geology. For his geological work alone, William Smith will always be remembered with honor; but to civil engineers his fame rests also on the way in which he applied his geological knowledge to his civil engineering work, work of no little importance a century ago. He indeed allowed his "faith" as a geologist to benefit his "works" as an engineer.

A Partial List of Publications Dealing with Geology as Applied in Civil Engineering

The bibliography of Geology as applied in civil engineering is not extensive. The following list contains particulars of those books, papers, and articles in which the subject is treated generally which have come to the attention of the author. Although foreign-language books are listed, papers in the English language alone are noted with one notable exception. Publications that deal with special aspects of the subject are not included; many of these will be found listed at the ends of succeeding chapters and in Appendix D. This compilation must be far from complete, but it is presented as a guide for those interested in examining the literature of the subject. For most of the titles of foreign-language books and for one or two of the other references, the author is indebted to the director of the Engineering Societies Library, New York, and to the library staff of the Institution of Civil Engineers, London. The author will welcome particulars of relevant publications not included in the list now printed:

Books

Elsden, J. V.: "Applied Geology," Quarry Publishing Co., Ltd., London, 1898.

Fox, C. S.: "Civil Engineering Geology," The Technical Press, Ltd., London, 1923.

Fox, C. S.: "A Comprehensive Treatise on Engineering Geology," The Technical Press, Ltd., London, 1935.

Penning, W. H.: "Engineering Geology," Baillière, Tindall & Cox, London, 1880. The author would pay special tribute to this little book which, although written over fifty years ago, can still be read with profit.

Ries, H., and T. L. Watson: "Engineering Geology," John Wiley & Sons, Inc., 5th ed., 1936. This well-known book is published by the same publishers also in an abridged form as "Elements of Engineering Geology," 2d ed., 1930.

SHAND, S. J.: "Useful Aspects of Geology," Thomas Murby & Co., London, 2d ed., 1934.

SHERLOCK, R. L.: "Man as a Geological Agent," H. G. and G. Witherby, London, 1922.

SORSBIE, R. F.: "Geology for Engineers," Charles Griffin & Company, Ltd., London, 1911 (2d ed., George Bell & Sons, Ltd., London, 1938).

BOOKS IN FOREIGN LANGUAGES

BRAUNS, D. A.: "Technische Geologie," Schwetsche, Halle, 1878.

KRANZ, WALTHER: "Die Geologie im Ingenieur-Baufach," Stuttgart, Enke, 1927.

NIVOIT, EDMOND: "Géologie appliquée a l'art de l'ingénieur," Baudry, Paris 1887–1889.

REDLICH, R. A., K. TERZAGHI, and others: "Ingenieurgeologie," Julius Springer, Berlin, 1929.

ROVERETS, G.: "Geologia; teoria-pratica-applicazioni," Milan, Hoepli, 1931.

SINGER, MAX: "Der Baugrund; prakitische Geologie fur Architekten, Bauunternehmer und Ingenieure," Julius Springer, Vienna, 1932.

WAGNER, C. J.: "Die Beziehunger der Geologie zu den Ingenieurwissenschaften," Spielhagen, Vienna, 1884.

PRODUCTS OF THE WORLD WAR, 1914–1919

It seems probable that one of the factors responsible for the attention devoted to Geology as applied in civil engineering during the last two decades was the attention that had necessarily to be devoted to this branch of applied Geology during the years of the World War. The following publications have therefore a special significance; all are unusually interesting, that by Dr. Gregory remaining one of the best introductions to Geology that can be placed in the hands of a civil engineering student, apart only from the natural emphasis on military matters.

ANONYMOUS: "The Work of the Royal Engineers in the European War 1914–19: Geological Work on the Western Front," Secretary, Institution of Royal Engineers, Chatham, England, 1922.

BROOKS, A. H.: The Use of Geology on the Western Front, *U. S. Geol. Survey Prof. Paper* 128d, 1921.

GREGORY, H. E.: "Military Geology and Topography," Yale University Press, New Haven, Connecticut, 1918, published under the auspices of the National Research Council, Division of Geology and Geography.

BOOKS ON ECONOMIC GEOLOGY

Books dealing with Economic Geology in its various general aspects naturally include some comment on the applications of the science to civil engineering work; the following two references are cited as typical of many:

GREGORY, J. W.: "Elements of Economic Geology," Methuen & Company, Ltd., London, 1927 (see Chaps. 19–21, Part IV).

LEITH, C. K.: "The Economic Aspects of Geology," Henry Holt & Company, New York, 1921 (see Chaps. 6, 19, and 20).

PUBLICATIONS IN PERIODICALS AND THE PROCEEDINGS OF ENGINEERING SOCIETIES

*BERKEY, C. P.: The Recent Development of Geology as an Applied Science, *Am. Philos. Soc. Proc.*, **72**: 25 (1933).

* Dr. Berkey has many other publications on similar detailed subjects, reference to some of which will be found in Appendix D.

BOYLE, ROBERT: "Geology in Relation to Civil Engineering" (Glasgow Technical College Civil Engineering Society), John Smith & Son, Ltd., Glasgow, Scotland, 1910.

CLARK, D.: The Applications of Geology to Civil Engineering, *California Jour. Mines Geology*, **29**: 161 (1933).

*CROSBY, I. B.: Engineering Geology of the Passamaquoddy Project, *Jour. Boston Soc. Civil Eng.*, **25**: 9 (1938).

DAWKINS, BOYD: On the Relation of Geology to Civil Engineering, James Forrest Lecture, *Inst. Civil Eng. Min. Proc.* (*London*), **134**: 254 (1898).

DOLMAGE, V.: Geology and Geophysics in Engineering, *Canadian Eng.*, **67**: 19 (1931).

ECKEL, E. C.: Engineering Applications of Geology, *Eng. Rec.* (*New York*), **67**: 667, 711 (1913).

ECKEL, E. C.: Engineering Geology and Mineral Resources of the T.V.A. Region, *T.V.A.* (*Tennessee*) *Geol. Bull.* 1 (1934).

EKBLAW, G. E.: New Demands of Engineering on Geology in Illinois, *Illinois Acad. Sci. Trans.*, **23**: 384 (1931).

EKBLAW, G. E.: Some Problems of Engineering Geology in the Vicinity of Chicago, *Illinois Acad. Sci. Trans.*, **25**: 142 (1934).

FOX, C. S.: Some Engineering Aspects of Geology, *Jour. Inst. Eng.* (*India*), **2**: 1 (1922).

HILLS, E. S.: The Geology of Bridge Foundations, Reservoirs, and Dam Sites, *Commonwealth Eng.* (*Australia*), **21**: 195 (1934).

HOTCHKISS, W. D.: Geology and Civil Engineering, *Civil Eng.* (*New York*), **7**: 760 (1937).

JOHNSON, D. W.: The Scope of Applied Geology and Its Place in the Technical School, *Econ. Geology*, **1**: 243 (1906).

*KNIGHT, B. H.: Geology as Applied to Road and Building Problems, *Jour. Inst. Municipal and County Eng.* (*London*), **63**: 1486 (1937).

LAPWORTH, H.: Applied Geology in Municipal Engineering, *Jour. Inst. Municipal and County Eng.* (*London*), **41**: 199 (1915).

LAPWORTH, H.: The Principles of Engineering Geology (two lectures to students), *Inst. Civil Eng. Min. Proc.* (*London*), **173**: 298 (1908).

LEGGET R. F.: Geology and Civil Engineering with Special Reference to Canada, *Eng. Jour.* (*Montreal*), **17**: 431 (1934).

LEIGHTON, M. M.: Glacial Geology and Engineering in Illinois, *Illinois Acad. Sci. Trans.*, **19**: 246 (1926).

LEIGHTON, M. M.: Modern Geology and Its contribution to Engineering, *Jour. Western Soc. Eng.*, **28**: 491 (1923).

MORSE, W. C.: Geologic Conditions Governing Sites of Bridges and Other Structures, *Mississippi Geol. Survey Bull.* 27, 1935.

*MORTON, E.: Some Problems in Engineering Geology, *Water and Water Eng.* (*London*), **31**: 153 (1929).

READE, T. MELLARD: The Advantage to the Civil Engineer of a Study of Geology, *Liverpool Eng. Soc. Trans.*, **10**: 36 (1888).

REINECKE, L.: Relations of Geology to Industry, *Geol. Soc. South Africa Proc.*, **33**: 23 (1931).

*RIES, H.: Importance of Geology to Civil Engineering, *Eng. Jour.* (*Montreal*), **12**: 3 (1929).

* These geologists have several other publications on similar detailed subjects, reference to some of which will be found in Appendix D.

RUNNER, D. G.: Fundamentals of Practical Engineering Geology, *Roads and Streets* (*Chicago*), **77**: 309 (1934).

SEMMES, D. R.: The Value of Applied Geology, *Texas Eng. Exper. Sta. Bull.* 20, 1918.

SMITH, W. D.: Tropical Engineering and Geology, *Philippine Jour. Sci.*, **18**: 221 (1921).

SOPWITH, T.: On the Construction and Use of Geological Models in Connection with Civil Engineering, *Inst. Civil Eng. Min. Proc.* (*London*), **1**: 161 (1841).

TWINEM, J. C.: Geology for Engineers, *Maine Tech. Exper. Sta. Paper* 11, 1931.

WILLBOURN, E. S.: The Relationship of Geology and Civil Engineering in Malaya, *Quart. Jour. Eng. Assoc.* (*Malaya*), **3**: 26 (1935).

YORKE, D.: Economics of Geology as Applied to Engineering, *Surveyor* (*London*), **80**: 139 (1931).

To the foregoing list must be added brief mention of the publications, in the journal of Czecho-Slovakian engineers (*Tech. Rev.*, Dr. Ed. Glegr and Sons, Prague), of Dr. Quido Zaruba-Pffeffermann, most of which are accompanied by an abstract in either French or English. One paper, in 1932, was entitled Geological Investigations in Civil Engineering, Dr. Pffeffermann's latest being on Landslides in the Pass of Lysa and Their Significance in Route Construction [**46**, No. 1 (1938)].

CHAPTER VI

PRELIMINARY AND EXPLORATORY WORK

> If a small fraction of the time spent on refined calculations and very accurate designing were spent on watching the preliminary boring operations, and in examining the foundations when work was in progress, the benefit would probably be many times greater than any advantage from the greatest attention to refinement of calculation in design.
>
> W. L. Lowe-Brown.[6.1]

Before a civil engineer can design any projected work, he must know something about the material on which his structure is to be founded or in which his work is to be carried out. This will necessitate some examination of the site of the work before any designing is started and a thorough investigation of the site before detailed designs are prepared. Truisms? Admittedly, but truisms that have often been neglected in the past, sometimes with disastrous results and almost always with consequent monetary loss. This last statement might be characterized as too sweeping were it not possible to confirm it indubitably by many examples from engineering practice, a few of which are mentioned below. To the civil engineer with even a slight appreciation of Geology, neglect of what is to him the obviously essential course of making thorough and complete preliminary investigations before embarking on construction will appear to be folly. He might with propriety compare such an omission to a surgeon operating without diagnosis or a lawyer pleading without any prior discussion with his client.

To the engineer who has not studied Geology, these analogies might appear to be far-fetched, for in many cases the nature of the surface of the ground and experiences on similar adjacent works might seem to him to be warrant enough for making certain assumptions as to subsurface conditions on which designs can be based. It may be that some such reasoning has led to the neglect of preliminary explorations in the past. Similarly, it is probable that unquestioning reliance on the information given by test borings alone, considered without relation to the local geology, may have sometimes resulted in engineers being faced with unforeseen difficulties in construction which have led them to distrust expenditure on such preliminary work. All too often, however, it has been the difficulty of obtaining the necessary funds before the start of construction that has prevented engineers from carrying out the exploration of a site that they knew to be necessary. This results from a shortsighted policy on the part of some owners which engineers may find hard to understand.

Discussion of this attitude, and of other factors that may have furthered the neglect of subsurface explorations on civil engineering works in the past, will not be profitable. Most civil engineers today recognize the supreme importance of preliminary work of this nature; any readers who may entertain doubts about its value will not have to read far beyond Chap. VIII to meet with evidence that should set all their doubts at rest. But all engineers may have to deal with owners who have not had an opportunity of visualizing the essential nature of exploratory work. The following section therefore presents data that may be of use in demonstrating the economic value of such preliminary expenditure.

6.2. Economics of Preliminary Work.—*Economics* may be too technical a word to apply to the following brief considerations of preliminary work; it will serve, however, to describe a discussion of the cost of such work relative to the total cost of a project and of the savings that may be effected and the difficulties that may be avoided by such a precautionary measure.

Preliminary work, as here intended, is held to include geological surveys, geophysical surveys, test pits, borings and drillings, and any further exploratory measures that may be called for in special cases. Detailed costs for test drilling and boring operations are suggested later (on page 96), but they are of little use in connection with over-all exploratory costs. Examples of the total expenditure on such work for complete projects are not easy to obtain, usually because the totals are so small that they cannot suitably be featured in the summaries of cost that form most valuable parts of published descriptions of engineering works. It has been possible, however, to obtain some figures for interesting examples of tunnel, dam, bridge, and building construction; these are given in the following brief statements:

Pressure Tunnel for Water Power

The Bridge River Tunnel of the British Columbia Electric Railway Co., Ltd. British Columbia, Canada, completed in 1930 in connection with water-power development. This tunnel is described on p. 130; it is included in this record as the preliminary work involved geophysical exploration.

Total cost of tunnel, including concrete lining throughout, surge chamber, etc., but not overhead, interest, and engineering charges $2,227,000 (£445,400)

Cost of preliminary work, including geological and geophysical surveys and test drilling..... $7,575 (£1,515)
or 0.3 per cent of the cost of the tunnel

References: See p. 130, but the preceding figures were kindly supplied by Mr. E. E. Carpenter, consulting engineer to the British Columbia Electric Railway Co., Ltd.

Pressure Tunnel for Water Supply

The First Narrows Pressure Tunnel, Vancouver, British Columbia, Canada, of the Greater Vancouver Water District, carrying water to the city area under the First Narrows of Burrard Inlet, a tidal estuary. The tunnel is driven through sandstone and shale and is 3,069 ft. long. It is connected to the main aqueducts at the surface through two shafts, each about 400 ft. deep, penetrating glacial drift, sandstone, shale, and conglomerate.

Total cost of tunnel, *including* engineering and inspection charges and interest during construction..... $1,249,641.85 (£249,928)
Cost of preliminary investigations and bedrock explorations..... $91,540.17 (£18,308)
or 7.3 per cent of the total cost of the tunnel

Special Note: This interesting case is included, since it is the most unusual that has come to the attention of the author in respect to the extensive preliminary work. Preliminary investigations extended from 1926 to 1930; several consultants reported on the available sites; and in all, about 50 test holes were put down to depths in some cases more than 700 ft., total footage being about 13,000. Despite the difficulties of the site and the unusual depth of the shafts, the work was completed successfully.

Reference: Powell, W. H., First Narrows Pressure Tunnel, Vancouver, B.C., *Eng. Jour.*, p. 253, June, 1934, in which an interesting analysis of drilling costs is given.

Water Storage Dam

The Boulder Dam, of the U. S. Bureau of Reclamation, located on the Colorado River in what is known as Black Canyon, 30 miles southeast of Las Vegas, Nevada. The dam is a mass concrete arch-gravity structure, 45 ft. thick at the top, 660 ft. thick at the base, and 726.4 ft. high from the lowest point in the foundations to the crown of the roadway. It is founded on andesite flow breccia and andesite tuff breccia.

Total cost of dam, excluding interest, engineering overhead, and the associated powerhouse and plant..... $76,770,396.12 (£15,350,000)
Total cost of all geological and exploratory preliminary work, at the site selected and other sites, including also consulting geologists' fees, $826,081.00 (£165,000)
or about 1.1 per cent of the total cost of the dam

References: Several references to this great dam will be found listed in Appendix D in connection with Chap. XIII. The preceding figures were kindly supplied by Commissioner John C. Page of the U. S. Bureau of Reclamation.

Steel Arch Highway Bridge

The Iskander Bridge over Perak River, Perak, Federated Malay States (Public Works Department, F.M.S.). The Bridge is 928 ft. long between abutments, providing a 30-ft. roadway. It consists of seven two-hinged steel arches, supported on mass concrete caisson piers and abutments, generally founded on rock (granite and limestone).

Total cost of bridge and approach roads, including engineering and inspection charges but not interest	£175,000	($875,000)
Cost of abutments and piers	£76,883	($384,415)
Cost of steel superstructure (erected)	£51,217	($256,085)
Cost of preliminary investigations and test borings	£3,150	($15,750)
or 1.8 per cent of total cost		
and 4.1 per cent of cost of piers and abutments		

Reference: Coales, F. G., and C. H. Clarkson, The Iskander Bridge, Perak, Federated Malay States, *Inst. Civil Eng. Min. Proc. (London)*, **240**: 342 (1935).

Steam Power-station Foundation

The Battersea Power Station of the London Power Co., Ltd., London, having an ultimate capacity of 500,000 kw. and an initial capacity of 160,000 kw. It was constructed in 1929–1932. For details of foundation-bed conditions, see page 492.

Total cost of constructional work only, for initial installation, excluding interest, consultant's fees, etc.	£2,141,550	($10,707,750)
Cost of foundations for main power station buildings	£226,370	($1,131,850)
Cost of trial holes and boreholes	£3,500	($17,500)
or 0.2 per cent of total cost		
and 1.5 per cent of cost of main building foundations alone		

Reference: Berry, C. S., and A. C. Dean: The Constructional Works of the Battersea Power Station of the London Power Company Ltd., *Inst. Civil Eng. Min. Proc. (London)*, **240**: 37 (1935).

These figures, for works of such differing character and in such varied parts of the world, show clearly that the cost of preliminary work relative to the total cost of a civil engineering project is small indeed. It is believed that the figures thus presented will be found to be typical for normal engineering work. If the costs thus presented are compared with other percentages which usually figure in civil engineering estimates, their significance will be appreciated. Special thought might perhaps be given to the comparison of this 1 or 2 per cent for the total cost of underground investigations with the item of "10 per cent for contingencies" which is so frequently featured in civil engineering estimates. For some types of work, and especially marine work, contingencies must be allowed for in any estimates of cost; but in general, this broad item must often be used to include for just those uncertainties which thorough subsurface exploration will assist in eliminating. Viewed in this light, the cost of preliminary work may not therefore appear to be so high as may at first be thought by an owner when faced with the necessity of spending some money in order to find out if he can even build the main project he is planning.

Further argument may be needed to persuade a dubious owner. If cost data are not sufficient, actual examples from engineering practice may be drawn upon to complete the picture that the engineer wishes to present. This book will be found to contain many examples that might thus usefully be quoted, but for convenience a few may be cited here in summarized form.

Considering, first, cases in which adequate preliminary work was *not* carried out, there may be cited the case of a tunnel the contract for which included a large bonus payment for every day gained in completion before a certain fixed date. The engineers based their estimate of construction time on a calculated rate for excavating the tunnel through the quartzite exposed at the two tunnel portals. The tunnel line pierced an anticline, however, the quartzite overlying easily excavated shale through which the tunnel was very quickly driven; the contractor's bonus was so large that it led to a legal battle. An even more serious case was that of a dam for water supply which was to be built in a valley as an earth-fill structure with a concrete-core wall carried to rock at an assumed maximum depth of about 60 ft. below the valley floor. This figure was derived from test borings which were stopped at what was assumed to be solid rock but which later proved to be boulders of a glacial deposit. The core wall had finally to be carried to a depth of 196 ft. below the surface, and the cost of the dam was increased correspondingly by a very large amount. As an example from bridge engineering practice, Messrs. Jacoby and Davis[6.2] mention a case in which a bridge pier was built upon a surface of hardpan in the river bottom, but no examination of this was made due to the difficulty of taking borings in the river current of 5 miles per hour. When constructed, the bridge pier sank out of sight, causing the loss of the two adjacent spans of the bridge and a number of human lives. When an examination was made later, it was found that the hardpan was only a thin stratum overlying a deep layer of soft clay.

These examples show what trouble can be encountered through neglect of preliminary investigations. What of the constructive contributions that such work can make to civil engineering practice? Many are the examples that could be quoted, although the actual monetary savings effected are usually hard to assess. The successful completion of many notable tunnels, such as those under the River Mersey in England or those constituting the Catskill Aqueduct for the water supply of New York in the United States, which will be referred to in more detail in Chap. VIII, provide striking evidence as to the value of preliminary geological studies and underground investigation. In like manner is evidence supplied by Boulder Dam, already mentioned, and the vast number of smaller dam structures successfully founded on strata previously thoroughly investigated. From the records of bridge construction, there might be mentioned, with similar interest, the San Francisco-Oakland Bay Bridge, the foundations for which are referred to in some detail in Chap. XII; despite their unusual size and the record depths to which they were carried, their construction agreed closely with what was anticipated as a result of the preliminary exploration on which the sum of $135,000 (£27,000) was expended.[6.3]

A somewhat unusual example may be given to illustrate the value of underground exploration for building foundations. The Lochaber Water Power scheme in Scotland has its powerhouse located at the end of a steel pressure pipe line 3,240 ft. long which descends from the slopes of Ben Nevis. The discharge from the Pelton wheels enters a tailrace about 3,000 ft. long excavated partially in rock tunnel, partially in open rock cut, and partially in open cut through drift deposits. As originally planned, the powerhouse was to have been much nearer to the River Lochy, with a tailrace only 100 ft. long, the length of pipe line consequently being 6,200 ft. Further investigations were carried out, however, in the form of an extensive program of test borings all over the available sites for the powerhouse and along corresponding routes for the tailrace. In all, 1,040 ft. of borings was carried out at a total cost of £1,340 ($6,700). Results so obtained were carefully plotted, mainly to locate contours on the rock surface underlying the surface deposits, all borings having been carried at least 3 ft. into the rock to ensure that boulders had not been encountered. On the basis of this information, 16 complete schemes for the powerhouse location were prepared, and their over-all costs estimated, the final decision being for the arrangement already described, for which the combined cost of tailrace, pipe line, and powerhouse excavation (on the average 40 ft. deep) was a minimum. A substantial saving was thus effected over the original plans which amounted to many times the cost of the exploratory work that made the saving possible.[6.4]

6.3. Geological Surveys.—With the necessity for adequate preliminary investigations thus demonstrated, thought may be given to the nature of such work and to its relation with Geology. This relation is a fundamental one, the first requirement always being a geological survey of the area in question to a scale and a degree of accuracy commensurate with the extent of the area and the nature of the work to be built.

In general, preliminary investigations and exploration will be undertaken with two objects in view:

1. The determination of all underground conditions at and near the site that may affect the work in view, these being generally the local geological structure and the distribution of underground water.

2. The determination of the character of all materials that may be encountered during the progress of the work.

Usually these two objects can at least be partially achieved by means of a detailed geological survey of the site. There are, of course, extreme cases in which the entire area may be drift covered, thus preventing accurate geological mapping of the rock surface. Building sites are often of this nature, but the larger types of civil engineering work, such as tunnel, bridge, and dam construction will usually be so located that some rock outcrops are available for study. Instances may sometimes occur of the other extreme condition with no drift occurring at all in the area to be surveyed. As a rule, therefore, it will be possible to make at least a partial geological survey of the area in which civil engineering work is to be carried out.

This survey work will follow the lines suggested in Chap. IV and need not therefore be detailed here. Emphasis may, however, be placed upon the desirability of making use in the first place of all the printed geological information available whether in the form of large-scale maps, published memoirs, or geological papers. As guides to sources of such information, Appendices B and C have been prepared. For other than elementary geological survey work, it may be necessary to have available an engineer specially trained in geological work, and for any areas that include any unusual features the services of a geologist will be essential.

It should not be difficult for an engineer to appraise the general nature of the local geology at a site, provided that he entertains an appreciation of the importance of the science in his work. He should have little difficulty in obtaining advice when necessary with regard to his geological problems from the respective geological survey, the staff of a local university, or a consulting geologist who has specialized in engineering work. He will probably be able to assist, either personally or through his staff, with the necessary geological survey work, and his cooperation will certainly be needed in placing before the geologist a succinct picture of the work to be carried out and the specific questions relating to subsurface conditions which are consequently of importance.

The results of the survey usually will be presented in the form of a geological map of the area, possibly accompanied by geological sections and a report on special features that demand attention and the materials to be encountered (including, if the engineer has so requested, details of construction material available). The geological map will serve to show the engineer something of the underground structure at the site of his projected work; and on the basis thus provided, he will be able to prepare preliminary plans of the work to be constructed. By considering the possibilities thus presented, he will be able to suggest in consultation with his geological adviser the detailed information that he will require in order to prepare his final plans, and in this joint manner the necessary program of exploratory work can be prepared. It is also advisable, in some cases, to obtain geological advice before construction begins with regard to the geological features affecting property that adjoins the construction site in order to determine the possibility of valid claims for damages caused by construction. When construction operations are to include much blasting, such a preliminary survey together with a detailed photographic record of the initial state of all adjacent property can often prove to be most useful in combating the claims that inevitably arise whenever householders and other "laymen" feel their property vibrate after blasting operations.

Finally, preliminary geological studies will enable the engineer to be acquainted, at least in general terms, with the materials that he may expect to encounter in his underground operations and also, if he has requested it, with what natural construction materials there are available in the vicinity of the proposed site. A geologist will usually be able to suggest also the extent of weathering that may be expected at the surface of the upper rock strata to be encountered, the probable effect of water on rocks that may eventually be exposed to it, and similar information derived by the application of his knowledge of petrology to the particular problems to be faced by the engineer.

6.4. Underground Exploratory Work.—It is but rarely that a geological survey alone, no matter how accurate it may be, will provide sufficiently reliable information with regard to underground conditions for civil engineering work. Data as to underground water, to mention but one feature, cannot always be obtained accurately from surface observations, and in many cases this information is a vital part of that required by the civil engineer. More often, however, the geological survey will be incomplete in respect to detailed matters owing to the existence of superficial deposits of unconsolidated material. As a general rule, therefore, underground exploratory work in the form of test pits or borings will have to be carried out in addition. It cannot be too strongly emphasized, however, that *all underground exploratory work must always be considered as supplementary to and conditioned by previous considerations of the local geological structure.*

Brief consideration of the possibilities of complex underground structures existing beneath a relatively simple-looking ground surface will confirm the validity of this suggestion. It is of even more than usual importance in country that has been subjected to glacial action where glacial drift now covers an original rock surface which may be entirely unrelated to present-day topography. Even without the existence of glacial conditions, however, disastrous results can and actually have occurred owing to reliance being placed on the results of test boreholes that had not been correlated with the local geology.

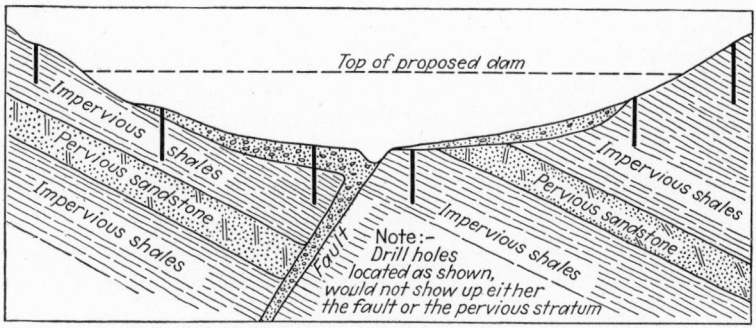

FIG. 6.1.—A geological section showing how a fault might be undetected by test drilling.

Consider as an example (not taken actually from practice but typical of many a river valley) the conditions shown in Fig. 6.1. Cursory surface examination of exposed rock in the immediate vicinity of the proposed dam would disclose outcrops of shale only. Test borings might then be put down to confirm what these rock outcrops might appear to suggest; and if they were located as shown—as well they might be if the local geology was not considered—it will be seen that they will give an entirely false picture of the subsurface conditions across the valley, possibly with disastrous results. How could Geology assist in such a case? Any geological survey made of this dam site would include at least a general reconnaissance of the neighborhood, and it is almost certain that, either by observation of the outcrops along the river bed or through some peculiar features of local topography, the fault would be detected. Even if this was not done directly, a detailed examination of the outcrops of shale on the two sides of the valley would almost certainly reveal differences between the two deposits, possibly minute, possibly even of their fossil contents, but enough to show that they were not the same formation, thus demonstrating some change in structure between the two sides of the valley.

Figure 6.2 shows another case, similar to that just discussed, in that casual surface examinations and the use of the information given by the boreholes shown would be misleading, in this instance because of the existence of folds in the strata, the outcrops of which are drift covered. Figure 6.3 illustrates the necessity, where the strata involved dip

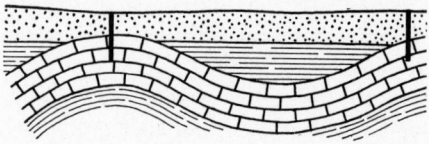

FIG. 6.2.—A geological section showing how a fold might be undetected by test drilling.

steeply across the site being investigated, for correlating the strata encountered in one drill hole with those pierced by the adjacent hole. If this were not done, and hole 3, for example, had been put down only as far as the point marked *A*, the existence of the fault would have been undetected, unless found from surface geological investigations.

A geological survey, even of a preliminary nature, will usually reveal whether or not an area has been subjected to glacial action. This information will always be particularly valuable in the prosecution of underground exploratory work, since it will at once suggest some of

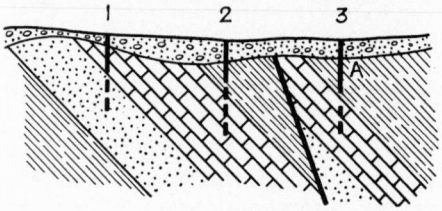

FIG. 6.3.—A geological section illustrating the desirability of correlating steeply inclined strata in test drilling.

the unusual subsurface features that may be encountered. Not only does the probable existence of boulders complicate the study of underground conditions in glaciated country, but also the frequent occurrence of buried river valleys. Those in the valley of the Rhine have already been mentioned; in Chap. VIII will be found reference to the buried valley that complicated the construction of the Mersey tunnels in England.

These examples are not mentioned to show that test borings and similar investigations are often faulty; such a conclusion would, in fact, be incorrect, because in all the cases cited the results from the boreholes may have been correct, but their *interpretation* was not, owing to the fact that the results may not have been correlated with the local geology. Arising from study of the cases described and many other similar instances, the following general rules with regard to underground exploratory work may be suggested. (For convenience, the word *boreholes* will be used to describe all such subsurface work).

1. No boreholes should be put down before at least a general geological survey of the area has been made.

2. Boreholes should, whenever possible, be located in relation to the local geological structure.

3. Boreholes should, whenever necessary, be carried to such depths that they will definitely correlate the strata encountered in adjacent holes by an "overlap" into at least one bed.

4. In exploring superficial deposits, one borehole at least should, whenever possible, be carried to rock.

5. In all cases of superficial deposits extending to no great depth, test borings should be taken into the rock for some specified distance, never less than 2 ft., but more than this if the nature of the work warrants the extra cost or if large boulders are liable to be encountered.

6. Unusual care must be exercised in putting down test borings in areas known to have been subjected to glacial action, especially with regard to checking all rock encountered during drilling (as in 5) and for the possible existence of buried river valleys.

7. The three-dimensional nature of the work must always be remembered. For example, three boreholes properly located will define exactly the thickness, dip, and strike of any continuous buried stratum.

It is not a part of the function of this book to present full details of the methods adopted in the various types of underground exploration, but this discussion of preliminary work would not be complete without at least a brief survey of the methods available. This is presented in the following notes and in Table E, a tabular summary in which suggestions are advanced with regard to costs. This is based upon, but a modification of, a table included in a most interesting booklet by Mr. H. A. Mohr, a publication of the Graduate School of Engineering of Harvard University, in which will be found details of most of the generally accepted methods of subsurface investigations.

6.5. Exploratory Test Borings.—Methods adopted for penetrating unconsolidated materials and solid rock are naturally different, even though they may often have to be employed in the same hole. It is convenient to describe the former as *test borings* and the latter as *test drillings;* test pits are sometimes used as an additional precaution in

TABLE E.—METHODS OF SUBSURFACE EXPLORATION[1]

Name of method	Materials in which used	Method of advancing hole	Method of sampling	Approx. cost as given below	Value for foundation-bed investigations
Test pits and test caissons	All soils: if groundwater is encountered, compressed-air caisson or water-lowering process will be necessary	Usually by hand excavation in lined pit or beneath caisson	Samples taken by hand from original position in ground	$10 and up, per foot (£2.0.0 and up)	*Most valuable*, materials can be inspected *in situ*, and adequate samples of all soils are easily obtained
Pricking	Shallow soils overlying rock	Steel bar forced into ground until rock is reached	.....	.....	Limited to locating rock surfaces at shallow depths
Hand auger boring	Cohesive soils, and some granular soils above water level	Augers rotated until filled with solid material and then withdrawn	Samples obtained from material brought up by auger	$0.75 to $2 per ft. (3*s*. to 8*s*.)	Satisfactory for shallow investigations for roads, small building foundations, etc.
Wash borings or wash-sample borings	All soils except the most compact (will not penetrate boulders)	Washing inside a driven casing	Samples recovered as sediment in wash water	$0.50 to $1.50 per ft. (2*s*. to 6*s*.)	*Almost valueless* and *dangerous* because of possibility of misleading results
Dry-sample or core-sample borings	All soils except the most compact (will not penetrate boulders)	Washing inside a driven casing	Open end of pipe or special spoon forced into soil at bottom of hole	$0.50 to $1.50 per ft. (2*s*. to 6*s*.)	*Most reliable* of inexpensive methods; enables accurate soil classification to be made, and penetration tests, if necessary
Undisturbed sampling	All soils except the most compact (will not penetrate boulders)	Washing inside a driven casing, but usually one of 4 or 6-in. diameter or even larger	Special soil-sampling tool forced into soil at bottom of hole	$5 to $8 per ft. (£1.0.0 to £1.12.0)	Best available method; enables undisturbed samples to be obtained for laboratory study
Core drilling	All solid materials—rock and boulders	Rotating power-driven coring tools with diamond, shot, or steel-tooth cutters	Cores cut out and recovered from holes	$3.25 to $10 per ft. (13*s*. to £2.0.0)	Best method of studying nature of rock and boulders
Well-drilling and rotary-drilling methods occasionally used					
Geophysical exploratory methods treated separately					

[1] This table is based upon, but is a modification of, Table I of Mr. H. A. Mohr's paper entitled Exploration of Soil Conditions and Sampling Operations, Harvard Univ. *Graduate School Eng. Studies Bull.* 208, 1936–1937.

superficial deposits. Exploratory work will therefore be considered under these three headings.

The purpose of underground investigations must be clearly appreciated at the outset. Only in special cases is the purpose merely to find out where the hidden surface of rock lies. As a rule, it is necessary to find out also the nature of the material penetrated before rock is reached, not only for purposes of construction but also for the preparation of designs, especially when foundations are to be constructed in the unconsolidated material. For important foundations, and when rock excavations have to be carried out, the nature of the rock penetrated by drilling must also be investigated. Preliminary work, therefore, usually aims not only to provide data as to underground boundaries between essentially different materials but also to procure samples of the material penetrated for laboratory analysis and testing. In view of the increased attention being paid to the testing of sands and clays, it is today often of the utmost importance to obtain these relatively undisturbed.

FIG. 6.4.—Open-type soil sampler in use, testing highway subgrade material.

(Reproduced by permission of the Alabama State Highway Department, through the courtesy of Mr. J. L. Land, Materials Engineer.)

The simplest type of exploration in unconsolidated material is that done by probing with a steel rod; in shallow ground, depths to rock may sometimes be explored in this way, but the method is extremely limited in its application. The next stage involves the use of a soil *auger*, a boring tool described by its name, which may be obtained equipped with many special devices for penetrating different kinds of material and for procuring samples that, although not undisturbed, are often satisfactory for preliminary testing in a laboratory. The auger bores the hole, samples of cohesive material being obtained by withdrawing the tool and cleaning off the material adhering to it. Holes through sand have to be lined with pipes, but special drills are available for this type of work, and devices with which samples may be obtained. A soil auger set, complete with the usual extra fittings and capable of boring to 30 ft., can be obtained for about $100 (£20); it weighs less than 100 lb. when packed for carrying. It is strange that more use is not made, at least for small-scale work, of this most useful equipment.

This thought persists when consideration is given to the most usual type of boring adopted by engineers, generally called *wash boring*. A hole is excavated by washing inside a pipe casing which is forced into the ground as the hole is made; material excavated is trapped in the

FIG. 6.5.—Soil boring and sampling apparatus (soil sampler indicated by arrow). (*Reproduced by permission of the Director of Building Research, Dept. of Scientific and Industrial Research, England.*)

washtub and there examined and identified (if possible). Wash borings may be useful in some cases to show where a hidden rock surface is located; but as a means of exploring unconsolidated material, they are almost useless and sometimes worse than useless, as giving misleading information. They reveal the material penetrated only in a thoroughly disturbed state, generally with much of the finer material washed away; they give no clue to underground water conditions and no samples of any use at all for testing. Although they can sometimes give a generalized *soil profile* when carried out by experienced men, they are dangerous and should be avoided as far as possible.

A simple modification of wash boring provides a much more satisfactory method. A cored hole is sunk by washing, but the washing is stopped at intervals (determined by the nature of the ground) in order that a pipe or special sampling device may be lowered into the hole, forced into the undisturbed material at the bottom of the hole, and withdrawn with a sample enclosed. Washing out then continues until another sample is required. The method still suffers from the disadvantage that it may not disclose underground water conditions, the boring of holes in the dry with an auger being the only certain way of checking up on this matter through the medium of test boreholes. By using special sampling devices (described in Chap. XX), samples that are relatively undisturbed may be obtained from the bottoms of test holes made in this way and thus made available for laboratory testing.

Remarkable examples of test-boring work of this nature have occurred during recent years. In the exploratory work that preceded the construction of the San Francisco-Oakland Bridge, undisturbed samples were obtained along the line of the bridge from depths up to the record maximum of 273 ft. below low water level, as a result of which accurate data were obtained which enabled the unusually deep foundations for this great structure to be designed with accuracy and confidence. On the east coast of the United States, another interesting test-boring program was carried out in 1935–1936 in connection with the Passamaquoddy tidal power scheme, then projected by the United States Government. This work was all in exposed coastal waters and was hampered by severe winter climatic conditions, tidal currents in excess of 8 knots, and a tidal range of 28 ft.—a combination of adverse conditions that could hardly be equaled. Despite this, however, test sampling was carried out, and core drilling into the rock over a large area of the sea near the coast, in conjunction with geological investigations, enabled foundation-bed conditions for the structures then proposed to be accurately investigated. The deepest hole put down during these interesting operations was in 120 ft. of water and through 164 ft. of unconsolidated material above the rock which was then core drilled. An accompanying photograph shows the elaborate rigging required for this work, which was carried out by the United States Army Engineers with Mr. John Sweeney in immediate charge.[6.5]

Naturally, there are available methods other than those mentioned, notably those developed in connection with well drilling. Those already cited, however, are the most usual. No mention has been made of some of the "short-cut" methods available, such as driving a long pipe into the ground and then forcing out the plug of material so obtained or, alternatively, driving two structural shapes (two channels, for example) into the ground together, withdrawing them, and then separating them to examine the "core" of material so obtained. Although useful as giving a rough guide to underground conditions,

FIG. 6.6.—Working platform, and rigging, for deep-water drilling carried out by United States Army Engineers for the Passamaquoddy tidal power scheme, November, 1935, off the coast of Maine. Top of the 160-ft. spud, working in 120 ft. of water, is seen at low tide from the drill scow to which it was attached.

(*Reproduced by permission of the Chief of Engineers, U. S. Army, through the courtesy of Mr. John Sweeney.*)

these methods cause such disturbance of the strata that they penetrate that the results are of doubtful value.

6.6. Test Pits.—A further means of investigating the nature of superficial deposits is by means of test pits—shafts excavated in the deposits to a size that a man can dig conveniently in them. In practically all cases, these must be lined with timber, the design and placing of which must always be carefully checked. Clearly, there is a limiting depth to which test pits may reasonably be carried, so that their main use is for the investigation of relatively shallow depths of unconsolidated material and particularly for the study of surface deposits of gravel, sand, or clay for use in construction. They have the advantage that the deposits penetrated can be easily examined in place, and undisturbed samples can be obtained, when necessary, from the bottom of the pit. They will disclose underground water conditions in the material penetrated but clearly are restricted in use to operations on dry land. For any extensive exploratory work, test pits will usually prove to be more expensive than test borings; but on small jobs for which a drilling output would have to be specially procured, they may sometimes prove economical. An average figure for the cost of labor and materials for lined test pits up to 30 ft. deep, in Canada, has been found by the author to be $7 (£1.8.0) per foot.

When test pits are used to investigate surface deposits of construction material, it will be necessary to take regular samples of the materials encountered. Sampling is a matter that does not always receive the attention that it deserves in engineering work, so it may be mentioned that it is not just the haphazard collection of small quantities of material that some would appear to imagine it to be. Two objects may be kept in mind when sampling:

1. Obtaining samples typical of average conditions for all material to be investigated.

2. Obtaining samples representative of maximum and minimum characteristics of the material.

When investigating material to be used in construction, both objectives must be kept in view, although the former will usually be the controlling one unless the material is very variable in composition. Samples must always be taken at regular intervals, in materials of the same appearance; and from a large number of these, thoroughly mixed at the site, one or more average samples should be taken. These may then be used for testing, but they *must always* be accompanied by at least one unmixed sample, selected as typical for the stratum of material being studied, in order to check up on the correctness of the average samples obtained. (This procedure applies, of course, only to materials that can easily be mixed and not, for example, to stiff clay from which a number of individual samples must be selected.) All samples should be most carefully marked with full data as to the location from which they have been taken; few comparable situations are more exasperating than to have a good set of samples the identification of which cannot be made.

6.7. Test Drilling.—When rock surface is reached, either in a test pit or test boring, a change in method is necessary if the rock has to be penetrated. This is usually the case if only to check up on the nature of the rock near the surface and to insure that solid rock has been reached and not a boulder. Various methods are available, but reference need here be made only to core drilling in which a cylindrical hole is drilled around a central core which is periodically broken off from the bedrock, at its lower end, and removed from the hole for examination. A rotary drilling machine is used for this purpose, fitted with a special bit equipped with black diamonds (*diamond drilling*), chilled steel shot (*shot drilling*), or removable steel cutting teeth. The choice of equipment will depend to some extent on the nature of the rock to be encountered; the size of hole will depend on the drilling tool available and on the anticipated depth of hole. Core drilling is a highly specialized operation, necessitating skilled workmanship and experienced supervision for its successful performance. Percussion drilling is another means of penetrating rock, but it is not so widely used for exploratory work.

When drilling in some types of soft or disintegrated rock, it will sometimes prove difficult to obtain complete sections of core, and estimates will then have to be made, based on the operator's observations and his experience, with regard to the exact nature of the material penetrated. The amount of core recovered will naturally vary with the type of rock penetrated. Mr. W. Simpson gives the following averages: quartzite, 90 per cent; granite, 85 per cent; sandstone, 70 per cent; limestone, 60 per cent; shale, 50 per cent; slate, 40 per cent.[6.6] These figures are naturally to be regarded as a guide only to what may be expected.

The action of a core drill will not, in general, affect the bedrock through which the hole is drilled, diamond bits especially boring a clean, smooth hole. Consequently, it is possible to utilize cored holes in rock for purposes other than just providing a core of rock for identification. When the rock is to be subjected to water pressure (as in the case of a dam foundation), drill holes can be capped and filled with water which is then kept under observation while pressure is applied to it, and note taken if the hole "holds water" or not. An interesting investigation can sometimes be carried out in drill holes of relatively large diameter (about 4 in. and over) by the use of a periscope device, equipped with an electric light below the inclined mirror. When this is lowered into the hole, the appearance of the rock walls can be examined with a fair degree of certainty. Finally, whenever possible, drill holes in rock should be plugged in such a way that when active construction starts at the site, they may be opened up again and used as grout holes. Drill holes are usually put down vertically, but they can be driven at a considerable inclination to the vertical for special investigations. On page 151 will be found mention of an interesting example of the use of inclined holes in connection with the crossing of the Catskill Aqueduct Tunnel beneath the Hudson River.

FIG. 6.7.—Diamond drilling at a prospective dam site in northern Ontario, Canada, showing log cribs and staging from which the work is being carried out.

(Photograph by R. F. Legget. Reproduced by permission of the Power Corporation of Canada, Ltd., Mr. L. C. Jacobs, Construction Manager.)

A most interesting development in connection with test drilling in rock on some of the large civil engineering projects carried out in recent years has been the use of cored holes large enough for a man to be lowered into them. Holes of this size have been frequently used in the past in other branches of work, and their value in civil engineering practice has been quickly recognized. They are drilled with machines of the "calyx" type, with diameters varying up to 36 in., this size being convenient for use by average-sized men. The method of drilling and removal of the cores is similar to the practice followed for smaller holes, although breaking off the core from the bedrock is sometimes a matter of difficulty necessitating special wedging devices or the use of blasting. The holes are of avail only if leakage of ground-water into them is small enough to be taken care of by a small pump; special precautions must always be taken to keep a supply of fresh air at the bottom of the holes, especially if blasting has been used for core removal.

When drilled to the requisite depth and cleaned out, the holes may be inspected by the geologists and engineers in charge (with the aid

FIG. 6.8*a*. FIG. 6.8*b*.

FIG. 6.8*a*.—Looking down into one of the 36-in. Calyx-drill holes bored into the foundation strata at the site of the Grand Coulee Dam.

FIG. 6.8*b*.—Close-up view of Calyx core (36 in. in diameter) from foundation strata of the Grand Coulee Dam.

(Reproduced by permission of the Commissioner, U. S. Bureau of Reclamation, Washington, D.C.)

of portable lights), lowered down in suitable cradles. In this way, the surrounding rock can be carefully inspected exactly as it occurs in place, boundaries between beds can be investigated, fissures studied closely, and a thorough and complete exploration of the rock can be made, with certainty and convenience. If the holes are drilled after grouting of the foundation beds has been carried out, the efficacy of the grouting operation can thus be checked up, this being a most valuable feature in view of the inevitable uncertainty regarding the penetration of grout. The cost of holes of this size may be considerable, but it is commensurate with the great advantages that they present for underground investigation.

The use of these large-diameter holes in civil engineering work appears to have developed principally in the United States of America, although they have been used successfully in several other countries. They have been put down mostly in connection with dam-foundation work, as for the Grand Coulee and Norris dams; Fig. 6.8*a* and Fig. 6.8*b* show their use at the Grand Coulee dam site. A particularly interesting application was at the site of the Prettyboy Dam, constructed to impound water for supply to the city of Baltimore, Maryland. The site is physically a good one, the valley being quite deep, but geological conditions required unusual precautions in the excavation of the cutoff trench below the base of the main structure. This called for the placing of about 190,000 cu. yd. of concrete. The rock formation beneath the dam is mainly mica schist with some limestone, gneiss, and intruded quartz, and it has been twice subjected to earth movement. Faulting was therefore to be expected. The exposed rock had undergone weathering to a considerable degree. It was realized that extreme care would have to be taken in blasting for the excavation of the foundation bed. As a preliminary operation, it was decided to drill several deep shafts, 36 in. in diameter, excavated with Calyx core drills, which permitted the consulting geologists of Johns Hopkins University, under Dr. J. T. Singewald, Jr., to be lowered into them so as to study the formation of the schist in place. By this means, the geologists were enabled to prepare accurate sections showing the position of all faults and large seams as well as the direction of their strike and dip. Disintegration of the schist on the hanging-wall side of a major fault, traced in this way, caused a considerable increase in the volume of excavation. Furthermore, it was in this way possible to obtain truly representative samples of the schist wherever desired, and the study of these samples constituted an interesting investigation in pure Geology.[6.7]

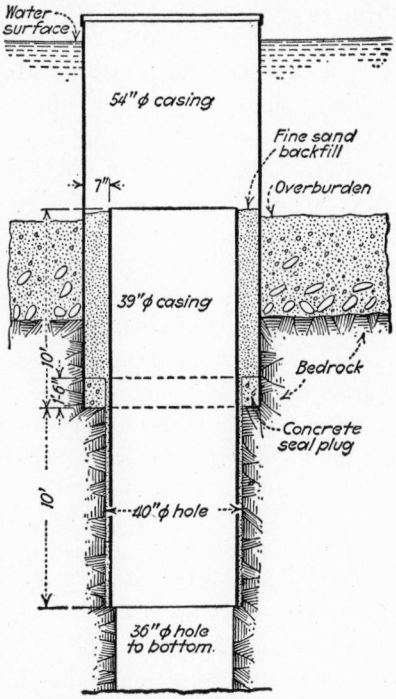

FIG. 6.9.—Telescoping casing and cofferdam as used for carrying out 36-in. diameter test-hole drilling at the site of the Watts Bar Dam, Tennessee River, by the Tennessee Valley Authority.

(*Reproduced by permission of the Editor, Eng. News-Record.*[6.8])

Another particularly interesting application of large-diameter drill holes was developed by the engineers of the T.V.A. in making explorations of an unusually complex underground structure on which the Watts Bar Dam is to be founded. This involved drilling large holes at various locations across the river at the dam site selected. This was done by means of the cofferdam arrangement sketched in Fig. 6.9, which worked admirably, depths up to 68 ft. of penetration into the rock being attained, permitting complete inspection of the sections thus revealed.[6.8]

6.8. Special Exploratory Methods.—The exploratory methods already described include all those generally adopted in normal practice. There are available special methods and devices which are sometimes applied in unusual cases, and brief mention of some of these must be made. As noted, test pits can be taken only to certain limiting depths, using ordinary construction methods and the usual timber lining. Occasions arise when visual inspection of deeper unconsolidated strata is necessary or when information is required not only as to the nature of such material but also as to how it will affect the sinking of caissons, designed to be part of permanent work. In such cases, special test caissons are sometimes constructed which may be sunk to the depth required, under compressed air if necessary, thus giving accurately the data required. Another special feature of the exploration of dam foundation beds is the use of test tunnels. The object of these is similar to that of test shafts, with the added advantage that they can be used to explore rock strata (as under river beds) which would be inaccessible by vertical shafts. Their cost limits their application to dam projects of unusual magnitude; some examples will be mentioned in Chap. XIII. Finally, there has developed in recent years a most interesting application of certain principles of geophysics to subsurface exploration, known variously as *geophysical surveying* or *geophysical prospecting* (the latter due to extended use in connection with mining). So important is this work, and of such potential value in civil engineering, that it is considered in some detail in Chap. VII.

6.9. Conducting Exploratory Work.—It will be obvious, even from this brief discussion, not only that the accuracy of underground exploratory work is of supreme importance but also that the conduct of the work often requires exceptional skill and wide experience on the part of the men in charge. In engaging men to undertake exploratory work, therefore, experience and reliability are essential qualities to be looked for. It is often advantageous to carry out all work of this nature by direct administration in view of the indefinite extent of the work when it begins; if this is possible, and the engineer is able to employ experienced men, it constitutes probably the most satisfactory procedure. Many large engineering organizations, such as highway and public works administrations, maintain special test-boring divisions for carrying out all their regular exploratory work. On other civil engineering work, however, carried out by consulting engineers or by engineering offices that are not able to maintain regular boring crews, it will be necessary to engage an outside test-boring contractor for this work. The necessary contract documents require special care in preparation, to allow for a wide degree of flexibility not only in the extent of the work but also in the location of test holes in view of variations that may be revealed as the work proceeds. It will be clear, therefore, that under no circumstances should a test-boring contract be awarded on a "lump-sum" basis. Unit prices must be secured, probably with a guaranteed minimum number of holes and total depth of drilling or boring, with modified prices for operations in excess of certain specified limits. Although the usual practice of calling for tenders will probably have to be followed, the award of the contract should not be made on the basis of price alone, but due regard should be paid to the experience and reliability of the respective contractors who have tendered and the test-boring and drilling equipment that they have available, details of which should be required with the tenders.

6.10. Recording Exploratory Work.—All exploratory work is of value only if a complete and accurate record of the results is obtained for the use of the engineer in charge and his advisers. This is so obvious that there would seem to be little need to emphasize the necessity for obtaining accurate records. Not infrequently, however, it will be found that all recording is left in charge of the test-boring foreman, supervised by occasional visits of an engineer, and this is often done when a considerable amount of money is being spent on the exploratory work. No matter how experienced and how conscientious a foreman may be, he should not be required to perform two tasks, not only because he will not be well versed in record work (as a general rule) but also so that an independent observer may check with him as to the results that he is obtaining.

It may safely be said, therefore, that, as a general rule, a member of the engineer's staff, himself a qualified engineer, should always be present throughout *all* exploratory work to watch its progress and keep the necessary records. If a geologist can undertake this task, so much the better, but this will be possible only when a large field geological staff is available. If a geologist is being consulted about the work for which the testing is being done, he should be enabled to visit the site of the exploratory work at least occasionally, while it is in progress, so that he may see the results for himself and discuss them with the men in charge.

When it is realized that the success or failure of an entire construction project may depend on the accuracy of records of exploratory work, this insistence on their importance may be appreciated. Records to be kept will naturally depend on the type of strata being encountered and the exploratory method in use, but it may be suggested that the following details should always be recorded:

RECORDS REQUIRED FOR EXPLORATORY WORK

Date.
Times of operation.
Names of operators.
Details of equipment.
Exact location of pit or hole (tied in to a control survey).
Elevation of ground surface.
Elevations of all changes in strata.
Elevations at which samples are taken.
Elevation of groundwater if encountered.
Elevation at which a hard bottom is reached.
Depth to which core boring is taken.

A special example of a boring record is given in Fig. 6.10.

Serial Number	Original Number	Date	Location	Purpose of Boring	Cost

Department securing boring	Elev. of Surface	Depth to bedrock	Penetr. in weathered zone	Penetr. in bed rock	Total depth	Depth to ground water	Cased to	Use of Dynamite	Type of drill used	Condition of Core	Size of Core	% of core Recovered	Originally described by	Geologists' report

Elev. C & GS	Graph	Depth	Progress	Core Recov.	Character of overburden or Condition of rock	Formations: Old name	Formations: New name	Petrographic variety	Dip	Aver. length core pieces	Evidence of faulting	Loss of water in boring	Remarks: Inc. structure, dikes, special minerals, porosity, unusual drill behaviour, relations at contacts, oxidation, etc.

FIG. 6.10.—The heading to a standard record sheet, for compiling full particulars of test borings and drillings.

(Reproduced by permission of the Municipal Engineers' Society of the City of New York, through the courtesy of Mr. John C. Riedel.)

Considering, now, the second objective of subsurface investigations —study of the materials encountered—it is clear that an accurate record of the nature of all strata must be kept. This record will be supplemented by the samples of each stratum that are secured and by the cores obtained in core drilling. Securing samples of unconsolidated materials for laboratory test purposes will be considered in Chap. XX. It may be noted here that whether samples are required for testing or not, they should be most carefully secured and should be stored in containers (glass for preference, to permit visual examination without removal) that are airtight so that the moisture content will not change. The hard, dried-out samples of clay that are sometimes seen in connection with boring results are often worse than useless, as they disclose nothing about the natural condition of the clay.

Rock cores must also be carefully stored, the most convenient way of doing this being to use special core boxes—flat wooden boxes divided into narrow compartments each wide enough to hold one core and, for

FIG. 6.11.—Calyx drill core (36 in. in diameter) from foundation strata of Grand Coulee Dam, laid out for inspection, with allowance for core losses.
(Reproduced by permission of the Commissioner, U. S. Bureau of Reclamation, Washington, D.C.)

convenience, an even number of feet long. Into these compartments, the cores are placed as they are obtained, care being taken to place them in compartments corresponding exactly in position to the location of the core in the holes so that gaps will have to be left periodically to allow for the inevitable core losses. Large-diameter rock cores cannot, obviously, be kept in storage boxes, but arrangements are usually made to have the various sections of core, as they are secured, laid out in order and in line adjacent to the hole so that they may readily be examined by those interested.

Finally, a prime requirement of all records of exploratory work is that materials encountered should be accurately described. When it is possible to examine samples of all the materials after the hole has been made, accurate descriptive notes as the work proceeds are not imperative, provided that all samples and cores have been properly correlated with the progress records. It will be a distinct advantage, however, if accurate terminology can be used in the day-to-day records. The engineer who is keeping them should therefore be familiar at least with main rock groups and the distinctions between the various grades of unconsolidated material. The latter should always be described by the use of the appropriate geological term—gravel, sand, silt, or clay, or a mixture of two or more of these—together with as accurate a notation of the physical condition of the material as possible, *e.g.*, hard packed or very loose. Terms that are essentially popular, such as *hardpan*, or local in application, such as *pug*, should be avoided. If an exact measure of the state of the material is desired, this must be obtained by using a penetration device of some type.

6.11. Interpretation of Exploratory Results.—Underground explorations such as have been described in preceding sections can be fully effective only if they are correlated, without delay, with the results of geological survey work at the site being investigated. This requirement provides another convincing argument in favor of having an engineer in constant attendance at all exploratory test work, since the correlation can best be carried out while the work is in progress. The obvious, and most useful, means of combining the results of the two methods of investigation is to have drawn up tentative geological sections, based on survey work, along lines on which test holes are to be put down, the records of these holes then being plotted to scale on the section as work proceeds. It may be necessary to use a distorted scale for this plotting; but provided that this fact is not lost sight of and that a definite scale *is* used to relate the boreholes and sections with one another, the resulting record can be easily interpreted. By means of this simple device, it will be possible to keep constant check on the progress of holes, to stop them when they have gone far enough, to locate new holes in order to clear up doubtful points revealed by the section, and generally to see that no effort is wasted and no necessary information left unobtained. A singularly valuable method of illustrating the results of test boring is to have a model prepared, using colored sticks to represent the various boreholes. Figure 6.12 is an interesting example of such a model.

6.12. Utilization of Exploratory Results.—Just as the completion of the initial geological survey of a site will usually enable the civil engineer to prepare preliminary plans for a project, on which will depend the program of exploratory work to be carried out, so will the completion of the exploration program enable final designs to be prepared in all detail. The combination of the results of the exploratory work and the geological survey, in all but exceptional cases, will enable the engineer to obtain a complete picture of the underground structure at the site in so far as it will affect his plans. He will thus know to what limitations his design is subject; he will be able to calculate with a fair degree of accuracy the amount of material that will have to be excavated; and he will know what natural construction

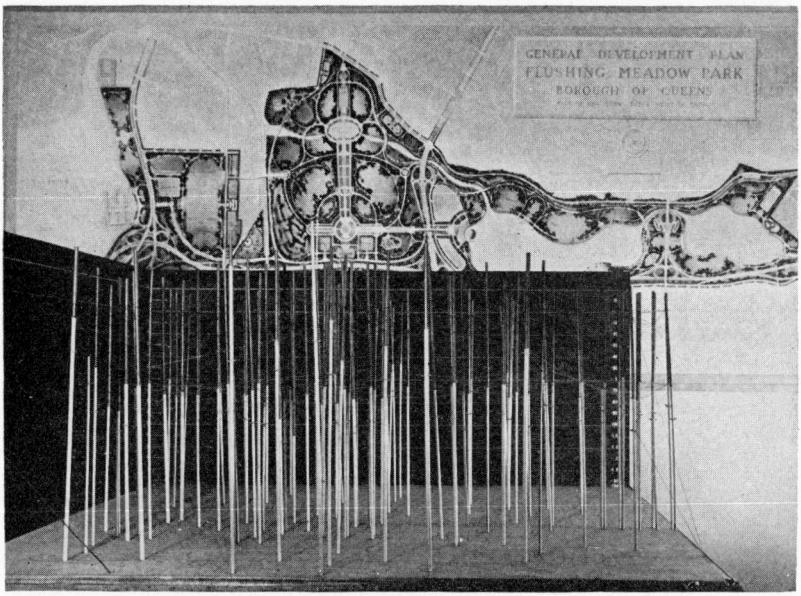

FIG. 6.12.—Diagrammatic model of test borings on the site of the New York World's Fair 1939, made by World's Fair staff and based on test borings put down by them and under the direction of Messrs. Moran, Proctor and Freeman, consulting engineers, New York.

(Photograph by New York World's Fair. Reproduced by permission of New York World's Fair, through the courtesy of Mr. Carlton S. Proctor.)

materials are available within easy reach of the site. Thus will he utilize the findings of his preliminary work, embodying the results in his contract plans and specifications.

In many specifications, however, and on many contract plans will be found something similar to this clause:

Drawing X contains details of borings that have been made at the site of the work, but their accuracy is not guaranteed; and intending contractors are required to take, before they tender, the borings that they may deem necessary to satisfy themselves as to the accuracy of the information regarding local conditions conveyed by the plans and specifications.

If this qualification is considered at all seriously, it will be seen to evidence a surprising paradox in that the engineer has presumably based his entire design upon the information that he suggests the contractor should not use in case it may be wrong! The clause is probably a carry-over from the days when preliminary investigations were not always comprehensive and when the methods available did not permit even the engineer to place great reliance upon the results obtained. Today, however, the clause is an anachronism. If there is any need for each of the contractors who intend to tender on a contract to take individual sets of trial borings, then the engineer's design cannot be assumed to be free from doubt, and the expenditure of all the money necessary to take the borings may be doubly wasted. On the other hand, if the engineer's design is based on accurate exploratory results, there is usually little need to include the clause. To complete the necessary comment on this strange feature of contract documents, it may be suggested that court records will not always be found to support the intention of the clause, *viz.*, to put the onus of anticipating satisfactory foundation conditions on the contractor. In this connection, it has been stated editorially in *Engineering News-Record* (July 7, 1938) that

> The established view of the courts on the matter roughly appears to be this: The owner is responsible for unforseen costs to the contractor when the engineer's borings are found to contain inaccuracies or fraudulent misrepresentation. The owner also is responsible when the engineer does not reveal to the contractor his complete record of preliminary investigations even when he has reason to doubt their accuracy.
>
> On the other hand, the owner is not responsible for the fact that incomplete borings do not reveal hidden ledges of rock in an earth bank, buried cribs along old waterfronts, and the like. . . . Lower courts have penalized owners when unforseen difficulties have caused the contractor to sue, but only under exceptional circumstances have higher courts ruled that the risk of such discovery is the owner's rather than the contractor's, provided always that there has been no concealment by the owner.

Some provision, however, must be made for the possibility that underground conditions when opened up during construction may not be found to be exactly as suggested by preliminary exploratory work, despite the care with which this has been carried out. One of the most satisfactory solutions to this difficulty appears to be that featured in the standard form of contract used for construction work carried out for the Federal Government of the United States of America.[6.9] This reads:

Changed Conditions.—Should the contractor encounter, or the Government discover, during the progress of the work subsurface and/or latent conditions at the site materially differing from those shown on the drawings or indicated in the specifications, or unknown conditions of an unusual nature differing materially from those ordinarily encountered and generally recognized as inhering in work of the character provided for in the plans and specifications, the attention of the contracting officer shall be called immediately to such conditions before they are disturbed. The contracting officer shall thereupon promptly investigate the conditions, and if he finds that they do so materially differ the contract shall, with the written approval of the head of the department or his duly authorized representative, be modified to provide for any increase or decrease of cost and/or difference in time resulting from such conditions.

This appears to be a reasonable provision and may suggest how the possibility of changed conditions can be dealt with in other specifications.

Another solution to this very valid difficulty is that adopted by certain public engineering organizations, *e.g.*, in the case of dam construction, of clearing the site of the dam and preparing the foundation bed by direct administration and then awarding a contract for the construction of the dam structure alone, about which there should be few serious disputes. This method, however, is applicable only to certain cases, and it is by no means free of disadvantages. It may be suggested, therefore, with propriety that the best of all methods for avoiding contractural disputes regarding foundation-bed conditions is to have the preliminary and exploratory work so well carried out that there remains, when construction begins, small chance of encountering serious variations from assumed subsurface conditions.

6.13. Conclusion.—This chapter ends as it began, with a renewed suggestion of the great importance in all civil engineering operations of adequate preliminary geological investigations and underground exploratory work. The services of geological advisers of wide experience can be secured without difficulty; methods and equipment of wide scope and great utility can now be utilized; estimates can readily be prepared to show the relatively small cost of this work; and many records can be found in the annals of the art to demonstrate the futility of its neglect and the great value of its correct utilization.

As an encouraging reminder of the extent to which exploratory work is carried out in these times, there may be cited such a case as that of the Shasta Dam, being built (started in 1938) by the United States Bureau of Reclamation on the Upper Sacramento River in California as a part of the great Central Valley Project. It is reported that before tenders for the work were received, "over a mile of tunnels and adits, two miles of diamond-drill holes, 180 ft. of shafts, and 190 ft. of 36 in. calyx drill holes had been driven" at the site, the local rock being "andesitic in character, made up of metamorphosed lavas and tuffs."[6.10] An even more striking example may be cited from French engineering practice. The preliminary investigations for Le Sautet Dam in the Valley of Le Drac, constructed in a deep and narrow limestone gorge, included 1,650 ft. of exploratory *tunnels*, in addition to much other exploratory work. From these tunnels, 20,000 ft. of holes was later drilled for the grouting of the foundation beds.[6.11]

These are examples from the list of monumental modern civil engineering projects, but equally effective examples could be cited from the great number of ordinary-sized jobs with which the average civil engineer has to deal. And it is these jobs that constitute the greater part of civil engineering practice; to them, also, can preliminary work carefully and accurately carried out prove to be of great value. More than twenty years ago, Sir Maurice Fitzmaurice, in his presidential address to the Institution of Civil Engineers, dealt in an interesting way with the principal matters that cause civil engineers trouble in the ordinary practice of their profession. "The first one," he said, "in my opinion, is the small knowledge we are able to obtain of the conditions which exist below the surface of the earth on which everything has to be founded." Progress has been made since those words were spoken; and although certainty in this branch of civil engineering work will never be fully attainable, it does appear that by a judicious application of Geology, coupled with accurate and constantly supervised subsurface exploratory work, many troubles may be avoided, many difficulties may the more easily be overcome, and desirable economy and efficiency in design may be achieved.

Suggestions for Further Reading

MOHR, H. A.: Exploration of Soil Conditions and Sampling Operations, *Harvard Univ. Graduate School Eng. Studies Bull.* 208, 1936.

SIMPSON, W.: "Foundations: Methods and Appliances," Constable & Company, Ltd., London, 1928.

Also the chapters dealing with subsurface exploration in leading textbooks on foundations.

Technical journals will be found to contain articles dealing with test boring and drilling; the following is an unusually comprehensive example:

SANBORN, J. F.: Exploratory Boring Needed as a Guide to the Development and Construction of Public Works, *Eng. News-Record*, 88: 1024 (1922).

See also MCCULLOUGH, C. B.: Highway Bridge Surveys, *U. S. Dept. Agr. Tech. Bull.* 55, February, 1928 (reprinted 1938).

CHAPTER VII

APPLIED GEOPHYSICS AND CIVIL ENGINEERING

> Bishop Watson compares a geologist to a gnat mounted on an elephant, and laying down theories as to the whole internal structure of the vast animal, from the phenomena of the hide. The comparison is unjust to the geologist. . . .
>
> LORD MACAULAY, "Critical Essay on History."

Few branches of scientific work have advanced more rapidly during the twentieth century than the application of what are known as *geophysical methods* to underground investigation. The name of the science of Geophysics is itself of recent origin, the word *geophysics* being less than one hundred years old. The science has been defined as being "concerned with the constitution, age, and history of the earth, and the movements of the earth's crust."[7.1] It will thus be seen to be closely associated with the science of Geology which is more particularly concerned with the constituents of which the earth is made. Geophysical investigations are therefore related to the physical properties of the earth, the term *physical* being used to include magnetic and electrical properties, in addition to density, elasticity, and rigidity. The methods employed will inevitably be utilized in connection with the measurable properties of the earth's crust, and in this way has arisen the use for utilitarian purposes of instruments and methods which might naturally be supposed to be of interest mainly to the scientist.

At the beginning of the present century, geophysical methods of investigation then in use were thus restricted to scientific research; but as the methods were developed, so were their utilitarian possibilities appreciated and explored. Experimental work was gradually initiated solely with utilitarian ends in view; and in the short space of less than two decades, geophysical methods have left the sheltered environment of the scientific laboratory and become an important and often closely guarded part of commercial endeavor on the part of mining and oil-producing companies and also of specialist organizations that undertake geophysical survey work, or *geophysical prospecting*, as it is frequently and not inaccurately called. Today, patent office records testify to the ingenuity and industry that are being applied to the further improvement of geophysical instruments and operations. Descriptive literature is abundant, from advocates of specialized geophysical methods on the one hand, anxious to publicize details of their work, and from the technical staffs of industrial corporations on the other, giving full details of results obtained but sometimes not a great deal about the methods employed.

This intense activity is confined generally to the mining and oil industries, and it is reflected in the specialized technical journals and societies serving them. The work thus being done is concerned with the investigation from the surface of the ground of underground conditions, and it is therefore of considerable interest to civil engineers. In the last decade, some of the methods have actually been applied to civil engineering work. These applications have been at least partially successful, and they are increasing in number. They suggest great possibilities and are therefore already worthy of serious study when an engineering project to which they may usefully be applied is under consideration. The civil engineer should consequently at least be familiar with the underlying principles of the main methods, their possible applications, and their limitations; this chapter is an attempt to provide the necessary introduction.

What follows is based on available published records of the application of geophysical methods to civil engineering work. Outlines only of the main methods available will be presented, in view of the usual employment of specialist operators; the engineer who wishes to follow the matter further will find listed a number of general references at the end of the chapter and in Appendix D.

All geophysical methods are concerned with the physical properties of the earth's crust, and so of the constituents of the crust. If, therefore, two rock bodies adjoin at or near the ground surface and have distinctly different *physical* properties, as distinct from their geological, properties, geophysical methods should be able to differentiate between the two rock masses. This idea is the basis of all geophysical work. To mention specific examples:

> . . . deposits of magnetite have a greater magnetic susceptibility than the enclosing rocks and may be located by means of a magnetic survey, massive chalcopyrite deposits are better conductors and may be discovered by electrical measurements, while a mass of iron ore has a greater density than the surrounding rock, so that a careful examination of the gravitational field in its vicinity may disclose the presence of this heavy body.[7.2]

The success of the various geophysical methods in detecting with accuracy the boundaries between rock masses will be indicated in succeeding sections. It must first be emphasized that geophysical methods constitute only another exploratory aid to geological surveying and that as applied to civil engineering they must never be regarded as anything more than this. They will not disclose more than will a good set of test bore and drill holes and usually not so much, and they can never be used without specific and constant correlation with geological information. In fact, a preliminary geological survey is necessary before the methods can be applied at all with any certainty of success, since they necessitate knowledge of certain general conditions of the local geology in order that their interpretation may be effective. The most favorable condition is when a large deposit or rock stratum occurs under the cover of a shallow superficial deposit, the physical characteristics of the two being markedly different. This condition, in connection with civil engineering work, occurs most frequently as a deposit of glacial drift overlying a solid rock body, and it is to the study of the geology of such deposits, especially as regards the depth to bedrock, that geophysical methods have generally been applied in civil engineering practice.

The methods available may be grouped conveniently into four divisions, each of which is considered separately in the following sections.

7.2. Magnetic Methods.—The phenomenon of terrestrial magnetism has been known for a long period, and it is not surprising that magnetic methods of investigation are the oldest of all geophysical methods for studying underground conditions. On the earth's surface, there is a magnetic field relatively constant in direction and strength. This can be measured by a magnetic needle. If this is freely suspended on a vertical axis, it will come to rest in a direction known as the *magnetic meridian;* if suspended on a horizontal axis, it will come to rest at an angle with the horizontal called the *magnetic dip,* if the axis is perpendicular to the magnetic meridian. The earth's magnetic field is not absolutely constant, varying slightly at any locality in a regular manner with a slow seasonal change and a rapid daily variation; it may also be affected irregularly by magnetic storms. In addition, the field may be found to vary in an unusual manner as between one location and an adjacent point owing to the presence underground of some material possessing the properties of a permanent magnet. Minerals of iron most notably possess this property; although certain other minerals and rocks are slightly magnetic, their magnetism is so weak that they cannot be detected by any of the normal magnetic methods.

De Castro, in the seventeenth century, appears to have been the first to realize that hidden masses of magnetic material might cause local magnetic anomalies and that these masses might be detected by determining the magnetic changes that they cause. Since that time, instruments have generally been developed for measuring these anomalies, based essentially on the simple magnet; the more usual types are known as *magnetic variometers*, and they may be of either the vertical or the horizontal type. In place of a magnet, a rotating coil may be used to measure the vertical or horizontal magnetic force, the induced current being measured by means of suitable electrical devices. The modern type of magnetic variometer is a highly complex instrument, and it cannot be used in the field by other than highly trained specialists; but despite this, magnetic methods of investigation are capable of relatively wide use in their restricted field.

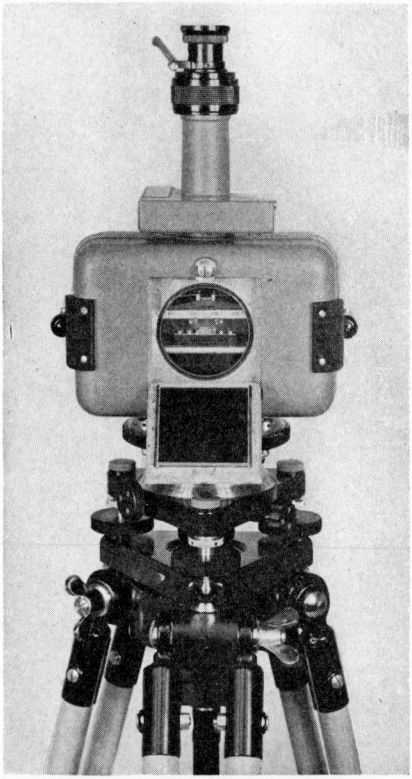

FIG. 7.1.—A Watts magnetic vertical variometer, as used in geophysical work.
(Reproduced, by permission, from an exhibit in the Science Museum, S. Kensington, London.)

7.3. Seismic Methods.—Earthquakes have been studied by scientists for a long period, but scientific methods have been applied to their investigation only in the last century and a half. As early in the development of this branch of study as 1846, Robert Mallet suggested that artificial earthquakes could usefully be created for experimental purposes by exploding gunpowder on the land or sea floor and that by this means earthquake waves could be studied. This idea is the basis of geophysical seismic methods, the fact that the vibrations set up by earthquakes, either real or artificial, do not travel at the same speed in different media enabling the existence of a change of medium to be detected. Although the idea was suggested at so early a date, and some experiments were later conducted along the lines suggested, it was not until the early part of the twentieth century that satisfactory instruments were devised for measuring and recording the vibrations reaching the point of observation. Since then, the original instruments have been improved, and new ones developed, so that seismic methods of underground investigation are today widely used.

Artificial earthquakes are produced by blasting with powerful explosives, high-strength gelatins and dynamites being most suitable; the charges should be buried in order to obtain best results. The waves so set up in the ground—*elastic earth waves*, as they are sometimes known—are of two types due, respectively, to longitudinal and to transverse vibrations. The latter may be principally in a vertical direction or alternatively horizontal. Other minor types of waves may be generated, but they are subsidiary to the two main types. Of these, the longitudinal waves travel faster than the transverse waves and so will be the first to reach the point of observation. Both types of waves travel through different types of rock with different velocities, and they will be refracted as they pass from one medium to the other. It is this fact that is the basis of geophysical seismic methods of investigation, records of observed vibrations being taken at different distances from the location of the explosion, and the results correlated with known facts about wave travel in different media.

If two rock strata are to be investigated, the upper one allowing waves to pass through it at low velocity, and the lower being a correspondingly high-velocity medium, a charge of explosive would be detonated at some convenient location in the upper stratum. Waves would then be dissipated in all directions through this stratum, and

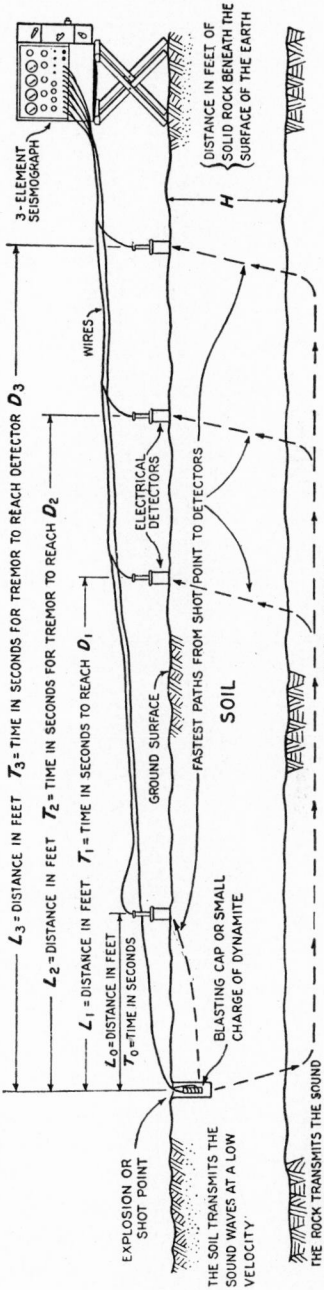

FIG. 7.2.—Diagrammatic section illustrating the principle of the seismic refraction method of subsurface exploration.
(*Based, by permission of the U. S. Bureau of Public Roads, on a drawing in Public Roads, June* 1935, *reference* 7.12.)

some would reach the observation points that had previously been set up at different distances from the explosion point, by thus traveling directly through the low-velocity medium. Some waves will reach the surface of the high-velocity lower bed. It has been found that some of this energy will travel along the boundary surface at a velocity equal to that which the waves would have in the lower stratum. It is being continually diffracted back to the ground surface through the upper layer, and eventually some will reach the observation points already set up for the purposes of the investigation. The relation of the time of travel by this route to that of the waves which proceed directly through the upper stratum will clearly depend on the relative velocities in the two media, the depth of the upper layer, and the relative positions of the observation points. It will be clear that a point can be reached at which the second path of the wave will become the quicker; and if simultaneous readings are taken at all observation points, it should be possible to find the position of this point (or, actually, this circle) within reasonable limits. This determination is the object of this type of seismic investigation.

A reflection method has also been developed, in which observations are taken at a convenient distance from the point of explosion of the time required for waves to travel downward to the surface of the lower stratum and then to be reflected therefrom to the point of observation. The practical difficulties of this method are considerable, being mainly in connection with eliminating the recording of the waves reaching the observation point directly, so that the very much weaker reflected waves may be noted. The difficulties are being overcome, and the method is credited with some remarkable results and is being widely used.

The instruments used for recording the vibrations reaching the points of observation are specially designed field-model seismographs. They are of two main types, the mechanical and the electrical. In the former type (now largely superseded by the latter), the necessary magnification of the very small vibrations is achieved by mechanical or optical methods; and in the latter, this is done by electrical means. The essential part of both types of seismograph is a heavy mass suitably mounted by a nonrigid type of suspension in a box or cover. The inertia of the mass will tend to keep it in a state of rest, whereas the cover will tend to move in accordance with vibrations reaching it through the ground, the instrument thus measuring its movement relative to that of the mass. Usually, a device is included that permits a continuous photographic record of the vibrations to be obtained by means of a clockwork mechanism, time intervals and the instant of firing the explosive charge being also marked on the same record strip. Normally, several seismographs are used, as already indicated; in the case of electrical instruments, the detector parts can be used at each observation point, the records being transmitted electrically to a control instrument on which all the results are photographed simultaneously.

FIG. 7.3.—An electrical seismograph (three-element) setup for field use.
(*Reproduced by permission of the U. S. Bureau of Public Roads, from Public Roads, June,* 1935, *reference* 7.12.)

7.4. Gravitational Methods.—The fundamental law of gravitation was announced by Isaac Newton in 1687; it stated that the force of attraction between any two bodies is directly proportional to the product of their masses and inversely proportional to the square of the distance between them. Originally, the law was associated only with large bodies such as those of the solar system. Scientists soon turned their attention, however, to the attraction exerted by large mountain masses. Bouguer appears to have been the first to experiment in this direction, at Mount Chimborazo (in South America) in 1737–1740, using a pendulum at a height of 9,000 ft. and again at sea level; similar work was done in Scotland by Maskelyne in 1775; and in the following century by various experimenters. Results then obtained were not as expected, but they were not explained until many years later following a further series of important observations in India. As a development from the study of these, there was finally formulated the principle of isostasy, which is "the basis of all accurate work on the variation of gravity at the earth's surface. [It] implies that above some depth below the surface neighboring vertical columns of the crust contain the same mass."[7.3]

Determinations of the value of gravity at different locations are still made by means of pendulum-type instruments, although of special design and construction to permit of their use in the field as well as in the laboratory. Many interesting gravity determinations, for example, have been carried out from submarines by Dr. Meinesz.

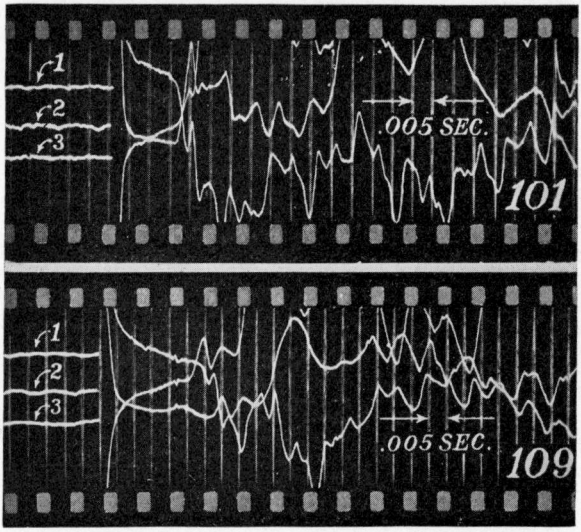

FIG. 7.4.—Typical photographic record strip from three-element electrical seismograph. (*Reproduced by permission of the U. S. Bureau of Public Roads, from Public Roads, June* 1935, *reference* 7.12.)

With such instruments, an accuracy of about one part in a million can be attained, but this is not sufficient to enable the variation of gravity between closely adjacent stations to be measured. That variations do exist is known from the different values obtained by gravity determinations, carried out at different points. The changes that can thus be measured by pendulum instruments have been found to be due to extensive buried masses of relatively dense material, some of which have been located in this way.

This method of detecting underground structures has been extended by the use of another type of instrument of greater sensitivity, known generally as the *Eötvös torsion balance*. This was first developed by Baron Roland von Eötvös in 1888 after a similar instrument used by Cavendish in 1797, although this had actually been designed by the

Rev. John Mitchell at the University of Cambridge, England, in 1762, who died before he could use it. As originally designed, the Eötvös balance consisted of a "single horizontal light aluminium beam, to the extremities of which equal masses of platinum were fixed. This system was carried on a delicate suspension wire, the torsion of which acted in opposition to the tendency of the earth to rotate the beam

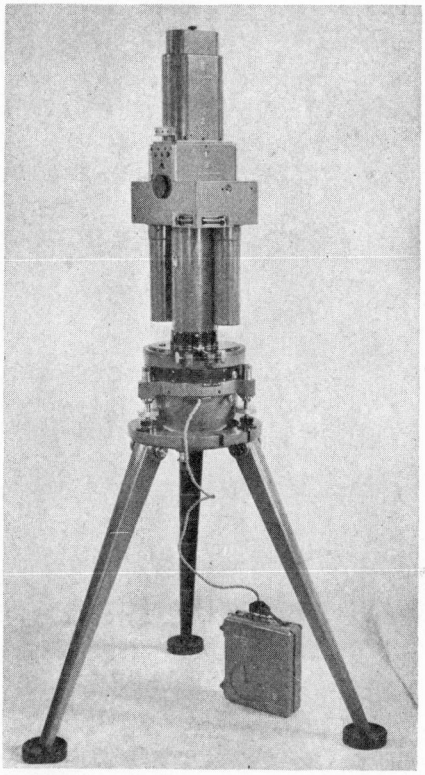

FIG. 7.5.—A Süss Eötvös torsion balance of the automatic photographic-registration type.
(Reproduced by permission of The British Geophysical Agency, London.)

into one definite orientation."[7.4] Another model of the balance has one of the masses suspended from one end of the beam by a fine wire, somewhat shorter than the length of the beam. Other models have been developed, some utilizing a light wheel instead of a beam, but the operation of all is along similar lines. If a torsion balance is set up in an area in which gravity varies from one point to another, possibly as a result of the presence of a buried body of rock, one of the masses on the balance will be subjected to a greater horizontal pull than the other, and so the suspended system will rotate until the external couple is balanced by the torsion in the wire, the degree of rotation being measured. The instruments are extraordinarily sensitive, being capable of measuring a change in the value of gravity of

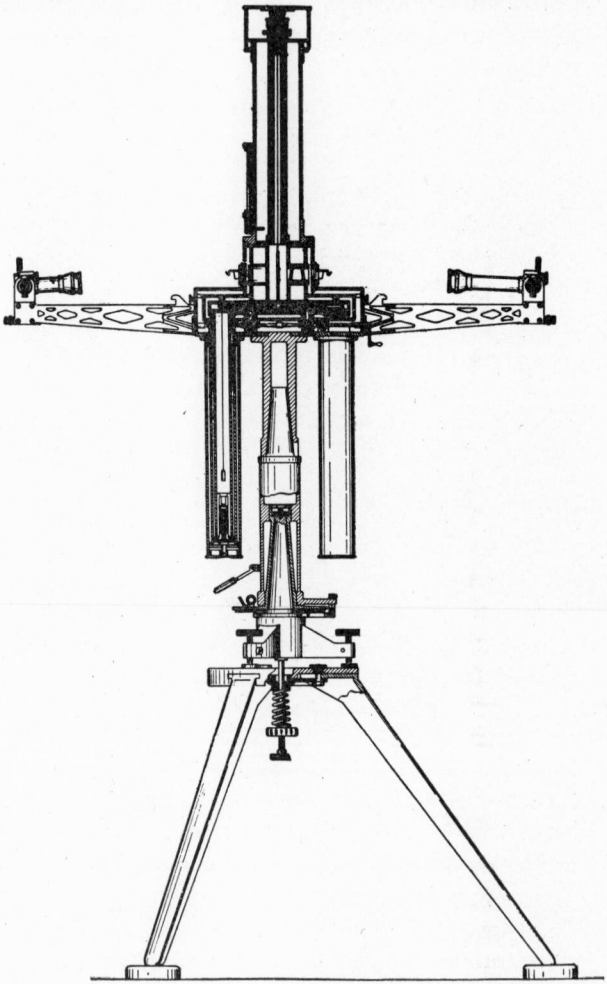

FIG. 7.6.—Diagrammatic section through a Süss Eötvös torsion balance of the hand-operated visual type.
(Reproduced by permission of The British Geophysical Agency, London.)

one million millionth part of its total value. Gravimeters are also used in a similar way.

As with other geophysical methods, a torsion balance can be used to detect the existence of two adjacent rock strata of different densities;

clearly the best results will be obtained if the change in density occurs near the ground surface. Not only will changes in rock density affect the instruments, but also any change in level of the adjacent ground, and even the presence near the point of observation of obstructions such as buildings or even trees. In addition, special care has to be taken in areas covered by glacial drift, since any large boulders included in the drift may affect the balance almost as much as the structural feature being investigated. The use of the instruments is consequently a most delicate matter, but it is possible by taking a series of observations to eliminate the effect of features other than that which has to be investigated. Although so sensitive, torsion balances are available in field models and have already been very extensively used for prospecting work.

7.5. Electrical Methods.—The various materials that constitute the earth's crust possess electrical properties of wide variation, the two most commonly used in geophysical work being conductivity and its reciprocal, resistivity. The differences that exist between the conductivity of different rock types are unusually large, the range of variation being much greater than for any of the other properties so far discussed; this may be a partial explanation of the wide use to which electrical subsurface prospecting methods have been put. Some sulphide minerals, as a result of chemical changes, occasioned possibly by groundwater, have set up in them small electrical currents which circulate in the adjoining ground. A good deal of experimental work has been done in investigating such currents, but the method is not generally applicable, because of its specialized nature. The more usual methods of investigation depend on passing a current through a section of the earth's crust between two (current) electrodes placed at a fixed distance apart and exploring the nature of the ground adjacent to or between them by means of two or more (potential) electrodes inserted into the ground at specially selected points.

The basic electrical theory underlying the several developments of this system is not new, but it was not until the beginning of the present century that successful field methods were evolved. Daft and Williams, working in England, appear to have been the first to devise a satisfactory equipotential method, which has since been improved and extended by other investigators.[7.5] By means of suitable equipment, lines of equal potential are traced out on the ground, and the shapes so obtained are compared with the perfectly symmetrical distribution that would be obtained in homogeneous ground, the differences between the shapes being an indication of the presence of buried material having different electrical properties from that at the surface.

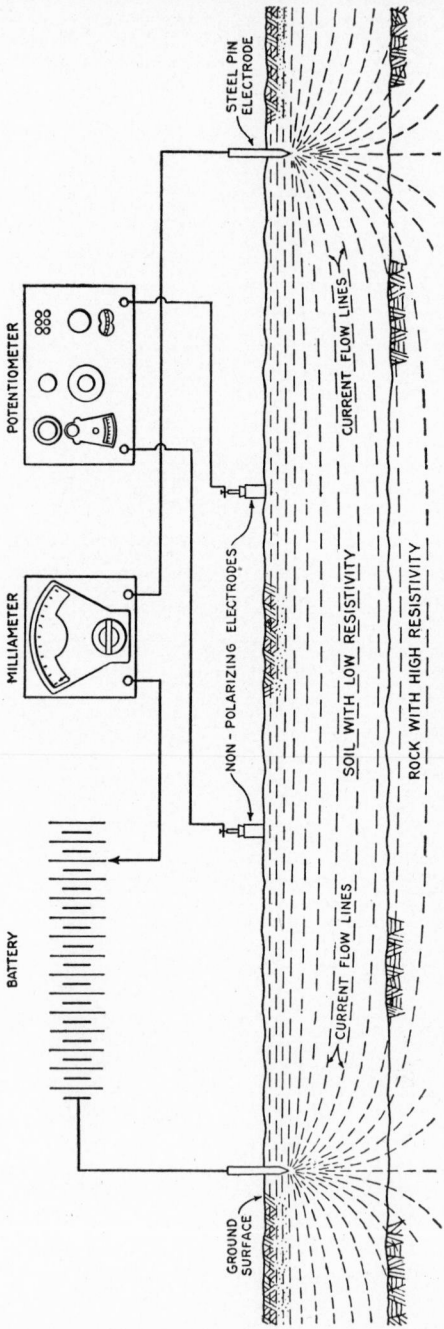

FIG. 7.7.—Arrangement of apparatus for subsurface exploration by electrical resistivity measurements. (*Based, by permission of the U. S. Bureau of Public Roads, on a drawing in Public Roads, June*, 1935, *reference* 7.12.)

Alternatively, the resistivity of sections of the ground may be determined in a similar way. Practical difficulties at first prevented the general use of this method, but these were gradually overcome, notable advance being made in 1912 and 1925 by Schlumberger and by Gish and Rooney following the work of Wenner, working independently in different countries. A third group may be classed generally as electromagnetic methods; some of these are related to the reception of Hertzian waves, measured at different points, the variation in intensity being an indication of geological changes in the adjacent ground. Other electromagnetic methods are based on the induction of currents in the ground under the action of alternating magnetic fields and measurement of the distortions from results that should be obtained in homogeneous material.

It will be seen that these methods cover a wide field of electrical theory, with which the civil engineer will be generally unfamiliar. He will therefore have to allow specialists in this class of work to apply the methods for him. Some of the methods in use are, indeed, of a proprietory nature the exact details of which are not generally known. The methods most usually applied in civil engineering practice, however, are related to resistivity measurement, and so a word or two may be added with regard to them. It may first be noted that if the methods are to be fully effective, it is necessary that a general conception of the local geology be available before they are applied. Particularly is this true with regard to the possible presence of groundwater, the resistivity of certain types of rock varying appreciably with their water content and also with the salinity of the water. A general mode of operation is to measure the average resistivity of a volume of earth by gradually increasing the distance between the potential electrodes and plotting the results so obtained in the form of a curve relating resistivity and electrode spacing. If the curve has an upward trend, it will suggest that resistivity is increasing with depth, whereas an abrupt change in curvature will indicate, if it is known that no other complicating features are present, a change in the nature of the material underground at a depth approximately equal to the distance between the electrodes at which the change occurs. It is of some interest to note that the instrument known as the "Megger," which will be familiar to some civil engineers as that used for testing the resistance of buried grounds in power station construction, can be used for measuring ground resistivity by means of a special adaptation. Results obtained with it compare favorably with those obtained by other methods. Another development of interest and of great utility in special circumstances is the use of resistivity methods in wells or drill holes, known usually as *electrical coring.* Three insulated conductors are lowered down the hole, their ends at different depths, the deepest being one of the current electrodes, and the other two the potential electrodes. The recording instruments are read at the surface; and by means of special operating devices, great speed can be achieved, up to 1,000 ft. of hole having been examined in an hour.

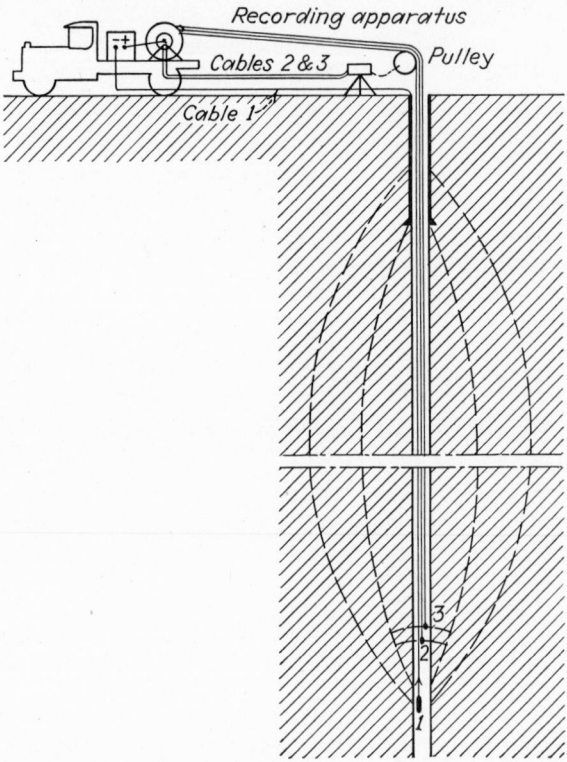

FIG. 7.8.—Diagrammatic section illustrating arrangement of apparatus for electrical coring.

7.6. Applications in Civil Engineering.—The foregoing notes outline briefly the main methods followed in geophysical prospecting and the basic ideas to which they are related. It will be seen that all the methods, in their practical form, are developments of the twentieth century, and some of relatively recent years. Despite this, geophysical methods applied to civil engineering have already achieved notable results. Some of these are outlined in the following notes, descriptive of tunnel, dam foundation, and excavation work.

The Bridge River Tunnel was completed in 1930 for the British Columbia Electric Railway Company, Ltd., of Vancouver, Canada.

The tunnel is 13,200 ft. long, passing through Mission Mountain at a depth of 2,400 ft. below the summit; it was constructed in connection with hydroelectric development, being located 110 miles to the northeast of Vancouver. A geological survey of the surrounding area was made, before tenders were called, by Dr. V. Dolmage of Vancouver, the structure disclosed being as shown in Fig. 7.9. Driving the tunnel

TABLE F.—A SUMMARY OF GEOPHYSICAL METHODS OF EXPLORATION
(*After Schlumberger*)

Methods	Normal conditions of use	Limitations
Gravimetric:		
Pendulum	Big regional surveys	Insensibility
Torsion balance / Gravimeter	Deep tectonics / Salt formations	Relief of the ground
Magnetic:		
Magnetometer	Magnetic rocks	Unreliability of magnetism
Seismic:		
Refraction / Reflection	Deep tectonics / Exact depth determinations	Use of explosions
Electric:		
Study of electric field	General surveys	Unreliability of resistivity
Resistivity map	Superficial surveys	
Map of potentials	Deep surveys	
Study of magnetic field	Dip of beds, etc.	Shallow depths only
Miscellaneous:		
Hertzian waves	Conductive masses	Very few applications as yet
Radioactivity	Radioactive rocks	

started at the two ends, and work at the northern end proceeded without trouble. About 1,200 ft. in from the southern portal, however, bad conditions were encountered as a very pronounced fault zone was met. The greenstone was found to be finely crushed and extensively altered to talc and serpentine, the whole zone being highly water bearing, exerting so much pressure on the timber supports that a short length of this was completely crushed. The probability of encountering such material had been foretold by Dr. Dolmage and confirmed by test borings, but it was necessary to find how wide this zone was going to prove. This was done by the application of the resistivity method used by the Schlumberger Electrical Prospecting

Methods.[7.6] It was found that the high water content of the fault zone had a resistivity much different from that of the surrounding hard greenstone, and it therefore proved possible to determine the width of the zone on three trial locations from the tunnel center line. The second of these was adopted, the tunnel being diverted by 150 ft. from its original location through this fault zone. As a result of the geophysical survey, it was estimated that the troublesome section of

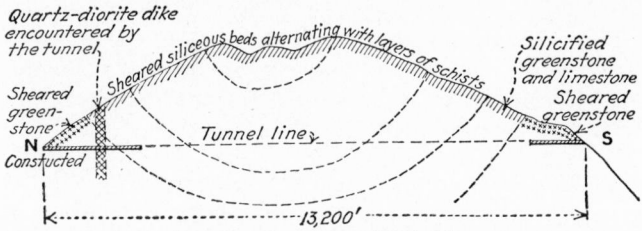

General section on tunnel center line.

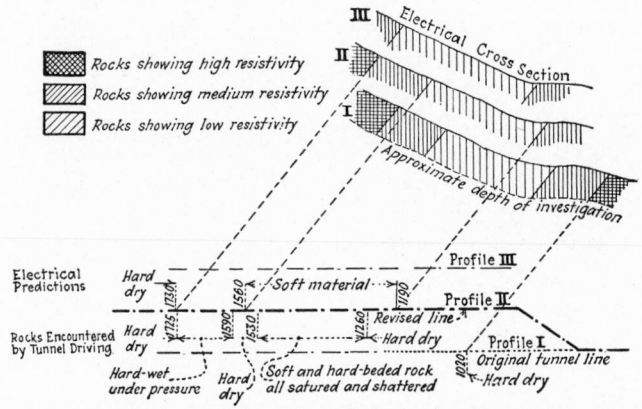

Detailed plan and section of water-bearing section, as revealed by geophysical exploration.

FIG. 7.9.—Bridge River Tunnel, British Columbia.

(Reproduced by permission of Mr. E. E. Carpenter, consulting engineer, British Columbia Electric Railway Co., Ltd.)

tunnel would be found to be 370 ft. wide. Excavation proved it to be about 270 ft. wide, with another short section of water-bearing rock some 60 ft. farther on. The value of this determination, to the engineers in charge of the tunnel work, will be obvious, since construction methods could be devised for driving through this relatively short stretch of water-bearing material with the knowledge that a return of satisfactory conditions could be expected as soon as it was penetrated.[7.7] The construction of other tunnels has been similarly assisted; geophysical explorations have even been carried out in connection with the proposed tunnel under the Strait of Gibraltar.

Geophysical methods applied to dam foundation problems have been principally concerned with determining the depth to the surface of solid rock through superficial deposits. Thus, what is believed to be the first application of geophysical methods to civil engineering work in North America was made in 1928 at the site of the Fifteen Mile Falls water power station of the New England Power Association on the Connecticut River near Littleton, New Hampshire. An interesting description of this work is available.[7.8] Briefly, the local bedrock of pre-Cambrian schist with occasional intrusions of granite was overlain by glacial deposits of boulder clay, gravel, and sand. The presence of boulders made test-boring operations expensive, and so an electrical resistivity method of investigation was employed for investigating the two available sites. A buried valley was known to be present beneath the glacial deposits, and its presence was detected under an overburden of 300 ft. Other positive results were obtained. Test boring and drilling was carried out as a check; and for those holes which were related to the previous electric prospecting, the depth thus predicted varied generally between 69 and 118 per cent of the true depths. Similar work has been done at the sites of proposed dams on the international section of the river St. Lawrence in Ontario, drilling in this case revealing an average error of only 6.6 per cent. The cost of this work has been stated to be $60 (£12) per determination, average progress being slightly less than three per day.[7.9]

In Europe, some progress has been made in applying geophysical methods to the detailed study of rock strata being used as the foundation beds for dams. MM. Lugeon and Schlumberger carried out experiments at the site of the Sarrans dam, located on a tributary of the River Lot, La Truyère, in the department of Aveyron, France. The dam is of the gravity type, slightly arched, 345 ft. high, containing 600,000 cu. yd. of concrete; it is founded throughout on granite. Drill holes and exploratory tunnels were used extensively in the investigation of foundation bed conditions, throughout a period of several months, this work disclosing several zones of weathered granite, transformed into a granular mass resulting from the alteration of the feldspar and mica. A geophysical electrical survey of the ground around the dam site was later carried out, and some interesting results were obtained, notably in a correlation of a decreased value for the resistivity with a known zone of moist crushed granite from which it was concluded that "the resistivities studied from the surface actually bring into evidence the direction of geological features which otherwise could be discovered only by drilling exploration or excavation work."[7.10]

Advantage was taken of the existence of holes drilled into the foundation rock below the dam cutoff wall for the purpose of grouting to carry out some "electrical coring." The resistivities thus measured in the several holes were compared with tests made on the permeability of the rock obtained from the cores, the permeability of the rock being clearly a vitally important factor in the design of the dam, and a fairly definite correlation was noted. The investigators expressed their conclusions thus:

It may be said that the experiments at La Truyère show that there exists a definite relationship between the specific electrical resistivity of the granite

FIG. 7.10.—Ground comparator set up to determine depth of overburden above rock level at prospective dam site in New York State.
(Reproduced by permission of Geophysical Explorations, Ltd., Mr. S. F. Kelly, President.)

and its degree of weathering, as well as its permeability to water. This relation seems probable at first sight, since the specific conductivity of a rock (which is the reciprocal value of its resistivity) is approximately proportional to the amount of absorbed water and to the percentage of dissolved salts contained therein. Since granite, when it is weathered and pervious, must contain, *a priori*, a larger percentage of water than compact granite, it must in consequence be more conductive than the latter.

After carrying out 500 measurements, the authors were able to ascertain that the granite on which the Sarrans dam is built shows electrical resistivities larger than 5,000 ohm-m^2/m. in the zones where it is compact and below that figure in the zones where it is weathered and pervious. Measurements carried out in the drill holes, on the other hand, showed a definite relation between the resistivities and the coefficients of absorption of the water.[7,10]

Although these results are only tentative, they suggest that geophysical methods of investigation may well prove to be of even greater service in civil engineering work than has yet been realized in general practice, when coordinated with adequate geological survey work and exploratory and confirmatory drill holes.

Electrical methods of investigation will be found to be more frequently mentioned than others in examples of applied geophysical work, probably because of the low cost of the apparatus and the freedom from interference by local bodies and surface irregularities. Gravimetric determinations of underground structures have been carried out, however, with notable success. Many examples could be cited from oil field records; but as typical of work akin to that involved in civil engineering operations may be mentioned that carried out by members of the staff of the British Geological Survey in tracing the preglacial valley of the River Kelvin, near Glasgow, Scotland, now drift covered in some places to a depth of 300 ft. It was found possible from the results of the survey work with a torsion balance to reconstruct the topography of the buried rock surface and to calculate depths to rock and rock slopes that agreed closely with data obtained from the available boring records.[7.11] Seismic methods have been used by the United States Bureau of Public Roads in connection with its studies of road foundations.

Finally, mention may be made of the steadily growing use of geophysical methods, both seismic and electrical, for the investigation of shallow surface deposits, as in exploration work for sand and gravel deposits and in the classification of highway excavation before the material is moved. Several states of the United States now include this work as a part of the regular operations of their highway departments, and the United States Bureau of Public Roads has been experimenting for some years with methods of promise. Interesting figures testifying to the success of this work are available.[7.12] Schappler and Farnham report that in 1931, without the aid of geophysical methods, an error of 12.92 per cent was made in estimating the quantity of solid rock on a selected number of highway projects; whereas on a similar number of similar jobs in the following year, utilizing geophysical methods, the error was reduced to 1.04 per cent. One of the states utilizing electrical prospecting methods for locating road material (sand and gravel) is Minnesota; it is reported that more than 100 prospective sites were tested for this purpose in a little more than a year and that not more than 2 per cent of holes dug subsequently have failed to verify the predictions of the electrical data.[7.13]

These figures must not be held to discredit geological surveying, for the geophysical work cannot be carried out without the benefit of preliminary geological investigation. They serve only to show what a valuable aid geophysical methods can be in subsurface exploration when conditions are suitable for their application. In the case of the investigation of shallow deposits, the usual resistivity methods can be followed in areas where well-defined deposits of sand or gravel may be expected. The success of the methods depends on the fact that the resistivity of sands and gravels is very high when compared with weathered topsoils, silts, and clays. The difference is due to dissolved salts in the water content of the deposits, since the ability to conduct

FIG. 7.11.—Automobile truck and field equipment used for geophysical exploration of sand and gravel deposits by the Minnesota Department of Highways.
(*Reproduced by permission of the Minnesota Dept. of Highways, Mr. J. T. Ellison, Chief Engineer, St. Paul, Minn., U.S.A.*)

electricity appears to be dependent on the presence of water, perfectly dry solid rock, for example, being practically a nonconductor.

7.7. Geophysical Methods and Groundwater.—This last remark leads to the question that naturally arises in the mind of an engineer when first he comes to realize the value of applied geophysics: Can geophysical methods be used successfully to determine the existence and position of groundwater? The question arises since it relates to what is often the one matter that cannot be foretold in preliminary work from geological investigations alone, test boreholes often being necessary to check up only on groundwater conditions. Evidence that may suggest an answer is steadily accumulating; already it can be said that, in general, the use of geophysical electrical methods may help in accurate groundwater determination.

It has been established that the resistivity of a rock is dependent mainly on two factors, the porosity of the rock and the salinity of the solution held in the pores of the rock, which may be the groundwater in the case of open-textured rocks. Thus gravel, even when water bearing, is sometimes found to have a very high resistivity, this being explained by the possibility that the water present may be exceptionally pure and free from electrolytes. In general, resistivities will be found to be high for dense impervious rocks and low for porous water-bearing rocks. Bruckshaw and Dixey[7.14] give the following typical values for resistivity (*i.e.*, the electrical resistance of a cubic centimeter of material when the current flows parallel to one edge):

Fresh crystalline rocks	From 20,000 to 1,000,000 ohm. cm.
Consolidated sedimentary rocks	From 1,000 to 50,000 ohm. cm.
Recent unconsolidated formations	From 50 up to 5,000 ohm. cm.

This variation in resistivity provides a clue to the possibility of the direct determination of groundwater by electrical prospecting methods. As waterlogging will affect the resistivity of any one material, the upper surface of groundwater occurring in a uniform medium will—in effect—be the boundary between two media with different resistivities, the wet and the dry material, respectively. This simple ideal case will rarely occur in practice; but by a development of the idea that it represents and its extension to three- and four-layer conditions, it presents possibilities for direct groundwater determination.

In some cases, satisfactory results have been achieved by a careful use of this method correlated with detailed knowledge of the local geology, but other trials have been equally disappointing. A geophysical investigation of special importance was carried out in 1934 in parts of the South Wales coal field with a view to ascertaining groundwater conditions in abandoned mine workings.

> The Depth-Resistivity methods [were found to show] fairly reliable indications as to the position of changes in the nature of the strata . . . [but the report on this work (dated 1935) suggested that] . . . although these first trials have not proved successful in detecting with certainty the presence and extent of the water-logged areas, the problem is of such importance, both from the point of view of safety in mines and of economic efficiency, that some further attempt must inevitably be made to develop a method of geophysical sounding for such areas.[7.15]

Work in this direction is steadily proceeding in many parts of the world (notably in U.S.S.R.); and in the meantime, continued application of geophysical methods generally is developing the second way in which they can assist in groundwater determinations. This is by revealing, at a cost generally lower than that of comparable test holes, the local geological structure from a consideration of which may be determined the probability of groundwater being present. An obvious example is the revelation, by geophysical methods, of a depression or buried valley in the rock surface under porous superficial deposits, in which groundwater might be expected to collect. The area that can be covered in a given time by geophysical methods when once the equipment is available in the field, as compared with the test holes which may be bored or the test pits dug, is so great that the resulting economy in water-finding operations will be obvious. Work of this nature is no different from the usual geophysical prospecting work in connection with civil engineering projects; but because of its association with water supply, in which the average man is directly interested, there is a tendency to consider such work as of a special nature. It is the interpretation of results so obtained that is, in a sense, a specialty necessitating a fine appreciation of the geology of the entire neighborhood and of hydrological principles governing the local groundwater distribution.

Notable work has been done, using geophysical methods, in finding groundwater in several states of the United States, in South Africa, in Palestine, and in U.S.S.R. In South Africa, savings of 66 per cent in the cost of drilling successful water boreholes have been reported, and doubtless similar economies have been experienced elsewhere.[7.16] Further advance may confidently be anticipated in this branch of applied geophysics, all of which will be of benefit either directly or indirectly to civil engineers in many branches of their own work.

7.8. Conclusion.—It has been estimated that the oil industry in the United States alone expends about fourteen million dollars (almost three million pounds sterling) annually on geophysical exploration, having almost three hundred crews in the field at a time engaged on geophysical work.[7.17] Similar and equally surprising figures could be quoted with regard to other aspects of modern geophysical work, descriptive literature, for example, probably including well over a thousand publications every year.[7.18] It is small wonder, therefore, that geophysical methods have attracted attention in civil engineering circles—and with some justification, as examples already cited will have made clear.

The development is so recent, however, that there is a real danger of popular enthusiasm for geophysical work being translated into injudicious appreciation of its place in civil engineering. It is useful therefore, to recall that the requirements of underground investigation, in civil engineering and in oil prospecting (for example) are essentially different, the accuracy and certainty demanded in the former being incidental in the latter to the discovery of oil. With this qualification must always be associated the thought that, in civil engineering, geophysical methods are no more than a special means of exploring subsurface conditions, generally in association with trial drill and boreholes and *always* as additional to preliminary geological surveying which must, indeed, always be carried out before the methods can be applied. All the methods are subject to definite geological restrictions —rocks of essentially different physical character must be in contact, and the strata encountered must be fairly uniform in respect to their physical character. And, finally, all data obtained as a result of geophysical prospecting must be studied and utilized only when properly correlated by a geologist with the maximum available information regarding local geological conditions. Despite all qualifications and necessary restrictions, geophysical methods yet remain a powerful and useful tool available for the use of the civil engineer in his preliminary and exploratory work; they are almost certain to achieve increasing importance in this field in the not too distant future.

Suggestions for Further Reading

The literature of the subject is already extensive but is mainly in the form of technical papers on special branches of geophysical work. The following titles will provide a general survey of the field:

BARTON, D. C.: Applied Geophysics in America, *Econ. Geology*, **23**: 651 (1928).

EDGE, A. B. B., and T. H. LABY: "Principles and Practice of Geophysical Prospecting," Cambridge University Press, Cambridge, England, 1931.

EVE, A. S., and D. A. KEYS: "Applied Geophysics in the Search for Minerals," Cambridge University Press, Cambridge, England, 2d ed., 1933.

JONES, O. T.: Geophysics, James Forrest Lecture, 1935, *Inst. Civil Eng. Min. Proc. (London)*, **240**: 689 (1936).

SHAW, H.: "Applied Geophysics," *Sci. Mus. Pub.*, H. M. Stationery Office, London, 1936.

Also the several volumes entitled "Geophysical Prospecting," published by the American Institute of Mining and Metallurgical Engineers.

Several excellent bibliographies are also available, notably those published periodically by the U. S. Geological Survey in the form of small volumes of abstracts. A general list of publications will be found at the back of "Methods in Geological Surveying," by Drs. Greenly and Williams, Thomas Murby & Co., London, 1930.

Typical of papers dealing with the application of geophysical methods to civil engineering work are the following:

CROSBY, I. B., and S. F. KELLY: Electrical Subsoil Exploration and the Civil Engineer, *Eng. News-Record*, **102**: 270 (1929).

STIPE, C. G., and S. F. KELLY: Geophysical Methods Aid Construction Work, *Civil Eng.* (*New York*), **7**: 264 (1937).

As geophysical work is of so specialized a nature, it may be useful to record the names of firms that undertake geophysical fieldwork as applied in civil engineering. The following are representative companies:

Canada: Geophysical Explorations, Ltd., 330 Bay Street, Toronto, Ontario.

Great Britain: The Geophysical Prospecting Co., Ltd., 9–11 Copthall Avenue, London, E.C.2.

United States: Geophysical Explorations, Ltd., 17 Battery Place, New York. Heiland Research Corp., Denver, Colorado. International Geophysics, Inc., 1063 Gayley Avenue, Los Angeles, California.

CHAPTER VIII

TUNNELS

Ezekias fortified his city, and brought in water into the midst thereof: he digged the hard rock with iron, and made wells for water.

Ecclesiasticus 48:17 (probably the earliest reference to the famous water-supply works for the city of Jerusalem).

Tunneling must have been one of the earliest constructional activities of man. It may be that natural tunnels and other results of water action such as "swallow holes," frequently found in limestone and similar formations, first suggested to early man the idea of an artificial passage through rock. The rock or cliff dweller may have been an example to other human beings in distant ages; but however came the inspiration, excavated underground dwellings and temples, the first tunnels, are to be found in many of the ancient civilizations that have been investigated in recent years. Excavated passages beneath the earth were early utilized for purposes other than residence, drainage being one of the first of the more general utilitarian purposes for which tunnels were specially constructed, often in connection with early mining and quarrying work. Water supply and road construction also necessitated tunnels as recognized features at an early date in the history of man.

Today, tunnels are used for the same purposes. Transport is facilitated by them, and water tunnels are now often used in the generation of water power as well as for the provision of water supply. Exceptional uses are to be found, but tunnels generally can still be classed today as they were two thousand years ago, either as aqueducts or as viaducts. It is indeed strange to reflect that this ancient branch of construction has changed but little in achievement, despite great advances in method, throughout the centuries. A tunnel is and can be a tunnel only, having a certain length and a certain cross section. The long Alpine tunnels are indeed magnificent achievements, but so also are many of the longer tunnels of the pre-Christian era, more especially when the construction methods then in use are considered.

8.2. Historical Notes.—Brief mention of some ancient feats of tunneling may therefore rightly lay claim to a place in even the most cursory review of the art. Although Geology as a separate branch of science was not recognized in these early years, yet brief reference here to a few of these tunnels is instructive and of interest. Examples of tunnels of some sort are to be found in almost all ancient civilizations, notably in Egypt where some of the rock-cut galleries of the early Egyptian kings are over 750 ft. long. In Malta may still be seen underground temples and gathering places, at least five thousand years old, hewn out of solid sandstone with flints which must have been brought to the island from the mainland. These early tunnels were generally built through the softer types of solid rock, but some were excavated through unconsolidated materials and so required immediate lining for stability. A notable example was the tunnel under the river Euphrates, probably the first submarine tunnel of which any record exists. It was 12 ft. wide and 15 ft. high and was built in the dry, the river being temporarily diverted.

The Romans were preeminent in early tunnel construction, alike because of their achievements and because of the improvements in method that they effected. They appear to have introduced the use of fire into tunnel construction, utilizing the principle (known to others earlier) that a heated rock if suddenly cooled will crack to some extent and so make excavation easier. They also most probably employed vinegar instead of water as a cooling agent when working in limestone and similar rocks, utilizing the acid nature of the vinegar to disintegrate the rock chemically as well as physically. The sufferings of the slaves who had to apply such methods can hardly be imagined. It is known also that the Romans utilized intermediate vertical shafts and even inclined adits in the construction of their longer tunnels, notably of the tunnel built for the drainage of Lake Fucino, driven through limestone for a distance of 3½ miles, being 6 by 10 ft. in theoretical section. It is said that 30,000 men were employed for 11 years on its construction. Volcanic tufa was another of the softer rocks pierced by these intrepid builders, one notable tunnel through this material being that which gave the road between Naples and Pozzuoli passage through the Ponlipio hills; it was 3,000 ft. long and 25 ft. wide.[8.1]

Many other examples from the past might similarly be quoted, but only one more will be mentioned. Lake Copais is now an area of dry land, fertile and well cultivated, of about 100 sq. miles in area, located near the east coast of Greece. It lies generally about 300 ft. above the level of the sea from which it is separated by a steep range of hills. The area was drained in 1886 by a French company. After drainage had lowered the water level, it was found that the ancient inhabitants (of the prehistoric Minoan period, 2000 B.C.?) had drained it previously. The ancient works then discovered led to some ancient drainage canals and to some of the *katavothras*, natural tunnels in the form of large fissures through the local limestone rock. Unfortunately, although these interesting natural tunnels were so effectively used in the past, the largest of them is now blocked up, and so a modern drainage tunnel has had to be constructed. To the north of the drained area in the pass to the sea at Laryma, there may be seen 16 shafts which appear to have been put down, by the prehistoric workers mentioned, with a view to tunnel construction, although the work had not proceeded far. The whole region is of great interest to the geologist and the civil engineer.[8.2]

The Middle Ages saw no advance in the technique of tunnel building, and even the introduction of gunpowder (first used in tunnel work in 1679–1681 at Malpas, France) had little immediate effect. It was the extension of canals and the introduction of railways that finally initiated the great advance of the last hundred years during which most of the tunnels now in use have been constructed. Although so rudely built and simple in conception, ancient tunnel works were inevitably dependent upon geological considerations—not only in design but also in construction, as in the possibility of the use of vinegar and the specially hard nature of flints and corundum being utilized for cutting tools. Simultaneously with advance in tunneling technique has gone the development of Geology, and so today the two are intimately associated.

8.3. A General Note.—Of all the activities of the civil engineer, tunneling is without question that to which the study of Geology can most fitly and usefully be applied, geological problems alone affecting design and construction methods when once general location and basic dimensions are determined. Except in the case of special tunnels such as those partially driven through an artificial deposit of clay (as in the case of the Sixtieth and the Fourteenth Street tunnels in New York) and those isolated examples built as tunnel linings prior to fill being placed over them (as in the case of the Golden Circle Railroad in the Cripple Creek mining district of Colorado, where the line was carried in this way across a future refuse dumping ground), all tunnels are driven through a part of the earth's crust. Accurate location and construction methods depend therefore on the rock through which the tunnel is to be driven; the necessity for lining must usually be determined by the behavior of the rock when exposed to the air and possibly water. A thorough geological investigation before construction begins is, therefore, of paramount importance in all tunnel work.

Basically, accurate geological sections along all possible routes available for the tunnel are the first requirement, knowledge of the rock characteristics to be encountered being the second and of little less importance. With this information at hand, the engineer can decide whether or not the construction of the proposed tunnel is a practical and economical possibility. He can determine with some semblance of accuracy what cover of rock must be allowed between the tunnel line and the ground surface in the case of a water tunnel in which the water is to be conducted under a pressure head (hereinafter called a *pressure tunnel*). The necessity for, and design of, an artificial lining to the tunnel can be generally determined, and the likelihood of grouting of the rock adjacent to the tunnel being necessary can usually be foretold from such information. Finally, such geological advice will give to the engineer at least some indication (often quite accurate) as to the percentage of overbreak that he is likely to encounter when construction is under way and for which allowance must be made in his estimates. To the contractor for tunnel work, geological data are equally vital, his entire construction program and construction methods depending on the material to be encountered, whereas his main hazards such as underground water are determined solely by geological conditions. On at least one major tunneling project, preliminary geological studies suggested the possibility of finding valuable mineral ores within the tunnel, the sale of which might have paid part of the construction cost. Unfortunately, not a trace of ore was found during excavation of the tunnel!

These several aspects of the application of the results of geological investigation to tunnel work will be considered in some detail and illustrated by examples from practice. There are some general observations which first demand attention. Clearly, the ideal condition to be met with in tunneling is to encounter one material only, and that easily excavated, containing no water-bearing fissures and being unaffected when exposed to air. Rarely is anything approaching such an ideal material met with, the London clay through which most of the London tube railway tunnels penetrate in part being perhaps as close to the ideal as is obtained anywhere. Tunnel locations cannot normally be changed for any other than economic considerations, and so no search for suitable tunneling material is normally feasible, rock conditions at a particular location having to be accepted by the engineer and explored thoroughly in order to render design and construction as certain as possible.

Archaean or Pre-Cambrian rocks, the oldest types geologically, will generally be found hard and slow to excavate, construction in such formations being consequently relatively expensive. Palaeozoic rocks, on the other hand (younger, geologically), are usually the most simple in which to tunnel, consequently affording more economical construction. Formations of recent origin increase construction difficulties, tunneling becoming generally harder as the formation becomes younger, recent sand and gravel deposits being particularly awkward to drive through. Novel methods have been adopted in tunnel work for penetrating such materials, the practice of freezing the water in water-bearing strata being perhaps one of the most ingenious. So far as is known, this method has been used only once in tunneling (a tunnel for pedestrain traffic built at Stockholm, Sweden, in 1884–1886 by a Captain Lindmark, through coarse gravel sand and clay, 80 ft. of which was driven by freezing), although Siberian miners are said to have long used the method in their mining operations. Grouting or cementation is a most valuable process for this class of work, as will presently be described, and an even more recent development has been the introduction of a chemical solidification process for water-bearing ground.

As in every other type of civil engineering work, economic considerations generally govern tunnel design and construction. Here, too, Geology is of avail, since not only will the materials to be penetrated affect the actual cost of construction, but they will definitely affect the maximum speed at which progress in excavation may be anticipated. From such estimates, detailed comparisons of the costs of various alternative routes for the work proposed can be prepared. In the case of the Kinlochleven hydroelectric scheme in Scotland (33,000 hp. with a head of 1,000 ft.), such a study led to the adoption of an open reinforced concrete conduit, following the contour of the hillside, as the main aqueduct; in the case of the neighboring Lochaber scheme (120,000 hp. with a head of 800 ft.), the main aqueduct is a tunnel, the selection of this route being based on a similar and extensive economic study, utilizing preliminary geological information.

8.4. Preliminary Work.—Normally no deviation from general survey methods will be necessary in obtaining the requisite geological sections along possible tunnel routes. By their nature, however, tunnels will usually be located in such a way that accurate preliminary correlation of the strata to be expected in the tunnel with surface conditions will be difficult. Subaqueous tunnels, for example, can be checked only from indications on adjacent dry land; tunnels under cities will pierce strata covered by built-up city areas; and tunnels through high mountain ranges will always be at great depths below the surface in some places—over 7,000 ft. in the case of the Simplon Tunnel in Switzerland. In practically all such cases, it will be impracticable, if not impossible, to put down the usual exploratory drill holes, which will generally serve to confirm the deductions of geological survey work. It is for this reason that tunnel construction records (to be mentioned later) are of vital importance.

Clearly, the first task of the geological adviser on a tunnel project is to indicate whether he considers that the geological conditions anticipated will be favorable for tunnel construction. It is then the task of the civil engineer concerned to decide whether or not the tunnel can be constructed economically or if at all. It is of interest to note that in connection with the Simplon Tunnel, it has been stated in public that

> . . . had the geologists been quite accurate in their preliminary investigation and reports, and had they truly and correctly anticipated the dangers and obstacles that were eventually met with in soft rock, the "Great Spring" or river of cold water, the high temperatures and hot springs, and the "creep" or lifting of the floor, no one would have dared to undertake the contract, and the tunnel would never have been constructed.[8.3]

Perhaps it is as well for human welfare that in this instance preliminary geological advice was not perfect! It should be added that the geological structures met with in the European Alps are of the most difficult type known to the science.

As has been noted, this tunnel is in places over 7,000 ft. below the surface of the ground, piercing high ranges of the Alps (the rocks encountered being gneiss, mica schist, and some limestone, or "sugar marble"), so that it is not surprising that preliminary predictions were not accurate; they did, however, give a reasonable conception of what rock would be encountered. An even more serious result of incomplete preliminary information occurred during the construction of the Lötschsberg Tunnel, from Kandersteg to Gopperstein, between October, 1906, and September, 1911. The tunnel is 9.04 miles long, and it was believed that it would penetrate solid rock (granite) throughout. Unfortunately, about 2 miles in from one portal, the drilling broke into an ancient glacial gorge, now filled with detritus and followed by the Kander River, and in an instant 8,000 cu. yd. of material rushed in, and 25 men lost their lives. The heading had to be bulkheaded off, and the course changed; only in this way was the tunnel eventually finished, finally being half a mile longer than had been anticipated.[8.4] Had geophysical methods of exploration then been available, trouble might have been avoided.

These examples serve to demonstrate that in the case of deep tunnels, even expert geological investigation will not always reveal difficulties to be encountered. Another example of serious trouble in tunnel work occurred in the construction of the Las Raices Tunnel,

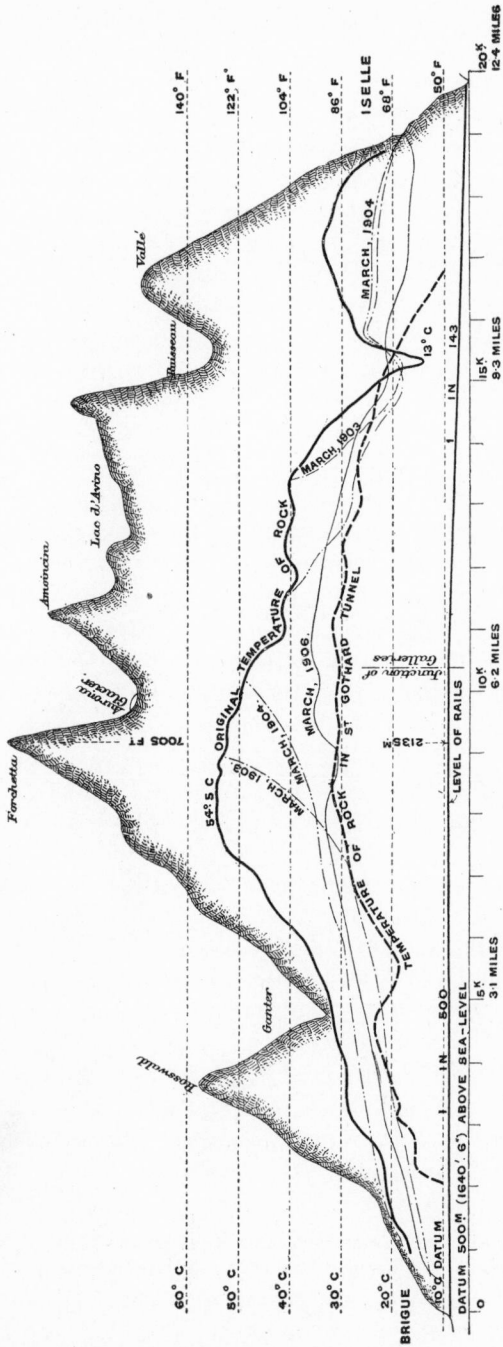

FIG. 8.1.—Section through the Alps along center line of the Simplon Tunnel, showing temperatures encountered (and also in St. Gothard Tunnel).

(*Reproduced, by permission of the Council of the Institution of Civil Engineers, from reference* 8.3.)

Chile, for the Transandine Railway. It is 14,885 ft. in length, piercing a section of the Andes. Surface indications (and there were few rock exposures above the tunnel) suggested that it would be driven solely through porphyrite, and so no trouble with water was anticipated. Driving disclosed porphyrite rock, generally firm, in some places cracked and fissured with bands of clay, some infiltration of water occurring. Practically without warning, a break-in occurred at about 1,750 ft. from one face, at a junction of a full section of the tunnel and a smaller leading section, a large amount of mud pouring into the tunnel through quite a small hole, trapping 42 men. Fortunately, they were eventually rescued, by means of a special rescue tunnel (the construction of which was a remarkable piece of work). The significance of the failure geologically, however, was that the break-in of this 2,616 cu. yd. of material caused a hole to form in the ground above the tunnel. This earth movement greatly facilitated the investigations of the failure that were carried out. It was then found, mainly from the presence of a boulder of granodiorite, that the tunnel had tapped and drained an ancient glacial bed which had been filled in with river deposits. The official report stated that "The accident could not have been foreseen for a rivulet which joins the River Agrio, 150 meters upstream of the tunnel mouth, forms a waterfall, the lip of which is 30 meters higher than the roof of the tunnel, this seeming to show that the thickness of rock over the tunnel was ample." Again, it will be seen what trouble can be caused by underground conditions due to glacial action of the past.[8.5]

Diamond drilling along the line of a proposed tunnel is made a supplement to preliminary geological investigation whenever possible. The results of this exploration are positive so far as they go; but their application to tunnel work must be considered in conjunction with the geological survey work also available. Cooperation of this nature has prefaced the construction of many important tunnels, among which are several of those that serve the great metropolitan area of New York. So important are these tunnels, and of such interest, that a brief description of some of them will be useful.

8.5. Tunnels for New York.—At the outset, perhaps, it should be said that some of the early experiences in tunneling in the New York area were not very fortunate. The first Hudson River tunnel was started in 1874, but apparently no exploratory work had been done in advance; the work met with repeated misfortunes and was several times interrupted, being completed only by the year 1908. Another example of early work, in which test soundings were utilized but apparently not with geological correlation, is that of the East River

Gas Tunnel, constructed between 1891 and 1894 for the purpose of taking three gas mains across the East River to the city of New York. About 10 pipe soundings were made in the two river channels under which the tunnel was to penetrate, and these indicated a solid rock bottom throughout, on which assumption a contract was let and work started. Trouble with water was soon encountered, however, and eventually the tunnel ran into a vein of soft decomposed rock. Construction methods had to be radically changed, compressed air and a shield having to be used before the work was finally completed, somewhat naturally not under the original contract.[8.1]

Just after the turn of the century, the Board of Water Supply of the City of New York initiated one of the greatest water supply undertakings that the world had yet seen, the Catskill Water Supply project, which even in this day of great projects is still a monumental piece of engineering. It consisted essentially of the construction of two large impounding reservoirs in the Catskill Mountains, to the north of New York, and of an aqueduct 110 miles long to convey the water so obtained to the city area, together with other associated works some of which are themselves of considerable magnitude. Two large pressure tunnels were also constructed beneath the city for the distribution of the Catskill water, City Tunnel No. 1 being completed in 1917, and City Tunnel No. 2 in 1936. Demands of water have been so great that the city is now (1939) constructing a duplicate water supply system, known as the Delaware River system, the main aqueduct of which will be 114 miles long, connecting with the distribution system already in existence.* These great works have been under the direction of Messrs. J. Waldo Smith, Thaddeus Merriman, and W. E. Spear successively Chief Engineers to the Board of Water Supply. They have been aided by many distinguished consultants, only three of whom can be mentioned here—Professors W. O. Crosby, J. F. Kemp, and C. P. Berkey, all famous as geologists, who were appointed in 1905 and 1906 to advise on all geological matters affecting the works. Dr. Berkey (in 1939) is still actively engaged on similar work for the city and many other authorities, although his two colleagues died some years ago.

From the very start of the work, in 1905, the prosecution of this great enterprise has been featured by the closest possible cooperation between geologists and engineers. In the words of Dr. Berkey:

* It is of interest to note that in connection with this new project 93,283 ft. of test borings were put down at a total cost of $373,182.18 [*Civil Eng.* (*New York*), **8**: 581 (1938)].

In this project virtually nothing was taken for granted. Every new step was the subject of special investigation with the avowed purpose of determining the conditions to be met; and, when these were determined, the plan of construction and design of the structure were brought into conformity with them. In this manner specifications could be drawn with sufficient accuracy to avoid most of the dangers, mistakes, and special claims commonly attending such work. Very few features or conditions were discovered in construction that were not indicated by the exploratory investigations, and such as were found proved to be of minor significance and were cared for at moderate expense.

When it is considered that the country traversed by the aqueduct (as is shown so clearly by the accompanying geological section along the whole route) "exhibits so great a variety of natural features and physical conditions that virtually every individual section of this aqueduct presented special geological problems" and that the dam sites were generally featured by great deposits of glacial drift with ancient river valleys thus covered from view (as shown in Fig. 20.19), this remarkable result will be more fully appreciated.

Fortunately for the engineering profession, the results of this cooperative work were made generally available by those in authoritative charge of the work through the medium of technical books, papers, and articles contributed to engineering journals. The record thus presented is a fascinating one and of great value to all engineers who may be faced with problems even indirectly comparable to those involved in this work for the city of New York. It is impossible to attempt to summarize any of these descriptions; in Appendix D, therefore, a list of the more easily accessible publications is given.[8.6] Special mention must be made of the general description of the entire project by Mr. Lazarus White ("The Catskill Water Supply of the City of New York") and also of the paper by Dr. Berkey and Mr. J. F. Sanborn, presented to the American Society of Civil Engineers, in 1923. The latter is already an engineering classic; it is safe to say that the practice that it describes is largely responsible for the change in attitude of the civil engineering profession toward Geology which has taken place since the turn of the century, certainly in North America.

For those interested in a brief account of the geological features affecting engineering work in and around the city of New York, reference may be made to "Guidebook 9," prepared for the Sixteenth International Geological Congress held in 1933. From this interesting handbook, two of Dr. Berkey's statements have been quoted, and the following quotation is taken—a description by Dr. Berkey of one of

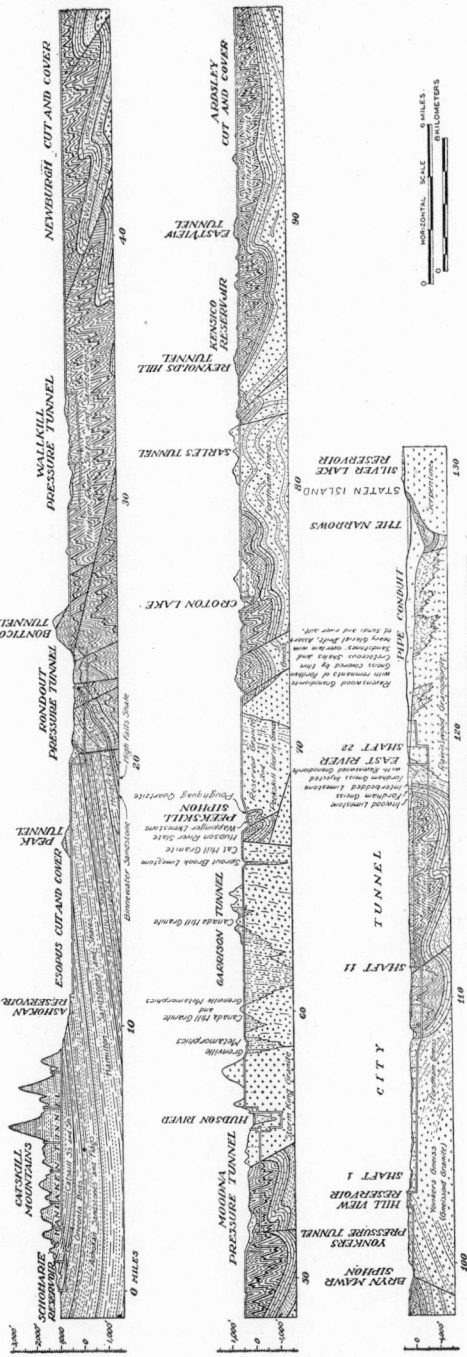

FIG. 8.2.—Geological section along the Catskill Aqueduct of the city of New York.

(*Reproduced, by permission, from "New York City and Vicinity," ed. by Dr. C. P. Berkey, Guidebook 9 of XVI International Geological Congress.*)

the most unusual parts of the Catskill Aqueduct, the Hudson River crossing, the general location of which can be seen in the geological section (Fig. 8.2):

On the basis of preliminary field studies, therefore, the Storm King crossing was selected because it appeared, from field evidence, that at this place a tunnel could be driven in a single formation of essentially the quality of a granite without encountering a fault beneath the gorge. . . .

It was commonly believed, however, that the Hudson River followed a great fault throughout much of its course and that this was the chief reason for its straight course and apparent independence of other structural control. In the preliminary studies therefore, three important questions had to be determined before suitable specifications could be drawn—first, the depth of

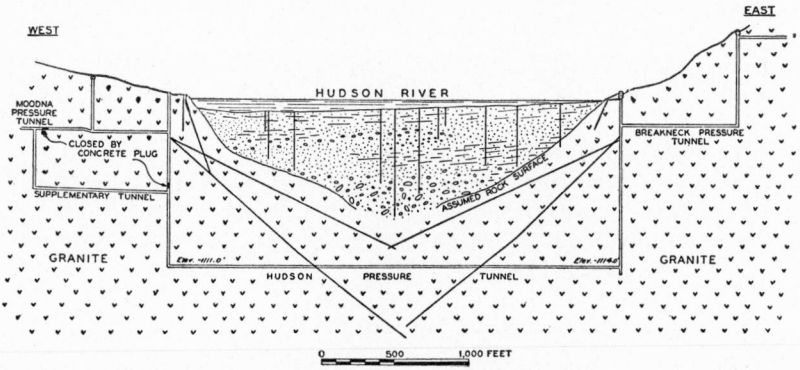

FIG. 8.3.—Geological section across the Hudson River, New York, at Storm King, based on exploratory borings for the Catskill Aqueduct (indicated in the section).

(Reproduced, by permission, from "New York City and Vicinity," ed. by Dr. C. P. Berkey, Guidebook 9 of XVI International Geological Congress.)

the channel, so that the grade of the tunnel could be fixed; second, the kind of rock to be penetrated, so that cost could be estimated; third, whether any special weaknesses or dangers were likely to be encountered, so that suitable provision for them could be made.

When explorations were undertaken at this particular site great difficulty was encountered in determining the depth of channel, and very much greater depth was found than was anticipated from previous geologic knowledge. . . .

The difficulty was increased because the Hudson carries a heavy river traffic. The boring equipment had to be anchored in midstream in the main traffic channel, where there was always danger of interference. Furthermore, the gorge is filled with mixed material in which large boulders are numerous. It was necessary to begin a boring with very large casing (18-inch) so that reductions could be made in passing through these obstructions.

After more than a year, borings to bedrock were successfully made on both sides of the river, finding granite on both sides. The gorge was found to be at least 500 feet (152 meters) deep for a width of 3,000 feet (914 meters). How much deeper it might be out in the center of the gorge no one could tell. A boring placed in mid-channel finally succeeded in getting to greater depth, reaching 765 feet (233 meters) without touching the rock floor. . . .

This fact, together with the slow rate of progress being made by the borings, finally led to the adoption of another plan for testing the ground beneath the river. The final working shaft on each side of the river was put down to a depth of about 200 feet (61 meters), and at that level a room was cut in the side wall, where a diamond drill was placed. This was set up to drill at an angle so as to penetrate the ground beneath the gorge out under the river. The first two borings, one on each side, were set to reach a depth in the center of 1,400 feet (427 meters) below sea level. These were successfully run in sound rock, and it was then decided to drill two others at a smaller angle, cutting the central ground at a maximum depth of 950 feet (290 meters) below sea level. In these borings sound rock, of granite type, was found continuously across the gorge. By this time the boring in the river had reached its maximum depth of 765 feet (233 meters) without touching bottom. With these data in hand, it was considered unnecessary to carry explorations further. . . .

When construction was finished the conditions uncovered in the Hudson River pressure tunnel were essentially as indicated by the exploratory and other investigations. The Storm King granite is continuous across the gorge. There are no great faults in the gorge at this point. The rock is sound and presented no special construction problem. . . .

This extensive program of exploratory work is typical of all the work done by the Board of Water Supply. On the 18-mile City Tunnel No. 1, for example, 46,000 lin. ft. of exploratory borings, were put down, and more than 50,000 ft. were drilled for City Tunnel No. 2, all being put down in conformity with geological advice. The results of the borings are utilized effectively on contract plans for construction as well as for the purposes of engineering design. Figure 8.4, which is reproduced with the permission of the Chief Engineer of the Board of Water Supply, shows a typical contract sheet, the attention to boring results on which will be readily observed.

Finally, there may be mentioned the Wards Island Sewer Tunnel, constructed in 1935–1937, in which preliminary borings, although carefully taken, did not disclose the extremely soft, almost fluid consistency of some decayed and disintegrated rock deep underground which finally necessitated a complete change of grade in the river section. The tunnel was constructed to deliver sewage from interceptor sewers on Manhattan Island to the new treatment plant on Wards Island in the East River, and the original plans were made for a depth of 297 ft. This proved too shallow, the tunnel finally being put through at a depth of 510 ft. below water level. The conditions

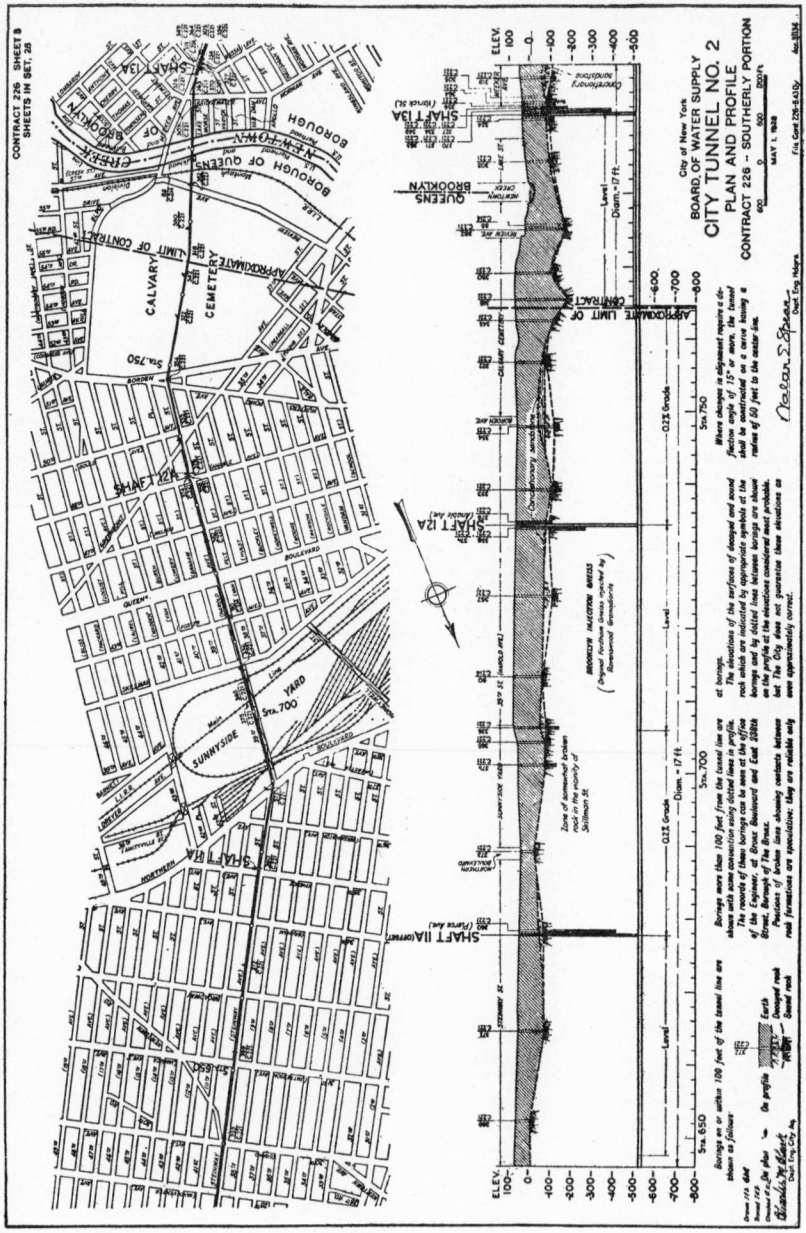

FIG. 8.4.—A typical contract drawing for a New York water-supply tunnel.

(*Reproduced by permission of the Board of Water Supply, New York, Mr. Walter E. Spear, Chief Engineer.*)

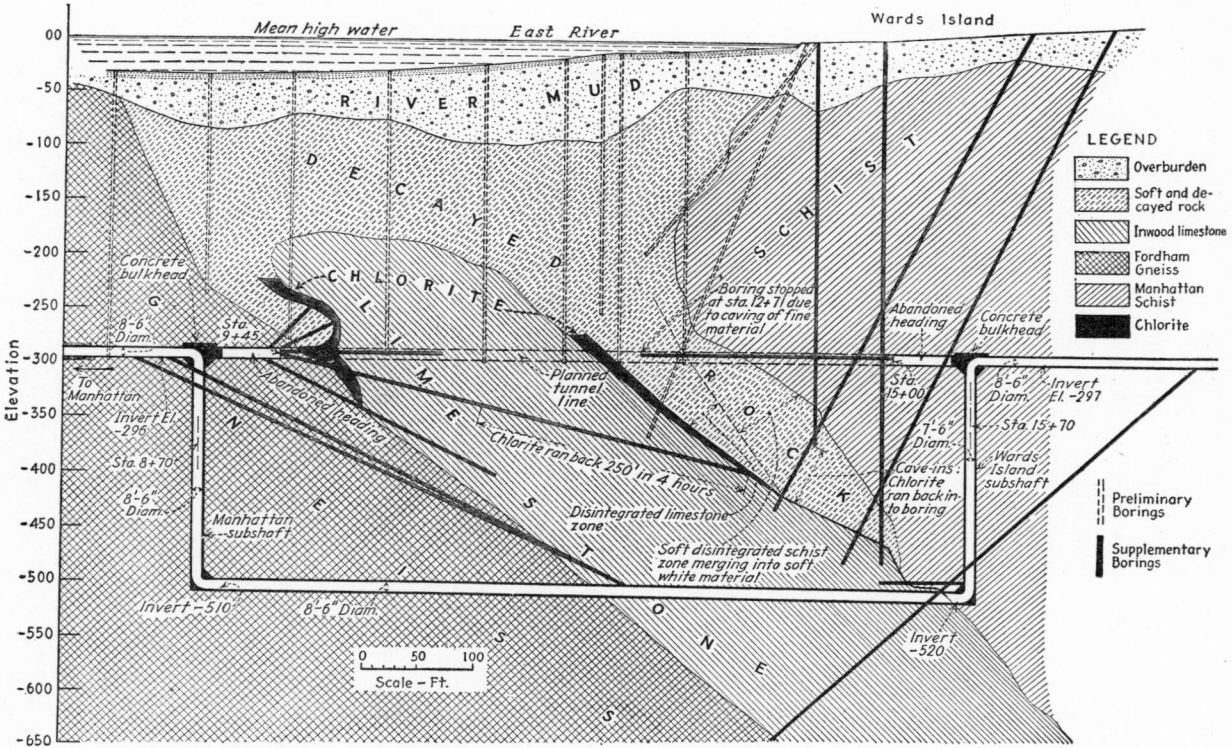

FIG. 8.5.—Geological section on center line of Wards Island Sewer Tunnel, New York, as revealed by test drilling, showing necessary relocation of tunnel in view of chlorite seams.

(Reproduced by permission of the Editor, Engineering News Record, from reference 8.7.)

encountered are clearly seen in Fig. 8.5, which shows also the location of the drill holes originally put down and the additional holes drilled after the work was temporarily closed in December, 1936, following the discovery that the seam of soft chloritic mass (of the consistency of fine mud) made further progress at that level practically impossible. One of the original borings had passed through this material; but as the hole had been drilled from the river, it had not been possible to secure such reliable samples as usual, and its wholly unsubstantial quality was underestimated. The cross section illustrates how the secondary drill holes disclosed satisfactory conditions at lower grade which enabled work to be resumed in March, 1937, the headings being holed through two months later. Dr. Berkey was the consultant for this work also, which was carried out for the Department of Sanitation of New York City, Mr. Walter D. Binger being Deputy Commissioner in charge of engineering, and Mr. Richard H. Gould, chief engineer.[8.7]

8.6. Tunnel Shapes and Linings.—The lining of tunnels can naturally be considered under two general classifications corresponding to the two general uses to which tunnels are adapted. In the case of road and rail tunnels, a lining may be necessary to support the pressure exerted by the material in which the tunnel is excavated; it may likewise be necessary to cover up the exposed material on the tunnel perimeter to protect it from atmospheric influences (and possibly the action of locomotive or automobile exhaust gases and vapors). Water tunnels, on the other hand, must often be lined to present a smooth surface to the flow of the water through the tunnel, thus reducing friction losses to an economic minimum; it may also be necessary to provide an impervious lining to prevent leakage of water into the surrounding rock if this is at all fissured or porous. Pressure exerted by surrounding material must be considered and, in the case of pressure tunnels, the pressure exerted by the water inside the tunnel. Finally, the material penetrated must be protected, if necessary, from the action of continued exposure to water and possibly air. These distinctive functions of linings are emphasized here, since sometimes they appear to be confused. Brief consideration will show that water tunnels are generally lined irrespective of the material through which they are driven, while traffic tunnels may or may not need such additional attention before completion. The few tunnels constructed for the passage of canals through high ground are governed by the characteristics of both types of lining.

Methods of lining are not within the scope of this book. It may, however, be mentioned that in soft ground, the use of segmental cast-iron or pressed-steel plates or precast reinforced concrete blockwork is now almost universal as the lining for resisting ground pressure (having succeeded the brickwork once extensively used). If necessary, such linings can have a concrete surface coating applied to them for presenting a smooth water surface. In the case of solid rock tunnels, timber framework used in construction may be left in place as a semipermanent lining in bad ground; it may be replaced by masonry or surrounded by a solid concrete lining. Mass or reinforced concrete linings are now almost universal for covering up exposed surfaces liable to disintegrate. The special linings for pressure tunnels are discussed in the next section.

The determination of the final shape of a tunnel cross section will also be affected by the material penetrated. In the case of water tunnels, a circular section is the most economical from the hydraulic point of view, but considerations of construction methods generally result in a horseshoe section with an inverted arch bottom. In soft ground and in poor rock, structural considerations of the design of lining will generally necessitate the use of a full circular section. Tunnel roofs are usually designed as semicircular arches, even in quite sound and solid rock, the remainder of the section (in the case of traffic tunnels) being proportioned from economic considerations. An interesting variation from this latter criterion is given by the Roman road tunnel already mentioned, through the Ponlipio Hills, which has its cross section in the form of a pointed arch 25 ft. wide, 22 ft. high at the center of the tunnel, but 75 ft. high at the ends, the increase having been intended to improve the illumination of the tunnel. As the total cost of a tunnel varies almost directly with its cross-sectional area, the importance of the correct determination of this will be clear. Economic considerations control this, but, as has been indicated, in some cases the nature of the material to be penetrated will override basic financial factors.

As the cost of lining a tunnel may amount to one quarter of its total cost, the importance of this feature of design will be obvious. Preliminary geological considerations will therefore be of great value in the determination of the design of the tunnel cross section. In igneous and metamorphic rock, known to be solid, a lining will probably be unnecessary for a traffic tunnel. Should disintegrated or shattered rock be encountered, most treacherous conditions may have to be countered, not only requiring permanent linings but possibly giving trouble during the temporary support of the exposed material during construction. In the case of the Simplon Tunnel, for example, much difficulty was experienced at a distance of about 2¾ miles from the Italian face, due to pressures exerted by decomposed calcareous mica schist, even 16-in. rolled steel beams used as temporary strutting being buckled. Quick-setting cement provided the final solution to the problem then presented, although at a cost of £1,000 ($5,000) per yard of tunnel. In other places, rock in the tunnel floor was forced up; but in sections driven through solid rock, even at the maximum depth below the surface, no deformation occurred, the lightest type of masonry lining proving adequate.[8.3]

Sedimentary rocks have produced most tunnel-lining problems, and for a variety of reasons. As a result of their mode of origin, sedimentary rocks are liable to change in character within relatively short distances; their original depositional characteristics render them often susceptible to changes in pressure; and the combination of these two factors renders them frequently of an unstable composition, easily and quickly affected by exposure to air and water. Furthermore, the bedding planes of sedimentary formations naturally affect the stability of any tunnel section bored through them, the angle made by the line of the tunnel with dip and strike of the beds being a most important feature. The underground water carried in fissures and along bedding planes will also affect the design of a lining. All these factors can be estimated, to some degree, from preliminary geological considerations.

8.7. Pressure Tunnels.—The design of pressure tunnels presents to the civil engineer some of his most serious difficulties in tunnel work. Three separate problems have always to be faced. In some way, the engineer must ensure that the material surrounding the water will be impervious. Some means must be secured for withstanding the pressures set up by the unbalanced head on the water passing through the tunnel. Finally, as a lining of some type is almost invariably used both to satisfy these requirements and to reduce frictional resistance to the flow of water, the design must take into consideration the possibility of the tunnel being empty, with groundwater tending to exert a considerable pressure on the outside of the lining. The engineer could naturally design a pressure conduit (of steel or reinforced concrete) that would meet all the requirements called for by the foregoing conditions while neglecting entirely the existence of the surrounding rock. Such a course would be far from economical. True engineering is the attainment of the economic solution to the problems faced, and so in the case of pressure tunnels the civil engineer seeks the cooperation of the geologist so that the best advantage can be taken of the rocks to be encountered in reducing the lining used to its economic minimum.

The three distinct problems must be emphasized, since they each have a separate bearing on the final solution although sometimes confused in discussion. It may be suggested that, in general, all important pressure tunnels are lined and that today such linings will almost certainly be of concrete. The section of a pressure tunnel will approximate a circle (being generally horseshoe shaped for construction convenience), and so the lining, when under pressure, will be generally in tension. Mass concrete has almost negligible tensile strength, and when under tension, even when reinforced, it will open up in minute cracks. If it is to be impervious to the water it is retaining, as it must be in the majority of cases (since the surrounding rock will not always be completely impervious to water), the lining must clearly be made from the finest quality concrete, placed carefully in position and well compacted in the forms. This is essentially an engineering matter, but it is stressed here, since the rock excavated from the tunnel will probably be used, if possible, as aggregate, and a geological opinion will have to be given before construction starts as to its suitability for this purpose (see Chap. XIX). The effect of the prolonged contact of portland cement concrete with the rock to be encountered must also be considered. A prediction as to how the rock will break will aid in the preparation of estimates of lining placing costs.

The structural design of the lining is governed by the anticipated maximum water pressure, from which can be calculated the total bursting pressure exerted on the tunnel lining. This has to be resisted by the lining and surrounding rock. At tunnel portals and where the *cover* of rock (the minimum distance from the tunnel perimeter to ground level) is small, the latter cannot be reckoned on at all, and the lining must be designed on standard lines as a reinforced concrete conduit, neglecting entirely the effect of the surrounding rock. When the rock cover is appreciable, and especially when it can be selected to suit design requirements, the geological nature of the surrounding rock is naturally of importance. Various methods have been suggested for arriving at the allowance to be made in design for the strength of the superimposed rock, varying from the minimum assumption that the actual weight of the column of rock vertically above the tunnel is all that can be assumed to resist movement due to the water pressure. Admittedly, any selection of the allowance to be made will be empirical, but a knowledge of the state and nature of the rock strata to be reckoned on in such calculations will tend to rationalize the decisions finally made.

It is important that the function of the rock be appreciated. The pressure on the water carried by the tunnel is transmitted equally in all directions; it is therefore uniform all around the tunnel perimeter, neglecting the weight of the water. This pressure, in the case of a pressure conduit, must be resisted by an equal and opposite pressure developed by the tensile strength of the material forming the conduit. If a tunnel lining is not designed as a conduit, the internal pressure must be resisted partially or totally by the rock around that section of the perimeter backed by solid rock of limitless extent (*ABC* in Fig. 8.6). Around the remaining section of the perimeter *CDA*, the resistance of the rock may be determined by the strength of the rock along planes such as *AP* and *CQ*. These conditions are ideal, assuming that the surrounding rock is homogeneous. In practice, such conditions will never obtain; and preliminary geological studies will enable some at least of the modifying features to be considered. Bedding planes may intervene; if these are filled with clay, these clay bands will form definite planes of weakness. The dip of the rock may provide a zone of weakness more serious than the shearing planes mentioned. The rock may have been faulted near the line of the tunnel, and the resulting fractured rock near the fault may have a serious effect on the tunnel. It may even be folded and thus in a state of initial tension liable to be seriously disturbed if a tunnel is excavated through it. Finally, it must be emphasized that, even assuming that the lining (with the aid of grouting) does bear solidly on the surrounding rock, it is not supported in this way on an unyielding base, since all rocks are to some degree compressible, and the consequent distortion may overstress the lining and cause serious rupture.

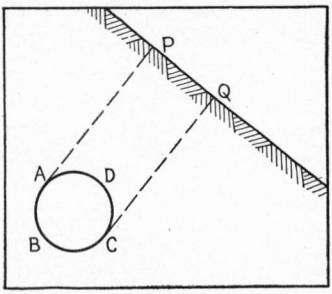

FIG. 8.6.—Diagram of pressure tunnel in relation to rock cover.

Civil engineering literature fortunately contains many most useful references to pressure-tunnel design and construction, notably in connection with tunnels that have failed. One of the most recent of these, and one of great interest, bears particularly on the preceding paragraph. The city of Sydney, New South Wales, Australia, constructed a pressure tunnel under a major part of the city area in connection with its water supply, during the years 1923–1928. The tunnel is 10 miles long and was designed initially with an internal diameter of 10 ft.; it is constructed throughout in the local Hawkesbury sandstone, of Triassic age, of varying character, fissured and jointed and marked by current bedding. Before placing the completed tunnel into service, the new Engineer-in-Chief to the Sydney Metropolitan Water, Sewage and Drainage Board arranged to have sections of the tunnel tested. This was done, and the lining was ruptured, to a varying extent, in all but one of the sections tested, although the test pressures were not unduly high. The failures occurred, generally, (1) where bad rock and fissures had been encountered, (2) where the cover of solid rock above the tunnel was theoretically inadequate. The test water was found issuing at the surface of the ground, and from the accurate test readings taken it was found that part of the lining and the surrounding rock had moved laterally ½ to ¾ in. in the direction of the axis of the tunnel and at an inclination conforming to that of the bedding plane of the sandstone. As a remedial measure, the whole tunnel had later to be lined with a solid and continuous steel pipe.[8.8] The published description of the work is a fascinating engineering document and repays much study; although faulty structural design may have been responsible for the failure, this was possibly associated with a lack of appreciation of the true nature of the sandstone rock in place and also when used as aggregate for the concrete of which the lining was formed.

8.8. Construction Methods.—To attempt to deal briefly with tunnel construction methods in such a chapter as this is to be forced on to a course between the Scylla of a statement of somewhat obvious facts and the Charybdis of a detailed exposition of drilling, blasting, and excavation, features belonging more properly to a volume devoted solely to tunnel construction. Yet the attempt must be made, since, as in other features of tunnel work, construction methods depend primarily on the geological strata to be encountered. Broad classification of the material to be met with will determine generally the type of construction to be adopted. The exact nature of the material, together with local practice, will determine the detailed variations of the main method in use, as, for example, the use of top, bottom, or central headings in rock tunnels and the use of the English, Belgian, German, or other schedule of excavation in the case of soft-ground tunnels. These considerations all depend on the geological nature of the materials met with, although methods will be determined by experienced construction men rather than by geologists. Special construction problems will almost invariably necessitate specialized investigation, and attention will therefore be directed to a few of the more important unusual conditions that may be encountered.

In the case of very deep tunnels, underground temperatures may sometimes cause inconvenience and affect adversely the progress of tunnel excavation. The study of such temperature variation can proceed only on the basis of actual experience, and it is therefore not possible to predict accurately the conditions to be met. It may be recorded that the temperature inside the Lochaber Tunnel in Scotland (during construction) rarely varied appreciably from 54°F. In the Simplon Tunnel, on the other hand, a considerable range of temperatures was experienced, as is shown in Table G. Mine workings have given some useful data on this aspect of tunnel work. Exhaustive tests have been carried out on this matter at the Lake Shore Gold Mine in northern Ontario, Canada, which show that the rock temperature there rises 1°F. with every 163.4 ft. of depth below surface. In Great Britain, a corresponding figure has been found to be 1°F. for every 60 ft. of depth.

TABLE G.—ROCK TEMPERATURES IN LEADING EUROPEAN TUNNELS[1]

Tunnel	Mt. Cenis	St. Gotthard	Arlberg	Simplon	Lotschberg
Length, miles	8.0	9.3	6.4	12.3	9.0
Elevation of portals, feet above sea level	4,162 3,765	3,756 3,638	3,995 4,271	2,080 2,253	4,002 3,936
Elevation of "crown"	4,248	3,788	4,297	2,312	4,067
Max. elevation of mountain range in profile	9,673	9,384	6,658	9,315	9,568
Max. cover	5,425	5,596	2,361	7,003	5,501
Max. rock temp., °F	85.1	87.4	65.3	132.8	93.2
Tunnel ventilation, cubic feet air per second, up to	247	35 to 71	106	1,236	388 to 883
Date constructed	1857–1872	1872–1882	1880–1883	1898–1906	1906–1911
Rocks encountered	Limestone, calcareous schist, gneiss, and schistose sandstone	Gneiss, mica schist, serpentine, and hornblende (fissured)		Slate, gneiss. and calcareous mica slate	

[1] Reproduced, with some additions, from I. SCHOKLITSCH, "Hydraulic Structures," Vol. 1. p. 15, by permission of the publishers, the American Society of Mechanical Engineers, and the J. R. Freeman Trust Estate, New York, 1937.

It has been during the construction of the long tunnels that pierce the European Alps that the most serious difficulties due to abnormal temperatures have been encountered. Table G has therefore been prepared, based (by permission) on one featured by Dr. I. Schoklitsch in "Hydraulic Structures." From the statistics therein presented, some idea can be obtained of the construction difficulties that had to be faced. Dr. Schoklitsch suggests that

. . . the variation in temperature along a tunnel depends on the thermal conductivity of the rock which itself is not constant for any given rock but varies with the moisture content and the dip of the strata. . . . It has also been found that the rock temperature in tunnels is, as a rule, lower than expected when the water breaks through from above [and vice versa] . . . abnormal, rapid rises in rock temperature can be regarded as danger signals.[8,9]

This last suggestion of this eminent European engineer must conclude consideration of this aspect of tunnel construction, important and interesting though it is, since there are other construction problems that arise frequently on the shorter tunnels more often encountered in the usual course of civil engineering practice. These are naturally most varied. They may even arise from the proximity of other under-

FIG. 8.7.—Breakneck Tunnels, New York Central Railroad, New York state. Old tunnel on left; portal of new tunnel under construction on right. Note proximity to syphon-shaft house of the Catskill water-supply system of New York, on extreme right.

(Reproduced by permission of the New York Central Railroad Co., Mr. J. W. Pfau, Chief Engineer.)

ground works, the safety of which must be safeguarded. For example, an unusual problem had to be faced during the construction of the new Breakneck Tunnel on the New York Central Railroad, in the state of New York. An existing double-track tunnel, extensively used and only partly lined, had to be increased in size, and a new double-track tunnel constructed, parallel to the existing bore, with a 30-ft. rock wall in between them. The material penetrated was a hard granite gneiss. The northern portals are close to one of the siphon shafts carrying the Catskill water supply from the deep-water tunnel under the Hudson River. Clearly, the excavation work had to be carried out in such a way that there should be no danger of the siphon shaft being affected by the excavation methods used. The compact nature of the gneiss would result in vibrations from explosions, if used, being transmitted without much damping over considerable distances. The excavation of the rock was therefore all carried out by the drilling of 2⅜-in. holes, 6 ft. deep, all around the final outline of the tunnel

FIG. 8.8.—New Breakneck Tunnel, New York Central Railroad, New York state. View showing close drilling adopted as excavation method because of proximity of aqueduct syphon shaft.

(Reproduced by permission of the New York Central Railroad Co., Mr. J. W. Pfau, Chief Engineer.)

section, spaced at 3-in. centers. The ⅝-in. gaps between each hole were then broached, after which the isolated core of rock was broken up and removed by standard methods. The operation was successfully completed, the finished appearance of the rock surface being as shown in the accompanying photograph.[8.10]

As indicated when linings were discussed, material may be encountered that is affected by exposure to air. Certain types of clay and shale are particularly susceptible to the influence of the atmosphere, and a common practice (as followed, for example, in Toronto tunnel work in Canada) is to spray the shale with a thin coat of cement as soon as it is exposed. Sometimes more elaborate methods have to be utilized, as in the case of the construction of the P.L.M. Railroad tunnels on a new line between Nice, France, and San Dalmazzo, Italy. Rock consisting of almost pure anhydrite was encountered in the Col de Braus tunnel for a length of about one kilometer, and in the Caranca tunnel for a few hundred meters. Water reaching the anhydrite, in the former tunnel, from adjacent jurassic limestone,

FIG. 8.9.—Equipment used for spraying exposed shale with cement, in duplicate water-supply tunnel for the city of Toronto, Ontario, Canada. Cement-coated shale is to be seen at extreme right.
(*Reproduced by permission of Mr. R. C. Harris, Commissioner of Works, City of Toronto.*)

caused an increase in volume of the exposed rock of about 30 per cent. The serious nature of the trouble was increased by the fact that the work was halted from 1914 to 1919. When excavation and lining were finally completed, the water which had not been completely drained continued its action and a portion of the masonry lining was ruptured seriously. Aluminous cement was used for its reconstruction, and an elaborate drainage system was installed; after an interval of ten years, it appears that a satisfactory result has been obtained. This experience was utilized in the later construction of the Caranca tunnel. Immediately after mucking, the rock was covered with coal tar; the lining was built in with the least possible delay, using aluminous cement; and after its completion, coal tar was injected under pressure behind the lining. Completed in 1921, the tunnel has given satisfactory service since.[8.11]

8.9. Overbreak.—Unless the specification covering the construction of a tunnel under contract is most carefully drawn up, with due regard for all the geological conditions anticipated, it is almost certain that the payment for overbreak will be one of the matters that cause most trouble at the conclusion of the contractor's operations. The term *overbreak* is one now generally used in civil engineering practice to

FIG. 8.10.—Rupture of masonry in the Col de Braus tunnel, S.N.C.F., between Nice and Coni, France, due to swelling of the anhydrite through which the tunnel was driven.
(*Reproduced by permission of Le Chef du Service de la Voie et des Batiments du Sud Est, S.N.C.F., Paris.*)

denote the quantity of rock that is actually excavated beyond the perimeter which is previously fixed by the engineer as the finished *excavated* tunnel outline. It will be realized that in the case of a lined tunnel, not only may the contractor receive no payment for rock excavated beyond this line, but also additional concrete filling will be necessary to fill up the resulting cavity. To engineer and contractor, therefore, the question of overbreak is an important one.

A close study of records and descriptions of early tunnels suggests that overbreak was not seriously considered in the early days of modern construction, possibly because many tunnels were then paid for on the basis of a unit length of completed tunnel. In modern work, payment is more usually made per unit volume of excavation and per unit volume of concrete or other lining. Under such circumstances, it will be readily seen that overbreak must be kept to a minimum; and as the quantity will depend primarily on the nature of the rock penetrated, given equally good driving methods, preliminary geological investigations are of great importance in estimating what overbreak will be.

In soft-ground tunnels, the term is rarely met with, since such works can generally be taken out exactly to the neat line of excavation, construction equipment (shields, lining plates, etc.) being designed to fit to this line. Close-grained igneous rock will usually break closer to the

FIG. 8.11*a*.—Lochaber water-power main tunnel, Scotland; view near heading 3E, showing the effect on excavation of drilling at right angles to the strike of the mica schist.

(*Reproduced by permission of Mr. W. T. Halcrow, Messrs. C. S. Meik and Halcrow, Consulting Engineers, London.*)

theoretical section than sedimentary or metamorphic rock, although certain types of chalk and compact sandstone will also give good sections. As overbreak will depend on the detailed nature and structural arrangement of the rock strata to be encountered, it cannot often be determined with great accuracy. When, however, the nature of the rocks to be met with has been determined by preliminary geological investigations, a combination of the previous practical experience of the tunnel engineers and the geologist's intimate knowledge of the characteristics of the rock types will usually enable a close estimate of the *way* in which the rocks will break to be made. It will then be possible to prepare the relevant parts of the specification so as to conform as closely as possible with the situations that will be met with as the work proceeds. The following represents a typical way of dealing with this matter, the items suggested referring to that part of a specification dealing with payment for excavation:

1. No points of solid rock must project beyond a line, called the *neat line* of excavation, fixed at a distance from the inside perimeter of the finished tunnel section equal to the minimum thickness of lining required.

FIG. 8.11*b*.—Lochaber Water Power Main Tunnel, Scotland; view in heading 5W showing the effect on excavation of drilling along the strike of the mica schist.

(*Reproduced by permission of Mr. W. T. Halcrow, Messrs. C. S. Meik and Halcrow, Consulting Engineers, London.*)

2. No flat areas of rock of more than, say, 3 sq. ft. must occur within a specified minimum distance from the finished perimeter, usually a few inches more than the minimum lining thickness.

3. All excavation and concrete lining will be paid for up to but not beyond another tunnel perimeter called the *pay line*, although all cavities beyond this line must be carefully filled, generally with concrete of an inferior mix.

The fixing of this pay line is usually the most difficult part of the cross-section tunnel design. It is the volume of excavation that has actually to be taken out beyond this line that is called the overbreak. Clearly, the engineer will want to fix the line so that a contractor will not be faced with the possibility of much overbreak, a feature that would be reflected in the unit prices tendered for excavation by those contractors who appreciate the significance of Geology and by disputes over claims made at the end of the job by those contractors who did not. The pay line will have to be fixed taking into consideration the nature of the rocks to be met; the way in which they will tend to break; the necessity or otherwise for temporary timbering and whether this will have to be incorporated in the finished lining or not; and, last but not least, the anticipated dip and strike of the rocks to be met with in relation to the axis of the tunnel.

Two photographs are reproduced showing finished cross sections of the Lochaber Water Power Tunnel in Scotland, as excavated. The first shows the shape of section obtained when the drilling was almost perpendicular to the strike of the mica schist, the marks of the drill steel being clearly visible, overbreak here being very small. The second view shows the shape of section obtained when the drilling was approximately along the strike, the section being almost rectangular owing to the way in which loosened blocks of the schist have fallen away. These views illustrate more vividly than can any words the dependence of overbreak on structural geological conditions as well as upon the nature of the rock that is penetrated. Each case must therefore be considered in relation to the special combination of circumstances that exist at the site. Actual experience can serve only as a general indication of what may be expected; with this in mind, Table H is presented as a guide, based (by permission) on a table in Messrs. Creager and Justin's "Hydro Electric Handbook."

8.10. Groundwater.—The presence of groundwater is often the main source of trouble in tunnel construction. It introduces at once the necessity for drainage facilities from all headings; and when the tunnel grades cannot be chosen to facilitate such drainage, the additional trouble and expense of pumping is necessitated. If water is present in any appreciable quantity, it will impede construction work of any kind; if the ground is soft and liable to be seriously affected by the drainage of water through or from it, the use of compressed air for all tunneling operations will become essential.

It is of the greatest importance, therefore, to have as accurate information as possible about the groundwater conditions liable to be encountered, before construction starts. It may safely be said that no major civil engineering operations that are to be carried out below surface level should be started before something is known about the

TABLE H.—STUDIES IN OVERBREAK IN EXCAVATION FOR TUNNELS[1]

Shape	Dimensions of neat excavation		Areas, sq. ft.		Per cent over-break	Nature of material excavated
	Height	Width	Neat Sec.	Exca-vated		
B	12 ft. 11 in.	13 ft. 10 in.	150	192	28.0	Rock—diorite, much jointed and slickensided
B	12 ft. 2 in.	13 ft. 6 in.	133	164	23.2	Rock—hard massive granite, numerous fissures
B	10 ft. 0 in.	10 ft. 8 in.	87	126	44.7	Rock—granite, generally coarse grained
B	12 ft. 2 in.	13 ft. 6 in.	131	161	22.8	Rock—black graphite slate
B	12 ft. 2 in.	13 ft. 6 in.	130	173	33.1	Rock—schist ranging from rotten to solid and massive
A	10 ft. 6 in.	10 ft. 6 in.	82	119	45.1	Rock—gneiss, micaceous
B	12 ft. 11 in.	13 ft. 10 in.	145	187	29.0	Rock—a complex quartzite and diorite
C	10 ft. 6 in.	9 ft. 7 in.	79	116	47.5	Conglomerate Roxbury pudding stone
C	11 ft. 5 in.	11 ft. 3 in.	101	144	42.6	Rock—felsite, very hard and flinty
A	13 ft. 2 in.	13 ft. 2 in.	148	222	50.4	Rock—limestone, hard and white
B	14 ft. 0 in.	15 ft. 0 in.	185	212	14.0	Rock—mica schist with occasional quartz veins

Shapes: *A*, circular; *B*, horseshoe; *C*, horseshoe with pointed-arch roof.

[1] Reproduced, by permission, from a more complete table featured on p. 504 of W. P. Creager, and J. D. Justin, "Hydro Electric Handbook," John Wiley & Sons, Inc., New York, 1927.

level and flow of groundwater at the site. It is equally important to keep careful check on all water that is met with during construction, not only with a view to determining its source and therefore the possibility of sealing it off in some way but also to check so far as possible the course that the water follows in order to ensure that it is causing no serious undermining or cavitation. In deep-tunnel work, plutonic water (water that has not come from the atmosphere or the surface of the earth) may be encountered, in which case no such precautions can be taken. During the construction of the Simplon Tunnel, for example, alleged plutonic water was encountered, at one point a hot spring and a cold spring of water issuing into the tunnel within one meter of each other, although differing by many degrees in temperature. In contrast to this, a test made during the construction of the Moffatt Tunnel, in Colorado, may be mentioned; calcium chloride was dumped in a lake 1,400 ft. above the tunnel line, and traces of the chemical were found in the water entering the tunnel no more than two hours later.[8,12]

The more general factors affecting the presence of water in underground rock structures are considered in some detail in Chap. XVII. These hold true for general tunnel work although often complicated and sometimes presenting special features. The problem presented by the presence of groundwater in tunnel work is the reverse of that usually to be solved, in that the water is not wanted in the tunnel, and its flow must therefore be stopped if this is at all possible. It is rarely possible to seal off an underground water flow at its source, assuming that the latter is known, although this has been done recently in the Severn Tunnel, England. Grouting is therefore generally the first recourse of the engineer, provided that the quantity of water encountered is such as to render economic the expenditure of this additional money and also that it can be definitely stated, from knowledge of the geological formation concerned, that the grouting will be effective.

A separate note follows on *grouting*, in which an interesting application to tunnel work is described. There may usefully be mentioned, in addition, the experiences gained in connection with driving the two tunnels under the river Mersey at Liverpool, England, to which brief references have already been made. The Mersey is slightly less than one mile wide at its mouth, where it separates the cities of Liverpool and Birkenhead, its banks being largely taken up with the great system of tidal docks necessitated by its 30-ft. range of tide. So great has been the traffic across the river at this point that a tunnel crossing was discussed as early as about the year 1800. In 1885, there was opened the Mersey Railway Tunnel, which still provides for an important electric railway service. And in 1934 was opened the Queensway Vehicular Tunnel, the largest subaqueous road tunnel in the world. The local geological formation is the Bunter sandstone, a red sandstone of the Triassic system, the tunnels lying almost wholly in the middle of three Bunter beds. The sandstone is hard, porous, massively bedded, and jointed; the strike is approximately north-south, and the dip between 2 and 5 deg. Faulting is a characteristic of the local geological structure, and a Liverpool geologist, Mr. G. H. Morton, suggested during a lecture that he delivered in the year 1861 that a fault would be found below the bed of the Mersey. The bedrock is covered throughout the local district with glacial drift, boulder clay, and sands and gravels, and these deposits extend across the river bed. In 1873, Mr. T. Mellard Reade, a distinguished Liverpool civil engineer and geologist, predicted that a buried river valley belonging to preglacial topography would be found under the present bed of the Mersey.

Construction of the Mersey Railway Tunnel started in 1879. Boreholes were taken in connection with the work, and Sir Francis Fox, the engineer for the work, talked to Mr. Mellard Reade, who stated in a paper that he maintained the correctness of his forecast of the buried valley

> . . . notwithstanding that I was given a section showing a series of borings taken by the first promoters of the tunnel, which was supposed to prove a rocky bed all through. Fortunately the engineers who carried out the work, forewarned, took further borings, with the result that the *level* of the tunnel was lowered. The trial headings found rock all the way through, but in the actual construction of the tunnel the bottom of the buried channel was cut through by the roof of the tunnel for about 300 feet, and I had the satisfaction of seeing the verification of my prediction without disaster to the undertaking. If the levels had not been lowered the result might have been disastrous failure.

The contractors met the buried channel suddenly and withdrew all their men from the workings, but the clay cover held, and work was eventually resumed and successfully completed. Mr. Reade's concise and modest statement illustrates vividly the value of this application of his studies into the glacial geology of the Liverpool district.

The fault predicted by Mr. Morton was also found, and both features were also encountered in the modern vehicular tunnel. When this was started in 1925, Professor P. G. H. Boswell undertook the extensive geological investigations that were necessitated by the increased size of the new tunnel (44 ft. internal diameter), its location seaward from the Mersey Railway tunnel, and the natural desire for economy in location and construction. Construction of the earlier tunnel assisted in these investigations, but a special study had to be made of the anticipated bottom level of the buried valley. This hinged on the elucidation of the fact that the old drainage system flowed in the *direction opposite* to that of the present river. This enabled the level of the new tunnel to be kept reasonably high, the actual minimum clearance between the tunnel excavation and the bottom of the valley being only 3 ft. at one spot! The old valley, although filled with water-bearing sand, was covered with impervious boulder clay, and so safety could be assured; it was the existence of this clay, which there came down to the rock surface, that enabled the railway tunnel to be completed even though it did pierce through the rock surface. It was not practicable to put down boreholes in the river (owing to the range of tide and shipping), and so all exploratory work was done by drilling *ahead* of the working face, up to 150 ft. away. Many of these holes were used later as grout holes. Despite its extensive nature, the total cost of all the preliminary and exploratory work amounted to less than one fifth of one per cent of the total cost, a remarkably good investment if the full record of the work is studied, as it can be in the publications listed in Appendix D.[8.13]

A singular difference between the two tunnels is related to groundwater. The lining of the railway tunnel was not made watertight, so that even today about 5,000 (imperial) g.p.m. have to be pumped out of its drainage sumps, the water having the composition of the Mersey and coming through the fissures and joints in the sandstone. It is *not*, however, related to the size of the tide, but it does vary sympathetically with the rainfall. Pumping had to be carried out during the construction of the vehicular tunnel, and this permitted many interesting observations with regard to groundwater in the vicinity of the tunnel to be made; particulars of some of the interesting results will be found in Professor Boswell's paper. The lining of the new tunnel is watertight, and the surrounding rock was grouted so that, after its completion, groundwater conditions reverted to their previous state.

In the case of soft-ground tunnels, grouting may be equally effective but only for such material as gravel through which the grout can penetrate sufficiently. Freezing has already been mentioned as another method (although of limited application) for sealing off troublesome underground water. A recent innovation, applicable to coarse-sand strata, is the Joosten chemical solidification process which consists in impregnating the strata being penetrated alternatively with silicic acid solution and a

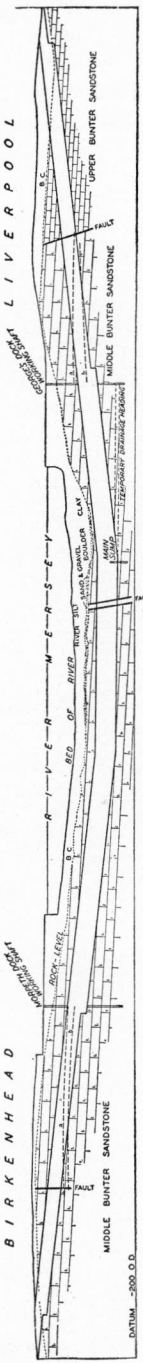

FIG. 8.12.—Geological section along center line of the under-river section of the "Queensway" vehicular tunnel under the River Mersey, Liverpool, England. (*a-a*, probable water level after construction pumping stopped. *b-b*, water level under conditions of maximum pumping.)

(*Reproduced by permission of Prof. P. G. H. Boswell and the Liverpool Geological Society, from reference* 8.13*b*.)

special salt solution, the interaction of the two causing the deposition of silica gel which eventually forms with the sand a solid type of special conglomerate insoluble in water and remarkably stable. Another method has been to use the special well-point installations utilized in open excavation work, with the well points jetted ahead of the working face which was thus kept dry by the usual pumping process through the screens in the points. Finally, there may be mentioned the more drastic procedure of actually lowering the underground water

FIG. 8.13.—"Queensway" vehicular tunnel under the River Mersey, Liverpool, England; view showing final phase of rock excavation, and permanent lining in place. Scale may be judged by the size of the two men in the center background.

(*Reproduced by permission of Messrs. Mott, Hay, and Anderson, Consulting Engineers, London.*)

table to such an extent that the ground through which the tunnel has to be driven is above this level and so in the dry.

An early example of the application of this solution is almost classical in nature, and so it may be briefly described. The Kilsby Tunnel near Rugby, England, proved an almost insuperable obstacle to the completion of the original railway line between London and Birmingham (now on the London Midland and Scottish Railway main line); it was eventually completed in 1838 at a cost of £300,000 ($1,500,000) as against an estimated cost of £90,000 ($450,000) and only at the expense of some lives. Preliminary exploratory work missed striking a pocket of water-laden sand and gravel of the inferior oölites in the lias shale; but when the tunnel works encountered it, extraordinary difficulties were met with, as the deposit proved to be a veritable quicksand; at one time, construction work had to be stopped. But Robert Stephenson was the engineer, and that remarkable man solved the problem in a characteristically satisfactory manner. Shafts were sunk, and steam pumps erected on the line of the tunnel. The following extract from his own report is of interest:

> As the pumping progressed the most careful measurements were taken of the level at which the water stood in the various shafts and boreholes; and I was soon much surprised to find how slightly the depression of the water-level in the one shaft influenced that of the other, notwithstanding a free communication existed between them through the medium of the sand, which was very coarse and open. It then occurred to me that the resistance which the water encountered in its passage through the sand to the pumps would be accurately measured by the angle or inclination which the surface of the water assumed towards the pumps, and that it would be unnecessary to draw the whole of the water off from the quicksand, but to persevere in pumping only in the precise level of the tunnel, allowing the surface of the water flowing through the sand to assume that inclination which was due to its resistance. . . . The simple result, therefore, of all the pumping was merely to establish and maintain a channel of comparatively dry sand in the immediate line of the intended tunnel, leaving the water heaped up on each side by the resistance which the sand offered to its descent to that line on which the pumps and shafts were situated.[8.14]

This account was made in a report to the London and Westminster Water Company in 1841 and apparently forgotten until rescued by Professor Boyd Dawkins and quoted in his interesting James Forrest Lecture in 1898. It is a striking example of scientific observation, establishing the nature of the *cone of depression* set up by pumping groundwater, and serves to illustrate vividly how scientific data can usefully be applied in civil engineering work.

Consideration has been given to the handling of underground water in so far as it affects normal tunnel construction. Special problems in connection with such underground water arise, of which the following is an interesting example. The Boyati Tunnel, 8.33 miles long, 7.54 ft. wide by 8.15 ft. high, with a horseshoe section, is a part of the new works for the supply of water to the city of Athens, Greece. Constructed between 1926 and 1931, it gave rise to many construction difficulties, all unforeseen, since no preliminary information was available and no borings were taken before work started. The trouble encountered soon demonstrated the need of borings, and therefore 12-in. holes, stepped in to 10, 8, and 6 in., were put down on the line of the tunnel at intervals of about ⅔ mile (the large size of the hole making it possible to use them for checking the tunnel survey and for leading electric power cables into the headings). The materials penetrated included compact marble, compact mica schist, a disintegrated chlorite schist (which smashed the concrete block lining two hours after setting and necessitated four sets of timbering before the squeeze was stopped), and crevassed and cavernous limestone. The latter was water bearing, and this water added to the troubles already encountered. Finally, at a distance in from the north portal of 13,700 ft.,

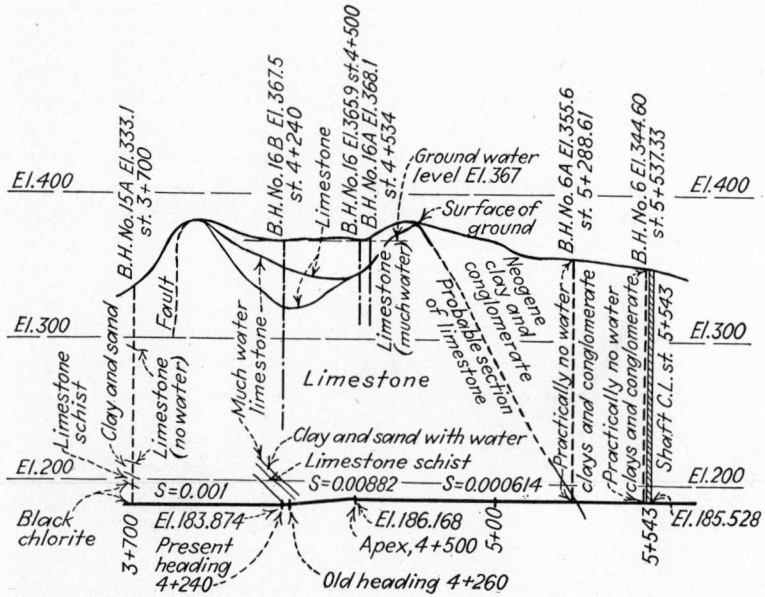

FIG. 8.14.—Section along center line of Boyati Tunnel, Greece, in water-bearing area. (*Reproduced by permission of the Editor, Engineering News-Record, from reference* 8.15.)

the water inflow was so great that the heading had to be bulkheaded off; the water pressure (estimated at 250 lb. per square inch) was so great that it forced a large sandbag bulkhead over 3 ft. back into the tunnel. Borings had disclosed the dip of this crevassed limestone as inclined toward the line of the tunnel in the direction of tunnel driving, with a superincumbent bed of clay and conglomerate which was practically dry. It was therefore decided to tackle the tunnel by driving through the limestone, working into the bed rather than from "underneath" it, and for this purpose a special shaft was sunk on the line of the tunnel, wholly in the clay and conglomerate. This was first joined up with the south heading, and then driving was resumed into the limestone. Excessive water pressure was again encountered, and a detour had to be made from the original tunnel line, but the radical change in attack suggested by geological considerations was probably the keynote of the ultimate successful completion of this notable work.[8.15]

8.11. Grouting.—Grouting, or cementation, has been mentioned already. It has been applied extensively in tunnel work, mainly, but

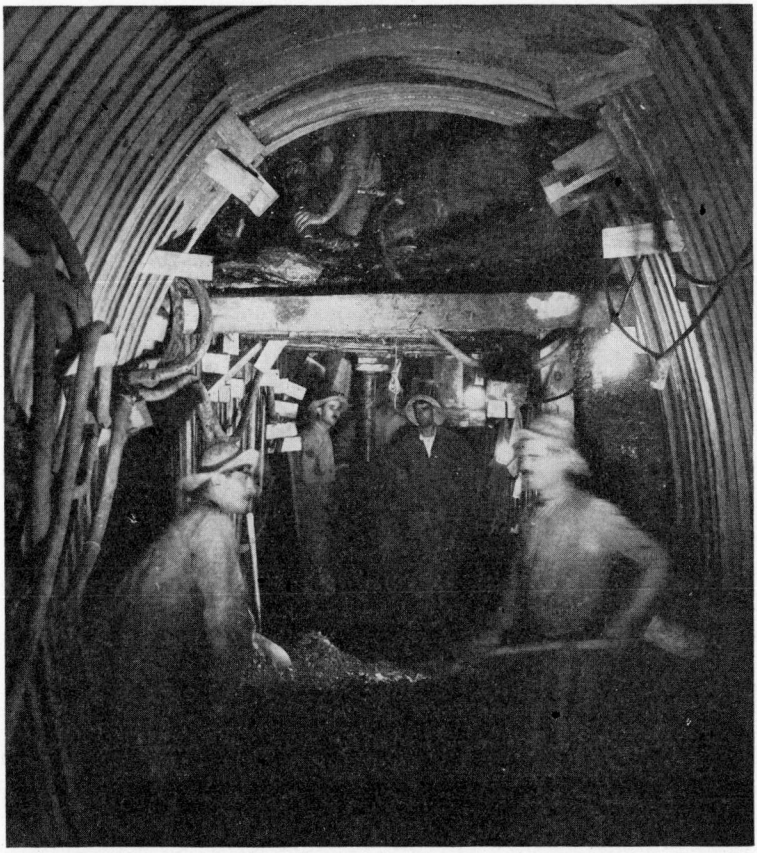

FIG. 8.15.—The Boyati Tunnel, Greece; a view showing enlargement of the tunnel section proceeding, after water had been drawn off—advance heading can be seen in background.

(*Reproduced by permission of Messrs. Ulen and Co., New York.*)

not exclusively, for reducing the flow of underground water into tunnels. Its application and the necessity for its application if anticipated before construction therefore depend on the geological nature of the strata to be encountered. The methods adopted in its application in tunnel work are similar to those generally used, a cement mixture being usual, although sometimes (as in the case of the well-known François process) special chemical additions are injected first to ensure cementation of fine cracks in the rock being grouted.

Fissures or joints must clearly exist in the rock if it is to be grouted satisfactorily. Fissures or general porosity of the rock are presupposed by the necessity for grouting, excessive inflow of water, structurally weak rock unable to support itself across the tunnel arch, or material liable to disintegrate when further exposed to the atmosphere resulting from these characteristics. In all these cases, if the anticipated trouble is determined beforehand, as is generally possible by careful geological investigation, grouting can prove an effective remedy and prevent serious trouble. The construction of tunnels through water-bearing strata such as coarse gravel can be materially simplified in the same way.

Leakage of river water into the famous Severn Tunnel of the Great Western Railway, England, has been practically stopped by the application of a specialized cementation process, many years after its construction. The construction of the new Mersey Vehicular Tunnel was greatly facilitated by means of the same process. Similar precementation in water-bearing gravel facilitated the construction of some of the inclined escalator shafts for the London tube railways. In one section of one of the P.L.M. railroad tunnels in France (mentioned on page 163), the use of tar for injection was successfully adopted as a means of coating voids behind the tunnel lining and preventing further disintegration.

The work carried out in the Severn Tunnel may be mentioned in more detail. The tunnel was started in 1873, but progress was slow; in 1881, the river water broke into the tunnel workings, passing through open marl beds. These were successfully plugged up by depositing clay from schooners sailing in the river above the breaks. Operating conditions in the tunnel have always been troublesome, and the Great Western Railway Company in the years 1924 and 1929 had to follow earlier practice and dump clay above obvious danger spots. After prolonged investigation, aided materially by the geological record kept by the original contractor (see page 79), it was decided to grout the natural strata behind the brick lining throughout a critical section of the tunnel. This work was entrusted to the François Cementation Company, Ltd., which carried out two contracts, using about 6.5 million pounds* of cement in 2,100 holes, totaling 9,500 ft. in one

* All weights, apart from general approximations, are given in pounds in order to avoid confusion between the short ton and the British long ton, examples having been selected, in so far as possible, to be representative of both British and North American civil engineering practice.

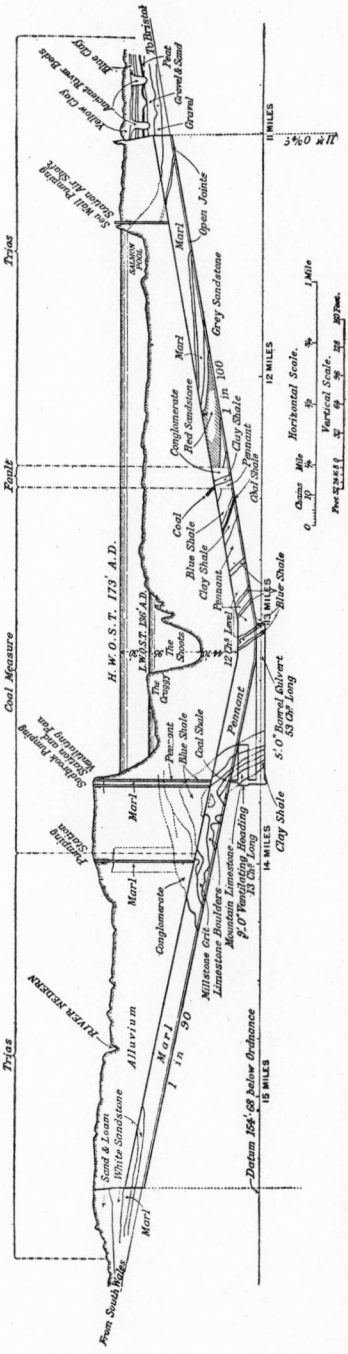

FIG. 8.16.—Geological section along center line of the Severn Tunnel, England (Great Western Railway); cementation carried out from Bristol end (R.H.S.) as far as the red sandstone, near mileage 12.

(Reproduced by permission of the Great Western Railway Co., Mr. R. Carpmael, Chief Engineer.)

contract, and 12 million pounds in 2,400 holes, totaling 10,500 ft. in the second. The existence of large voids between the brick lining and the surrounding strata, even beneath the invert of the tunnel, was proved by the travel of the cement. The river bed above the tunnel was patrolled at low tide during the grouting operations; but in only one instance did cement actually travel to the surface, where it formed a protective covering over the mouth of a fault in the marl.[8.16]

8.12. Construction Records.—In all tunnel construction work, it is of the greatest importance that accurate, complete, and up-to-date geological records be maintained from the start of construction. This can be done most conveniently by means of a geological map of the tunnel route and a geological section along the line of the tunnel. The exact nature of the rock excavated must be observed after every round has been fired, and the direction of strike and dip of the rock recorded regularly. All such data should be carefully and continually compared with the geological section along the tunnel predicted from preliminary investigations before construction started and with similar data obtained at other headings in the tunnel. In addition, it must be compared with surface topography and geology above the tunnel and in its vicinity, in order that as complete a picture as possible may constantly be available for study. In this way, a close check can be kept on the relation of observed to anticipated formation, and any unusual departure from the latter can be checked up immediately it is discovered, and its implications investigated. Alternatively, if difficulties are encountered underground due to fissures or faulting, surface conditions may serve as a guide to future similar troublesome spots when once a correlation has been established.

The geological record so obtained is of great interest and importance in another direction entirely—to the geological surveyor, as giving him information obtainable in no other way and, if the tunnel is to be lined or used as an aqueduct, at no other time. Such records can therefore be seen to be of supreme value to the community as a whole, and all civil engineers in charge of tunnel work should therefore invite the attention of the director of the local geological survey to any section available for inspection and should furnish to the survey a copy of the final section obtained. The courtesy, involving no expense and but little trouble, should likewise be extended when possible to the head of any university geological department in the immediate neighborhood. The discretion of scientists is such that no engineer need ever hesitate to take the step here suggested.

In addition to being of vital importance to pure Geology, construction records often have practical application and general utility.

Should any trouble develop in the future operation of the tunnel, the cause may often be traced to some unsatisfactory condition of the surrounding ground if this information can be obtained from the records available. When tunnels, and especially pressure tunnels, are being examined after continued periods of use, a geological section will always help to conserve time and increase the value of the survey by indicating the sections that need special attention. If a railway tunnel is to be increased in cross section, an accurate geological record of the original work will be of inestimable value. And if there should unfortunately be litigation with regard to the construction of a tunnel, as is sometimes caused by overbreak settlements, a geological record of the tunnel line will often be a deciding factor in a court of law.

As an example of a tunnel on which close attention to Geology during construction was an important feature, the main tunnel of the Lochaber Water Power scheme (Scotland) may be cited. The tunnel extends from the main reservoir for the project, Loch Treig, to the side slope of Ben Nevis (Great Britain's highest peak) overlooking Fort William, on Loch Linnhe, being slightly more than 15 miles long with an effective average diameter of about 15 ft. The tunnel passes through rock strata which, from the geological standpoint, constitute one of the most complicated parts of Great Britain. Many types of rock were encountered, from shattered mica schist to granite and baked schist of excessive hardness (some requiring three sets of drill steel in each hole drilled into them). Prior to construction, a survey of the route selected for topographical reasons was made by Dr. E. B. Bailey, F.R.S., now Director of the Geological Survey of Great Britain, who for many years past had made a close study of the rocks to be encountered. The preliminary information then obtained was confirmed to a remarkable degree by the geological section revealed by the excavation. During construction, Dr. Bailey made a complete survey of the tunnel, collecting many rock samples, and in this way the detailed mapping of the surface above has been completed. Several interesting facts, not shown by surface outcrops, were discovered; the public records were thereby enriched. Most of the staff of the Resident Engineer, Mr. James Brown, had a working knowledge of Geology and so were able to observe and record the varying geological features encountered in the course of their regular engineering work. From among the many interesting features of the work, one may be mentioned. At the Loch Treig end of the tunnel, excavation proceeded in soft mica schist, much disturbed by faulting and lines of movement. Seams of clay were met with in this section, and the rock adjacent to these was soft, special timbering and cast-iron lining having to be used. After passing two of these disturbed areas, it was noticed that they corresponded to the lines of cliffs on the hill above the tunnel; accordingly, succeeding lines of cliff were surveyed, and the presence of bad rock in the tunnel predicted thereafter to within a few feet of where it actually occurred. The value of such work to the contractors needs no elaboration.[8.17]

8.13. Conclusion.—Many more interesting examples could be quoted from the records of tunnel practice, the availability of this

FIG. 8.17.—Minneapolis–St. Paul intercepting sewer tunnels; view showing jetting of sandstone rock and pumping of sand slurry as an excavation process.
(Reproduced by permission of the Minneapolis–Saint Paul Sanitary District, through the courtesy of Mr. C. C. Wilbur, Consultant.)

interesting material being probably due to the fact that every tunnel job is unique at least in some one respect and often in several. Descriptions of tunnel construction consequently figure prominently in civil engineering literature, a fortunate feature in connection with any study of the application of Geology to civil engineering work in view of the dependence of all tunnel work on geological conditions. This dependence will probably be clear from the examples that have already been quoted, and the constructive contribution that the science has made to the art will have been obvious.

As a final note, reference may fitly be made to a few cases in which even the nature of the material encountered has enabled the use of special methods during construction. In connection with tunneling in soft ground, for example, the particular type of clay encountered in Chicago, Illinois (and Winnipeg, Canada) permits its removal in large slabs, "cut off" the working face by means of specially shaped power-operated knives. In great contrast to this work of peace may be mentioned the unusual procedure developed during the excavation of mine tunnels in the chalk of northern France during the World War, 1914–1919. In order to soften the chalk before excavation and to reduce the noise of the underground working, auger holes were bored into the working face and filled with vinegar, a strange reversion to the practice of the Romans![8.18]

Finally, it may be mentioned that during the construction of sewer tunnels for the twin cities of Minneapolis and St. Paul, Minnesota, some tunneling work was carried out in a local white sandstone formation. Although this rock has a good appearance, it can be cut with a high-pressure water jet. This means of cutting was adopted to break up the working faces, using a $^3/_{16}$-in. jet at 300 lb. per square inch pressure. The rock quickly disintegrated to the state of fine sand, and the resulting slurry was pumped, by means of centrifugal pumps, along the tunnel and up the working shafts to settling bins at street level. And with this reference to solid rock being pumped through a 4-in. pipe line, perhaps this chapter should be brought to a close![8.19]

Suggestions for Further Reading

Most of the standard textbooks on tunnel construction contain sections in which geological features are considered. The subject can most usefully be studied by reference to detailed descriptions of actual tunnel projects; the papers listed under this chapter heading in Appendix D are a representative collection.

An unusually complete tunnel construction record is to be found as an Appendix to the Report of the Chief Engineer, Annual Report of the Metropolitan District Water Supply Commission, for the year ending Nov. 30, 1935, Public Document No. 147 of the Commonwealth of Massachusetts. Entitled "Geology of the Quabbin Aqueduct and Reservoir in Central Massachusetts," it was prepared by Frank E. Fahlquist, Senior Civil Engineer, and Dr. C. P. Berkey, who was consulting geologist for the works described.

CHAPTER IX

OPEN EXCAVATION, EMBANKMENT FILLS, AND RETAINING WALLS

Mr. Sopwith called the attention of the meeting to the valuable Geological Sections presented by the railway cuttings, and other engineering works now in progress; this was particularly the case on the North Midland Railway, where the crops of the various seams of coal, with the interposing strata, were displayed in the clearest manner, developing the geological structure of the country which the railway traverses.

At the Institution of Civil Engineers (London), February 2, 1841.[9.1]

Open excavation work in civil engineering may be conducted with one of two main objectives. Materials may be excavated in order either to form a hole for some purpose, in which case the disposal of the material excavated is a matter of no great significance, or so that the materials may be used in some way—as building stone, gravel, and sand for concrete work or as ordinary fill. The quarrying of stone of all types is now a specialized line of endeavor; it will therefore be briefly considered separately, since it has problems peculiarly its own. Excavation of gravel, sand, and other material required for use is generally a straightforward process when once the material required has been located, the excavation process rarely introducing unusual problems. As the location of such deposits is dealt with elsewhere in this book, no special consideration will therefore be given to the excavation of unconsolidated materials for use as materials of construction.

The digging of a hole in the ground—for open excavation work is frequently so regarded—appears superficially to be a relatively simple matter. All experienced engineers know otherwise. The problems introduced are sometimes intensified by their very simplicity. The economics of engineering designs involving open excavation may be initially more indefinite and in consequence more involved than in any other branch of design work. Excavation methods and rates of progress are clearly dependent on the material to be encountered and its geological structure, the two factors that alike determine also the finished cross section of the excavation. All too often, these vital factors do not receive the study they deserve. The design and construction of embankments have generally to be based on empirical data in a similar way to open excavation work. Unforeseen problems are not the same hazard as in excavation work, but the history of fill placement bears testimony to the fact that problems of serious moment are not unknown. These are often associated more with the effect of filling on existing strata than with the nature of the fill material itself. They will also be considered in this chapter as being intimately associated with the main features of open excavation work.

The pressure on retaining walls is a leading subject of research and discussion in civil engineering circles and has been for a great number of years. It will frequently be found that the application of the results of research work and of complicated formulas based on ideal soil conditions to practical design work in which materials far from ideal have to be considered is something to which little attention is paid. The fact that the pressure on the back of any retaining wall is due to the pressure induced by the material that it retains and not to the pressure given by an ideal sand or clay is obvious. Yet it is a fact sometimes so patently neglected that it may well serve as an introductory note to some comments on the adequate appreciation of geological data in such design work.

9.2. Open Excavation.—Open excavation consists essentially in removing certain naturally formed material, within specified limits and to certain definite levels, in the most expeditious and economical manner possible. To do this, it is desirable to know with some degree of accuracy the nature of materials that have thus to be handled, their relative structural arrangement, their behavior when removed from their existing position, the possibility of water being met with during excavation, and the possible effect of the excavation operations on adjacent ground and structures. On some civil engineering projects, these preliminary requirements are not fully answered before excavation begins, with the result that money is lost and trouble encountered. This neglect of preliminary investigation work in open excavation is to some extent understandable: such work looks easy, and the unit prices generally charged are so low that it is often thought that a few cubic yards more or less will make little difference. The few can easily become dangerously many.

Preliminary investigations and constant check on excavation work as it proceeds are therefore clearly necessary. The preliminary work will follow the lines already suggested, with the addition of such special investigation as is found to be needed. The resulting geological maps should show rock contour lines where possible or the contour lines for any stratum so different from the surface material as to affect excavation methods and progress. The geological sections prepared will assist in visualizing the structure of the mass of material to be excavated. The usual care must be taken to investigate the possibility of groundwater being encountered. The sections prepared across and along the site to be excavated will show generally the relation of the hole to be made with adjacent strata and structures, and possible damage may in this way be foreseen. Complete and detailed records of all geological features encountered during the progress of excavation work are just as necessary as in other types of civil engineering work, since they will act as a constant check on the validity of information deduced from preliminary exploration and surveys and often serve as a warning of possible unforeseen future troubles. Similarly, such records will be of interest to geologists and may reveal structural details of great value. A close watch should be kept for fossiliferous strata, and, whenever possible, permission granted for fossil beds to be studied by interested geologists.

The preparation of limited areas of excavation for all special foundations, as for buildings, bridges, and dams, will be considered to some extent in the chapters devoted especially to these subjects. Considering general excavation work, the dip of the strata to be encountered is a factor of major importance. If the strata are horizontal, or approximately so, excavation work will be relatively straightforward, and side slopes can be determined with some degree of certainty. If the bedding is appreciably inclined, excavation methods will be affected, and hazards possibly increased. The side slopes must be so selected as to be in accord with the natural slopes given by the bedding. Faults that cross the area over which excavation is to be carried out may cause serious trouble; in soft ground, it may be difficult, if not impossible, to detect them before digging starts, hazards thus being increased.

Finally, all excavation work is influenced primarily by the nature of the material to be excavated. Broadly speaking, this can be classified as unconsolidated material ("earth") and rock. It will be convenient if the general problems of excavation, already indicated, are considered under these two headings. Note may again be made of the great importance of adequate definitions in contract specifications of materials that have to be excavated, since the difficulties to which reference was made on page 76 most frequently occur in connection with open excavation. The suggestions already advanced for overcoming these difficulties are therefore of particular importance for this class of work.

There are naturally construction projects on which the open excavation work is through material varying from rock to soft unconsolidated deposits; as one of the most instructive descriptions of unusual excavation work in recent civil engineering literature deals with such a case, it will be briefly outlined. This was the opening out of the Cofton Tunnel on the Birmingham to Gloucester section of the London Midland and Scottish Railway in England in 1925. The tunnel was constructed in 1838–1841 and forms a part of the route taken by this line of railway in crossing a fairly high ridge of land immediately south of Birmingham. Modern traffic requirements had made necessary its opening up and reconstruction as a cutting. The

FIG. 9.1.—Opening-out of Cofton Tunnel, England (London, Midland and Scottish Railway); a "hanging wall" slip in the west slope on a fault plane.
(*Reproduced by permission of the L. M. and S. Rly Co., Mr. W. Wallace, Chief Engineer.*)

strata shown up in the cuttings on either side of the tunnel were soft false bedded sandstones with thin beds of marl (north) and thick beds of hard sandstone and beds of tough marl (south), so that to the civil engineer they can be classified generally as neither rock nor earth. No records of the construction of the tunnel and the strata passed through were available, but the geological survey map of the district showed a major fault traversing the tunnel. Troubles due to earth movements were therefore anticipated, and some of a serious nature were encountered very early in the work. Before the final 80 yd. of tunnel was demolished, a careful geological survey of the surrounding country was made in order to attempt a deduction of the hidden structure yet to be encountered. The survey was considered in conjunction with the records of dip and details of faults recorded from the start of the work. This investigation carried out in the midst of construction, although incomplete, was yet of sufficient value to be of great service. The work was finally completed satisfactorily but only after surmounting great difficulties. The evidence tended to show that the tunnel had originally been built instead of an open cutting because of a realization of the troubles to be encountered. Examination of the geological survey map showed that if the original line had been located either 200 yd. west or 300 yd. east of its actual position, little or no trouble due to the Longbridge Fault would have been experienced. The paper describing the work is one of those rare civil engineering publications dealing with unusual difficulties in construction and their surmounting and as such is worthy of special study. It is interesting also to note that at least one member of the staff of the Geological Survey of Great Britain was in constant attendance at the work.[9.2]

9.3. Economics of Open Excavation.—The end of a tunnel is often the beginning of an open cut; the location of the change of section is always a matter demanding careful consideration by the designing engineer. The basic criterion, but not the only one, will be the relation of the respective costs of construction of tunnel and open cut; geological structure is intimately associated with such a study. Tunnels having already been dealt with, little need be said about the estimated cost of their excavation. For an equivalent open cut, the volume to be removed per foot of length is naturally much greater, but the unit cost of excavation will be lower, since the work can be carried out in the open, without underground hazards and with excavating machinery that could not be used underground. Calculation of the actual quantity to be removed in an open cut will be mentioned in a later section, side slopes and rock dip being important factors. No equivalent to the lining of a tunnel will normally be necessary, if the side slopes can be trimmed off to a satisfactory angle, although if they have to be restricted, retaining walls at the foot of the slopes may involve a greater cost than tunnel lining. Based on figures thus obtained, comparative estimates of cost for tunnel and cut can be calculated; and by applying these to locations with increasing depth to grade, a section can be selected as being that at which the change in type of construction can be made most economically.

In many cases, this result will be the final one; but in all investigations of this type, true economy has not been studied unless some consideration is given to geological structure. Geological sections along the center line of the proposed cut and at right angles to it (at several stations) are an essential, together with a detailed material study. The relation of the cut proposed can then be seen in relation to underground drainage, the dip of the strata, and possible unusual underground structural relationships; important modifications of design may have to be made in consequence.

Consider such a section as that sketched in Fig. 9.2. In the case of the tunnel, drainage of underground water need not be considered, as no water will reach the tunnel; but in the case of the cut, drainage will certainly present serious problems. Water will be continually seeping out at contact A; and if the quantity of water is sufficient, it may tend to wash out some of the sand on to the face of the slope. In order to escape, the water must run down the face of the shale, thus tending to weaken it. There might even be some travel of water along bedding planes of the shale, lubricating them and giving rise to dangerously unstable conditions leading possibly to serious slips. Alternatively, through this unrestricted drainage of the underground porous bed, unstable conditions might be induced in the clay stratum above with consequent possibility of serious earth movement.

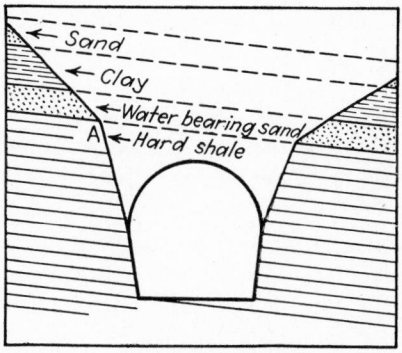

FIG. 9.2.—Simple geological section illustrating (diagrammatically) a possible relation between tunnel and open-cut excavation.

In all such cases—and they are by no means rare—standard calculations of relative economy may not therefore present a true picture. In the particular case illustrated, it will be advisable to extend the tunnel construction beyond the theoretical economical limit in order to avoid possibility of the difficulties indicated, even at the expense of increased first cost. Cases will occur in which no such simple geological structure as is sketched will be encountered. Faulted ground, for example, is alike a menace in solid rock and in soft material; and although causing difficulties in tunnel work, it may result in greater difficulties in open excavation, mainly because of the greater freedom given to material on fault planes to move in large masses with consequent increase of unnecessary excavation. The complications resulting from excavation in faulted ground which arose during the opening up of the Cofton Tunnel, described in the last section, serve as a good example, the factors there influencing ordinary economic criteria being clear.

In the case of open excavation work, unassociated with tunnel construction, over-all dimensions and depths to finished grade will normally be determined by considerations other than excavation costs, so that no question of relative economy will normally arise. Thorough preliminary investigations are necessary all the same, and these may possibly reveal special conditions which do affect the economical aspect of the work undertaken. For example, subsurface explorations may show that by a slight change in location troublesome material can be avoided. Finally, geological conditions will have a bearing on the necessity for finishing off side slopes either with or without retaining walls, when there is no restriction as to how an excavated area is to be

FIG. 9.3.—Simple geological section illustrating (diagrammatically) how the dip of rock strata may affect excavated cuttings.

bounded. The local geology will affect generally the economic cross section of the cut desired in this case, since the structural relationships of the strata to be exposed will affect the nature of side slopes quite apart from the nature of the material. A point to be specially watched in this connection is the question of drainage, since the adoption of simple and relatively inexpensive drainage systems can readily lead to appreciable modifications in the side slopes adopted.

9.4. Open Excavation of Rock.—Excavation of large masses of rock in the open may be subdivided into excavation in the dry and excavation under water. Most of the following comments apply equally well to both branches of work, although underwater work is generally of a less accurate and exact nature than that carried out in the dry. The problems that may be encountered are, in the main, similar to those met with in rock tunnel work. They necessitate preliminary determination of the structural arrangement of rock strata, the nature of rocks to be encountered, possible presence of water, and the resistance of exposed rock to weathering influences. The structural arrangement of rock strata will affect considerably the side slopes to be adopted in design, since if the general dip is *toward* one side of the cutting, there will be a tendency for shatter rock to fall off the ends of successive layers, unless trimmed well back; a reverse effect will naturally be found on the opposite side of the cutting. Structural arrangement will also affect underground water problems in determining the general lines of water movement and also, in connection with the porosity or impermeability of beds, the levels at which water may be encountered. Angles of dip and other special structural features such as fault planes will obviously have a considerable influence on excavation methods.

Of great importance is the nature of the rock to be excavated. By nature is meant not merely the general classification of the rock but a vivid conception of the actual properties of the particular type of rock to be encountered all over the area to be excavated. The value of general classification is by no means unimportant—indeed, it is a first essential—but each class of rock can vary so considerably that, for civil engineering purposes, it must be further described by some indication of its physical properties. Thus, sandstone can vary from a hard and compact rock to material that is little better than well-compacted sand; granite can prove to be one of the hardest of all rocks and yet may be found in such a state of decomposition that exposure to the atmosphere will lead to immediate disintegration. All preliminary geological information must therefore be checked thoroughly, and the general description of rock types, obtained from borings, considered in relation to the physical state of the rock samples so found. Similarly, it is essential to know with some degree of accuracy, before work begins, how the rock to be encountered will stand up after long exposure in order to obviate possible future accidents and keep maintenance charges at a minimum.

If due allowance is made for exceptional variations in physical properties, the following notes will serve as a general guide. The stability and permanence of igneous and metamorphic rocks when exposed to the atmosphere can generally be relied upon unless they are badly weathered. If such rocks are badly fissured, jointed, or shattered by any local faulting, the greatest caution must be observed in finishing off slopes. Especially is this true for work carried out during severe winter weather, as in Canada, where frost action may temporarily consolidate local disintegration. Sedimentary rocks require more caution and judgment when being considered in such work. The presence of clay in any form, even as a shale which may be exceedingly hard when first exposed, must be regarded with suspicion, particularly if the bedding planes are inclined to the horizontal at any appreciable degree. The action of the atmosphere may soon reduce this material to an unstable state, with consequent extra trouble and expense. When clay is known to be present, ample allowance should be given in side slopes as designed, and the economics of protecting the exposed faces with light retaining walls should be investigated. Sandstone and limestone, if in a firm and solid state, will stand with vertical or almost vertical faces; but both types of rock vary from a sound solid state to material that can be crumbled in the hand, so that every case must be considered on its own merits.

The provision of adequate drainage facilities is naturally an important feature of all rock excavation work. It will be provided in one of the standard methods available in civil engineering practice, but geological considerations may affect it to some degree. For example, in the construction of a new railway in Salvador and Guatemala (by International Railways of South America) many cuttings had to be made through a local conglomerate called *talapete*. The side slopes in all these cuttings had to be left as *steep* as possible in order to carry off quickly the torrential tropical rains that occur in this region, since their intensity would otherwise lead to erosion of the conglomerate.[9.3] Similarly with all types of shale, drainage must be so arranged as to remove surface water as quickly and surely as possible in order to minimize its action on the rock. Limestone may also be affected by excessive contact with water.

For underwater rock excavation two methods are in general use: underwater drilling and subsequent blasting with explosives, and alternatively the use of a floating rock breaker. The former method is in almost universal use for rock excavation in the dry. In both cases, but especially when a rock breaker is to be used, accurate knowledge of the bedding of the rock to be removed and of the nature of the rock will prove to be of great value not only in planning construction methods but also in estimating construction schedules and in specifying the work suitably in relation to possible overbreak. Many remarkable examples of underwater rock excavation testify to the difficulty of the work and its uncertainty.

One of the earliest as well as one of the most remarkable major examples of underwater rock work was the removal of Flood Rock, a rocky ledge in Long Island Sound, New York; 80,000 cu. yd. were removed in this operation in 1885; the work had to be carried on in later years, a modern channel being completed only in 1920. Another interesting piece of work was carried out in New York in 1922 when a subaqueous reef in the East River was removed to improve navigation. The work had to be carried out under the water of a tidal river with a swift current and to a final depth which was only 15 ft. above the roof of a busy subway tunnel. So accurately was the drilling and blasting of the crystalline Manhattan schist carried out, however, that the project was successfully completed without damage to the tunnel or interference with subway traffic.[9.4] A feature of special interest was the extensive use of field seismographs to check up on the extent of the vibrations caused by the blasting. This idea can often be of great service whenever blasting has to be carried out in built-up areas. The author knows of a large rock excavation job which had to be carried out in the vicinity of an important hotel. At the end of the job, the usual claim for damages caused by blasting was received, but it was not pressed when it was divulged that the contractor had rented a room in the hotel and there maintained a vibration record throughout blasting operations which showed that the vibrations were not so serious as those caused by heavy traffic.

The importance of the nature of the rock to be excavated in underwater work is well illustrated by experience obtained during rock-dredging operations in the harbor of St. Helier, Jersey, one of the Channel Islands. A rock breaker equipped with a 22,000-lb. ram was used for breaking the rock of the local sea bed which varied from a syenitic granite to a true diorite. In the harbor area, it was badly fissured; but outside the harbor entrance channel, a purer and sounder diorite was encountered; the contractors estimated that the latter material cost five times as much to break as the former, even though the two were found in such close proximity.[9.5] The type of rock may also dictate the excavation method to be adopted, as is shown by a recent dredging contract carried out at Sunderland, England. Over 100,000 cu. yd. of rock were removed, working to a maximum depth of 44 ft. without the use of blasting or a rock breaker. This was possible since all the rock was a loosely bedded limestone, the beds being from a few inches to 2 ft. in thickness, and it was broken up by the dipper of a large bucket dredger digging underneath the upper projecting stratum on the working face.[9.5]

Not only in subaqueous excavation work is the nature of the rock of importance but also in excavation carried out in the dry. The usual method of drilling and blasting may not always prove to be suitable when the exact condition of the rock to be moved is known, as instanced by the interesting work carried out in preparing the foundation bed for the Prettyboy Dam, in Maryland (described on page 320). A similar case occurred during the construction of the main spillway for the Fort Peck Dam in Montana, which is founded on local Bearpaw shale. This is a relatively soft rock, its excavation being possible without the aid of blasting. To excavate the 200,000 cu. yd. taken out, three different types of cutting machine were used—an auger which drilled holes 5 ft. in diameter; and two special adaptations of electrically operated coal-cutting machines, one of which cut vertical slots between holes, and the other for horizontal undercutting. Incidentally, the Bearpaw shale disintegrates when exposed to the atmosphere, for which reason a 4-ft. cover layer was left in place over the excavated area at the start of excavation, the finally finished

FIG. 9.4.—Coal-cutting saw making horizontal cut in shale, excavating for training wall foundation slab at spillway to Fort Peck Dam, Montana.
(Reproduced by permission of the Chief of Engineers, U. S. Army.)

surface being sprayed with a bituminous paint against which the concrete was immediately placed.[9.6]

Another example of interest is given by the excavation for the powerhouse substructures at the Wheeler Dam on the Tennessee River, U. S. A. Extensive preliminary exploration work disclosed that the rock at the site consists of an "alternating series of nearly horizontal layers of pure limestone and a cherty siliceous rock, the latter occurring in much thicker strata than the former." More than 500,000 cu. yd. of rock had to be removed to depths varying to 57 ft. below the existing stream bed, and it was deemed advisable to avoid any blasting close to the faces of the deep cuts in which the power units were to be placed, owing to the nature of the rock to be encountered. A system of close line drilling was therefore developed and carried out with wagon drills all round the site; the rock so enclosed was drilled and blasted following usual methods, except that no blasting was done within 30 ft. of the line-drilled faces, from which the rock broke away cleanly. The undisturbed face of rock was of great importance in the structural design of the powerhouse, and the elimination of all overbreak resulted in the method proving to be quite

FIG. 9.5.—Line-drilled excavation for turbine pits in powerhouse area of Wheeler Dam, Tennessee River, U.S.A.

(Reproduced by permission of the Tennessee Valley Authority, Mr. T. B. Parker, Chief Engineer.)

economical.[9.7] Many other examples could be given, but the purpose of all would be but to confirm the importance of geological features in rock excavation.

Ancient rock excavation work has not yet been mentioned, mainly because of its wide extent and of the familiarity of the average reader with at least some leading examples, such as the unusual excavations at Petra in the red sandstone which has made the beauty of this ancient city almost proverbial. Occasionally these works of the past are put to modern use, thus achieving an added interest. During the World War, 25 caves of various sizes were discovered in the Arras district of northern France. They were developed by New Zealand troops and proved to be the underground quarries from which tough unweathered chalk had been obtained in the seventeenth century for the rebuilding of the city of Arras. They were warm and comfortable and provided accommodation for 11,000 men. There were probably over a hundred of these caves in the area occupied by the British Empire troops, although all were not in use at the same time. They were ventilated by means of vertical or inclined 6-in. boreholes; to ensure their success, geological examinations were often made before these were drilled.[9.8]

9.5. Open Excavation of Unconsolidated Material.—Excavation in soft material may vary from work in good clean sharp gravel to digging in soft cohesive clay. Between the two extremes lie the many and varied substances classed generally as soils, earth, or some other such generic name, in nearly all of which cohesion is a property of some significance. It can be seen that in considering the mechanics of these materials, as distinct from solid rocks, conditions other than those considered in the ordinary strength-of-materials laboratory must be investigated. It is perhaps natural that although the ordinary physical properties of solid rocks have long since been investigated, by standard testing methods, it is only in comparatively recent years that *Soil Mechanics* (as this special study has come to be known) has developed as a separate branch of study. It is of such importance and it has advanced so rapidly in recent years that Chap. XX of this book has been devoted to the subject. The two main mechanical properties of soils (using this general term to indicate all unconsolidated material), internal friction and cohesion, are both of importance in open excavation work, the relative importance of friction varying inversely as the coefficient of cohesion for the material. Similarly, the relative significance of the presence of underground or surface drainage water varies with the coefficient of cohesion, but internal friction is affected by water to a much lesser degree.

From this general comment on the mechanical features of the stability of loose material, it will be seen that knowledge of the specific nature of the material is of the greatest importance in the determination of its probable behavior when excavation has proceeded. From a study of typical samples of the material, coupled with knowledge of its geological characteristics and of the possible presence of groundwater, a close estimate can be made of a practical angle of repose to adopt for the material and of the angle of internal friction (the two not necessarily being equal), side-slope determination being a major problem in open excavation work.

In preliminary considerations of open excavation work, the bottom grade level will be determined either by factors depending on other parts of the works involved or by general economic considerations. In both cases, the effect of the side slopes on the total quantity of material to be excavated, and therefore on the total cost, will be considerable. With flat side slopes, it is easy to see that in the case of ordinary railway cuttings for example, the volume of material to be excavated to form these side slopes, nonproductive excavation, as it might be called, can be larger than the volume of material to be moved from that part vertically above the level base of the cut, the material that must productively be moved. It is clear, therefore, that side slopes must be the steepest possible consistent with safety in order to keep to a minimum the volume of material to be excavated.

It is in the determination of such side slopes that geological considerations should play a conspicuous part. In the past, this has not always been done, side slopes used in design being based largely on the past experiences of those concerned with the work. On the basis of practical experience, tables of angles of repose for materials of various kinds have been built up and are in general use today. Although these tables undoubtedly have their use, unquestioning reliance on them is inadvisable. Since it is almost impossible to correlate the broad material classification in these tables with the detailed results of test borings, the procedure is without a logical basis and so is the more surprising in view of normal engineering method. It is true that the side slopes of many cuttings stand up quite successfully; it is also equally true that trouble has been encountered with the sides of many such cuttings and that large quantities of material have probably been unnecessarily removed from the side slopes of many cuttings. The unsound basis of this procedure is emphasized by the frequent use of round-figure tangential proportions (1:2 or 1:4) as a method of expressing side slopes, instead of the use of angles; a diagram can best demonstrate the weakness of this practice. The inevitable use of assumed angles of repose in preliminary designs may be admitted; but it is suggested that before the completion of any final designs, a full study should be made of the geological structure of the complete prism of material to be removed, not only to discover the geological arrangement of the material but also to investigate the effect of removal of the prism on adjoining geological structure.

Sands and gravels will usually be found to be unaffected by exposure to the atmosphere. Clays are different, and special attention must always be paid to the effect of exposure to the atmosphere on any clays that are to be uncovered during excavation. Moisture content of the clay is a critical factor in such determinations; it can profitably be studied in the laboratory if undisturbed samples are obtained for testing. It is in order to maintain unchanged the water content of a clay that the old construction "dodge" of covering up exposed steep faces of clay was evolved, a device that will often result in considerable saving in excavation.

9.6. Open Excavation in Water-bearing Strata.—Adequate surface-drainage arrangements are a first essential to success in all open excavation work. Consideration of the problems presented by the presence of groundwater must inevitably be made in conjunction with geological studies, if possible waste of money and time are to be avoided. Pumping will be the engineer's first thought in determining how to handle these problems. But the provision of the usual sump, for example, when excavating into water-bearing gravel, might be worse than useless. If preliminary investigations show that such water is moving, then a watertight cutoff must be provided before progress can be made. Interlocking steel sheet piling of a type with watertight joints, if it can be tightly driven into an impervious layer beneath, may provide a remedy. Alternatively, the job may be tackled with the aid of what is known as the *well-point system.* Various systems incorporating well points (a self-descriptive term) are available for this type of work, some involving the use of submersible pumps which are lowered into specially prepared sump holes, but the "Moretrench" system has certain special features, and as the author has personal knowledge of its use on some unusual jobs, it will be briefly described.

Basically, the system consists of a special pump and a number of well points for lowering the underground water table below the lowest excavation level, in this way predraining the material to be excavated and so converting it from the wet state, in which its behavior may be treacherous, to the dry, thus facilitating excavation progress and obviating water troubles. The lowering of the groundwater level is achieved by means of special well points with riser pipes about 2 in. in diameter. The well points are fitted with special jetting nozzles, and immediately above these is wrapped a triple layer of bronze screen mesh, the standard area being 350 sq. in. The well points are jetted into the ground, to a depth 5 ft. or more below bottom grade level and located close to the area to be excavated in such a way that they will not be disturbed when excavation is complete. They are spaced at certain distances apart so that the cones of depression around each well point intersect, resulting in the water level midway between points being lowered below any level at which it might cause trouble. All the points are connected with a header pipe leading to the pump which is set in operation when the complete system is ready for working. Successive cones of depression thereupon surround completely the area of excavation. If the pumps are properly installed and of sufficient capacity, they will prevent the rise of groundwater into the area, and excavation can proceed within the area in the dry. The system can be extended and modified in many ways, notably in the excavation of deep trenches, successive surfaces of depression being as indicated in Fig. 9.6. Bold and simple in conception, the system can naturally be fully effective only under

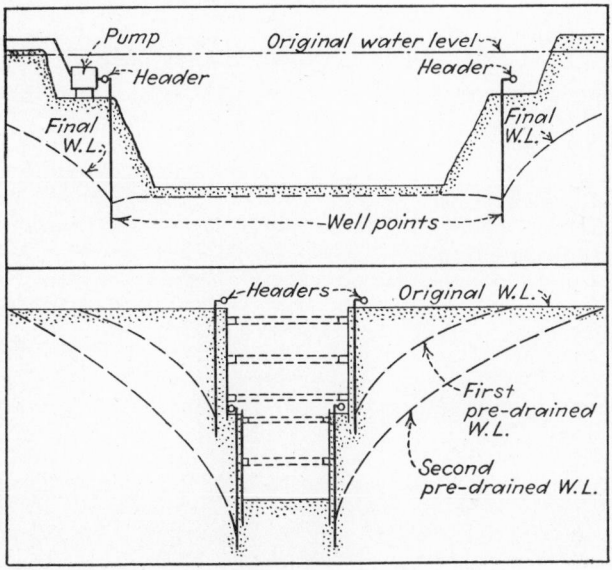

FIG. 9.6.—Diagrammatic sketches illustrating the use of well points for open and for trench excavation in water-bearing ground.

skilled direction and with a sure and certain knowledge of the geological conditions obtaining at the site, since the success of its operation depends on the travel of the groundwater in the strata encountered.

As an example may be cited a well-point installation on the construction of the Victoria Park Pumping Station, a part of the duplicate water supply system of the City of Toronto, Ontario, Canada. The buildings of this station cover an area of 87 by 297 ft.; and during their construction, excavation had to be taken to a depth about 35 ft. below the level of Lake Ontario the shore of which was only about 100 ft. away from one of the longer sides of the excavated area. Excavation proceeded through stiff clay for about 30 of the total depth of 46 ft.; small water boils then made their appearance. Auger borings were taken all over the site to check those originally taken, and a few thin water-bearing sand seams were encountered. A well-point system was therefore put in, well points being jetted down all around the area to be excavated to full depth. The excavation was then completed without water being encountered. It is of special interest to note that methane gas was "pumped out" by the Moretrench pump in addition to water, the gas being lit and burned; this gas was thought to have been the result of decaying organic matter entrapped with the sand deposits.[9.9] A similar installation had been made some years

FIG. 9.7.—Excavation proceeding on the site of the Victoria Park Pumping Station, Toronto, Ontario, Canada. Well points have been jetted into place along right-hand wall of steel piling, at the foot of which the header pipe can be seen. Steam shovel is working on what was previously waterlogged ground. (Dominion Construction Corp. Ltd., Contractors.)

(*Reproduced by permission of Mr. R. C. Harris, Commissioner of Works, City of Toronto.*)

before (1928) during the construction of the North Toronto Sewage Treatment Plant, an area of 170 by 250 ft. and consisting of swamp mud, clay, sand, and gravel, having its groundwater table lowered by 13 ft. during the progress of excavation. Other cases are on record in which the facility of working in the dry made possible by the use of the well-point system has enabled modifications of foundation design to be made in the direction of simplicity. The large Holland Plaza building in New York City was designed to be supported by a concrete pile foundation placed in a clayey quicksand. A well-point system dried up the excavation, and a heavy reinforced concrete mat foundation was used instead.[9.10]

At first sight, it would appear that the system would operate only in porous materials such as sands and gravels. It has been found, however, that in many types of clay it will give equally effective results. The quantity of water pumped out of clay is naturally very much smaller than in the case of the more porous material, and careful study has suggested that in such cases the suction caused by the pumping action causes a slight vacuum at the base of each point, this leading to an unbalanced pressure at the nearest exposed face, atmospheric

FIG. 9.8.—Excavation in swamp mud, predrained by a well-point installation (seen on right-hand side), for installation of 84-in. precast concrete pipe for outfall sewer from Jamaica Sewage Treatment Works, New York.
(Reproduced by courtesy of the Moretrench Corporation, New York.)

pressure outside preventing water from leaving the clay and, indeed, tending to hold up the material at a slope greater than its normal angle of repose. Figure 9.8 is a striking example of this type of work, the ground in which the reinforced concrete sewer is being laid consisting of 5 or 6 ft. of swamp mud overlying a mixture of fine sand mixed with mud. The ease of working in the dried material is obvious. This illustration is a useful reminder of the fact that the water content of soils alone is the cause of many of those peculiar physical properties described by such names as *gumbo* (clay), *mud* (often silt), and *quicksand.*

An alternative type of installation, extensively developed in Europe, consists of the use of submersible pumps lowered into specially prepared drainage wells. A water-lowering plant of this type was used during the construction of the Albert Canal in Belgium and also in the building of the King George V Graving Dock in Southampton, England. In this latter installation, ten 14-in. pipes gravel packed in 24-in. holes formed the drainage wells for dealing with the artesian water that was found by test borings to exist under part of the area now occupied by this great dock. Permanent groundwater relief pipes were installed in the completed structure.[9.11]

This most interesting aid to excavation is not the only method available for dealing with groundwater. Freezing as a construction tool was mentioned in Chap. VIII; when subsurface conditions are suitable, it also can be used in open excavation work. Suitable conditions for such an application must be definitely known before the project can be considered; they consist of regular strata of water-bearing material, water being present in such quantity that difficulty will be experienced in opening up the excavation in any other way. Somewhat naturally, the freezing method is especially applicable for deep excavation, where pumping would be difficult. Once the conditions are determined, suitably sized pipes are sunk (generally by jetting) all around the area and connected up to a refrigerating system which, when operating, gradually freezes solid the ring of material around the area of excavation, this ring acting as a natural temporary shell for the excavation work.

An interesting example was the excavation of large shafts used during the construction of a vehicular tunnel under the River Scheldt at Antwerp, Belgium, in 1931–1933. These shafts were 70 ft. in diameter and 87 ft. deep; and because of the construction method adopted, they were entirely free from bracing. They penetrated water-bearing sand with an interbedded turf layer. For the freezing process, 116 holes were bored, one set on a circle of 86-ft. diameter, and one set on a 78-ft. circle, the spacing between holes being 4½ ft., all being sunk to a clay stratum 90 ft. below ground level. A 6-in. pipe was sunk and sealed in each hole, and a 2-inch pipe open at the bottom was inserted in each of the larger pipes, thus forming the necessary circulating system for the brine solution which was used to lower the temperature of ground and water. It took four months to complete each of the requisite cylinders of ice and frozen ground, which were then maintained until the shaft construction was complete.[9.12] The method is a most interesting one but dependent for its success on the porous nature of the strata to be penetrated.

There is another method which has been successfully used in the excavation of exceptionally troublesome water-bearing strata, this being the chemical solidification of loose water-bearing ground to a degree where it becomes self-supporting. One proprietory method was discovered by Dr. H. Joosten in Germany, during work in potassium and salt mines where the presence of large accumulations of

FIG. 9.9.—Chemically consolidated sand and gravel, overlying clay, in excavation for an escalator tunnel at Monument Station, District Railway, London.

(Reproduced by permission of Messrs. Mott, Hay and Anderson, Consulting Engineers, London, through the courtesy of Sir Henry Japp, Works Director to the Contractors, Messrs. John Mowlem and Co., Ltd., London.)

brine was a constant source of trouble. By impregnating the ground with a solution of sodium silicate, it was found that both liquid and ground were converted into a solid mass; the method was therefore patented in 1925 and 1926 for general work. It is applicable only to loose porous material; it is analogous to cement grouting but is instantaneous in action and has the great advantage that it can be operated even in the presence of running water underground. Injection pipes are driven into the material to be consolidated, and a solution of sodium silicate is forced through the special point of each pipe at intervals, as it is driven in to its final penetration. As the pipe is withdrawn, a measured quantity of a calcium chloride solution is forced in, again at regular intervals; the reaction of the two solutions in the open pores of the material results in its consolidation, or *petrification*, as it is sometimes called. Sandy soils containing about 20 per cent of quartz are especially suitable for the application of the process; soils with a large clay content or an appreciable free lime content are unsuitable. The first application of the method in England was in April, 1932, when a trial length of a sewer diversion being carried out in connection with an extension of a London tube station was consolidated in this way. The surface of the London clay was about 2½ ft. below the crown of the sewer tunnel which was located in loose gravel. An arch of chemically consolidated gravel 2 ft. thick was therefore formed over the crown of the tunnel; this was later found to be so hard that small air hammers were required for its excavation.[9.13] Subsequently, the method was applied to other underground work in England and other countries. A particularly interesting application of a similar process was during the construction of the first section of the Moscow subway tunnel in the U. S. S. R.[9.14]

To describe these excavation methods has necessitated some deviation into the realm of civil engineering construction practice, but this was inevitable in order to present the methods adequately. What it is desired to emphasize is that all the methods described, and indeed any others that may be evolved for similar special work, are largely dependent on subsurface conditions, which must therefore be investigated in as careful and thorough a manner as is possible before work begins and which must be closely observed and followed as excavation work proceeds.

9.7. Embankment Fill and Dredging.—The construction of embankments produces problems in many respects similar to those which arise in excavation, notably the determination of a safe angle of slope to use in design for the sides of the embankment. Although this will be clearly shown up practically as filling proceeds, it is often necessary to have an accurate idea of the slope that must be used before construction begins, especially if the ground area to be covered by the fill is restricted, so that preliminary investigation and experimentation are often necessary.

The correct placing and consolidation of fill material in ordinary embankments, exclusive of water-retaining earthern dams, are matters of standard construction practice. Associated with them are the problems of settlement and of bulking—the relation of the volume of fill material after deposition to the volume of the same material before excavation. Knowledge of this ratio is essential in accurate design work; it is, naturally, dependent on the type of material being handled. Bulking will depend principally on the undisturbed state of the material to be moved, on any change in the moisture content of the material between excavation and deposition, and to some extent on the methods used to consolidate the material after it has been placed in final position. The second two factors are of special moment in the case of sand and sandy materials; the first, in the case of clays and materials with a large clay content. When rock is used as fill material and has to be obtained by quarrying, the bulking may be exceptionally high.

Settlement of an embankment, the first of the two detailed problems mentioned, may be the joint result of shrinkage of the fill material and actual settlement of the strata supporting the fill. The former figure will be related to bulking, but its accurate determination is difficult. Ingenious methods, including the use of vibrating rammers, have been introduced in recent years, notably in Germany, with the object of expediting the shrinkage of fills in order to permit their use without the usual long delay allowed for the fill to take up its final position.

The troubles that develop during the placing of fill material—and they have been frequent throughout all civil engineering practice—may be said to be due in general to one or both of two main causes. Construction methods may not have been properly adapted to the material being handled, and the strata on which the fill is placed may fail under the load to which it is subjected. The first of these causes is one that can readily be guarded against if due regard is had for the nature of the material being placed. Thus, except in the case of clean dry granular material, side tipping of an embankment may lead to trouble as a result of the fill binding together and taking up temporarily a slope greater than its angle of repose. On some works, side tipping is not allowed.

Failure of foundation beds is a rather different matter. If the weight of material contained in even an average-sized embankment is calculated and reduced to the form of a load per unit area of original ground surface, the result will be surprisingly high. Although decreasing toward the sides of the fill, this loading is directly comparable to that induced by concentrated structures, and the stress distribution in the foundation strata should be investigated accordingly. Progressive settlement will inevitably take place, unless the fill stands on solid rock, while the foundation beds take up their new loading. If the strata are not uniformly strong, failure of one bed may occur, and the fill may collapse. The investigation of fill stability thus becomes a part of the study of Soil Mechanics, both the nature of the fill and foundation materials and the mechanics of the combined structure having considerable bearing upon the success of the fill construction. Stability investigations can be made only when accurate data as to the local strata are available, and so the need for preliminary exploratory work is again made clear. So many fill failures, great or small, will be known to most engineers that space will not be taken to mention one here, instances such as those in which a new fill completely disappears below the original ground telling their own story to the engineer whose appreciation of Geology reminds him constantly of what may lie below a seemingly solid surface.

Filling of large areas, as distinct from relatively small embankments, is often economically carried out by pumping the fill material mixed with water. Settlement of the pumped material is usually the main problem, other than those of construction method; this is also studied as a part of Soil Mechanics. The colloidal material present in clays sometimes causes difficulty, but this can be foreseen if adequate preliminary tests are carried out. Dredging operations naturally necessitate a thorough knowledge of the material to be dredged, so that suitable plant may be made available; adequate preliminary exploratory work is again the only way in which this can be told.

Records of dredging practice contain descriptions of many interesting projects; only one will be mentioned, since it has a somewhat unusual geological interest. In order to prevent a repetition of the disastrous floods of 1926 and 1928, a combined flood protection and navigation project for waterways in Florida was initiated in 1930. Included in the project was the construction of 66 miles of levees around Lake Okeechobee. The material of which the levees were to be built was to be obtained from the bed of the lake. Much of Florida is geologically new country, the bed of the lake being still in process of formation. Bed materials vary widely, uniform deposits being exceptional; but under the recent deposits are strata of marl and limestone of varying thickness. The recent deposits often contain considerable quantities of sea shells, sand, and other material so finely ground as to be almost colloidal. From this brief description, it will be seen that the material was most unfavorable to dredging operations. Almost all the contractors who started on the work elected to use draglines for excavation, but later experiments were tried with a powerful hydraulic dredge and proved so successful that other dredgers were later brought on to the work. Special cutters had to be used at the end of the suction pipes, but good progress was made at low cost, and the resulting levees were more satisfactory than those made with the draglines. The marl and limestone layers had to be broken initially by drilling and blasting, but the dredging equipment handled all material thereafter, the relatively new deposits proving their geological youth by the way in which they could be disintegrated.[9.15]

9.8. Quarrying in Civil Engineering Practice.—Building stone used in civil engineering will usually be obtained from an established quarry, and it is not often that large-scale quarrying operations will be necessary in the course of normal civil engineering practice. The two most common types of work requiring special supplies of quarried rock are the construction of rock-fill dams and the use of rock for the construction of embankments or special structures such as mound breakwaters. If the quantity of rock required for projects of this or any other kind is appreciable, the civil engineer will be well advised to obtain the services of a quarry expert, as quarrying is an art, the best practice of which can be developed only after long experience. On some projects, however, civil engineers may often be called upon to open up small quarries that do not warrant the employment of an expert and an appreciation of Geology can then be of good service.

After the site of the quarry has been stripped, the exposed rock surfaces should be studied carefully, special attention being paid to the dip and strike of the strata, the presence of any unusual features such as folds, and particularly to the jointing of the rock. On the strength of the data so obtained, it will be possible to open up a working face in the most advisable way, advantage being taken of the direction of dip, strike, and jointing to facilitate both the blasting and the removal of rock, jointing being of particular importance. Methods of operation will depend on the amount of rock to be moved, the rate at which it is required, and the type of rock to be quarried. If a large output is not required, it may be desirable to utilize for drilling the regular job hand drills and jackhammers, in which case a series of low faces and narrow benches will probably prove most suitable. If the job is a large one, however, consideration should be given to the possibility of using larger drilling equipment, high faces, and deep snake holes which can be sprung before being finally shot, this leading to economy in operation.

The location and depth of drill holes and the necessary amount of explosive to be used are matters that will have to be determined accurately for every new quarry face opened up. The problems involved are not dissimilar from those involved in the normal excavation of rock in grading work; and as the United States Bureau of Public Roads has carried out field studies in connection with highway grading, some of the conclusions reached may usefully be cited.[9.16] A study of 71 different excavation projects showed over half to be featured by poor blasting, only 15 being classed as good in this respect. The consequent *average* increase in operating times for steam shovels was 42 per cent; and as the volume of rock handled was less than the dipper capacity, the average over-all efficiency was only 50 per cent. In Mr. Anderson's paper will be found suggestions for improvement of output, mainly related to drilling and blasting. The accompanying tables (I and J), which are reproduced by permission, summarize some of this

TABLE I.—AVERAGE AMOUNT OF EXPLOSIVES USED PER CUBIC YARD OF PAY MATERIAL, FOR VARIOUS ROCK TYPES[1]

Job No.	Explosive, pounds per cubic yard				Condition of material for moving with power shovel	Kind of material	Aver. approx. depth of holes, feet	Aver. approx. spacing of holes, feet
	Dynamite or gelatin		Black powder	Total				
1	0.04	....	0.85	0.89	Poor	Soft and seamy granite	30	25 by 18
2	0.55	....	0.98	1.53	Poor	Hard basalt	18	19 by 19
3	0.27	0.44	....	0.71	Good	Massive granite	14	10 by 8
4	0.01	....	0.55	0.56	Fair	Conglomerate	15	Variable
5	0.25	0.85	....	1.10	Poor	Stratified limestone	16	15 by 12
6	0.06	....	1.06	1.12	Good	Shale	13	12 by 12
7	0.02	0.03	0.45	0.50	Good	Disintegrated granite	10	Variable
8	0.15	0.16	0.67	0.98	Fair	Stratified limest. and shale	12	10 by 10
9	0.02	0.17	0.50	0.69	Good	Seamy basalt	21	15 by 15
10	[2]	1.90	....	1.90	Very poor	Hard granite	5	3 by 3
11	0.25	0.75	....	1.00	Poor	Very hard granite	14	8 by 8
12	0.55[3]	1.00	....	1.55	Good	Very hard felsite	13	7 by 7
13	0.14	0.59	....	0.73	Fair	Hard schist	12	9 by 9
14	0.10	0.50	....	0.60	Good	Foliated granite	10	10 by 10
15	0.18	0.75	....	0.93	Fair	Hard granite	16	8 by 8
16	0.15	0.87	....	1.02	Good	Very hard granite	16	8 by 8
17	0.10	0.47	....	0.57	Good	Soft granite	18	10 by 10

[1] Reproduced by permission of the U. S. Bureau of Public Roads, from *Public Roads*, February, 1932.[9.16]

[2] Near close of studies springing was begun, using approximately 0.30 lb. of 60 per cent dynamite per cubic yard for springing and 1.60 lb. for blasting.

[3] Part of this was used in burning the deeper holes to keep drills from sticking.

helpful information; they offer at least a guide to rock excavation methods in newly opened-up areas. It is particularly to be noted that as rock drilling is an expensive operation, 1 ft. of hole being about equivalent to 1 lb. of dynamite (in North America), economic practice will tend toward keeping drilling to a minimum.

TABLE J.—SUGGESTIONS FOR THE SPACING OF DRILL HOLES AND AMOUNTS OF EXPLOSIVE TO BE USED FOR THE FIRST TRIAL SHOTS IN EXCAVATING UNFAMILIAR MATERIAL WHEN HARD SHOOTING IS PERMISSIBLE[1]

Type of material	Spacing of drill holes as fraction of depth		Depth of hole below grade, as fraction of spacing along rows	Pounds of explosive per cubic yard of burden excluding springing	Use an explosive equivalent to
	Along rows	Between rows			
Rocks exceptionally difficult to shatter	½	½	½	1⅛	60 per cent straight dynamite
Most hard, dense, unweathered (ledge) rocks	⅔	⅔	⅓	1	50 per cent straight dynamite
All medium-hard, weathered or partly disintegrated rocks, those which shatter readily, and very hard shales	¾	⅔	⅓	1 to 1½	40 per cent straight dynamite or blasting powder if holes are dry
Ordinary shale, cemented clay and gravel, and similar material	1¼	1	¼	⅞ to 1	Blasting powder if holes are dry; otherwise 20 per cent straight dynamite

[1] Reproduced by permission of the U. S. Bureau of Public Roads, from *Public Roads*, February, 1932.[9.16]

9.9. Earth Pressure and Retaining Walls.—Coulomb was the first to formulate the laws of friction and cohesion which govern the pressures on retaining walls; his name has since then always been associated with earth pressure calculations. Rankine suggested his well-known formula at a later date, but it was specifically restricted to cohesionless material. Despite this, and possibly because of its simplicity, Rankine's formula has since been widely used in civil engineering, the angle of repose often being used in the place of the angle of friction. The effect of this formula was to give an earth-pressure diagram equivalent to a fluid-pressure diagram, a condition that experiment has now shown to be incorrect. Much research work has been performed in investigating earth pressure, and much speculation demonstrated, but the publication of Professor Jenkin's paper in 1932[9.17] and the researches of Dr. Terzaghi[9,18] seem at last to have cleared up definitely the problem of the actual pressures obtained from clean sand, whereas work that proceeds at present on clays may finally clarify even this most involved aspect of the same question.

With the mathematical computations necessary in determining earth pressures this book is not concerned; but as all the calculation methods involve an assumption as to the physical properties of the material to be retained by the wall, some comment on the geological aspect is called for. Adequate preliminary investigations of the natural strata are again a prime essential if the wall is to retain material to be left undisturbed, and of the fill material if the wall is to retain newly deposited fill. Samples that are typical should be secured and tested for their coefficients of cohesion and internal friction. The average figures so obtained may then be used in the respective earth-pressure formula to be adopted, not indiscriminately but taking into consideration local circumstances, the type of wall, and the likelihood of normal conditions being modified in the future. Geological conditions will serve as a useful guide to the probability and possibility of groundwater accumulating behind the wall and thus to the possible necessity for providing drainage behind and weep holes through the wall.

The procedure described is logical; and yet how often is it followed? In the case of small walls, it is understandable that frequently no money can be spent on extensive preliminary work, an omission that may not be serious in view of the inevitable factor of safety included in such designs. But in the case of large retaining walls, neglect of preliminary investigations may be serious. The fact that the bending moment induced in retaining walls varies approximately as the cube of their height serves to illustrate the validity of this suggestion. This fact is one not always realized by some engineers who might be surprised to find, for example, that a wall retaining 40 ft. of fill has to be twice as strong in resisting bending as a wall retaining 32 ft. of the same fill. For walls of large dimensions, therefore, it is imperative that adequate data as to the material that they are to retain should be available if truly economical designs are to be prepared.

Recent engineering practice contains at least one record of how the design of a large retaining wall was assisted by careful preliminary work. The Fifteen Mile Falls Dam of the New England Power Association, constructed in the years 1928–1931 on the Connecticut River in New Hampshire, has already been mentioned in connection with geophysical prospecting. The project has further special interest for civil engineers in that the dam structure, consisting partially of a concrete gravity structure and partially of an earth embankment, includes one of the largest retaining walls ever constructed. This is located at the junction of the two main parts of the dam and is therefore at right angles to the face line, as its function is to retain the end part of the earth fill. It is about 900 ft. long and reaches a maximum height of 170 ft.; the material that it had to retain was classified as *glacial till.* In view of the size of the wall, it was felt desirable to make some experimental check on ordinary theories of design; and so at the suggestion of the New Hampshire Public Service Commission, an investigation was carried out at the Massachusetts Institute of Technology. A special laboratory building was constructed to house a test bin 14 ft. square and 7 ft. deep at a total cost of about $50,000 (£10,000). The work done was directed by Dr. Terzaghi and proved to be a research undertaking of a monumental nature. The results obtained cannot adequately be summarized, but the final record made available to the engineering profession is worthy of special study by all interested in retaining wall design.[9.18]

Finally, the increasing use of interlocking steel sheet piling of various types for the construction of retaining walls is introducing special problems into retaining wall design and construction. The severe vibration set up by the driving of the piles, although relatively unimportant in sand and gravel, may produce strange results in clay, especially if the driving is under water. The author has encountered a small pier construction in which steel piles were driven into preconsolidated clay which naturally showed up in the preliminary wash borings as "very hard." When disturbed by the pile driving and consequent churning action, it changed character completely, becoming almost fluid in the neighborhood of the piles, and gave a greatly reduced passive pressure as resistance to movement. A large dredge was engaged at the same time as the pile-driving operations in excavating a channel in the same clay. When some distance from the pier, the steel cutter head of the dredge had to be used to break up the clay sufficiently for it to be pumped; but starting about 400 ft. away from the site of the pile driving, the effect of the vibrations was noticed in the progressively softer nature of the clay as it was excavated, the steel cutter finally being stopped and the clay removed by direct pumping.

The presence of boulders can be a serious impediment to the driving of steel piling; in the interpretation of the results of the preliminary test boring which is *always* necessary in this branch of work, an appreciation of the local geology can be of great assistance—knowledge as to the glacial nature of strata encountered, for example, suggesting the possible presence of boulders even if the test borings do not show any. The possibility of scouring taking place in front of a steel pile wall, the passive ground resistance at the toe of the wall then being reduced, can be the better estimated if attention is given to local geological characteristics. Furthermore, if the steel piling is to be driven through the full depth of unconsolidated material at a particular site, toehold being desired in the rock below, a study of the local geology will sometimes be the only means whereby a reasonable estimate can be made as to the nature of the rock surface and whether or not it will be weathered to such an extent that the piles can penetrate it at least for a little way. In this branch of work, therefore, as in others, not only will an appreciation of the significance of geological features be of service to the civil engineer, but in many instances a direct application of geological data will assist in both the design and the construction of his works.

CHAPTER X

EARTH MOVEMENT AND LANDSLIDES

> Nature has been compared with the old—and bad—type of expert witness. What we succeed in learning from her depends on the skill of our cross-examination. She will carefully conceal what we are most anxious to find out but will offer us instead a great deal of apparently irrelevant information.[10.1]

Excavation is often associated in the minds of engineers with some form of earth movement—landslides or rock falls—occurrences that so often interfere with the orderly progress of excavation work. The fact that landslides do occur is frequently accepted as a part of the natural order, certainly in some localities and with some classes of materials. In keeping with this attitude of mind, it is all too often the case that such earth movements come to be seriously studied only after they have occurred when damage has been done and when initial movement has interfered, perhaps seriously, with previously existing geological structure. Reference to landslides in literature, both classical and technical, are in keeping with this general position—even that passing reference in the works of Mark Twain to the farm in Nevada that slipped over on to the top of another farm, thus raising difficult problems of ownership! Few references, even indirect, are to be found in early writing, and so study of the subject does not lead to the interesting historical record that serves as a useful background to consideration of other branches of civil engineering work.

The general subject of earth movement, and of landslides in particular, is patently one of great importance, not alone because of the trouble and expense caused by unexpected earth movements during civil engineering construction but also because of possible loss of life through such movements of completed works or even untouched natural ground—catastrophes that civil engineering can sometimes avert. Any newspaper file will testify adequately to the validity of this statement, as hardly a month passes without a reference in the daily press to loss of life and property from this cause somewhere in the world. Fortunately, major catastrophes have not been frequent in recent years, but historical records show something of what has happened in this direction in the past by references such as that to the fall of Mount Grenier in A.D. 1248, which buried five parishes, and the slide at Rossberg, Switzerland, which took the lives of 800 people.

In its major aspects, the subject of earth movement is of paramount importance to geologists and geophysicists; earthquakes, earth tremors, and land subsidence are being studied and analyzed as never before. The civil engineer is more usually concerned with more detailed aspects of the subject and, in similar fashion, is pursuing with unusual vigor his research into the causes of landslides and the problems related to stability of earth slopes. His works, however, are affected to some degree by major earth movements in many parts of the world. The attempt will therefore be made in this chapter to review the whole subject briefly, in so far as it concerns the civil engineer and with special reference to the significance of local geological detail.

10.2. Stability of the Earth's Crust.—The basic factor in all considerations of earth movement is that the crust of the earth is composed of ordinary solid materials which react to the stresses induced in them generally in a similar manner to structural materials such as may be tested in a laboratory. This fundamental is one that will be found stressed elsewhere in this book; the risk of undue repetition is taken, since it appears to be a matter that is not always appreciated. If this essentially structural viewpoint is retained in considering the notes that follow, earth movement will be the better understood. Almost of equal importance is the fact that the constituents of the crust are under the influence of gravity. This statement is important. It is a reminder that the principal cause of all minor earth movements such as landslides and rock falls is the action of gravitational attraction functioning in the usual way. A mass of rock, for example, detached in some way from the bedrock of which it has been a part will not be held in position by unusual means if it is in unstable mechanical equilibrium. It may be expected to fall in just the same way as will any other solid body in similar circumstances. Similarly, if the underside of a loose mass of rock is inclined to the horizontal; and if that surface is well lubricated, then the mass will tend to move downward in just the same way as would any other similarly placed body move down an inclined plane.

Minor earth movements are, then, no matter of mystery but the inevitable result of instability of part of the earth's crust, and to a large extent they are subject to the ordinary laws of mechanics. This instability cannot always be foreseen prior to movement occurring; but gradually, by an extension of experience and by continued study of records of landslides and similar phenomena, general knowledge is increasing, with consequent general benefit. With instability admitted as the basic cause of movement, it will be asked what finally starts such movements. Construction operations are an obvious possibility; the great slides on the Panama Canal (see page 264) were the result of construction operations in connection with the excavation of the deep Gaillard Cut. Drains may become blocked, thus raising the level of groundwater tables and possibly affecting the moisture contents of unconsolidated materials which thus become unstable. Vegetation, acting as a binder of loose surface material, may be removed by either burning or stripping, thus loosening the material in which it grew. Unusual vibrations may be set up, as by blasting or gunfire—in these and many other ways may serious earth movement be initiated. Even earthquake movements can give rise to subsidiary minor earth disturbance, as in the case of the great rockfill dam 1,000 ft. high which was formed on the River Indus, India, in this way.

The cause of major earth movements such as earthquakes was referred to briefly in Part I of this book, as a part of the geological cycle. Reference may therefore be made to the brief note on page 27 and to books of reference listed at the end of this chapter for a discussion of this interesting geological subject. It is generally accepted that the cause of major earthquakes is a sudden yielding of a part of the earth's crust to strains set up in it due to a lack of balance or equilibrium in an adjacent area. The civil engineer is not primarily interested in the detailed study of such causes, being mainly concerned in designing his works, in regions subject to earthquakes, so that they shall be safe. The matter is mentioned, however, since the stresses thus set up in the crust of the earth by natural forces can be duplicated on a small scale by the operations of the civil engineer. The construction of a great dam, for example, subjects the local part of the crust to unusual stresses; the design of dam foundations has to take this factor into account, necessitating unusual care in preliminary geological investigations, as is made clear in Chap. XIII. Within comparatively recent years, there has occurred a most unusual demonstration of what these stresses can do to a local rock formation, relatively weak; and as no better example could probably be chosen to illustrate the suggestions made at the start of this section (that the materials in the earth's crust will react to stresses set up in them), it will be briefly described.

The Arapuni water power scheme is located on the North Island of New Zealand, about 120 miles south of the city of Auckland which it supplies with power; it was constructed between 1925 and 1932 under the direction of, and in the later portions by, the Public Works Department, all public power supply in New Zealand being under government control. The scheme consists essentially of a main dam, of the curved gravity type, 192 ft. high with a crest length of 305 ft. on a 250-ft. radius, which diverts the water of the River Waikato into an open headrace canal ¾ mile long, finishing with a spillway dam and penstock intake. Steel penstocks, built into rock tunnels, lead to the powerhouse the ultimate capacity of which is 200,000 hp. operating under a head of 175 ft. The geology of the Waikato valley is complicated, since the valley has been the scene of repeated volcanic

FIG. 10.1.—General view, from the air, of the Arapuni hydroelectric scheme, New Zealand, showing location of the crack the occurrence of which is described in the text.

(Reproduced through the courtesy of Mr. F. W. Furkert, late Engineer-in-Chief, P.W.D., New Zealand.)

activity as a result of which the course of the river and its gradient have varied widely and often. Old river courses have been filled with erupted matter, and consequent denudation has not always followed the old courses, with the result that several old river channels now occur at varying heights above the existing river bed in the vicinity of the Arapuni works, one of which was utilized as the main part of the headrace canal. The main rocks encountered were volcanic tuff and breccia. The vitric tuff is supposed to have been ejected from vents in the ground as incandescent dust and then to have fused together as it cooled except at the bottom of river valleys where the probable presence of water caused quicker cooling, resulting in a rock of the same chemical and mineralogical composition as the solid tuff but completely unconsolidated. The solid tuff may contain up to 30 per cent of water which may be held in ultramicroscopic pore spaces, as the rock is durable and moderately hard. Below this tuff occur varying beds of tuff and breccia, including a pumaceous

FIG. 10.2.—Arapuni hydroelectric scheme, New Zealand; view of the crack which developed in local rocks adjacent to the main intake structure.
(*Reproduced through the courtesy of Mr. F. W. Furkert, late Engineer-in-Chief, P.W.D., New Zealand.*)

breccia on which the main dam is founded. Below this last stratum, softer tuffaceous material was encountered extending to great depths. These volcanic deposits were uneven; they also proved to be so elastic "that the absence or presence of 10 feet of water in the gorge as the diversion tunnel was opened or shut, caused decided and opposite tilts to be registered on the seismograph in the powerhouse." On June 7, 1930, while the dam was retaining water and the powerhouse under load, a crack occurred in the local country roughly parallel to the flow of the river. It was widest where the spillway joined the

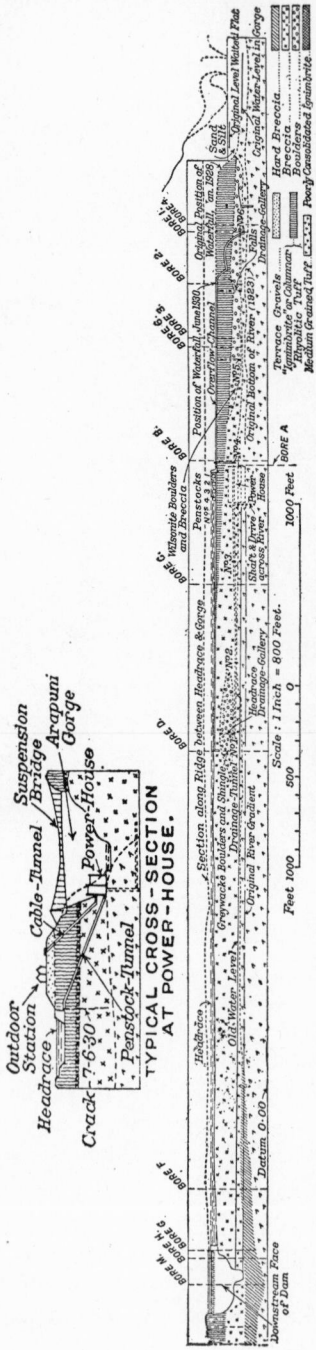

FIG. 10.3.—Arapuni hydroelectric power scheme, New Zealand; geological sections along headrace and overflow channel, and through power-house.

(*Reproduced by permission of Mr. F. W. Furkert, and the Council of the Institution of Civil Engineers, London, from reference 10.2.*)

penstock intakes, where it measured 2 in. across; it extended for about 2,000 ft. Observations then made showed that "the whole mass of country, about 2,000 feet long, 150 feet thick, and 400 to 800 feet wide, was bent over towards the gorge. . . . " The power station was shut down, and the lake drained; "as the lake fell, the country recovered its position, the cracks closing, except where jammed by drawn-in debris, and the powerhouse, suspension bridge, etc., regaining their original positions."

This most unusual occurrence was naturally investigated and studied closely by expert geologists and engineers. The consensus of opinion was that the movement was due to leakage of water from the headrace canal, affecting the adjacent volcanic rock to such an extent that movement took place. The remedial measures undertaken, therefore, centered on the provision of a waterproof lining to the headrace. After this was installed, apart from some minor troubles, the works functioned satisfactorily, and they do so today. In carrying out the remedial works, many unusual features developed. Thus, while the water level behind the dam was being lowered, gas consisting of 96 per cent nitrogen was found to escape from the rock in the headrace. The gas was unusual in that it contained no oxygen, and its volume may have reached the surprising figure of 50,000 cu. ft. (instead of the 25,000 cu. ft. mentioned in the published description of the work, according to information received privately from Mr. F. W. Furkert). Much grouting was carried out; and in drilling the necessary holes, water was frequently lost, a fire hose being turned into one hole, finishing at a depth well below river water level, without any trace of escaping water being found in the vicinity. In other holes, great difficulty was experienced due to the drills encountering the roots of vegetation which had been growing on ground formed of early volcanic deposits when covered by later volcanic material. Timber was secured from holes and appeared to be quite sound; "and though in the opinion of geologists it had been there for 10,000 years, the bark was still intact."

This quotation and those preceding are from the paper prepared by Mr. Furkert, who was Engineer in Chief throughout the whole course of the work, describing this trouble that developed at Arapuni and the remedial measures undertaken. It is a fascinating engineering document, as will be clear from even this brief summary, and it is an inspiring record of the successful surmounting of seemingly insuperable difficulties. The evidence thus presented by the rock movement at Arapuni shows clearly how the works of the civil engineer do affect the earth's crust. Although in most other cases the stresses set up in natural formations do not cause actual rupture of bedrock, the experience at Arapuni is a timely reminder of the strains that inevitably accompany such stresses, in rock as in other solid materials.[10.2]

10.3. Earthquakes.—Turning now to major earth movements due to natural causes, those described as earthquakes are the most common and the most serious. The earthquake that devastated Tokyo, Japan, on September 1, 1923, will be within the memory of most readers; it destroyed 140,000 lives and did damage estimated at three thousand million dollars. The earth movement that causes the quaking may be either volcanic or tectonic, the latter type being the more common. In many instances, the permanent deformation of the earth's crust can be observed. Thus the San Francisco earthquake of April 18, 1906, caused movement along 270 miles of the local San Andreas rift, a great fracture zone which runs almost parallel to the Californian coast line of western America. A maximum horizontal movement of 21 ft. was afterward found, the illustration on page 328 showing how this affected two dams that crossed the rift line. No vertical movement took place; but in the Yakutat Bay (Alaska) earthquake, of 1899, a vertical movement of 50 ft. took place over a large area, completely altering the local topography and creating new waterfalls on local watercourses.[10.3]

Earth movements of this type and magnitude are generally restricted to certain parts of the world that have come to be known

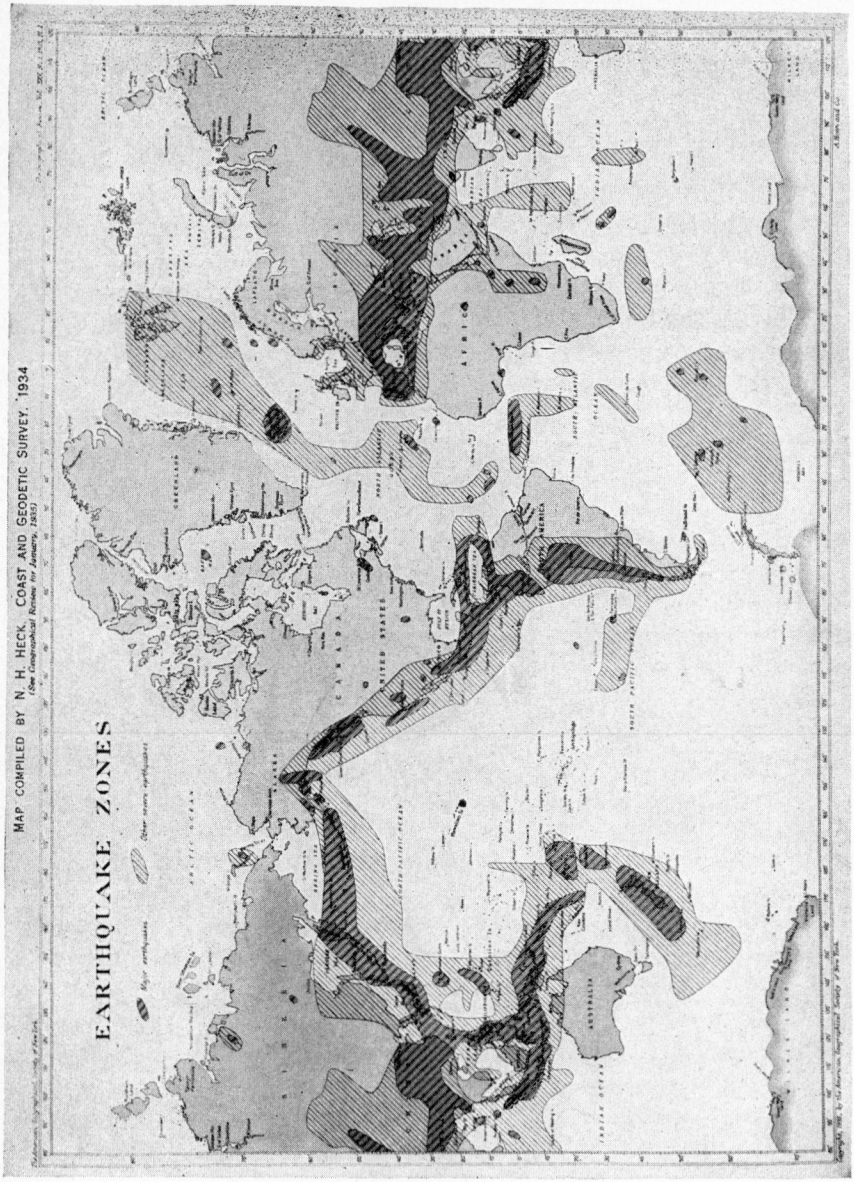

FIG. 10.4.—Earthquake zones of the world.

(*Reproduced from the map by N. H. Heck in the Geographical Review*, vol. 25, No. 1, 1935, *by permission of Captain Heck and the American Geographical Society of New York.*)

as *seismic areas.* These are indicated in a general way on Fig. 10.4. Seismic areas coincide generally with areas in which active volcanos are to be found, but both phenomena may be due to the fact that the areas in question are usually close to the borders of continental land masses and so may be zones of weakness in the earth's crust. Earthquakes are not restricted to the main seismic areas, earth tremors probably occurring in most parts of the world. Even in such an unsuspected area as eastern Canada and the bordering New England states, an earthquake occurred as recently as February 28, 1925; it was remarkable for its low intensity over a large area, but it caused much local consternation and did some damage to structures in the vicinity of Quebec and along the banks of the St. Lawrence River. Reinforced concrete grain elevators at Quebec were badly cracked, as was also the ground alongside; columns had been wrenched off their foundation bolts. There was no direct loss of life, but several deaths through shock. The incident served to focus attention on earthquake studies in North America; its significance is due in part to another earthquake which is recorded as having occurred in the same region on February 5, 1663, probably in the vicinity of Three Rivers, causing many landslides which made the water of the St. Lawrence River muddy for a month.[10.4]

It must be emphasized that the actual earth movements such as have been described are not the result of earthquakes but, on the contrary, are the cause of the quakes that follow them. The great masses of material involved in the movements naturally set up dynamic effects of magnitude. In the New Zealand earthquake of 1929, several observers witnessed the earth's surface undulating to such an extent that a ripple seemed to pass along the ground under observation. The main result, however, is the initiation of earthquake waves (or, more accurately, vibrations) inside the crust of the earth. These travel great distances, and it is due to them that a localized movement of earth can have disastrous effects over a wide area. The waves are recorded by means of seismographs, delicate instruments similar to those used in seismic prospecting but housed in special buildings and far more sensitive. Modern seismographs are, indeed, so sensitive that the instrument in the Bidston Hill Observatory near Liverpool, England, regularly records the rise and fall of the tide in the adjacent Liverpool Bay and Estuary, which cause movement of the rock on which it is founded. *Seismology* is the term applied to the study of earthquakes and their effects. Seismologists have distinguished three types of waves set up by earth movements, two of which travel through the earth, and the third at the surface. All three can be distinguished on the seismograms obtained from seismographs, and by means of suitable calculations the origin of the waves, or *epicenter*, can be determined.

The area affected by earthquake waves is often extensive; in the case of the Assam, India, earthquake of 1897, it is estimated that an area of 1,750,000 sq. miles was thus influenced. Amplitudes and periods of vibration vary; the factor of most significance in engineering is the horizontal acceleration that the vibrations tend to produce. Various scales have been suggested and used to denote the differing intensities of earthquake shocks; the Rossi-Forrel is one of the best

TABLE K.—DR. F. OMORI'S SCALE OF EARTHQUAKE INTENSITIES[1]

Maximum Acceleration, Feet per Second per Second	General Effects
1	Sufficient to induce people to run out of buildings. Produces cracks in badly constructed masonry, buildings, and factory chimneys. Wooden structures crack; furniture is overturned; and water becomes turbid
2	Pendulum clocks stop. Walls of wooden buildings crack. Old wooden buildings get out of plumb. Tombstones are overturned
4	Factory chimneys damaged, and badly built brick buildings destroyed. Timber buildings mostly destroyed, and wooden bridges damaged. Fragments of rock break away from the hillsides
6½	All factory chimneys ruined, most ordinary brick buildings and some wooden ones destroyed. Cracks 3 in. wide appear in the ground. Wooden bridges partially destroyed
8	All ordinary brick buildings very seriously damaged. About 3 per cent of wooden houses destroyed. Railway permanent way twisted. Fissures 1 to 2 ft. wide along river banks. Landslides occur
13	50 to 80 per cent of the wooden buildings totally destroyed. Embankments badly damaged. Railway permanent way considerably contorted. Steel bridges thrown down, and wooden bridges badly damaged. Wide fissures in the ground, whence jets of water and sand may emerge. Landslides general. Low-lying ground so distorted that flora die
Over 13	All buildings destroyed except those specially built to withstand seismic conditions. Enormous landslides result from the movements along the geological fault lines

[1] *Imp. Earthquake Investigation Comm. (Tokyo) Pub.* 4, pp. 138 *et seq.*, 1889, reproduced in *Inst. Civil Eng. (London), Min. Proc.*, 236: 309 (1933), and now reproduced by permission of the Institution.

known. Of more interest and use to engineers, however, is a scale that has been suggested by Dr. F. Omori; it is here reproduced as Table K.

In addition to a geological action being the cause of earthquakes, local geological features can have a marked effect on the local results of earthquake shock. This is evidenced by the fact that earthquake shocks are not felt in deep mines, even though they are located within seismic areas. Since the radiating vibrations travel at different speeds in different materials, it is to be expected that effects in rock and unconsolidated material will be different. This is found in practice, earthquakes causing much more trouble in areas of unconsolidated materials than in those with solid rock exposed at the surface. Greater amplitudes of vibration are possible in the former than the latter; and in consequence, greater accelerations are to be expected in the softer materials. In the San Francisco earthquake, the maximum acceleration recorded in marshy ground was about 10 ft. per second per second; in corresponding rock outcrops, a figure as low as 0.89 ft. per second per second was observed. Similar records have been obtained in other earthquakes. It follows that in civil engineering work in seismic areas, it will always be advisable to found structures wherever possible on solid bedrock. Another reason for this is that under the action of the transmitted vibrations, unconsolidated material will fracture and be displaced much more easily than will solid rock. Actual experience with buildings and engineering structures in severe earthquakes is in line with this suggestion. When the local geology is such that a rock foundation bed cannot possibly be used, foundation design must take into consideration the unusual forces that may be exerted upon the structure following an earthquake; a raft type of foundation, so designed that it will withstand upward forces in addition to the usual downward loads, is one kind that proves satisfactory, as have also continuous foundations carried on piles driven so as to rest on an underlying rock stratum.

In built-up city areas, the shock from an earthquake is liable to fracture water mains; and as fire is always a consequent hazard, water distribution systems have to be designed and constructed with special care in seismic areas. Underground conditions may be so affected that groundwater distribution will be changed; old springs may stop flowing, and new ones may be formed. Unconsolidated deposits that are below the surface may react in strange ways. Fill material will very often be affected, the intense vibration compacting it to an extent not possible by any normal means. In the New Zealand earthquake of 1929, it was noted that all made ground sank to some extent: one fill which had been in place for six years sank 3 ft. although only 22 ft. high, the grass on the surface not being displaced.[10.5] A final feature to be noted which is sometimes of most serious consequence is the effect that an earthquake shock can have at exposed vertical or sloping faces such as the steep banks of rivers or those of artificial cuts. The action of the vibrations on reaching a sloping face has been likened to the movement of the last billiard ball of a row that is struck sharply at the other end! Although hardly an accurate picture, this does assist in an appreciation of the severe landslides and rock falls that so often occur on such faces after passage of an earthquake shock.

The importance of all these features to civil engineering design and construction will be clear. In seismic areas, local regulations usually govern the allowance to be made in structural design for the possibility of earthquake shock, a common rule being that a structure must be able to withstand a horizontal force caused by a horizontal acceleration of one tenth that of gravity. This aspect of earthquake design is now being investigated experimentally by means of special *shaking tables* which simulate earthquake vibrations. In other directions, however, regulations will give the engineer no guide, and he will then have to study the past seismic history of the locality, to see if any special features have been revealed. Local geology will be a necessary complementary study. The advisability of getting foundation beds of rock has been noted. Similarly, level ground should be selected in preference to undulating or hilly ground. Marshy ground should be particularly avoided, and the use of fill material kept to a minimum. Close vicinity of civil engineering works to cuttings such as river beds, canals, and coasts should also be avoided so far as this is possible.

10.4. Ground Subsidence.—A feature that sometimes accompanies earthquakes is a vertical displacement of ground in addition to the main earth movement responsible for the quake. Thus in the New Zealand earthquake, to which reference has just been made, the town of Karamea, located on a deltaic formation, sank 2 ft. Other similar depressions of large areas of land are constantly taking place. As outlined in Part I (pages 25 to 27), this gradual movement is also a part of the general geological cycle; it is usually so slow that it cannot be observed, but some instances can be and have been studied. For example, it is now known that the area in which are located the Great Lakes of North America is slowly tilting, the rate being 5 in. per 100 miles per century. Although so small, this, as has been pointed out, would in 1,600 years cause the upper Great Lakes to discharge by way of the Chicago River into the Mississippi Valley instead of into the sea by the St. Lawrence![10.6]

There are a few localities at which an even faster movement has been detected. Only in comparatively recent years has it been possible, through the medium of precise leveling operations, to observe such movement. Indeed, it has sometimes been in checking precise leveling results, especially of inland areas, that the movement has been discovered. A subsidence of 15 in. in Utah was located in this way in the vicinity of Kosmo; it followed an earthquake which occurred on March 12, 1934.[10.7] Dr. O. E. Meinzer has noted ground subsidence in connection with his groundwater investigations, notably in Dakota and in the Goose Creek oil field, U. S. A., and was apparently the first

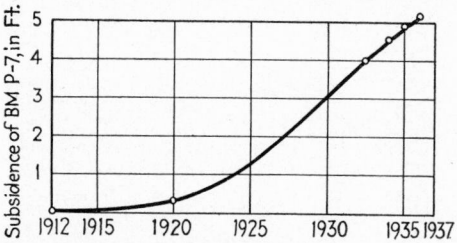

FIG. 10.5.—Record of the subsidence of the Bench Mark on the Hall of Records Building, San Jose, California.

(*Reproduced by permission of the Editor, Engineering News-Record, from reference* 10.9.)

to treat this matter in a scientific manner, showing how the subsidence could be accounted for by the abstraction of water from underlying beds.[10.8] It seems probable that this cause is responsible for one of the best known cases of subsidence, that in the Santa Clara Valley, about 20 miles south of San Francisco, California. The area affected is 200 sq. miles, and the maximum settlement (in 1937) amounted to slightly more than 5 ft., this occurring in the business district of the city of San Jose, with a population of 80,000. The subsidence was first noted during the progress of a survey in 1920, and it has been observed regularly since then. The results are already becoming serious, especially as the area involved includes a part of the shore line of San Francisco Bay, which has now to be protected by dykes.[10.9]

It is naturally in connection with coastal developments that land subsidence can be most serious. One interesting solution to the problems thus created was adopted in connection with the protection of a stretch of shore line in Venezuela. The area around Lake Maracaibo near Lagunillas in this South American state is extensively developed as an oil field. The ground is swampy, and so a drainage program was initiated in 1923, following which it was noticed that the ground was slowly subsiding. The rate of subsidence was found to be about 12 in. annually; so that before long, properties of the operating oil companies were in danger of flooding. Various methods of protecting the lake front were tried, but finally a flexible type of reinforced concrete retaining wall, with a sloping face secured by raking piles at the back, was evolved and constructed in 1932. The wall was subjected to a slight earthquake in 1933 but showed no signs of damage.[10.10]

There is a further type of subsidence which may be briefly mentioned, this being the purely local phenomenon sometimes referred to as a *sinkhole*. Such depressions are the result of superficial unconsolidated material subsiding into holes formed in underlying rock (generally limestone) which has been eroded in some way, often by solution in groundwater. Some of these holes are of unusual extent, Culpepper's Dish—a well-known feature in England—having a circumference of almost 600 ft., the sides sloping to a depth of about 150 ft. In Jamaica is a district known as the *Cockpit Country* because of the sink holes there found in the local white limestone, varying from shallow basins to pits 500 ft. deep. An interesting case occurred in connection with the disposal of sewage from the city of Norwich, England; it was conveyed to a farm at Whitlingham, and the land on which it was discharged was soon covered with holes from 3 to 5 ft. in diameter and of varying depth, due to the washing out of *sand galls* in the local chalk formation and consequent subsidence of the surface material.[10.11]

10.5. Landslides.—Minor earth movements will be considered, for convenience, under the two headings Landslides and Rock Falls, using the latter term to include sudden falls of solid rock, and the former to cover all other types of localized earth movement other than subsidence. Landslides are so common a feature and are often so serious in their effects that they have been the object of much study. Many official commissions have considered them, notably that of the Swedish Government, and two governments today—the Swiss and the Italian—maintain official organizations to deal with landslides and similar troubles. In many of these studies, attempts have been made to classify landslides in order to simplify their investigation, those of Heim and Terzaghi being perhaps the best known. In such a brief review as this, detailed classifications cannot be followed; reference will therefore be made to the main four of the five types of slide suggested by Dr. George E. Ladd in one of his most valuable publications on landslides.[10.12] These are:

1. Mud flows.
2. Slope readjustments.
3. Undermining of strata.
4. Structural slides.

The fifth type cited by Dr. Ladd is an unusual ejection of clay from old caverns during excavation; this is met with very infrequently.

All four types are associated with movement of material constituting a part of the earth's crust, movement caused fundamentally by gravity and taking place because of some inherent instability in the arrangement of the materials concerned. All types are dependent completely on the nature of the materials involved and on their relative arrangement, in other words on the local geology at the site of the slide. All types may occur naturally; they may all develop during the carrying out of some civil engineering work. If they occur naturally, they must be regarded as an inevitable part of the general geological cycle, in that they contribute to the erosion of parts of the earth's surface, thus taking their place as important factors in the processes that are developing topographical features of the world today. If they occur during civil engineering construction, they betray some interference with the natural stability of the part of the earth's crust in which work is being carried out. In all cases, although the exact cause of the slide may be difficult to determine, they will be due to either one or a combination of several natural causes which can be determined if investigations are pursued in the proper directions. It follows that although naturally occurring slides may be classed as "Acts of God," if by that term is implied a part of the natural order of the Universe, they cannot be so classified if the term is meant to suggest that the cause of the slide is a mystery. Finally, all types of slides do not suddenly develop their essential instability (if those due to earthquake shocks are excepted), the movement that develops being merely the indication that a critical point has been passed. It therefore follows that it should be possible to anticipate many landslides if the necessary preliminary investigations are made, that is to say, if the local geology is studied with unusual care. The fact that all too often such investigations are made *after* a landslide has occurred is probably the reason why the value of preliminary studies is not better appreciated.

Natural landslides may affect the work of the civil engineer in many ways. Those which have taken place in the past may be responsible for topographical features which the engineer has to deal with in his work. Many major slides have in the past blocked up river valleys, the resulting constrictions giving dam sites which appear at first sight to be admirable. In view of the nature of the material in slides, and its unconsolidated and often disturbed condition, sites of this kind may prove to be very far from ideal. Dams have, however, been successfully founded at such sites. In a paper by Mr. W. G. Atwood,[10.13] some interesting examples have been given, one of which —the Farmers' Union Reservoir, on the Rio Grande River in Colorado

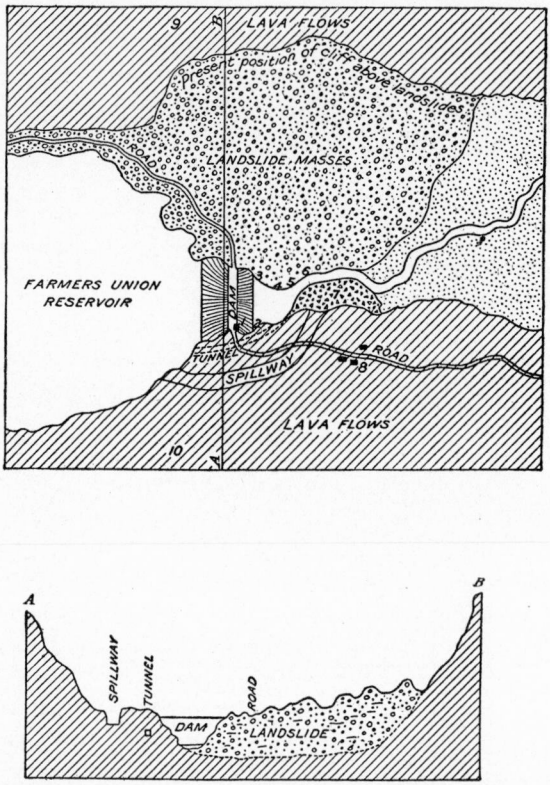

FIG. 10.6.—General plan and section of the site of the Farmers' Union Dam (earth and rock fill; concrete-core wall), Rio Grande River, Colorado, showing relation of dam to landslide mass.

(*Reproduced by permission of the Director, U. S. Geological Survey, from U.S.G.S. Bull.* 685, *reference* 10.13.)

—blocks off only one quarter of the width of the true valley in the local lava flows, the remaining three quarters being dammed effectively, although with small leakage, by a great mass of landslide material. The incidence of another great landslide mass upon a major dam location is mentioned on page 362, the Bonneville Dam site having been created by the deflection of the Columbia River by an ancient landslide. It will be clear, therefore, that recognition of ancient landslides is an important part of geological survey work. Local topographical detail will often be a guide, but the composition of the unconsolidated mass will be more certain. The rock fragments will be angular and not polished as in glacial deposits; all materials must obviously be the same as found in adjacent strata in distinction to glacial deposits; the disposition of the different materials in the mass will be irregular as a final distinguishing feature.

The motion of landslides, natural and those caused by engineering work, may vary from an almost imperceptible rate to what may be appalling suddenness. The great Gros Ventre slide in Wyoming, in which a river valley was filled for 1½ miles of its course by a natural

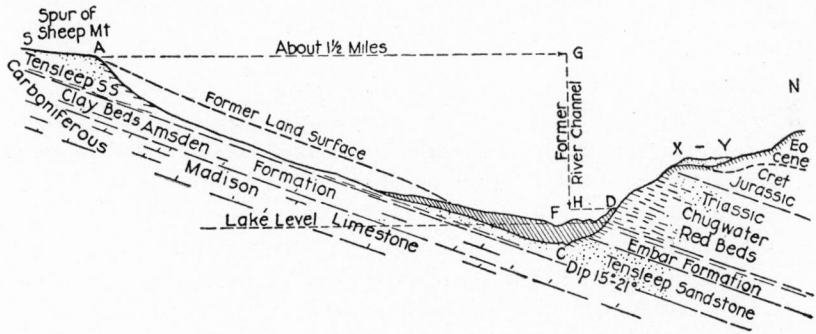

FIG. 10.7.—Geological section across Gros Ventre Valley, Wyoming, showing the landslide of June, 1925.

(Based, by permission of the American Institute of Mining and Metallurgical Engineers, on a section by W. C. Alden in reference 10.14.)

dam 1,000 ft. wide at the top with an average depth of 180 ft., the total volume being estimated at 50,000,000 cu. yd., was seen by eyewitnesses who stated that it occurred in less than 5 min.[10.14] It was later washed out by the natural flow of the river. Figure 10.7 shows a typical section through the main slide from which it can be seen that it was essentially of the structural type, slipping occurring on a bedding plane. By way of contrast may be mentioned those slides which are moving so slowly that their surfaces can be regarded as practically firm ground. An example is a part of the Canadian Pacific Railway in the vicinity of Field, British Columbia, Canada, which is reported to be built across the toe of the talus slope on the east side of Mount Stephen, of the Canadian Rockies, which is moving very slowly; it is only occasionally, however, that relocation of the tracks has to be made.[10.15]

Talus slopes, and similar deposits of eroded material which has fallen into place without the aid of water or glacial action, are frequently affected by slides, instability arising from the way in which the material has reached its present position. Glacial deposits and materials deposited from water will not usually be affected without the aid of some major change in their natural condition, such as mentioned below; deposits containing clay will then be found to be those which are principally affected, beds of pure sand and gravel rarely moving unless they are seriously interfered with by construction operations. Instability in these cases arises from the variable character of clays, especially if moisture content is changed, the physical properties of sands and gravels, on the other hand, being relatively constant whether dry or wet. Solid rock strata that have an appreciable dip and are featured by marked bedding planes, by excessive jointing, or by the essentially fissile structure of many of the metamorphic rocks will be found to be those most liable to give trouble through slides. Slickensides are often a potent cause of movement. Of common mineral contents, *serpentine* is that most liable to lead to trouble. It is an alteration mineral, sometimes found in massive form; it is greasy to the feel, slippery even when dry, and when wet can be most troublesome. It has been responsible for many serious slides. Another mineral that sometimes causes movements is *anhydrite* which will change to *gypsum* when exposed to the atmosphere by absorbing water; in so doing, it expands 33 per cent in volume, so that the resulting instability can readily be imagined.

The presence of treacherous materials alone is not sufficient to start a slide, but in combination with other causes they can add to the troubles that movement will create. Earthquake shocks are always a potent cause of movement, as already explained. The process of erosion is another common cause; it may act in one of several ways. Differential erosion of strata of varying stability may leave overhanging material of a harder stratum which will eventually break away and cause a slide. Erosion of the toe of a slope of unconsolidated material may remove the essential support from the material above which will start to move downward until stability is restored again. This will happen more easily on sloping bedding planes which are always a source of possible weakness when bordering on any natural or excavated sloping face. Fault planes constitute another frequent cause of slides, if so arranged that they isolate blocks of material which are thus left free to move; the work done in opening out the Cofton Tunnel (see page 185) is an example of slides due to this cause. The equilibrium of material in its natural position may sometimes be affected by the placing of unduly heavy loads upon it. This is often the case with excavation work, the dumping of excavated material too near to the cut being made being responsible for many slides, among which those on the Panama Canal may be numbered, although the slides there were due to a combination of causes of which this was only one. Erection of buildings on unconsolidated material will sometimes have the same effect.

Probably the most important cause of all is a change in groundwater conditions. This may be due to interference with natural drainage conditions, to excessive evaporation from normally damp ground, or to increase in groundwater due to excessive rainfall. This last is probably the most usual way in which groundwater conditions are affected; it is especially serious because excessive rainfall will also increase surface runoff, which may result in erosion of material at the toe of a slope and so intensify sliding tendencies. It is rarely that surface erosion, of itself, will lead to trouble, the landslides that often accompany this result of intense rainfall being generally due to a corresponding change in groundwater conditions. If, however, surface erosion strips off vegetation and leaves bare ground exposed to future rainfall, this may lead very quickly to an increase in groundwater and so to slides. The presence of water underground has three main effects: It will increase the weight of the material that it saturates; it will tend to lubricate any sliding planes, especially if these are formed by materials, such as shales and clay, that are affected by moisture; and it will tend to weaken many materials, including the weaker kinds of rock and unconsolidated materials with any clay content. The combination of one or more of these effects with other consequences of heavy rainfall will show why many slides occur in wet weather and why drainage is so often an effective remedy for sliding.

The presence of excessive groundwater is so generally realized as a potent cause of landslides that it may be difficult to imagine a slide being caused by drought conditions! Dr. Ladd has reported such a case, however, which led to the wreckage of two trains, although the volume of material in the slide was only about 400 cu. yd.[10.16] The slide occurred on the Richmond, Fredericksburg and Potomac Railway in Virginia, in 1933. It took place in a cutting 1,000 ft. long at a point on the west side where the depth of cut was about 80 ft. Excavation was through sedimentary deposits of fine material derived from local metamorphic beds but containing no clay; deposition had been irregular, and some induration had taken place, to an extent such that the railway engineers classified the material as rock, having found it difficult to excavate with a steam shovel and able to stand at a slope of about 60 deg. in the lower part of the cut, flattening off to about 45 deg. in the upper part. Three distinct strata were evident in the cut, the slide occurring from the third one and impinging on the tracks with the result noted. The strata were considerably jointed, and Dr. Ladd's opinion is that the slide was due to "the actual or incipient joint planes that parallel the west face of the cut [which] are the result of general shrinkage of the exposed beds because of excessive loss of moisture during a long dry period."

Landslides occur frequently and come so often into the purview of the civil engineer engaged in general practice that it has been thought

FIG. 10.8.—An aerial view of the Elysian Park Landslide, Los Angeles, California, November 26, 1937. (SPECIAL NOTE: This slide wrecked part of the Riverside Drive concrete viaduct, as can be seen. A study of the failure of the structure is to be found in *Engineering News-Record*, vol. 120, p. 273, New York, February 17, 1938.)
(*Reproduced by permission of Mr. Lloyd Aldrich, City Engineer, Los Angeles.*)

better to discuss them by means of the foregoing general notes rather than by the use of a few selected examples. The variety of slides is such that any examples so given can illustrate but a few of the many individual types; references are therefore given to publications that deal with slides much more fully than can be done in this short chapter, in which many instances are cited. In other chapters, slides are mentioned in other connections, notably the Panama Canal slides (page 264), the slides at the opening of the Cofton Tunnel (page 185), and mudflows of the Andes (page 259). The interest of the civil engineer in the Geology of landslides is not in the geological features as such but only in so far as these may assist in remedial or preventive work. Further examples will therefore be given in connection with a discussion of this important branch of civil engineering work.

10.6. Preventive and Remedial Work.—All engineering work directed toward either preventing landslides or remedying the results of slides must be based on the fact that the movement betrays instability of the materials involved. If the slides are of undisturbed material, the fact that they are a part of the normal geological cycle must not be lost sight of; this may mean that they cannot be stopped but only anticipated and possibly controlled. If the slides develop during construction, their cause can then be traced to a construction operation, and suitable remedial measures evolved, taking into consideration the further construction work still to be done. But considerations of the over-all stability of the entire mass of material involved must always be of cardinal importance. If due regard is paid to stability, such remedial measures as removal of material at the toe of a moving slope and the construction of either pile or crib retaining walls at the toe of a slide will not be adopted except possibly as temporary emergency measures. In a few isolated cases of very small slides, these measures may be effective, but the general result of toe removal will be to remove some of the resistance to movement and so to increase the extent of the slide instead of stopping it. The use of retaining structures will usually betray a neglect of the fundamentals of engineering design. Retaining structures should be designed for a predetermined load which they are to transmit to a foundation bed of known capacity. If constructed to stop a slide movement, unless the slide is of very small extent, they will have to withstand the load induced by movement of the whole slide—generally unpredictable with any accuracy—transmitting it, in many cases, to a lower part of the material affected by the slide! To illustrate the frequent failure of such measures to control slides would be to labor the point too strongly; examples will be found quoted in several of the references given. There is one possible exception to this general unsuitability of retaining walls, this being in the case of cuttings both banks of which are unstable and both of which can be held by retaining structures connected by solid struts across the bottom of the cutting. In some cases, this measure may prove economical; usually, however, it will be more satisfactory to achieve stability of the slopes of the cutting, thus dispensing with the need for such walls.

Three general preventive and remedial measures are available: consolidation of unstable material contributing to the movement, treatment of the slopes that are unstable, and correction of groundwater conditions responsible for the movement. In special cases, special methods may have to be adopted, such as the protection of the toe of a natural bank from erosion by a river, protection of a weak stratum where differential erosion is taking place, or removal of unusual loads adjacent to slopes that have contributed to movement; but the three methods listed are general in application and apply to all types of slides to some degree.

Consolidation of unstable material may be achieved by chemical means, by the use of cement grout in certain special cases, and by the use of freezing. The object of all processes is to render more stable material that is moving because of its own state, or to *solidify* it, if this term may be used in connection with the conception of a fluid or semifluid state of moving unconsolidated material. Consolidation methods are therefore especially applicable to slides of the mudflow type. These commonly take place from talus slopes; in mountain regions subjected to severe temperature changes, flows that are popularly called *rock glaciers* may develop as a peculiar kind of mudflow; and in regions featured by recent clay deposits and moderate or severe rainfall, clay slopes will often be found to flatten by means of slides that approximate flows in form. The east of Canada, and especially that part of the Province of Quebec to the immediate north of the St. Lawrence River, is noted for slides of this kind; there is, indeed, a small town actually called *Les Éboulements* (The Landslides) picturesquely situated in an area noted for slides. Apart from some interference with highways, these slides are not of great consequence; consolidation methods have not yet been applied to them.

Freezing has been successfully applied to the control of a slide that might have been serious had it developed further, this occurring during the construction of the Grand Coulee Dam for the United States Bureau of Reclamation.[10.17] The dam is located on the Columbia River, Washington, and when completed will be one of the major dam structures of the world. At the dam site, silt originally filled the valley of the river to a depth of about 500 ft., but the river has subsequently worn this down to a depth of between 40 and 50 ft. During this erosive process, the river channel has swung from side to side of the valley, with the result that slides of the silt have been frequent. Excavation in the dried-out river bed encountered the toes of many of these old slides and in this and other ways so disturbed the stability of the remaining silt that many slides occurred during construction. The silt was found to be a very fine rock flour, containing from 20 to 25 per cent of colloidal material; undisturbed the material stood up well, but when once moved it proved unstable on any slope steeper than 1:4 even when relatively dry. Drainage by wells and tunnels assisted in controlling slides, as did also slope correction; but near the center of the east excavation area, the bedrock was found to be intercepted by a narrow gorge, 120 ft. deeper than the average bedrock elevation, the silt in which created a slide of an unusually serious character. A 5-cu. yd. shovel could make no headway against the

FIG. 10.9.—Flow of silt in the east excavation area at Grand Coulee Dam, Columbia River, Washington, stopped by an "arch dam" of frozen material. Refrigeration pipes may be seen near the center, at top of photograph.
(*Reproduced by permission of the Commissioner, U. S. Bureau of Reclamation.*)

slide, and so it was finally decided to freeze an arch of the material so as to form a solid blockade of the gorge behind which excavation could proceed. This was done, with the aid of 377 special freezing points and a circulating ammonia-brine solution refrigerating system of 160,000 lb. of ice per day capacity. The frozen dam was about 40 ft. high, located on top of a concrete and timber crib structure 35 ft. high; it was 20 ft. thick and 100 ft. long, costing $30,000 (£6,000). Although freezing started when the slide was moving in to the excavation at a rate of 2 ft. per hour, the dam was successfully completed and was estimated to have saved its own cost in the excavation which did not have to be done; in addition, several weeks of very valuable time was saved in connection with work in the cofferdammed river bed.

Slope readjustment, the second main remedial measure noted, really amounts to doing under control and in a limited way what a slide will do automatically if it is allowed to take place; it will be recalled that the same term was applied to the second type of landslide, using Dr. Ladd's classification. As an extreme case, there may be mentioned the statement apparently made in 1894 by the Chief Engineer of the London and North Western Railway of England that

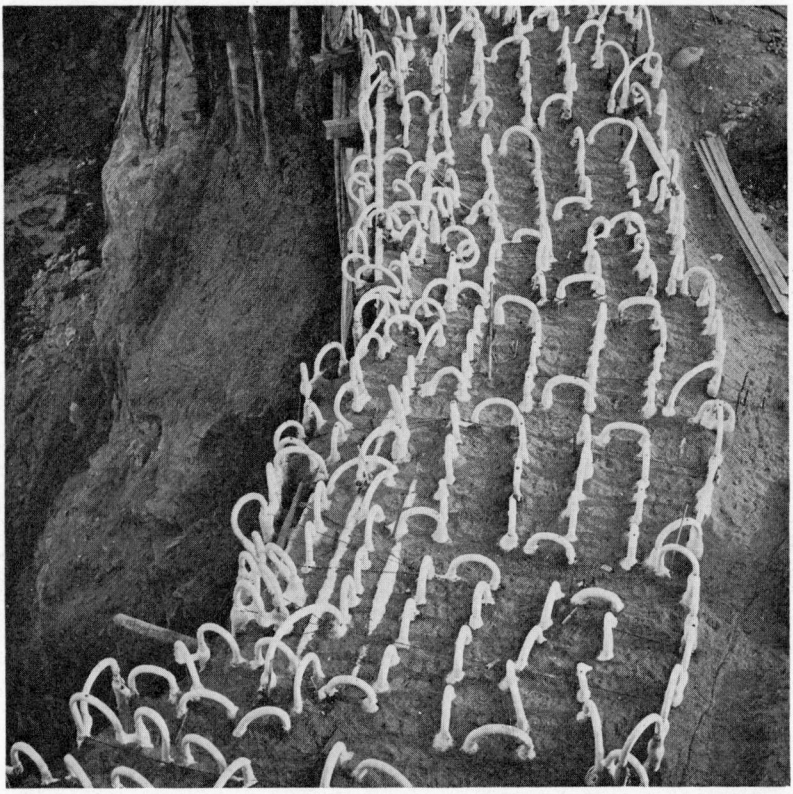

FIG. 10.10.—A close-up view of the "arch dam" of frozen silt at Grand Coulee Dam used to halt a flow into the east excavation area.
(Reproduced by permission of the Commissioner, U. S. Bureau of Reclamation.)

"there was neither a bank nor cutting between Euston [London] and Rugby that had not slipped at one time or other," this statement dealing with the first 100 miles of one of the main railway lines of England.[10.18] Railways have been very general sufferers from this type of slide, both in natural undisturbed material near to which railway lines have been located (as on hillsides, for example) and in respect to the cuttings and fills constructed to carry the roadbed. It is perhaps safe to say that no railway line has suffered more (although some of the South American lines may have suffered as much) than the Hill Section of the Assam Bengal Railway in India. It is 114¼ miles long and was opened to traffic in February, 1904; its construction was undertaken largely for military reasons, having been described by a viceroy of India as a "millstone round the neck of the Indian Finance Department." From the start of operation, slips and washouts gave trouble; in 10 years, about £100,000 ($500,000) had been expended in necessary maintenance, and the line had been closed on several occasions. In 1915, a further disaster overtook the line, due to excessive rainfall, culminating in a fall of 26 in. in 48 hr.! It was subsequently closed for 2 years while remedial work was carried out at a further cost of over £225,000 ($1,125,000). Disregarding the financial aspects of the work, all engineers can appreciate the heroic efforts of the engineers concerned with the maintenance of this line in the face of the difficulties that the quoted figures suggest. The difficulties were due in large part to the local geology, the local rocks being of the Tertiary measures, mainly alternating beds of carbonaceous shales and sandstones so affected by earth movements of the past that they possess little durability and break up into small fragments on exposure to the atmosphere. The shales vary from rock as hard as slate to material with the consistency of clay; the sandstones, from rock sand to first-class building stone. A record of some of the major difficulties in dealing with this material is fortunately available[10.19]; in it, many details of slides will be found of which the following may be noted as illustrating a notable case of slope readjustment—first by way of a slide, and subsequently as a remedial measure, coupled with other features of interest.

Figure 10.11 represents a cross section at a point on the line near to the south portal of the Chamartalla Tunnel; sketch *a* is as this was early in 1913. The slope of the cutting in the rock was at first taken out to a steeper slope than shown, and many falls of rock and earth occurred, with the result that the slope was cut back to that shown. Revetment was also built to protect the newly exposed shale from exposure to the atmosphere. The retaining wall shown at river level was the cause of much trouble. When rebuilt in 1899, its foundation was on hard shale at river-bed level, and no erosion was anticipated; but in 1902, and again in 1908, the river in flood flow undercut this wall to a depth of up to 12 ft., eroding the hard shale completely; the wall had to be rebuilt again. In 1913, a great mass of material slipped from the hillside above onto the tracks, the load being so great that the retaining wall was forced into the river, the whole railway forma-

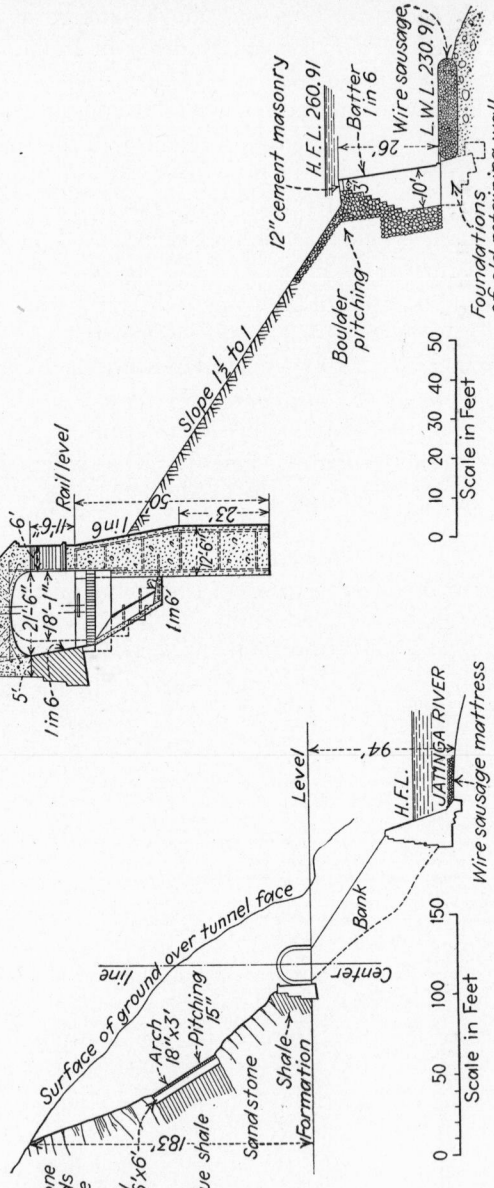

a b

Cross section south of Chamartalla Tunnel in 1913, before major slip. Covered way structure erected following 1913 slip; undamaged in 1915.

FIG. 10.11.—Landslide remedial work on Hill Section of the Assam Bengal Railway, India.

(*Reproduced by permission of the Council of the Institution of Civil Engineers, London, from reference* 10.19.)

tion being carried away for a length of 150 ft. Communication was restored temporarily within 15 days. Permanent reconstruction introduced many problems, but the final solution is shown in sketch *b*, the design of the covered way being so prepared that it would offer a minimum obstruction to further slides that were intended to pass over it. A sloped cushion of earth fill and handpacked stone was to be permanently maintained on top at the slope indicated, the slope below the track being trimmed as shown. The foot of the retaining wall was protected with wire sausage mattresses to secure it against future scouring. The piers for the covered way, as will be seen, had to be carried to a depth of 50 ft. below track level, in order that a solid foundation bed should be available, rock above this level being badly fissured. In 1915, when the major disaster to the line took place, a slip passed over the covered way without causing any damage. Fortunately, it is but seldom that engineers are faced with difficulties such as have just been recited, but the case is of great interest as showing how even such a serious situation can be remedied by judicious work on the affected slope.

Drainage was the third main method to be listed. Surface drainage is generally envisaged as that necessary in most engineering work, including that dealing with slides. In a few cases, surface drainage may be effective; generally, however, it will be useless in view of the fact that the critical location in a slide is the plane on which sliding is taking place—usually well below the surface except at the extremities of the slide. In some cases, surface drainage may even be harmful. When surface erosion is taking place, drainage to control the flow on the slope will assist in limiting the erosive action. When vegetation is lacking, the planting of appropriate shrubs and plants is often an effective remedial measure. Robinia trees (a pseudo acacia) have been used with success for this purpose in India; the Hottentot fig plant has also been used, its growth being such that its large leaves will effectively shield sloping ground from intense rainfall and also from intense sunshine which may lead to cracking of the surface.

To be effective in preventing or remedying slides, drainage must intercept the groundwater which is assisting to promote the instability. In some cases, a very small amount of water can cause extensive slides. Examples from actual practice have shown that such features as thin beds of sand, only 1 to 2 in. thick and only slightly water bearing, and even a 1-in. seam of decomposed coal, can lead to slides of the superincumbent material which may be of wide extent. Drainage can, therefore, be applied effectively only if full data as to the cause of the slides are available and full information as to the source of the groundwater that is contributing to movement. This information can be obtained by a judicious combination of geological survey work and test boring; needless to say, it is work that can be carried out far more easily before any movement has taken place than after a slide and to much better effect.

From the many examples of successful drainage work of this kind may be cited an interesting case from the records of the State Highway Commission of Indiana.[10.20] Southwestern Indiana is underlain with a deposit of fire clay which has been the cause of many slides and which contributed to that at Horton's Hill, 13 miles east of Evansville, in 1933. Relocation of the road at this location was impracticable because of the rugged nature of the ground around; the slide that eventually developed was complicated by an abandoned coal mine beneath the road. Figure 10.12 shows the cross section through the site of the slide which was obtained by the local highway engineers, following the usual practice of their department, after it had been found that the subsidence of the roadway was not due to collapse of an old mine roof, mud jacking proving ineffective in restoring the road to its original level. It was found that sliding was taking place on both fire-clay surfaces, aided by seepage from the hillside above the road and from the old mine workings. It was therefore decided to install a drainage system of the type and in the position shown in Fig. 10.12, the drain consisting of 174 ft. of perforated 24-in. steel pipe (specially lined because of the acid nature of the water from the old mine workings) connecting to 540 ft. of perforated 18-in. pipe leading to a side-ditch outlet. Manholes are provided for inspection and cleaning. The total contract price for the work described was $10,700 (£2,150). It proved successful; when the old mine workings were drained out, they were backfilled with gravel, to some extent, to protect the drainage system.

This example epitomizes successful drainage work in connection with landslides, except that it was carried out after a slide had occurred instead of before, as can often be done. The subsurface was adequately explored; the cause of the slide was found; the main contributing cause—leakage of groundwater—was dealt with by a carefully planned drainage system. These steps will feature most preventive and remedial work in connection with landslides although the exact nature of the method for dealing with the main cause of sliding may differ. When it is normal groundwater, special methods may some-

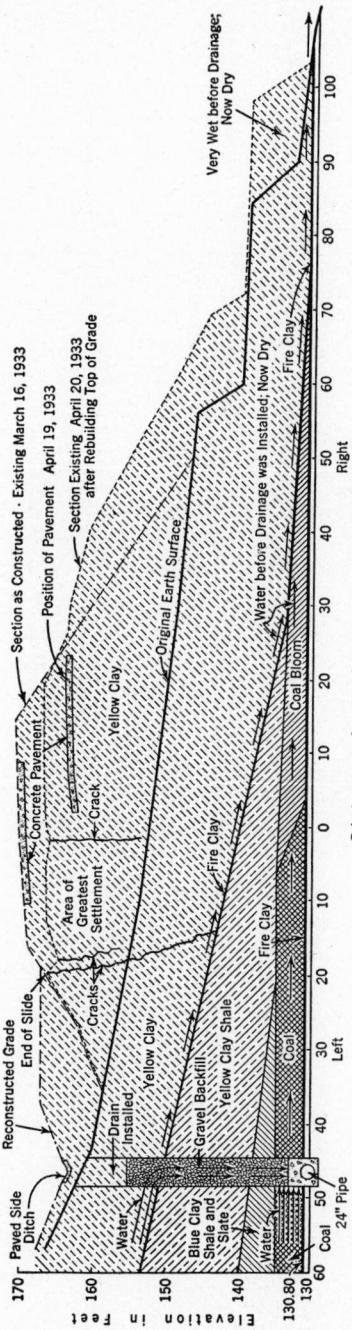

FIG. 10.12.—Simplified geological section through the Horton's Hill slide, Indiana, showing remedial measures adopted to stop sliding on fire-clay stratum.

(Reproduced, by permission of Mr. G. H. Allen and the American Society of Civil Engineers, from Civil Engineering, reference 10.20.)

times be adopted in addition to regular drainage systems. Well points (see page 196) may be used to advantage in some places. One large

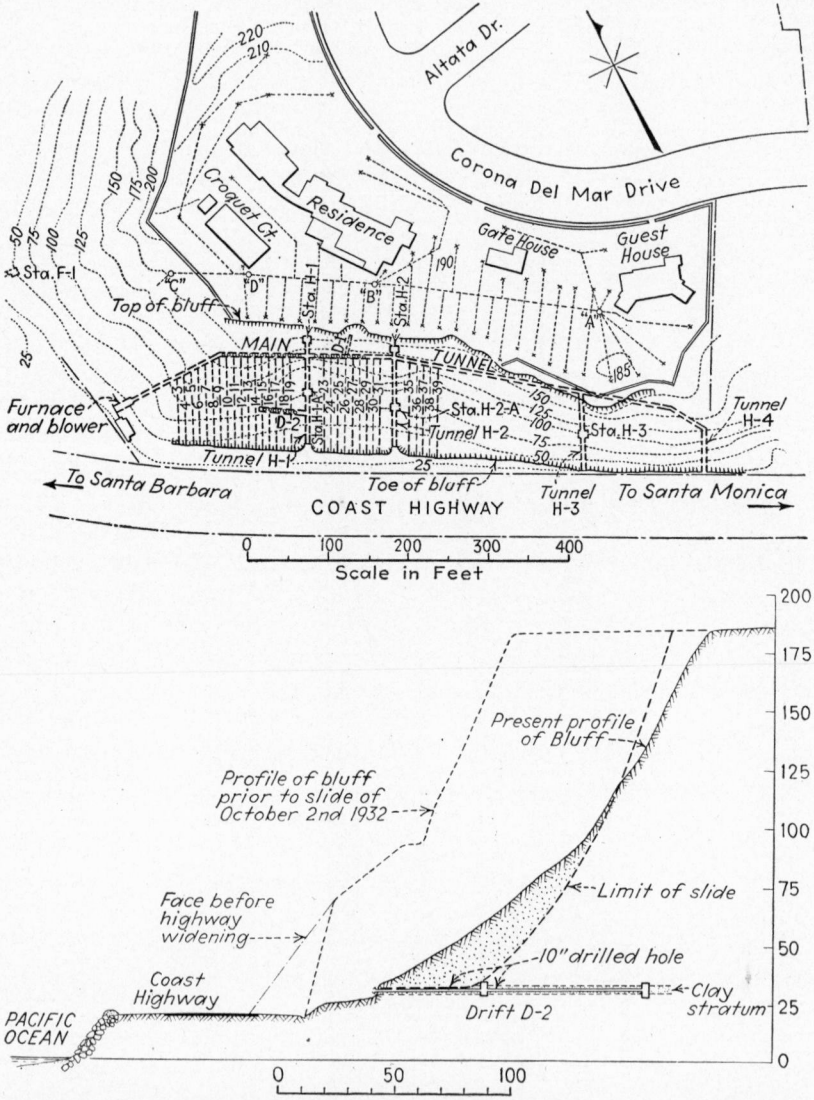

FIG. 10.13.—General plan and typical cross section of landslide area at the Palisades, west of Los Angeles, California, showing arrangement of hot-air drying installation.

(Reproduced by permission of Mr. R. A. Hill, Messrs. Quinton, Code and Hill-Leeds and Barnard, Consulting Engineers, Los Angeles.)

hillside in California is being maintained in its present state only by the constant operation of an extensive hot-air pumping system which circulates warm air through tunnels excavated in the clay bank in order to dry out groundwater that is constantly seeping toward the hill face.[10,21] This unusual scheme has worked so well that it is now planned to fill all the tunnels with gravel and discontinue drying, while leaving the general arrangement so fixed that drying can again be resumed in the future, should this ever be necessary.

10.7. The Stability of Earth Slopes.—The attainment of stability as a necessity in dealing with incipient or actual landslides has been stressed to such an extent that the question will naturally arise as to how this stability can be determined in the case of those sloping faces of unconsolidated materials which are so common a feature of civil engineering work. The sides of cuttings, the sides of deposits of fill material, and natural hillsides that have to be used in connection with civil engineering construction—these are the main types of earth slopes, the control of which can so often be a matter of great difficulty. Whenever such slopes are composed of material that does not occur in regularly bedded strata, the determination of stability becomes a matter of investigating the mechanical stability of a mass of unconsolidated material conforming to the outline of the slope under consideration. Once the physical properties of the material are known, the investigation becomes a mathematical computation of some complexity. As early as 1846, Mons. Alexander Collin investigated landslides that had occurred in France (along canals) in this manner, and he showed that the surface along which such slides will take place approximates a cycloid. This work appears to have been long forgotten, but it was mentioned by Mr. H. F. Reid in connection with the study of Panama Canal slides.[10.22] Further work was done by the Royal Commission of the Swedish Government to investigate the landslides that had occurred on the Swedish state railways. From these beginnings have arisen the extensive mathematical data and the several methods of investigation that are now available through the work of Soil Mechanics research workers. All methods necessitate some data on the physical properties of the materials concerned, and this information has been made available through Soil Mechanics experimental work, the properties of internal friction and cohesion from which a determination of the shear strength can be made being among the leading properties that are at present under study by research workers.

An extensive bibliography already exists on the stability of earth slopes, approached in the manner indicated. The application of all methods to actual practice must be based on a thorough knowledge of the local geology at the site considered. Calculations usually assume a homogeneous material; only from a consideration of Geology coupled with study of numerous samples can the accuracy of this assumption be judged. The extent to which Mathematics has already been applied to this branch of work makes this note especially important; it is probable that no natural material is ever completely homogeneous, so that the utility of extreme refinements in calculation must always be determined by what Geology can suggest with regard to the actual nature of the material involved. This comment is in no sense to be taken as belittling the value of all the work done on the stability of earth slopes; on the contrary, this note is included so that the methods may be appreciated as an important factor in slide control.

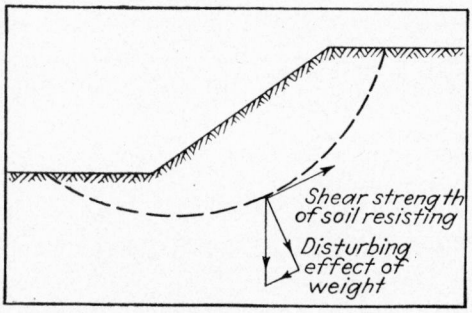

FIG. 10.14.—Diagram illustrating the basis for the calculation of the stability of earth slopes.

An accompanying illustration shows how the mathematical concept of a curved sliding surface is realized in practice.[10.23]

10.8. Rockfalls.—Rockfalls, in general, are more determinate and understandable than other types of minor earth movement. They coincide, to a great extent, with the fourth class of slides mentioned on page 225. They are due to a mass of rock becoming detached from surrounding bedrock in some way and thus being free to move downward if its position will permit. The loosening of blocks of rock will almost always be associated with such features as its bedding plane, joints, cleavage, or a local fault zone or plane. The immediate cause of loosening will be a change in the material adjacent to the planes of possible movement which will usually be the result of weathering. Natural rockfalls are evident at any steep mountain face, talus slopes being a direct result. Many are of great extent, but great rockfalls will generally take place in uninhabited mountain regions and so be of interest mainly to the geologist. One great rockfall was not so located. Figure 10.15 shows Turtle Mountain in Alberta, Canada, which will be seen to be composed of Palaeozoic

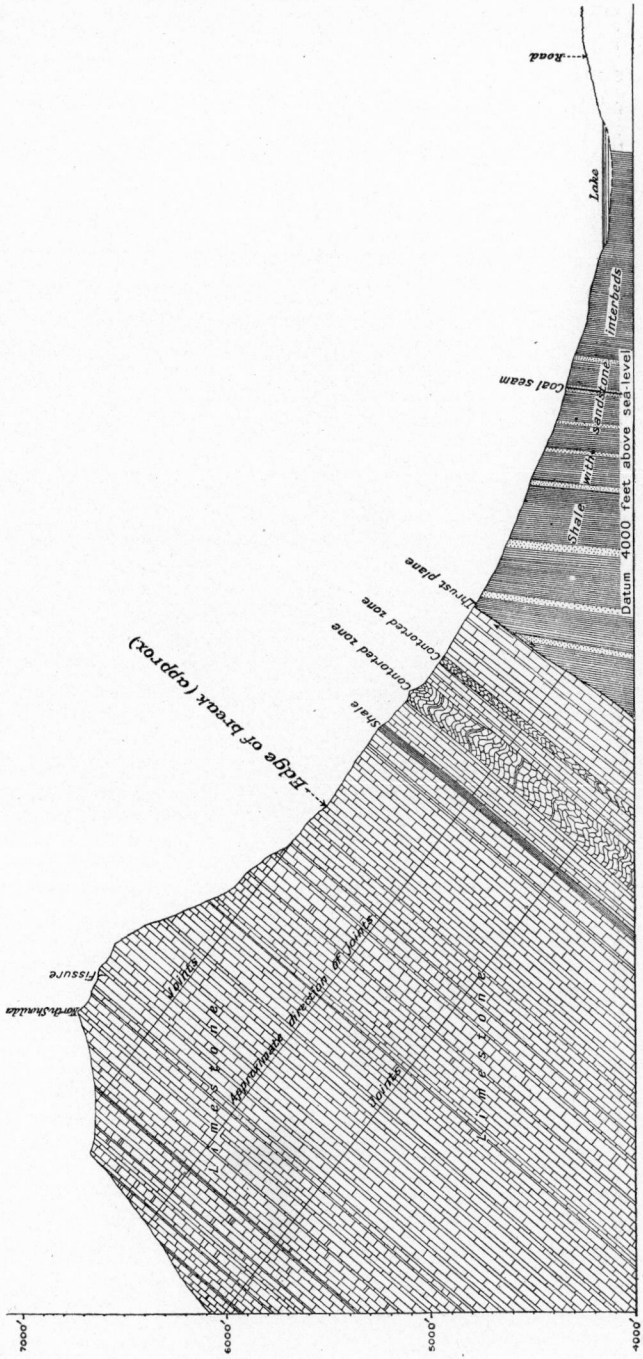

FIG. 10.15.—Geological section through Turtle Mountain, Frank, Alberta, showing relation of rockfall to the local geological structure.

(Reproduced by permission of the Geological Survey of Canada, from reference 10.24.)

limestones overthrust upon a syncline of Mesozoic shales, sandstone, and coal beds. In the early morning of April 29, 1903, the crest of this great mountain broke off, sliding to the valley below, taking about 70 lives, damaging much property, and burying nearly 7,000 ft. of the Crow's Nest Railway. The catastrophe was closely investigated by Canadian geologists and engineers; the general conclusion seems to be that the major cause was failure of the rock along the joint planes which are a prominent feature of the limestone formation.[10.24]

FIG. 10.16.—Reinforced concrete protective works at the Vriog Cliffs, near Barmouth, Wales, on the Great Western Railway.
(*Reproduced by permission of the Great Western Railway Co., Mr. R. Carpmael, Chief Engineer.*)

Fortunately, rockfalls of this magnitude are extremely rare in developed areas; the case is cited to illustrate how a structural feature that might not at first be suspected may lead to loosening of blocks of rock even on a large scale. Rockfalls of smaller extent are not uncommon along transportation routes that are adjacent to natural rock cliffs and in cuttings excavated through rock. Rockfalls can but seldom be prevented; although if they are on a major scale, the loose rock may possibly be anchored to solid bedrock by means of steel dowels if these can be placed without the necessary hole drilling still further loosening the rock. An example of this type of work was carried out successfully in connection with the foundation of the Polson Dam, Montana. Usually, if probable rockfalls can be detected, it will prove economical to remove the loose rock under control with the aid of skilled "rock men." If this cannot be done, then some form of protective works may be constructed so designed that rockfalls will not cause serious damage. This type of protection work is most common on transportation routes, the importance of rockfalls to railways being emphasized on page 258. The work illustrated in Fig. 10.11 is a good example of what can be done and of what has to be done in this direction.

FIG. 10.17.—Concrete guard walls which protect, from rockfalls, the Calderwood powerhouse of the Aluminum Company of America, Little Tennessee River, Tennessee.

(*Reproduced by permission of the Aluminum Company of America, Mr. J. P. Growdon, Chief Hydraulic Engineer.*)

Another example is illustrated in Fig. 10.16, which shows the reinforced concrete protective "tunnel" constructed by the Great Western Railway Company near Barmouth on the west coast of Wales. Vriog Cliff, seen in the photograph, rises to a height of 1,500 ft.; road and railway are benched into it as they are carried along this stretch of coast. On March 4, 1933, a train was swept off the tracks at this location by a slide that happened after a severe blizzard followed by a period of thaw. The slip took place at a reversed fault in the local Ordovician strata, consisting here of dark flags and slates with one or two beds of limestone and abundant volcanic ash. The protective works were constructed after a complete survey of the adjacent cliff in order to obviate any possibility of trouble from slides that may develop in the future due to continued weathering of the exposed rocks.[10.25]

Suggestions for Further Reading

Earthquakes

Naturally, many books are available that deal with earthquakes, their origin and effects and their influence on engineering design. The following titles are typical of modern earthquake literature and provide a useful guide to further study:

Davison, C.: "A History of British Earthquakes," Cambridge University Press, Cambridge, England, 1924.

Davison, C.: "Great Earthquakes," Thomas Murby & Co., London, 1936.

Freeman, J. R.: "Earthquake Damage and Earthquake Insurance," McGraw-Hill Book Company, Inc., New York, 1932.

Heck, N. H.: "Earthquakes," Princeton University Press, Princeton, New Jersey, 1936.

See also the publications of U. S. Coast and Geodetic Survey (several by Captain N. H. Heck) detailing earthquakes of the U. S. A. (*e.g.*, *Special Pub.* 149).

Landslides

One of the most complete modern treatments of landslides considered in relation to civil engineering work is:

Ladd, G. E.: "Landslides, Subsidences and Rock-falls," *Am. Ry. Eng. Assoc. Proc.*, **36**: 1091 (1935). This comprehensive paper includes a lengthy bibliography.

After this chapter was written, the following detailed study was published—also containing an extensive bibliography—to which reference may be made for a full treatment of the "mechanics" of landslides:

Sharpe, C. F. Stewart: "Landslides and Related Phenomena," Columbia University Press, New York, 1938.

CHAPTER XI

TRANSPORTATION ROUTES

> Now in the middle ages, to keep these roads and especially these bridges in repair, was one of the first calls on Godly piety—charitable concern for all travellers. Turn to your Litany and read: "That it may please Thee to preserve all that travel by land or by water."
>
> SIR ARTHUR QUILLER-COUCH.[11.1]

The history of transportation routes presents a vivid picture of the development of civil engineering—from the first efforts of primitive man to bridge a stream with a fallen tree trunk to the conception and construction of modern express highways with all associated works. Beaten tracks to drinking places were probably the first artificially constructed pathways. Communication between settlements was a natural development, and thus gradually arose the idea of main transportation routes. In early historical records, many references to practical road building are to be found. The road work of the Roman Empire was the culmination of such development, some of the Roman roads being still in use today. Many centuries had to pass before road construction regained the eminence that it held in Roman times. This last advance has not yet been halted and continues as the virile construction practice of today.

Simultaneously has gone the gradual development of transportation by water routes, primitive localized use of rivers having been extended into lakes and inland seas, finally including transoceanic traffic. Change has been seen mainly in the vessels employed, but this has inevitably caused a steady extension of the marine works necessary for berthing vessels and protecting them when at anchor. Dock and harbor engineering has thus an ancient history, even the most primitive seagoing craft requiring some artificial port works where natural facilities did not exist.

The last century saw the road challenged by the railroad as the main artery of transport; in this present age, the road has answered the challenge; and both now play their part in the general transportation scene. Railroad engineering was, in effect, an extension of road construction, limiting characteristics being more severe, thus involving engineering works of greater magnitude than for highways. Tunnels, for example, are a leading feature of railroad construction although not often encountered in road work. A final development of the last few years has been the advent of travel through the air, bringing its own special problems, which include the provision of necessary landing fields and airports.

The civil engineering works called for in this wide field of transportation are many and varied. Tunnels have been mentioned; river crossings of road and rail call for bridges of many types and sizes; uneven topography calls for cuts and fills often of considerable size. These are matters dealt with in general terms elsewhere in this book. There remain certain unusual types of work in connection with transportation which demand some attention. These will therefore be briefly reviewed, and the relation of Geology to the leading branches of transportation will be illustrated by some cases of interest.

11.2. Route Location.—Two main problems remain to be faced in undertaking road construction: (1) the selection of the location of the route to be followed and (2) the provision of a road bed and surface of reasonably permanent stability. In both problems, the final choice is largely dependent on economic aspects, but Geology may often have a considerable bearing on the final solutions. Considering first the matter of location: in flat country, this is largely a matter of convenience, the route being chosen so as to connect the places in question by as short a road as possible, making deviations if necessary to obtain improved river crossings. In hilly country, many alternatives can be considered. The ridge route up a valley, for example, may be chosen in preference to the valley route, the climb to the ridge of a valley more than compensating in convenience for the numerous river crossings necessary if the road keeps to the valley. The Roman method of laying out a road as straight as possible irrespective of natural features may have something to commend it on occasion. These details of location economics are mentioned only as indications of the general factors that must be given consideration in route selection. They will be affected indirectly by the local geology because of the topographical demonstration of the underlying structure. In addition, the nature of necessary cuttings, the availability of suitable material for embankment fills, the stability of sloping ground above and below a selected route, and the availability of suitable materials for construction are all factors influencing the final selection of a route which are directly associated with the local geology. It is usually imperative, therefore, that geological survey work be carried out together with general topographical route surveys. This applies equally to railways and roads; many early railroad surveys enabled pioneer geological reconnaissances to be carried out in addition to that required for route location. The survey methods to be employed and the interpretation of results are not unusual, but the application of geological study to route surveying is one of unusual importance. An interesting old paper on this subject is to be found in the *Proceedings of the Geologists' Association (London)*.[11.2]

11.3. Road-building Materials.—With a route chosen, the engineer's next problem is the selection of suitable materials for the construction of his highway, a problem that also arises when existing roads are to be reconditioned, both classes of work being covered by the following notes. The problem has a dual character—the provision of a suitable subgrade or foundation, and the construction of the roadway surface. The success of the Romans as road builders is largely attributable to the care with which they prepared and drained their subgrades before laying on them the paving-rock slabs that they used as their roadway. They recognized the value of a layer of sand as subgrade material and, where it did not exist, often transported it over great distances to serve as a base for the four masonry layers that constituted their standard type of construction. It was not until the seventeenth century that serious attention was again devoted to road materials, and this revival of interest in France had to await the study of Telford and McAdam, the great British road engineers, before it developed into what may be called modern road engineering.

Water-bound and Tar Macadam Roads.—In those early days, a pavement of natural stone was the only type of surface considered, water-bound macadam (named after the engineer) being the finally perfected type of this construction. The character of the rock used for such road structures is clearly most important, and this importance was recognized by as early a worker as McAdam himself. It is interesting to read his own notes on the matter, presented in a forthright manner bespeaking the practical man. Thus he states that:

> Flint makes an excellent road, if due attention be paid to the size, but from want of attention, many of the flint roads are rough, loose and expensive. Limestone when properly prepared and applied, makes a smooth solid road and becomes consolidated sooner than any other material; but from its nature it is not the most lasting. Whinstone is the most durable of all materials; and wherever it is well and judiciously applied, the roads are comparatively good and cheap.[11.3]

Among the principles that McAdam laid down, some of which have been disregarded in the construction of macadam roads, are the following: No stone should exceed one inch in any major dimension; all stone used should be clean; the foundation should not be solid rock or hard pavement (a matter on which Telford had other opinions);

the camber used should not be too great; and the rock when laid should possess a natural water bond.

McAdam's ideas were based on his practical experience; his selection of rocks suitable for road use was similarly empirical. Subsequent detailed study and microscopical examination of rocks available for road work has generally confirmed his original ideas; it has also extended them so that today it is possible to specify with a reasonable degree of accuracy material suitable for use as *road metal*, to use the old English term. The main geological requirements may be summarized as follows (quoting Professor P. G. H. Boswell)[11.4]:

1. The rock should be crystalline, and not too fine in grain, or splintery or flaky in fracture:
2. The ideal textures are equigranular and interlocking, especially ophitic and granophyric.
3. The rock should be both hard and tough, so as to resist attrition and not yield too much powder.
4. All constituent minerals should be of equal hardness. Quartz should preferably be absent on account of its greater hardness than other common rock-forming minerals and its yielding practically no cementing substance.
5. The constituent minerals should be such as to yield decomposition products of good cementing value.
6. The rock should be neither too fresh nor badly altered.
7. The specific gravity should be moderately high and the rock not too porous since its absorptive power depends upon its porosity.

These ideal requirements must naturally be considered in relation to the supplies of material available and to the economic aspects of the construction in view. Generally speaking, the intermediate and basic igneous rocks are those which satisfy the requirements best. The most widely used in Great Britain are the porphyries, the porphyrites and diorites, and the dolerites and basalts among the basic rocks. Granites and quartz porphyries are extensively employed, despite their high quartz content. Metamorphic rocks are not, as a rule, suitable for road work, the common foliated structure militating against their wide use. When, however, the structure of metamorphic rocks consists largely of equigranular crystals, they may be considered as similar in application to the igneous rocks. Of the sedimentary rocks only the hardest types of sandstone and limestone can be considered as the main constituents of a macadam roadway, the other types being too soft to stand up to hard wear. To this must be added a note with regard to the wide use of limestone as a base for the many tar and bitumen preparations now so widely used under the generic name tar macadam. The tar binder so cements the rock constituents together and cushions the wear on the road surface that limestones too soft for use alone can safely be selected for use with tar. The main criterion in these cases is the way in which the rock selected will allow the tar to bond on its surface, bonding being clearly a most desirable feature.

For the testing of rocks as prospective road-building materials, the use of thin geological sections in the form of slides suitable for microscopical examination is most convenient and reliable. The use of slides will enable the mineral composition of the rock to be determined; and when considered with the nature of the crystal structure, this will serve as a guide to the suitability or otherwise of the material being examined. In addition, there are certain standard physical tests which must be used to supplement any geological examination such as the determination of specific gravity, absorption of water, resistance to abrasion, hardness, toughness, and cementing value. A combination of geological investigation and physical testing will give a reasonably accurate estimate of the worth of the rock in question as a road constituent.

Concrete Roads.—The modern reinforced concrete road may be regarded as a logical development of the water-bound macadam road, cement now being used as a binder, the concrete aggregate being in effect the road metal of the macadam road. The geological requirements of concrete aggregate for general civil engineering work are discussed in Chap. XIX. For road work these still hold true, with the added requirement that the rock used for aggregate should be as highly resistant to abrasion as possible. Structural strength is required also but not usually of a high order, so that abrasive quality can be the main criterion in the selection of aggregate for road work. There are two main physical testing methods for investigating the suitability of rock for aggregate in concrete for road use, the Deval test and the Los Angeles abrasion-machine test. The former is essentially an abrasion test rather than an impact test; the latter allows the samples of rock to drop in addition to being abraded by contact with moving steel. The two test methods have been correlated, but neither can be used without reference to the mineralogical composition of the rock in question, requirements listed in connection with material for water-bound macadam road surfaces being generally applicable. For further information, see B. H. KNIGHT: "Road Aggregates, Their uses and testing," Edward Arnold & Company, London, 1935.

Stabilized Soil Roads.—It is in the development of secondary or minor roads that Geology has been of signal service to the road engineer, more especially perhaps on the North American continent, although the results there obtained have now a world-wide significance. American interest in this branch of engineering is a natural consequence of the size of the United States and of the relatively recent history of much of the land embraced in that great country. Not only are the countries of the Old World smaller in area, but they have long been populated and permanently settled. European roads, even in the more remote districts, are therefore well-consolidated tracks if not, indeed, metaled roads, the size of the national road systems having enabled comparatively high standards of construction on secondary roads to be achieved, though perhaps not always maintained, certainly during the last century. In America, on the other hand, all minor roads have had to remain on a different level; and even today, the magnificent major highways of the United States contrast strangely with tributary road systems when compared with similar features in Europe. The construction and maintenance of these secondary roads, however, is a stupendous task considering the area to be covered and the varying climate and topographic conditions encountered. The progress made can truly be termed remarkable. Research work on the problems of these secondary roads by the United States Bureau of Public Roads and the various state highway departments now presents a scientific background to this road work of great value.

Most engineers will know from experience how bad an earth road can be, especially in bad weather; no words need therefore be wasted in attempting to describe the initial conditions to be faced in such work. The ultimate goal of this branch of research work has been to devise simple methods for the construction of all-weather stable road surfaces from whatever local unconsolidated materials are available. The proper combination of suitable sand and clay can produce what is, in effect, a natural *soil concrete.* That methods have successfully been evolved for this purpose and are now in use is evidence of the work done.

It would appear that investigations initiated in 1906 by Dr. C. M. Strahan, then County Engineer of Clark County, Georgia, were among the earliest attempts to utilize in a practical application to road work some of the facts concerning soil behavior that were then just being observed. Dr. Strahan's first public statement was made in 1914. At the first conference of state highway testing engineers held in Washington, in February, 1917, the matter received considered attention, and from that time forward advance can be noted. It is largely as a result of these studies that the general study of Soil Mechanics has arisen, a matter now so important that Chap. XX of this book may be regarded as merely an introduction to it.

In view of this development, many of the fundamentals of soil study will be left for consideration in this later chapter. Soil surveys are there described, and the more general tests now used for investigating the physical properties of soils. The application of this scientific analysis to the composition of earth-road surfacings generally is concerned with an investigation of the possibility of the local materials providing a soil mixture having a suitable binder in appropriate quantity. It is but rarely that the material at the road site will be a naturally suitable soil mixture: it may be pure sand or, at the other extreme, a perfect clay. The problem has then to be faced of selecting a material, either natural or artificial, to mix with the local formation in order to give a surface of the desired stability with the further possibility of suggesting whether a surface treatment of some deliquescent material such as calcium chloride will or will not be advisable.

Investigation started in an empirical way, samples being taken from selected earth roads which under widely varying conditions displayed unusually good stability of structure and water-resisting properties. From the analysis of the materials so obtained, an appreciation of the desirable features in soil road surface mixtures was gained, and this formed the basis of future laboratory work and theoretical reasoning. It can now be suggested that gravel and coarse sand furnish structural strength and hardness; fine sand provides support for the coarse sand; silt acts as a filler to prevent the coarse material from moving; clay provides minute particles and colloidal material to cause connecting moisture films to have high cohesion; and the deliquescent salt, if used, will tend to prevent excessive loss of moisture during dry weather. There naturally exists an extensive background of experimental results on the basis of which new work is being continually initiated. Most of the state highway departments in the United States now maintain special soil-testing laboratories, and research work is being pursued in many university laboratories. Not only are unconsolidated materials alone being investigated, but soils stabilized with such materials as portland cement, heavy oils, bituminous material, and even molasses—all in the search for low-cost sound and stable road surfaces.

The general relation of all work of this nature with Geology will be clear; but in what detailed way can the application of Geology assist in its development? First in the assistance that an appreciation of Geology and the application of geological methods can give in the development of soil survey work. This is something more than just obtaining samples from specified locations along a soil profile; concurrent study of all local geological features will often assist in the interpretation of sample test results and suggest where more should be taken. Secondly, if all soils are studied and tested with due consideration being given to their geological origin and history and their mineralogical composition, the significance of physical test results will often be enhanced, and sometimes special information of use may be obtained. Both features will be referred to in Chap. XX. Two illustrative examples from highway practice may usefully be cited here. Figure 11.1 shows a cross section through a marsh between two lakes, now connected by a small stream. It is clear that the highway could

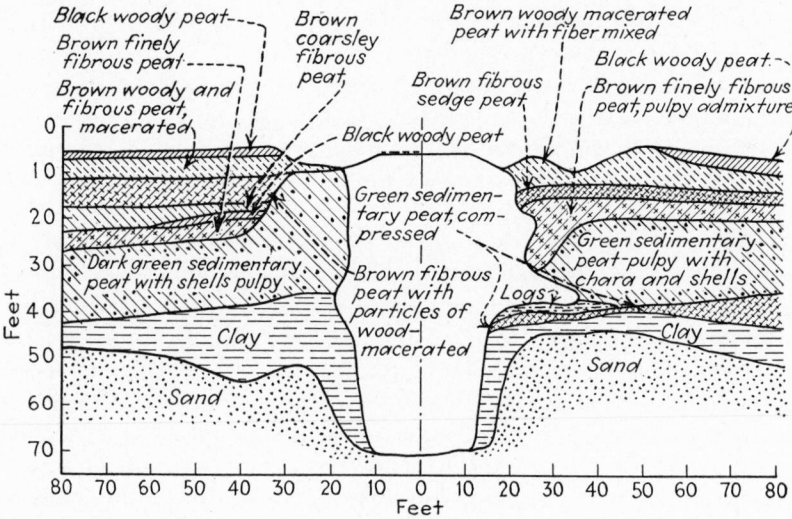

FIG. 11.1.—Cross section through an unusually deep fill; preliminary soundings would have indicated the local ground condition and so have permitted better location. (*Reproduced by permission of the U. S. Bureau of Public Roads, from Public Roads, reference* 11.5.)

not have been located more unfortunately, since it is directly above the ancient glacial valley that at one time existed between the two lakes. The existence of this could have been determined by a combination of test borings and geological investigation, and about 25 ft. of fill over this section of road would have been saved.[11.5] Many similar examples could be given. To illustrate how the geological nature of soils may influence the nature of a road surface, there may be mentioned the interesting practice developed in Australia (patented by Mr. L. R. H. Irvine of Sydney) of baking or calcining in place the adobe clay that is found in parts of New South Wales and Queensland to form a road surface. The clay is one that can be satisfactorily calcined by means of a wood-fired down-draft furnace of the air-gas producer type, the resulting product being bricklike in texture. The untreated clay (which softens when moist) is used as a binder in places where termite mounds (providing *antbed* material) are not available, the resulting road being stable in wet weather although achieved at relatively low cost.[11.6]

11.4. Road Construction.—Road construction consists essentially of preparing a suitable foundation bed for the surfacing to be adopted, in addition to constructing the surface pavement and the structures, such as bridges, required to carry the roadway over natural depressions. Clearly, the preparation of the roadway is dependent on the subsurface conditions existing all along the proposed route, an accurate survey of which is an essential preliminary to the completion of designs and the start of construction. It might be thought that sound and solid rock would provide an admirable roadbed; it does not always do so, since in many cases it is of varying strength and durability, the existence of joints often adding a further complication. The softer rocks such as chalk constitute fair roadbeds when dry, but they can be troublesome when wet. Clay is similar, being extremely unreliable when wet owing to the expansion of most types as they increase their moisture content. Probably the best of all roadbeds is provided by well-packed coarse sand and gravel resting on some type of solid substratum.

This ideal case is not often met with, and the engineer has to contend with actual ground conditions varying from fissured igneous rock to peat bogs. In practically all cases, drainage is of supreme importance in order to prevent waterlogging of the subgrade and consequent trouble. Drainage methods vary; they must inevitably be suitably related to the nature of the materials to be drained, the physical properties of which—in the case of unconsolidated materials—can be tested in the same way as can those of other soils. The underlying geological principle of keeping the water table sufficiently low so that gravity flow through the material can take place is applied in various ways, with "self-draining" porous road-bed materials by means of suitable cross-sectional road design, and with impermeable materials by the provision of suitable artificial drainage channels or the use of a layer of sand or gravel.

It is not often that the reverse process has to be undertaken, but this has been done in at least one instance. A concrete roadway in Ellis County, Kansas, was found to be settling unevenly, causing high joints between adjacent slabs. Investigations showed that the differential movement was due to changes in the moisture content of the subgrade soil occurring at joints and cracks. This was successfully corrected, and the concrete slabs releveled by introducing water into the subgrade material through an installation of well points jacked into place about 10 in. below the slab from the roadway shoulder.[11.7]

This interesting case is cited here to show how even one soil characteristic can affect road construction; a study of the frost boils that cause trouble in Canada and the adjacent parts of the United States would demonstrate the same thing, again pointing to the necessity for accurate subgrade surveys and adequate preliminary soil testing. In some cases, this preliminary investigation will reveal material so unsatisfactory that a road cannot safely be built across it.

FIG. 11.2.—Blasting operations in progress for fill consolidation over swamp land, near Ha-ha Branch, on U. S. Route 40, Maryland.

(*Reproduced by permission of the Maryland State Roads Commission, Mr. Nathan L. Smith, Chief Engineer.*)

Peat bogs, swamps, marshes, muskeg, and tidal flats are examples of such a condition; all are relatively new geological deposits which have not consolidated to any extent and which consequently contain a high percentage of water. Study of the local geology together with an adequate program of test borings will usually reveal the extent of these deposits, since they are not the "limitless sinkholes" sometimes popularly imagined.

In early civil engineering work, construction across such materials was generally tackled by dumping solid fill material along the route until it finally stopped settling or, alternatively, by laying down mattresses of brushwood or similar material to serve as artificial foundation beds for solid fill which was then deposited on them. The crossing of Chat Moss in 1827 by the original Liverpool and Manchester Railway in England, George Stephenson being the engineer, was one of the earliest notable examples of the use of this device. These ideas are sometimes used today; but with accurate data as to the extent of the poor material available, other methods can often be applied to better advantage. Replacement of the softer material with dry fill by means of jetting is one new method, jetting of peat with pressures up to 100 lb. per square inch being carried out underneath the dry fill, allowing the latter to settle as the disturbed peat is displaced laterally.

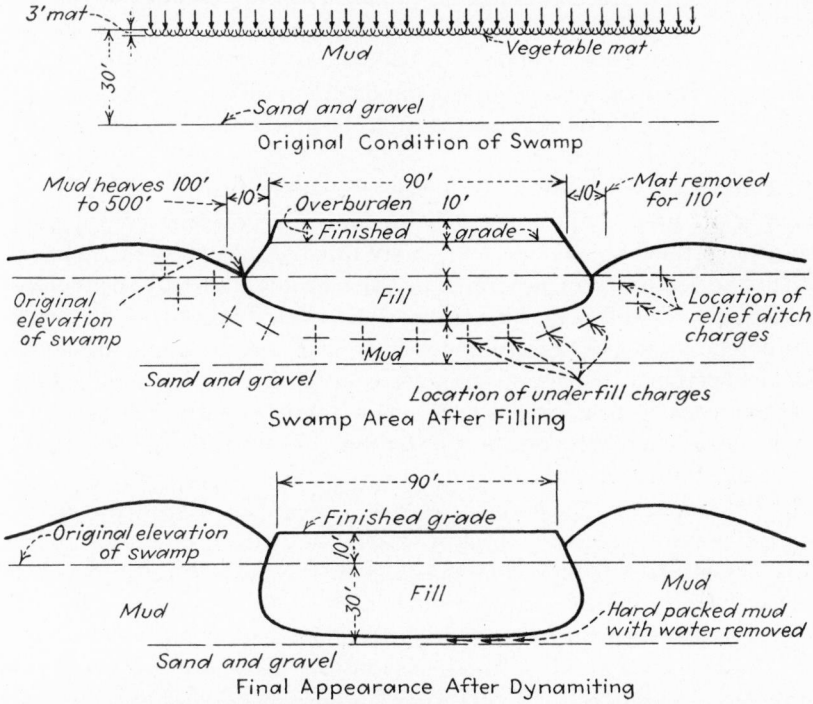

FIG. 11.3.—Diagrams illustrating the method followed in consolidating fill material over swamp land, during construction of U. S. Route 40, Maryland.

(Reproduced by permission of the Maryland State Roads Commission, Mr. Nathan L. Smith, Chief Engineer, and The Editor, Engineering News-Record, from reference 11.8.)

Displacement of unstable deposits by blasting is an even more widely used method, blasting with high explosives being carried out either beneath dry fill which has been dumped in place or else well down in the soft material in front of the fill deposit. An accompanying photograph shows a typical example of this work, the dependence of which on accurate data as to subsurface conditions will be clear.[11.8] The work illustrated, on the notable dual highway constructed in 1934 from Baltimore to Havre de Grace, Maryland, was necessitated where the route crossed swamp areas. The surface vegetable mat was first removed by blasting with 50 per cent nitroglycerin, and the fill was then advanced and blasted into position, a 40 per cent gelatin explosive having a slow heaving effect being used. This type of construction operation has been considerably developed in North America in recent years.

11.5. Railways.—Features of railway construction are discussed elsewhere in this book, as in the chapters on open excavation, tunnels, and the foundation of bridges. Roadbed construction is not unlike highway construction, problems of route location drainage and subgrade stability being common to both. Apart, therefore, from mentioning railway construction as an important branch of civil engineering to which geological studies can usefully be applied, little more need be said.

Railroad maintenance work calls for some further comment, in view of its great importance. Local failure, or obstruction, of other transportation routes does not usually involve the safety of travelers, since the trouble can generally be seen in advance and so avoided. In the case of railways, human lives are often involved, if the tracks cannot be kept clear and intact. Thus it is that track maintenance and inspection are such important parts of railway work. Bridge foundations constitute a special hazard; in Chap. XII (page 302) will be found some comments on this matter and a note on two accidents due to bridge failures. Tunnels always require special care in inspection, with due regard to the stability of rock exposed in unlined sections.

Embankments always require special attention, especially in the period just after their completion and in the vicinity of bridge abutments. Cuttings are similarly important, as was shown in Chap. X. The trouble in cuttings most readily imagined is that due to landslides of adjacent unconsolidated material; all too many occurrences of this type have taken toll of human lives in the past. Rockfalls can be just as serious; even minor falls can result in accidents due to the solid blocks of rock that will sometimes break off an exposed face. A train on the Western Pacific Railroad, United States, was derailed on November 11, 1937, with consequent loss of two lives, as the result of hitting a single 3-ton rock boulder which had fallen on to the track in the interval between the arrival of the train and inspection of the track just 30 min. previously.[11.9] Many similar instances could be cited. In view of the possibility of such occurrences in cuttings through rock, the stability of which is uncertain, ingenious safety devices have been evolved. One of the first of these was on the Callander and Oban line in Scotland, where it passes along the north side of the Pass of Brander, being there subjected to frequent rock falls. As long ago as 1881, a special 9-ft. wire fence was erected alongside of the line, connected to 14 signals located throughout the critical length of 3¼ miles, so arranged that if a falling rock cleared the fence it would not land on the track, whereas if the rock touched the wires one or more signals would be put in the "danger" position.

Another unusual source of trouble in connection with railway maintenance is due to what have been described as *mud runs*—streams of mud and rock debris washed down from the lower slopes of bare mountain sides by the runoff that follows torrential rains. These are generally confined to tropical countries, being especially severe in

FIG. 11.4.—Arque Bridge on the Antofagasta and Bolivia Railway, showing two of the original 40 spans (which have to be kept open) and the accumulation of rock debris above the original valley floor, about 3 meters below the track level.

(Photographed by and reproduced by courtesy of Mr. Hugh Gibson, Chief Engineer, Ferro-Carril de Antofagasta a Bolivia.)

sections of South America and in India. The extreme climatic conditions promote rapid disintegration of exposures of relatively weak rocks such as shales, and the corresponding concentrated rainfall acts as an efficient transporting agent. If mud runs reach bridge openings under a railroad designed to take normal stream flow, the consequent restriction of movement will result in blocking of the openings, and thereafter control of the runs will be a task of difficulty. It is reported that several sections of the Bolivia Railway have had to be relocated owing to this cause, in one place a town adjacent to the railway having had to be completely relocated because of the original site being buried deep under the unusual *detrital cone* formed by blockage of a large mud run at a bridge opening. This bridge, originally 10 ft. above stream-bed level, is now located in a cutting![11.10]

As mud runs are a natural phenomenon, they cannot be prevented, control being the only means of dealing with them. This has been effected in some cases by the construction of check dams in the valley well above the railway location, by which the velocity of the runs is reduced, thus resulting in deposition of the rock debris behind the dams.

11.6. Rivers.—Although rivers still constitute an important part of many inland transportation systems, they call for even briefer mention than railways, since their improvement and maintenance for navigation purposes is allied to other branches of river engineering work, all of which can be grouped under the heading of River Training Works, a subject dealt with in Chap. XV. It is there noted, and may be suggested here also, that all engineering work in connection with river training must be carried out with due regard to the physical features of the river as a whole and not merely in relation to the particular local problem involved. Thus it is that the *life history of rivers,* as the study of the physical geology of watercourses is sometimes called, is of great importance in civil engineering work of this kind. The combination of river navigation requirements with flood regulation and water-power-development works, as on the river Rhine in Germany and the Mississippi River in the United States, suggests how wide is the field in which river transportation may be featured.

11.7. Canals.—The extension of inland waterway systems and the penetration of narrow land barriers between seas by means of canals have attracted the attention of master builders throughout the ages. The canal connecting the Nile and the Red Sea is said to have been begun in the fourteenth century B.C.; work on the project was abandoned by Necho about 610 B.C. after 120,000 men had lost their lives in the excavation work; according to Strabo, it was finished by Ptolemy II, who is said to have constructed in it locks with movable gates. The many canals constructed in what is now Iraq are familiar to students of early history. Xerxes, in his war against the Greeks, constructed a canal across the isthmus of Mount Athos, trouble being encountered due to the sides (up to 50 ft. deep) being excavated at too steep a slope. The Greeks made several attempts to build a canal across the isthmus of Corinth; another attempt was made by Nero; but it was not until 1893 that this great rock cut, giving a channel 4 miles long, 70 ft. broad, and 26 ft. deep was completed.[11.11]

Many other instances could be mentioned, not a few of them being records of repeated but unsuccessful attempts to excavate the necessary water channel. Defeat was generally met with because of unusual features of the material to be excavated which necessitated modern construction methods and plant in addition to the man power which was always available in ancient days. Problems in canal construction have not been confined to work of the distant past. Mr. T. Mellard Reade in a paper presented in 1888 records the following interesting case[11.12]:

> Another well-known piece of stupidity was the attempt to make a canal between Loughs Mask and Corrib in the West of Ireland. This was one of the relief works planned during the great famine, and it is not known who was responsible for it. A canal between the two loughs would have been of some utility, but the strata in which they lie is carboniferous limestone much jointed, and the communication or drainage between the two waters is by underground channels. The engineers, Royal or otherwise, went bravely to work, the canal was cut in the limestone rock, the water turned in, when lo! it all quickly disappeared. The abandoned cutting may now be seen sparsely overgrown with grass.

Although exceptional, this case is not unique in all respects, as records of modern canal construction will show. Superficially, the excavation of a regular water channel along a selected route appears to be a simple operation. It is only when the details of design and construction are considered that the true complexity of the work becomes evident. Canal bed and canal banks should be impermeable; if permeable, the leakage must be predictable with certainty. Canal banks, if necessary, must be stable. Bridge foundations adjacent to the canal must be sound and secure; similarly, the foundation of all lock and control structures must be susceptible to rational design and expeditious construction. Geology is of fundamental importance in all these problems, as can be seen at a glance. Adequate knowledge of all geological features is therefore an essential prerequisite to canal construction.

The foundation of locks is of particular interest in view of the varying hydrostatic conditions to which the completed structures will be subjected. As an example of somewhat unusual conditions ingeniously utilized may be mentioned the main sea lock of the IJmuiden Canal, Holland, opened in 1930, a diagrammatic cross section through which is shown in Fig. 11.5. The local fresh water supply is obtained from groundwater trapped beneath the top clay stratum shown in the diagram. There was grave danger that if construction operations interfered with this underground reservoir, the supply might be interfered with in quantity and possibly in saline content by ingress of sea water. As a result of thorough subsurface exploration, the conditions shown in Fig. 11.5 were determined, and this permitted the complete enclosure of the space occupied by the lock—1,315 ft. long by 161 ft.

wide—by a wall of steel sheet piling. This course was followed as indicated in the diagram, and the work was successfully completed. It constitutes an outstanding example of the adaptation of local geological features as a definite part of a scheme of construction.[11.13]

The canal is but one of several notable canals constructed in relatively recent years. The new Welland Canal joining Lakes Erie and Ontario across the Niagara escarpment is another outstanding example. It is about 26 miles long, with a total drop of 326 ft. 6 in. achieved by means of only seven locks, each of which has a usable size of 820 by 80 ft. Throughout its course, it crosses 10 distinct geo-

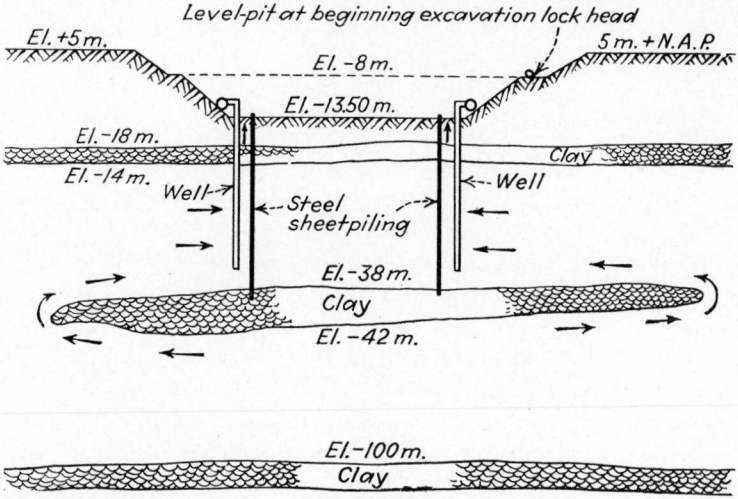

FIG. 11.5.—Subsurface conditions at the side of the new Ijmuiden Lock of the North Sea Canal, Holland.

(Reproduced by permission of Dr. J. A. Ringers, and the Editor, Engineering News-Record, from reference 11.13.)

logical formations of the Ordovician and Silurian measures including dolomites, limestones, sandstones, and shales. Of these, the Medina or Queenston shale gave unusual trouble during excavation, disintegrating as it dried out in the atmosphere. The practice developed, therefore, of so excavating it that a thin cover was left in place until just before concreting began, when it was finally removed after thus protecting the final excavated surface.

One more canal must be mentioned—a project the completion of which was one of the most notable achievements in civil engineering up to that time. The Panama Canal, joining the Atlantic with the Pacific Ocean, is now one of the most important waterways of the world. Located just north of the equator, it crosses the isthmus of Panama where the great continental divide dips to its lowest elevation, about 312 ft. above sea level. The route of the canal follows existing river valleys, crossing the divide in the great Gaillard Cut; it is 40.27 miles long. Its history is interesting. First projected in 1529 by Alvaro de Saavedra (a companion of Balboa's in the discovery of the Pacific and a lieutenant of "stout Cortez"), it remained for over three hundred years as no more than an international hope. Construction was started in 1882 by a French company which operated until 1889; a reorganized company restarted work in 1894 but was eventually bought out by the United States of America in 1902 under whose

FIG. 11.6.—A landslide on the Panama Canal, September 18 and 19, 1915, showing island formed overnight in 30 ft. of water.

(Reproduced by permission of the Governor, Canal Zone.)

auspices the canal was completed in 1914. The canal was officially opened on July 12, 1920.

Despite the magnitude of its engineering features, attention must here be restricted to the canal's geological interest. This is typified, perhaps, by the following data as to excavation carried out up to June 30, 1938:

	Cubic Yards
Excavated by French	29,908,000
Excavated by Americans	413,864,655
Excavation in Gaillard Cut	168,265,924
Excavation in Gaillard Cut attributed to slides	73,062,600

This last surprising figure is at least a clue to the difficulties encountered in the excavation of the great cutting penetrating the continental

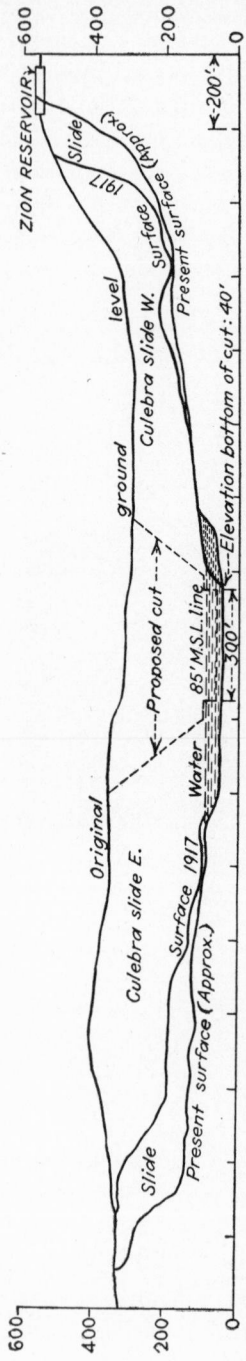

FIG. 11.7.—Cross sections through Gaillard Cut, Panama Canal, showing excavation originally proposed, and the actual shape of the cut in 1938.

(*Reproduced by permission of the Governor, Canal Zone.*)

divide. It is 8.75 miles long. The geological formations encountered during the construction of the canal are mainly sedimentary beds of shale and sandstone, irregular masses of intruded basaltic and agglomeritic dikes and plugs, and volcanic tuff. Their distribution is somewhat complicated and seriously affected by frequent faults. The greater part of the Gaillard Cut is through rock strata which, although variable, did not cause much trouble due to slides. In the Culebra section of the cut, however, a synclinal trough about one mile wide at canal level crosses the cut directly at its deepest portion. The trough is filled with a "fine-grained sandy clayey formation" called *cucuracha;* it is structurally weak, and by reason of its repeated failure, it led to most of the slides that interfered so seriously with the construction of the canal.

These slides began at least as early as 1884, just after the start of excavation by the French; they continued until long after the completion of the canal. One slide of over half a million cubic yards happened overnight; others were of considerable duration. They were of varied types and included upheavals of the floor of the cut, described thus by Colonel Gaillard:

> Most of the slides of the past year (1911–12) were breaks resulting from the failure of an underlying layer of rock of poor quality due to the pressure of the enormous weight which crushes the underlying layer, forces it laterally and causes it to rise up and heave in the bottom of the cut. The heaving at times is 30 feet.

An accompanying photograph shows such an upheaval in the form of an island which appeared overnight out of a depth of 30 ft. of water.

Figure 11.7 represents a cross section through the Gaillard Cut as it exists today and as it was originally projected. From a study of this, coupled with recognition of the structural instability of the *cucuracha* formation, it can be appreciated that the slides were due to the inherent weakness of this material in standing up to the loading to which it was subjected by progress of the excavation, intensified in some places by the extra loading due to the dumping of excavated material too near to the canal. The unsatisfactory nature of the *cucuracha* formation was reported on in 1898 by the French geologists Bertrand and Zurcher, and it has been suggested that their opinions were not utilized. It is therefore of significance to note that the Committee of the National Academy of Sciences (of the United States) appointed by the President of the United States to report on the Panama Canal slides, in 1915, following the closing of the canal in September, 1915, due to blockage by slides, included this statement in their final report:

> The Committee regrets that the United States engineers in charge of the canal excavation have not had the benefit, from the outset, of the best available technical advice in regard to the proper slopes for the canal banks, based on a thorough study of the rocks in the banks, and in regard to the character of the slides. Dr. Howe, who was attached to the canal force under Engineer John F. Stevens, 1906–7, before much progress had been made in excavation, was occupied mainly in the preliminary geological study of the canal route and in special problems which were pressing at that time. Dr. Hayes, the first geologist called upon to examine the Culebra slides, was sent to the canal in 1910, on the request of President Taft, but remained there for a short time only. On his recommendation a geologist [Mr. MacDonald] was attached to the canal corps and for three years rendered valuable assistance, although the great slides were already past control. He was the first to recognize the general character of the deep-seated deformation which characterizes the Culebra slides and explains the upheaval of the bottom of the cut, which was a feature of some movements.

The report of this Committee and its valuable appendices constitute a document of great engineering and geological interest and illustrate vividly the necessity for the application of geological study to all problems such as those which the construction of the Gaillard Cut involved.[11.14]

11.8. Docks and Harbors.—Harbor engineering has always been a leading branch of civil engineering work; it is also one of the oldest, having probably started around the shores of the Mediterranean Sea about 2,000 B.C. Many ancient harbors proved unsuccessful owing to the action of natural causes which may fitly be classed as geological. One of the most famous of ancient ports was Tyre, second port of the

Phoenician empire; it was founded on an island and included two harbors each protected by rockfill breakwaters. In 332 B.C., Alexander the Great destroyed the city after building a solid causeway to the island. This interfered with the local movement of coastal sand, and the channel soon filled up, the site of Tyre being now on a peninsula. The great Roman harbor of Ostia, featured by marine structures which even today command respect, finally silted up owing to sediment brought down by the Tiber, despite the ingenuity of the Roman engineers in designing structures to offset this sedimentation; its site is now 1½ miles from the sea.[11.15] Throughout the intervening 2,000 years, engineers have had to wage ceaseless effort in overcoming similar difficulties. If today the abandonment of a harbor is almost unheard of, this indicates no modification of the action of natural processes along seacoasts but only the availability of dredging equipment and other devices capable of dealing with the great volumes of sediments carried to sea by rivers or moved along the coast by the sea. This latter phenomenon is termed *littoral drift;* it is associated with coastal erosion and so is considered in some detail in Chap. XV.

Modern dock and harbor practice has included some of the most notable civil engineering works of recent years. The extensive dock system of Liverpool, England; the steady development of the port of New York; and the many and varied engineering works in the gulf and tidal portion of the St. Lawrence, Canada—these and many similar developments testify to the magnitude of dock and harbor works. Geological features affect all construction of this kind to some degree although usually in one or more of the ways described in the other chapters of this book. Thus a notable underwater geophysical survey was carried out in the harbor of Algiers, using an electrical method, in order to obtain data as to the position of a rock surface covered by superficial deposits.[11.16] For the construction of the rockfill breakwater at the new harbor of Haifa, Palestine, a large quarry had to be opened up—a striking reminder of the similar construction methods adopted at the adjacent harbors Tyre and Alexandria over 2,000 years before.[11.17] During the construction of the notable harbor works at Valparaiso, Chile, South America, many unusual features developed, one being an interesting illustration of sedimentation processes. The base of a large breakwater extension was formed by depositing sand from dredgers in depths of water up to 160 ft. on to a sea bed consisting of black clayey silt. Slides of the deposited material occurred owing to deformation of the sea bed, over 50 per cent of the material deposited being "lost" as compared to theoretical quantities. It was found during deposition that "everything of a light character —shells, mud and things of that sort—was washed out of it, and the sand that reached the bottom formed a material so hard that an anchor fluke would not enter it."[11.18]

Many similar instances could be quoted covering a wide field of geological interest. Occasionally, special problems arise, as in other branches of work; brief mention may well be made of a recent unusual case, since it has more than local significance. Transportation between England and France across the narrow English Channel is of a nature and a density equaled at few comparable locations. As early as 1800, a proposal to construct a tunnel was made; in 1867, the first definite plans appear to have been advanced, and thereafter public interest in the project continued. A great deal of geological investigation was carried out, 7,000 soundings across the channel being taken in 1875, for example, and almost 4,000 samples obtained from the sea bed. Exploratory shafts were sunk, and there exist today several trial headings which extend for some distance under the sea. The concensus of general geological opinion is that the proposed tunnel (39 miles long) should be constructed throughout in the Lower Chalk Measures, although the continuous existence of this formation across the channel can only be surmised. Mons. Fougerolles has worked out in detail a scheme for disposing of the excavated chalk by grinding it into a slurry and pumping it upward into the sea! The latest report on the project was made to the British Government in 1930; it was favorable, but the Government subsequently decided against the project not on any technical grounds but as a matter of policy.[11.19]

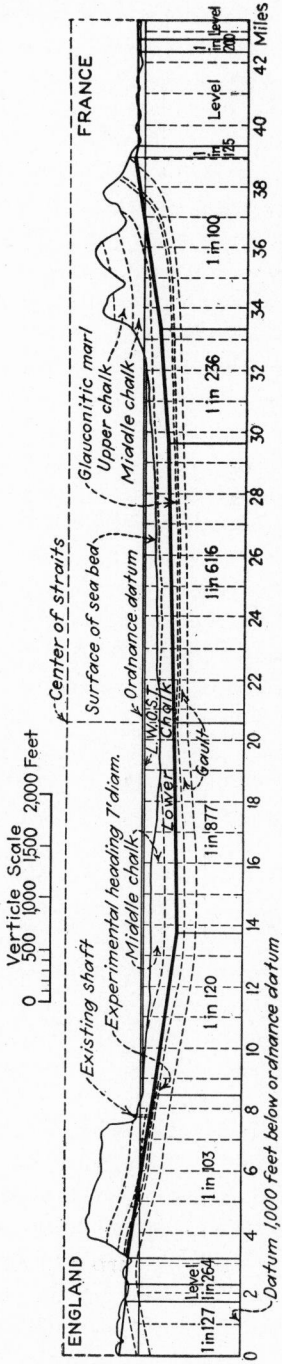

FIG. 11.8.—Geological section along line of proposed channel tunnel between England and France, across the Strait of Dover.
(*Based upon the plan accompanying the Report of the Channel Tunnel Committee* 1930, *by permission of the Controller of H. M. Stationery Office, London.*)

The Southern Railway Company of England thereupon decided to proceed with a train-ferry scheme connecting Dover with a port on the other side of the channel.

The ferry terminal at Dover was designed as an enclosed and watertight dock in which the train-ferry vessels could berth at all states of the 25-ft. tide, the water level being raised or lowered by pumps to the berthing level. The geological formation of the sea bed at the site consists of the Lower Chalk Measures. Work and borings previously carried out in the vicinity and the existence of the underwater headings for the proposed tunnel which was within a few hundred yards distance and which had been practically dry for 50 years suggested that the chalk would be found to be of a solid and homogeneous nature. It was therefore proposed that the work should be carried out in the dry within a cofferdam constructed of steel piling driven into the chalk. This did not prove possible, as the hardness of the chalk limited the penetration of the piles, but a slight modification of the design permitted the work to go ahead. When pumping of the enclosed area was started, it was "found that, although a head of from 10 to 20 feet could be sustained, the difference in pressure . . . caused an inflow of water through the sea-bed in the immediate vicinity of the works greater than the pumps could discharge." Usual methods of sealing were tried, but all proved unsuccessful. Small bags filled with permanganate of potash were placed by divers in fissures in the sea bed outside the cofferdam, the color showing up all over the enclosed area. Further consideration showed that increased pumping might enlarge the fissures and make matters worse. Special methods such as the use of freezing were contemplated but eventually the entire project had to be carried out in the wet instead of in the dry, the placing of large quantities of underground concrete proving to be a notable construction operation. Various explanations of the unusual state of the chalk were advanced, the most interesting of which is that the chalk encountered consisted not of solid chalk but of ancient rockfalls from adjacent chalk cliffs, consolidated by the passage above it of the littoral drift of gravel for which the adjacent coast is noted, over a long period of time. A drill hole put down some miles away for the construction of the Beachy Head Lighthouse showed solid chalk, although similar difficulties were encountered during construction.[11.20]

11.9. Airports.—The development of aerial navigation within comparatively recent years has necessitated similar progress in the location and design of airports the importance of which is apt to be forgotten in view of the ever popular appeal of "flying machines," to use an old but suggestive name. All leading cities today aim at possessing an efficient airport; but because of urban development, landing fields for city use are generally remote from central city areas. Toronto, Canada, with one of its airports on Toronto Island (see page 374) is a notable exception, an unusual local geological feature enabling the field to be located within about a mile of the center of the city. Airport design involves several branches of civil engineering work. Hangar buildings have already featured unusual structural designs; water supply and sewage disposal are always important; and the provision of highway access and utility services introduce engineering work of no small importance. Seaplane bases may include unusual wharf and slipway structures; they must always be so located that no possibility of silting up of the landing area can arise.

It is, however, in connection with the preparation of landing fields that local geology is of importance. The areas will have to be graded to a given level, drained, and provided with suitable runways. Design of the latter is comparable with highway design; similar materials are used for similar design requirements. Turf may sometimes be used to surface a field, in which case special attention will have to be given to the type of grass and to the spreading of a good top layer of soil if the local material is unsuitable for good growth. Drainage is the counterpart of all such work; it will often prove to be the most difficult part of airport work. The fact that the landing field has to be practically level constitutes a leading problem necessitating the closest attention to the gradients adopted for drains in order to keep trench excavation to a minimum.

Drainage systems can clearly be designed efficiently only with thorough knowledge of subsoil conditions all over the site to be used. A soil survey is therefore essential. This can often be carried out with the aid of hand-boring outfits. If grading is extensive, data as to ground material will serve a double purpose, in enabling accurate estimates for the grading work to be prepared. Once these data are available, the type of drainage system necessary can be decided upon. If porous materials underlie the site, the system can be simple; if clay or similar material is to be found, an elaborate system of field drains and main drains may be necessary. Installation of drains must be undertaken with great care, backfilling (of selected porous material) up to within 6 in. of the surface being essential. Surface material should be uniform over the field and therefore should cover all drains and refilled soft spots, the discovery of which in preliminary test borings can be one of the most important contributions of the civil engineer to this phase of transportation engineering. American figures suggest that one accident out of every eight in aviation is due to landing field defects; even in this branch of work, therefore, adequate preliminary investigation of ground conditions can be of real avail.

CHAPTER XII

THE FOUNDATION OF BRIDGES AND COFFERDAM PROBLEMS

Of alle werkys in this worlde that ever were wrought Holy Chirche is chefe. . . . Another blessed besines is brigges to make. . . .

RICHARD FORMAN, about A.D. 1458, in an old MS. now in Christ's Hospital, Abingdon, England (rhyme on the building of Abingdon and Culham bridges).

In discussing transportation routes, considerations were limited to normal sections of roadways and other traffic arteries. Where such routes meet natural obstacles that cannot be circumvented, tunnels or cuttings will be necessary. In a similar way, bridges form a necessary component part of most land transportation routes; and if for no other reason than this, their foundation would call for individual treatment to the extent possible within the limits of this book. The founding of bridge piers and abutments, however, is at once so special and so important a part of civil engineering work that of itself it calls for detailed consideration. Bridge construction is, moreover, such a widespread operation, falling at some time to the lot of most practicing civil engineers, that it merits more detailed study than some other specilized types of work.

Schoolboy memories (which may not be very pleasant) of the piled foundation of a Roman bridge across the Rhine must not be allowed to deflect attention from a subject even the history of which is full of interest. Of the many famous early bridges, London Bridge, over the Thames, England, is one of the best known. According to a third century Roman writer, there was a bridge across the river just above its mouth as early as A.D. 43, and records indicate the continuous existence thereafter of a river crossing of some type at this location. A reference of peculiar interest in this book is that which tells of King Olaf, "The Saint," coming in A.D. 1014 to aid King Ethelred of England against the Danes who held London. His fleet " . . . rowed quite up under the bridge and then rowed off with all the ships as hard as they could down stream [having secured ropes to the piles supporting the bridge]. The piles were then shaken at the bottom and were loosened under the bridge . . ." which gave way, throwing into the river all the defenders ranged upon it.[12.1]

Although such a barbaric use of bridge foundations fortunately has few modern applications, the further history of this bridge is of special significance in this study. The structure known so well through illustrations in history books appears to have been completed in the early part of the thirteenth century. Fynes Morrison, writing in 1671, states that it was founded on "pakkes of wool most durable against the force of water," but this is probably a garbled reference to the tax on wool which enabled the King to pay his share of the cost! The waterway of the river was reduced by this multiarched structure to a width of 195 ft., resulting in such swift rapids between the arches that many lost their lives in passing through, the old saying being that "London Bridge was made for wise men to go over and fools to go under." An act of Parliament passed in 1756 ordered all the buildings on the bridge to be removed and the two central arches rebuilt into one arch, work that inevitably diverted the main flow through this opening and set up serious scouring, which eventually led to the necessity for demolishing the bridge altogether. A modern structure now occupies a site parallel and adjacent to this famous crossing.[12.2]

This is but one of the ancient bridges the piers of which have given rise to trouble. Records are extremely scarce, unfortunately, but it can safely be said that scouring out of the foundation beds adjacent to bridge piers has been, in the past, a major cause of trouble. The piers of ancient bridges rarely failed because of excessive loading on the foundation beds, if only because of the limitation of span length imposed by the structural materials available. The two defects mentioned can be regarded as the two main possibilities of failure to be investigated in the design of bridge piers. Both may readily be seen to be essentially geological in character.

12.2. The Importance of Bridge Foundations.—As a necessary preliminary, the fundamental importance of Geology in the founding of bridges must be stressed. However scientifically a bridge pier may be designed, the whole weight of the bridge itself and of the loads that it supports must be carried ultimately by the underlying foundation bed. The piers and abutments of a bridge are relatively "uninteresting" to the keen structural engineer. Herein should lie, however, more than passing interest, for all too often the design of the foundations is left to the structural engineer responsible for the design of the superstructure. Many structural engineers appreciate the difference between the two types of design work and their relative importance. Some, however, may not—perhaps naturally, since the careful consideration of the materials available as foundation beds and the forecasting of the forces that may in time affect them are different from the determinate mathematical calculations relating to the arrangement of steel, reinforced concrete, or timber to be used for the superstructure.

It may be that in some cases an assumption as to the relatively small cost of foundations compared to the total cost of a bridge may militate against sufficient attention being given to their design. Actual cost records, however, show that the cost of foundations (piers and abutments) generally almost equals the cost of superstructure, even on large bridges, as the typical figures presented in Table L clearly demonstrate. Whatever may have been the cause, it cannot be denied that the importance of bridge foundation design has not always been fully recognized, thus betraying on occasion the basic

TABLE L.—SOME TYPICAL BRIDGE COSTS, ILLUSTRATING THE RELATIVE COSTS OF SUPERSTRUCTURE AND SUBSTRUCTURE

Bridge	Type	Cost of		Ratio $a:b$	Reference
		Superstructure (a)	Substructure (b)		
Cherry St. Bridge, Toledo, Ohio.	7 Concrete arches + Bascule	$ 545,482	$ 531,103	1.02:1	*Am. Soc. Civil Eng. Trans.*, **80**: 789 (1916)
Hell Gate Bridge, New York.	Steel arch, 977.5-ft. span	$2,000,000	$1,700,000	1.18:1	*Am. Soc. Civil Eng. Trans.* **82**: 882 (1918)
Sydney Bridge, Australia.	Steel arch, 1,650-ft. span	£3,202,000	£1,046,000	3.06:1	*Inst. Civil Eng. Min. Proc.*, **238**: 193 (1935)
Iskandar Bridge, Fed. Malay States.	7 steel arches, 160- to 148-ft. span	£51,217	£76,883	0.67:1	*Inst. Civil Eng. Min. Proc.*, **240**: 363 (1937)
Tees (Newport) Bridge, Middlesbrough, England.	Steel lift bridge, 259-ft. 6 in. span	£201,302 (excluding approaches)	£65,688 (2 piers)	3.06:1	*Inst. Civil Eng. Min. Proc.*, **240**: 598 (1937)
Chelsea Bridge, London.	Self-anchored suspension bridge	£149,000	£75,000	1.98:1	*Jour. Inst. Civil Eng.*, **7**: 420 (1938)
Thousands Islands Bridge, Canada-United States.	Group of structures, including 2 suspension spans, steel arch, etc.	An extreme case, in view of unusually favorable foundation bed conditions:→		5:1	*Civil Eng.* (New York), **8**: 408 (1938)

assumption of the superstructure designer—that his pier- and abutment-bearing surfaces will provide him with fixed and solid pedestals on which he can support the structure he conceives, without fear of serious movement.

12.3. Special Preliminary Work.—The first considerations in bridge location are generally those of convenience and economy, foundation bed conditions taking usually a subsidiary place, for the prime requirement of a transportation route is that it shall connect its terminal points by the shortest convenient route consistent with topographical configuration. When crossings of any natural defile have to be made, economical considerations usually limit the choice of site to that calling for the shortest possible structure. The bridge engineer is therefore not usually given much choice of location; in consequence, he has often to accept the foundation bed conditions at a site so determined and, if it is possible to do so, design for them a suitable foundation structure.

This limitation of site is one reason necessitating the most complete data possible with regard to the geological conditions at the site. A still more potent reason is that when once the construction of bridge piers is started, their respective locations cannot possibly be changed except in most unusual circumstances. More than the usual degree of certainty must therefore be attached to the design and anticipated performance of bridge piers and abutments. There is yet a further reason calling for this special care in preliminary investigations. Bridges, as a rule, are constructed to cross river or other valleys—depressions below the normal level of the ground which by their existence suggest some departure from normal geological structure. In districts that, in past ages, have been subjected to glacial action, it is not unusual to find an older river bed or other depression now completely hidden, well below the existing river bed, by either subsequent glacial or river deposits. Such a condition can have a serious effect on design, even if discovered. If the existence of a buried valley is not determined before construction begins, it can lead to untold difficulties. Again, river beds, from their nature, are prone to contain many types of deposits, including boulders; and if preliminary work is not carefully done and correlated with geological considerations, an extensive boulder deposit can easily be mistaken for solid rock. As an example may be mentioned a bridge near Cornwall, Ontario, Canada, which failed in 1898, with the loss of 15 lives, as the result of a pier having been founded on a boulder occurring in a layer of "hardpan" which had not previously been explored by borings and which proved to be only about 2 ft. thick. It was scoured out in the vicinity of one of the bridge piers, disclosing the clay beneath, the pier eventually tipping over and dropping two of the bridge spans into the St. Lawrence which it crossed.[12.3]

Another example has been given in a paper presented to the Institution of Civil Engineers describing the construction of the George's River Bridge, New South Wales, Australia. This paper is

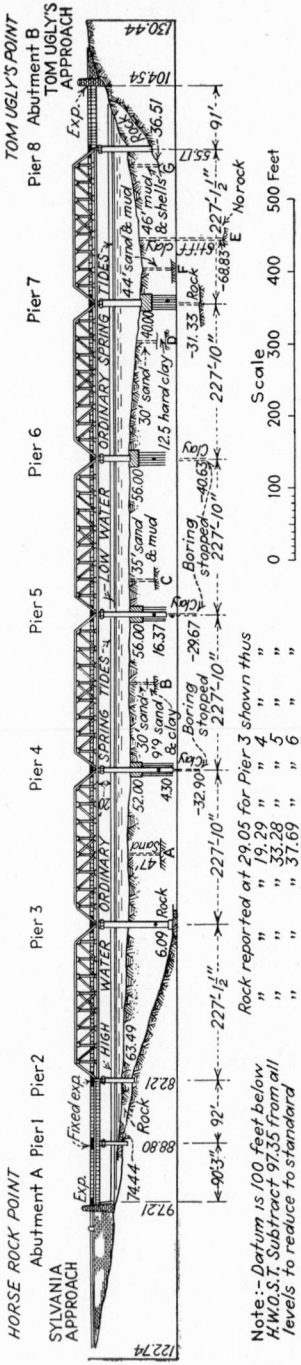

FIG. 12.1.—Elevation of George's River Bridge, New South Wales, Australia and section through foundation strata showing results of original borings and strata as revealed during construction.

(*Reproduced by permission of the Council of the Institution of Civil Engineers, London, from reference* 12.4.)

one of those valuable but all too rare records of difficulties encountered in civil engineering work from the study of which so much more can often be learned than from a description of work that has gone forward without any trouble. George's River flows into Botany Bay, 12 miles south of the city of Sydney. Work toward the construction of a toll highway bridge to replace existing vehicular ferries was begun in 1923. Borings were put down at three possible sites for the bridge, all by an experienced foreman, those at the site finally selected disclosing solid rock at depths below bed level varying between 35 and 47 ft. at regular intervals across a river section about 1,500 ft. wide, the rock at the sides of which was known to dip steeply. On the basis of this information, a through truss bridge of six main spans, supported on cylinder piers, was designed, and a lump sum contract awarded. During construction, it was found that rock existed, as expected, at two only of the seven main piers, additional borings taken to depths of up to 130 ft. failing to disclose any solid rock at all at the other pier sites and, what is even more strange, disclosing no stratum harder than "indurated sand." Construction had to be stopped, and designs changed, in consequence of which the bridge took five years to build instead of two and cost 27.6 per cent more than the contract price. Discussion of the paper naturally emphasized the rigid necessity of having borings most carefully watched by a trained observer, the remark of Dr. Lowe-Brown quoted on page 85 being made in this connection. The absence of geological references in both paper and discussion suggests that neglect of geological features may have been a contributary cause of the trouble that was experienced.[12.4] Although an unusual and possibly exceptional example, the construction of this bridge is a most telling reminder of the supreme importance of preliminary geological information in bridge design.

Another reason for devoting unusual care to the study of the geological conditions at bridge sites in all cases of river crossings is the fact that so much of the ground surface is hidden below water. Dependence has thus to be placed on the results of the underwater borings obtained, correlated with geological structural data secured on the adjacent shores. Where solid rock is encountered, this calls for no unusual attention, provided that the exposed surfaces of the rock show no signs of disintegration or weathering; but if any part of the foundation beds consist either wholly or partially of clay, then it is desirable —in some cases imperative—to obtain samples of the clay in as undisturbed a condition as possible. Suitable test-boring devices have been developed with which undisturbed samples of clay and other unconsolidated materials can be obtained, even through great depths of water. One such piece of equipment was developed especially for investigating underwater soil conditions at a bridge site—that of the San Francisco-Oakland Bridge in California. This bridge project is one of the largest yet undertaken; and as the design and construction of all piers was determined only after the most thorough preliminary investigations, the subsurface conditions will be briefly described.

San Francisco is separated from the district of Oakland by the entrance to San Francisco Harbor in the center of which Yerba Buena Island divides it into the East Bay and the West Bay, transportation across which has been maintained for many years by ferryboats. A long-projected bridge scheme finally reached the construction stage in 1933, the great structure being officially opened on November 12, 1936. Early in the planning of the scheme, the engineers in charge decided upon a program of borings and soil tests to determine the nature of the materials to be encountered and enable them (1) to ascertain the most desirable location for the center line of the bridge, (2) to determine the best location for individual piers, and (3) to select a logical basis for the design of the piers. Preliminary jet borings enabled a contour map of the underlying solid rock surface of the harbor to be prepared, including the West Bay Bridge area. With the aid of additional wash-pipe borings and diamond-drill core borings into the rock, a final design for the West Bay crossing was readily prepared,

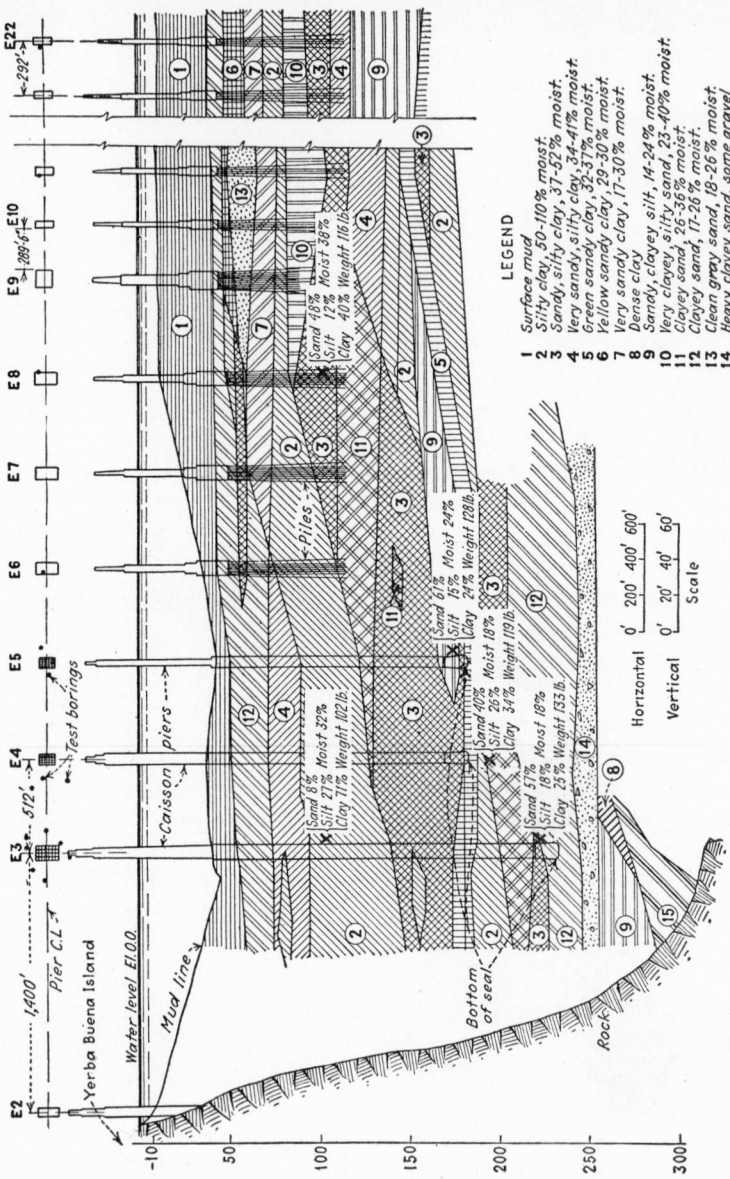

FIG. 12.2.—San Francisco-Oakland Bay Bridge; geological section along east channel section showing results of subsurface exploration and pier location.

(Reproduced by permission of the Editor, Engineering News-Record, from reference 12.5b.)

and piers located and designed, all founded on solid rock, constructed by means of caissons, the behavior of which could be accurately foretold.

The East Bay crossing presented quite distinct problems, since rock was not found by borings at practicable depths, and it became necessary to determine the nature of the overlying unconsolidated material. It was for this purpose that the special sampling device already mentioned was devised, cores so obtained being hermetically sealed without removal from the sampling tube on reaching the deck of the drill barge, to be tested later at the University of California. On opening up the containers, perfect cores were generally found, although in some cases a slight swelling was noticed, possibly due to the change in internal pressure in the sample, coming from its deep river bed. Material was obtained in this way from depths up to 273 ft. below water level. The section on the final center line of the bridge as finally disclosed by the consequent tests is shown in Fig. 12.2, from which the completeness of the subsurface survey will be evident; the use of such data in the design of pier foundations will be to some degree self-evident.

The soil samples were subjected to a rigorous and extensive series of tests, special attention being paid to the question of consolidation, in view of the loads to be carried by the piers. Special features that they disclosed were a confirmation of the fact that in beds of clays and silts, (1) the moisture content decreases with the depth below the mud line, and (2) the moisture content, for clays and silts at similar locations, increases with the fineness of the individual grains composing the mass. In view of the special circumstances of the site, study of probable settlement resolved itself into study of the results of direct consolidation tests and the interpretation of these into specific data for design. Actual loading tests on driven test piles were also conducted; and as a result of all the data so assembled, the nature of the piers could be finally determined.[12.5]

The example just described is in some ways unique, although many features of the preliminary investigations have general application. Adequate test borings, not only along the line of the selected bridge site but on either side of it; careful study of core specimens so obtained; and the correlation of this data with the geological structure of the adjoining dry ground should present a reasonably accurate structural picture of the foundation beds which will enable the designing engineer accurately to locate and to design the bridge abutments and piers. Finally, the necessity for taking all test borings deep enough below the surface of solid material (and especially of unconsolidated material) must be stressed. Loading from bridge supports are always relatively concentrated and often inclined to the vertical. It is therefore doubly necessary to be as sure as possible that no underlying stratum may fail to support the loads transmitted to it, even indirectly, by the strata above.

FIG. 12.3.—Central anchorage pier of San Francisco-Oakland Bay Bridge, anchoring two supsension spans; a pictorial representation showing subsurface conditions and construction details.

(Reproduced by permission of California Dept. of Public Works, Mr. C. H. Purcell, Chief Engineer.)

An interesting example of trouble due to this cause is furnished by the failure of a highway bridge over the La Salle River at St. Norbert, Manitoba, Canada, on October 31, 1920. The bridge was a single reinforced concrete arch, with a clear span of 100 ft., the spandrels being earth filled. The roadway was about 30 ft. above the bottom of the river, the height of the fill placed in each approach being about 20 ft. The bridge abutments were founded on piles, driven into the stiff blue clay exposed at the site and thought to overlie limestone bedrock, as a result of preliminary auger borings and the record of an adjacent well. Failure occurred by excessive settlement, the north abutment dropping 4 ft., bearing piles being bent and broken. Subsequent investigations disclosed the existence of a stratum of "slippery white mud" about 25 ft. below the original surface, failure of which to carry the superimposed load resulted in the bridge collapse. Local soils are sediments from an ancient glacial lake and usually overlie compacted glacial till, under which is limestone carrying subartesian water. The existence of this water complicated the underpinning of this bridge foundation, but the work was successfully completed, and the bridge restored to use. The existence of the white mud was previously unknown in the vicinity; it illustrates the uncertainty of glacial deposition. The occurrence has a special interest for engineers in that, although the bearing piles were driven into the compacted till ("hardpan"), settlement of the abutment occurred as the result of failure of soft material overlying this.[12.6]

12.4. The Design of Bridge Piers.—Although this chapter is not an introduction to bridge-pier design, some attention to design requirements is necessary in order to obtain an adequate appreciation of the relation of geological conditions to this design work. Generally speaking, there are four types of loading, one or more of which may have to be provided for in design: (1) vertical loads, possibly of varying intensity, from truss or girder spans or suspension bridge towers; (2) inclined loads, again of varying intensity and possibly varying direction, from arched spans; (3) inclined tensions, from the cables of suspension bridges; and (4) horizontal thrusts due to the pressure of ice, possible debris, and the flow of water impinging on to the piers and also due to wind acting on the bridge superstructure and piers. Combinations of these several loads will give rise to certain maximum and minimum unit pressures to be taken on foundation beds; from considerations of these results and of the nature of the strata to be encountered, the type of foundation can be determined.

The estimation of the load that the foundation strata at the site of a bridge pier will safely support is generally similar to the same operation for ordinary foundation work. The usual precautions must be observed with regard to strata below the surface that might fail under load. There are two unusual features to be noted, both being reductions which may be applied, under certain circumstances, to the calculated net load on the base area. The first is the allowance that can be made for the displacement by the pier of water and the natural material excavated; the second is the reduction that can be allowed for the skin friction on the sides of the pier in view of the usually large surface area exposed as compared to the base area. These two factors are obviously dependent on the nature of the strata penetrated. The estimation of the first is straightforward, but that of the second is generally a matter of past experience or of experiment during pier sinking. Table M shows a few typical figures as a matter of interest, although the correlation of the loading data shown with the nature of the foundation strata is not definite.

TABLE M.—SOME TYPICAL VALUES FOR SKIN FRICTION[1]

Foundation stratum	Foundation element	Skin friction, pounds per square foot in round figures	Reference
Mud and clay....	Steel and concrete piles	100 to 200	*Inst. Civil Eng. Min. Proc.*, **235**: 478 (1932)
Alluvial deposits..	Steel cylinders	200 aver.	*Inst. Civil Eng. Min. Proc.*, **225**: 244 (1927)
"Rotten" chalk...	Brick monoliths	300 to 500	*Inst. Civil Eng. Min. Proc.*, **224**: 254 (1927)
London clay.....	Cast-iron cylinders	450 aver.	*Inst. Civil Eng. Min. Proc.*, **239**: 156 (1934)
Clay.....	Concrete cylinder	700	*Eng. News-Record*, **88**: 822 (1922)
Gravel and clay...	Steel caissons	750 to 1,000	*Inst. Civil Eng. Min. Proc.*, **235**: 115 (1932)

[1] These values are illustrative only of the general nature of skin friction; for actual use, values should be accurately determined.

Throughout all determinations of safe foundation loads, the nature of the local strata does play an important part; it may even dictate the use of hollow piers, to reduce unit loads, or of such unusual structures as the open reinforced concrete framework abutment supports adopted for the Mortimer E. Cooley Bridge across the Manistee River in Michigan. This singularly beautiful bridge, consisting of two 125-ft. deck truss-steel cantilever arms, supporting a 50-ft. suspended span and balanced by two 125-ft. anchor arms, has its deck level about 60 ft. above the level of the ground on either side of the river, which consists of varying strata of unconsolidated materials which were accurately explored. In order to keep foundation loading on this material to a minimum, the open framework design shown in Fig. 12-4 was adopted and successfully used.[12.7]

When preliminary investigations have shown that the foundation bed material will have poor bearing capacity, consideration may be given to the use of artificial methods of consolidating such material in attempting to improve its bearing capacity. That this is no new expedient is shown by some of the older records of bridge building, *e.g.*, the account given by Leland (Antiquary to King Henry VIII), who wrote in 1538 concerning the Wade Bridge in England, that ". . . the foundations of certain of th' arches was first sette on so quick sandy ground that Lovebone (Vicar of Wadebridge) almost despaired to performe the bridge ontyl such tyme as he layed pakkes of wolle for foundation. . . . " Although this use of wool has been disputed, the record is interesting as demonstrating that some artificial means was used to improve bearing capacity.[12.1] Modern methods are described elsewhere in this book. Grouting is naturally the most common method, but chemical consolidation has been suggested for some bridge works. Another approach to the problem is to consider leaving in place the steel piling of the pier cofferdam in order to confine the foundation bed material and thus prevent lateral displacement, in this way increasing bearing capacity to some extent.

Among the special problems to be faced in bridge-pier design in the solution of which geological information can usefully be applied is that of predicting possible settlement of piers when loaded, in view of the serious effect that unequal settlement can have on certain types of bridge structures, notably continuous truss spans, vertical lift

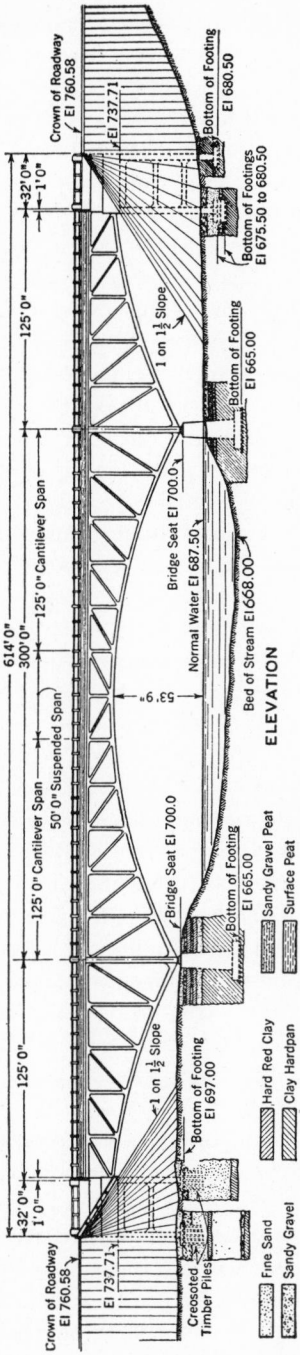

FIG. 12.4.—Elevation of the Mortimer E. Cooley Bridge over the Manistee River, Michigan, showing nature of foundation strata and special abutment structures.

(Reproduced by permission of Michigan State Highway Department, Mr. L. W. Millard, Bridge Engineer and the American Society of Civil Engineers, from Civil Engineering, reference 12.7.)

bridges, and fixed arches. An accurate knowledge of foundation-bed conditions, and the probable behavior under load of the material in these strata, will usually enable accurate predictions of settlement to be made. What happens when uneven settlement does take place is well illustrated by the failure of piers 4 and 8 of Waterloo Bridge, London, as a result of which the whole bridge has had to be taken down, and a new structure erected. Described by Canova as "the noblest bridge in the world worth a visit from the remotest corner of the earth," Waterloo Bridge was constructed from 1811 to 1817 to the designs of John Rennie, who took especial care to protect all his pier foundations—timber rafts on timber piles, bearing on gravel—against scour. Progressive settlement of the two piers mentioned became serious in 1923, the total settlement of pier 4 exceeding 2½ ft. and naturally causing an arching action between piers 3 and 5.[12.8]

This case is cited mainly because of its general interest; as an example it can hardly be called typical, since the conditions encountered were relatively unusual. In a general way, settlement may occur from one or other of the following causes: (1) displacement of part of the bed by scour; (2) lateral displacement of the foundation bed due to lack of restraint; (3) consolidation of the underlying material by elastic deformation of the material, by crushing or alteration of the water content; or (4) failure of an underlying stratum. Of these only the settlement caused by (3) can be called *controlled* settlement, the other three types being of a nature that may cause serious trouble to the structure. All types can be foretold, in the majority of cases, provided adequate preliminary information is obtained. Dangerous possibilities can therefore be avoided, and the controlled settlement accurately predicted, as in the case of the San Francisco-Oakland Bridge foundations just described.

Provision against unequal settlement of piers in the case of relatively small-span bridges has assumed considerable importance in recent years owing to the development of the rigid-frame type of structure, an essential feature of which is that the abutments are *unyielding*, meaning, practically, that any settlement of piers must be uniform. For rigid-frame structures founded on clay, the difficulties of which as a foundation bed will later be fully stressed, some interesting expedients have been adopted, one (at a Canadian National Railways Bridge at Vaudreuil in Quebec, Canada) having been to isolate the bridge foundation from the bearing piles, the load being transmitted through a specially tamped layer of crushed rock. This is an interesting return to an old practice (although for a new reason), the foundations of the present London Bridge, over the Thames in

England, bearing on clay through a similar layer of crushed rock and timber piles, installed almost a century ago.

Stresses set up in foundation beds during the bridge construction must also be carefully considered in design, especially those which may be set up during unbalanced loading caused by irregular construction scheduling. During the construction of the Broadway Bridge, Saskatchewan, Canada, in 1932, a three-span reinforced concrete arch structure, the concreting of the arches proceeded in varying stages, as a result of which the piers tilted when carrying the dead load of only one adjacent arch rib (even though this was supported on centering). Careful records were taken, and a maximum deflection of 0.60 in. recorded. In this instance, the result was not unexpected.[12.9]

The inclined tensions, mentioned as the third type of loading possible, are generally transmitted to anchorages in solid rock, in which case design resolves itself into providing for shearing resistance in the rock which, together with allowance for the dead weight of the anchorage, will be sufficient to balance the tensile forces in the bridge cables. Sometimes there occur cases in which the inclined tension in bridge cables has to be taken up wholly by concrete piers. The Île d'Orléans (suspension) Bridge, in Quebec, Canada, is one such example. The suspension span of this bridge is flanked by long approaches from the shores, the main cables being secured in anchor piers of which one is founded on rock but the other on sand. The stability calculations for these piers naturally had to take into consideration all stabilizing factors, while keeping the unit toe and heel pressures within the limit of the foundation bed material. Friction between concrete and rock and concrete and sand had therefore to be considered, as well as the shearing value of both strata. As a result of these studies, inclined H-beam piles were driven into the sand underlying one of the anchor piers, being left so as to project into the concrete of the finished pier, in this way giving the necessary increase in stability.[12.10]

Another interesting, but unusual, example is provided by the double suspension spans of the West Bay crossing of the San Francisco-Oakland Bridge, the two 2,210-ft. suspension spans joining and being anchored into a central anchorage pier 500 ft. high above rock level. The engineering design of this structure is of unusual interest; its peculiar interest in this discussion is that, under one of the most severe combinations of loading considered, a unit pressure on the rock of 79 tons per square foot may occur, a figure that could be contemplated with safety only after the very thorough preliminary studies made for this bridge project and already described (see Fig. 12.3).

As a final example, there may be mentioned the building of the new Burford Bridge, as part of a new by-pass road across the River Mole in Surrey, England, which is one of the most unusual contacts of special geological features with civil engineering practice known to the author. The bridge is a single reinforced concrete arch span of 80 ft., 100 ft. wide between parapets with specially selected brick facing. Many who know the lovely valley of this small stream, located some 25 miles to the south of London, will know also that its valley is as geologically interesting as it is scenically beautiful. The river flows over the local

FIG. 12.5.—New Burford Bridge on Mickleham by-pass Road, Surrey, England. Reinforced concrete domes over swallow holes Nos. 1 and 2 under construction.

(Reproduced by permission of Mr. W. P. Robinson, County Engineer, Survey, through the courtesy of the Ministry of Transport, London.)

chalk formation which has been so eroded in places by the action of the dissolved carbon dioxide in the river water as to have underground cavities large enough to receive the whole normal flow of the stream. Trial borings were put down at the site of the new bridge and carefully checked to see if any such *swallow holes* in the chalk were revealed. Two soft spots were located, and these proved to be solution channels of this kind, having almost vertical sides and filled up with alluvial matter. It was at first intended to secure adequate bearing for the road and abutments over the holes by driving reinforced concrete bearing piles into the material found in them, but further studies led to a more unusual solution which could be adopted with more certainty. Concrete domes were constructed over each of the holes, founded on circular ledges cut in the chalk around the top of the holes, the largest being 58 ft. in diameter with a rise of 8 ft. The holes were filled up to the undersides of the domes, the filling being then covered with waterproof paper and used as the lower form for the concreting of the domes. Each dome was furnished with an access shaft connecting to a manhole at road level by means of which the engineers may inspect the swallow holes from time to time to see that no dangerous undercutting or further erosion of the chalk is taking place.[12.11]

12.5. The Design of Bridge Abutments.—In addition to having to support, at least partially, loads that the piers of a bridge have to carry, the abutments of a bridge may and often do have to resist another load, that of earth pressure against the face and wing walls of the abutment structure. Design may therefore be a complicated matter, combining the difficulties of both bridge-pier and retaining-wall construction. In the case of abutments retaining artificially formed fills such as those of approach embankments, difficulties of design and construction can generally be foreseen and so guarded against. The careful placing of fill material, working away from the abutments; adequate provision for drainage, by the installation of cross drains along the lower part of the inner faces of abutment structures, connected to suitable weep holes which cannot become plugged up; and the supervision of all fill settlement are perhaps the leading matters for attention in this type of work.

Abutments located on sloping ground, as so many of them have to be, may present different problems; for in addition to having to retain the pressure of earth backing, often with considerable surcharge, they may be subjected to forces set up by the instability of the whole hillside slope; or, alternatively, their foundations may be made insecure by earth movement on slopes below. Again, the forces acting on them are far from being symmetrical, so that the balancing (and especially during construction) is a matter calling often for great ingenuity in design. A few examples will illustrate the type of problem presented and the dependence of abutment design on foundation bed conditions.

Bridges with solid rock providing abutment support will be familiar to almost all engineers, this being an ideal condition and one essential for certain types of arch designs. A famous English bridge illustrates in an interesting way a slight variation from the ideal case, one of the abutments of the Grosvernor Bridge across the River Dee at Chester being founded partially on rock and partially on sand. The bridge has a clear span of 200 ft. and a rise of 40 ft., having been built in the year 1833. Figure 12.6 illustrates the construction that had to be adopted at the north abutment, due to the fact that the outcrop of solid rock suddenly terminated and was succeeded by a deep stratum of loose sand, to which the arch thrust was transmitted by means of timber piles.[12.12]

When no rock is available, abutment designs must follow methods suggested in the case of pier design, subsoil conditions being thoroughly investigated and the foundation-structure design being determined after a full consideration of test results. In many cases of abutment designs on unconsolidated material of low bearing capacity, bearing piles will have to be employed to provide for the anticipated load. Such piles, unless driven at a batter, will offer small resistance to lateral movement; and for this reason, quite a few abutments have

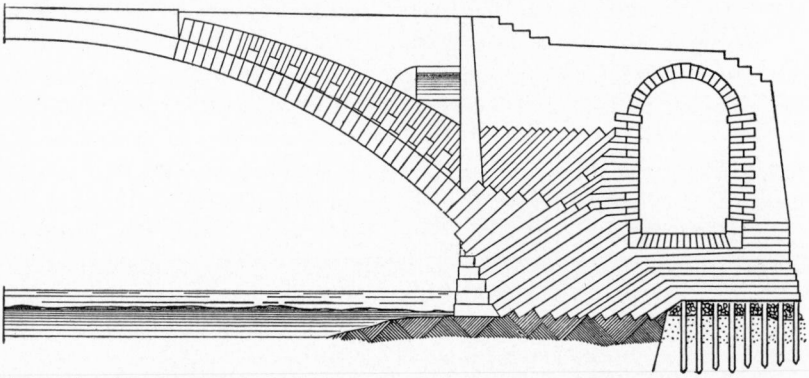

FIG. 12.6.—Cross section through abutment of the Grosvenor Bridge, Chester, England, showing variation of structure with foundation strata.

(*Reproduced by permission of the Institution of Structural Engineers, London, from The Structural Engineer, reference* 12.12.)

been moved inward from their original positions by the excess of earth pressure behind them over the stabilizing forces. Earth pressures and, similarly, the frictional resistance between foundation and foundation bed must receive the most careful attention in all such designs. In some cases, the only solution possible is to brace one abutment against the other by means of underwater struts. One example of this type was that of the Summit Street Bascule Bridge across Swan Creek in Toledo, Ohio, in which a series of braced concrete struts, supported on piles, 4 ft. thick and each 9 ft. wide, placed in 16 ft. of water, had to be installed after abutment movements had seriously affected the bridge operation.[12.13] Many similar cases could be cited, another well-known one being Bridge No. 16 over the new Welland Canal in Canada.

A strange case which may be mentioned as a matter of some interest is provided by the third of a series of three hinged concrete single-span arch bridges, constructed in Holland across the Meuse-

Waal Canal about 1925, each of which had a span of about 212 ft. All were founded on strata of sand and gravel, apparently identical in character, the maximum pressure to be exerted on the ground being the same in each case. When the centering of the third bridge was removed, it was found that the abutments were moving *backward*, horizontally, rotating about a horizontal axis and with some settlement. By means of jacks under the arches, half the movement was removed. As no reason could be found for the movement, the engineers had to design and construct a secondary *friction* abutment. This successfully prevented further movement, and the bridge has since maintained its correct position with no trace of movement. But the cause of the movement has never been discovered.[12.14]

Not only abutment design but even the design of bridge superstructures may be affected by the geological structure of the abutment site. Nowhere else has this been better demonstrated, perhaps, than in the case of the Kohala Bridge carrying the Rawalpindi-Kashmir Road across the Jhelum River on the boundary of the Punjab and Kashmir in North Western India. Originally constructed as a three-span girder bridge (130-ft. center span and two 90-ft. approach spans), the unstable condition of the hillside at the Punjab abutment caused movement of the girders toward the Kashmir side, the Punjab approach span being seriously damaged in 1929 and finally wrecked in 1931 through serious landslides occasioned by heavy rains. The site was examined in 1929 by Mr. W. T. Everall, Deputy Chief Engineer, N. W. Railway of India, who discovered a large fissure in the hillside about 200 ft. above the level of the bridge, this occurring in a slope of loose soil and rock detritus, indicative of future movement. In consequence, after the failure of the Punjab approach span, Mr. Everall had a special design for reconstruction accepted and carried into effect, cantilevering a 70-ft. span back from the first main river pier and connecting it by means of a light suspended span to the hillside road level. Suitable drainage work and stone pitching were also carried out. Although future earth movement has been well guarded against, should a landslide occur it is the intention that it shall displace the suspended span which can safely ride up and over the end of the cantilever span, thus causing little damage to the main bridge structure and being easily replaceable after movement has stopped. The accompanying photograph shows this feature clearly. Engineering ingenuity has here accepted geological forebodings and anticipated trouble in a particularly ingenious manner.[12.15]

An unusual and extreme case of settlement may finally be mentioned. The Pemberton Loop line of the London Midland and

FIG. 12.7.—The Kohala Bridge across the Jhelum River on the Rawalpindi-Kashmir Road, in northern India, seen from the right bank (Punjab side) after the completion of the scheme of reconstruction described in the text. Part of the old abutment can be seen in the lower right hand corner, and on the left is some of the protective stone pitching. The floating shore span is in place, supported by the cantilever extension of the main river span.

(*Reproduced by permission of the Chief Engineer, North Western Railway, Lahore, India*).

Scottish Railway crosses the Leeds and Liverpool Canal by Bridge No. 14 near Wigan, England. The ground adjacent to the bridge site consists of alluvial and glacial deposits overlying the Middle Coal Measures of the Carboniferous formation; the ground surface was originally a few feet above water level in the canal. Owing to extrac-

FIG. 12.8.—Abutment of Bridge No. 14 on Pemberton Loop line (L. M. and S. Rly) near Wigan, England, showing timber trestle support in place, on steel pile box foundation. Successive cap-stones can be seen at the corner of the brickwork abutment structure, showing settlement which has taken place.

(Reproduced by permission of the London, Midland and Scottish Railway Co., Mr. W. Wallace, Chief Engineer, and the British Steel Piling Co. Ltd., London, from reference 12.16.)

tion of coal from the strata immediately under the bridge site and the adjacent area, continual subsidence took place, the fields around gradually becoming an extensive lake. Correspondingly, the plate web girders of the bridge had to be jacked up, and the abutments built up under them to compensate for the settlement. By the beginning of 1935, the headstones on which the girders had originally rested had sunk 13 ft., being only just visible above the canal water level. Subsidence was still continuing, and its final extent could not be ascertained. The engineers therefore decided to construct timber trestles to replace the abutments (although they were to be left in place), the trestles being supported on two series of special boxes constructed of steel sheet piling, all piles being driven as deeply as possible. Jacking up could then easily be done from the trestles which are intended to serve until subsidence has ceased, when the brick abutments can be permanently rebuilt. The building of the steel pile boxes was carried out without interference to traffic and constituted a most interesting piece of construction.[12.16]

12.6. Some Special Features in Design.—In addition to those features of design which have already been mentioned, and the general requirements of structural design, there are a few special factors which have to be considered in certain cases of bridge foundations. The possibility of stresses due to earthquake shocks is one such matter to be considered in certain regions of the world. This subject has already been considered in Chap. XI, and allowance for seismic forces in bridge-pier design calls for little additional comment. An acceleration of one tenth that of gravity has been utilized for this purpose in certain bridges in the western part of the United States. The only complicating factor is the possibility of a large part of the pier being buried below ground level and not responding to the seismic forces exactly similarly to the surrounding mass. Suitable allowance can be made in calculations for this modification. Consideration of this major form of earth movement suggests mention of the obvious necessity for locating bridge center lines away from all fault planes that may be suspected of even slight movement.

The possibility of a change in permanent water level occurring at a bridge site is another matter sometimes requiring attention. An example of the serious effect of such a change, almost classical in nature, is the scouring around the pier foundations of the old Westminster and Vauxhall bridges over the Thames, London, in the early years of the nineteenth century following the removal of old London Bridge and consequent lowering of the water level in the stretch of the river immediately upstream. This opening up of the river channel restored normal flow conditions at the sites of the bridges mentioned, for which the foundations had not been designed, with the result that, in course of time, both structures had to be completely rebuilt.[12.17]

It will be useful to consider briefly the results of bridge construction across a waterway on the natural configuration and flow of the stream. Three main effects may generally be foreseen: (1) The construction of piers and abutments will generally decrease the effective cross-sectional area of the stream, thus inevitably increasing the velocity and raising the water level (usually very slightly) above the bridge; (2) the existence of piers as obstructions to the stream flow will set up eddies around the piers and may possibly institute cross currents in the stream, tending to change it from its normal course below the bridge site; and (3) the combined effect of increased average velocity and eddies may disturb the equilibrium of the bed material between piers and so lead to scouring. All these results represent interference with natural conditions and therefore call for the application of geological data in the engineering solution of the many problems that they bring up.

The first of the foregoing suggestions is of consequence only on rivers subjected to severe flood conditions. Geological factors may affect runoff calculations appreciably, as will be indicated later in Chap. XIV, and it is from these calculations that estimates can be prepared showing what hydraulic conditions the bridge will have to withstand. On many tropical rivers with low-lying banks, bridges have to be designed on the assumption that they will be completely submerged in flood periods; naturally, in all cases of this kind, the geological stability of the foundation bed for piers, abutments, and approaches must be assured before a submergible design can be entertained. An interesting example has recently been described, this being the bridge across the Nerbudda River near Jubbulpore in India. It is 1,222 ft. long, consisting of six 98-ft. and eight 46-ft. spans, all reinforced concrete arches; it is founded throughout on basalt. At flood periods, the bridge is completely submerged.[12.18]

The second and third suggestions are both related to that branch of civil engineering known as *river training work*. Work of this nature has frequently to be carried out in connection with bridge construction projects on rivers with relatively unstable beds, for the reasons covered by the suggestions mentioned above. If the river approaches toward such bridges have not been suitably trained, the result may be that the volumes of water passing between the several piers will not be the same: cases have been known in which the flow between pairs of piers has actually been reversed. It is not only on the great rivers that flow through alluvial plains that river training and associated problems occur. Scouring of a river bed may happen on even the smallest stream if its normal state is interfered with, as by the construction of bridge piers in midstream. It is safe to say that scouring of bed materials from around the foundation structures supporting bridge spans has been responsible for more bridge failures in the past than has any other cause. In olden days, there was more tendency to set up scouring action under bridge spans, because of the relatively small spans and consequent relatively large cross-sectional pier area as compared with water area, a contributory factor being that construction methods then possible did not permit of foundations other than timber piling being constructed much below low water level. Construction methods today suffer from no such limitations, and the general use of longer spans does not cause constriction of waterway areas; the problem still remains, however; and even in the records of modern bridge building, there occur cases of serious scouring out of foundation bed material, two of which are mentioned in the last section of this chapter.

12.7. Some Construction Requirements.—The design of bridge foundation structures can rarely be considered without specific relation to the problems of constructing the piers so designed. Limitations in the restriction of a watercourse during construction, requirements of navigation, requirements of traffic across the bridge in the case of reconstruction work, depths of water and tidal range, and the depth below water to foundation bed level are all features that must be considered in relation not only to structural stability but also to construction methods. Here, too, the geological structure of the underlying strata must be correlated with other design factors.

Construction of piers will generally be carried out by one of three main methods: the use of open cofferdams (working either in the dry or in water), the use of open dredging caissons, or the use of compressed-air caissons. There are, of course, other special methods available. The general arrangement of foundation strata will be a potent factor in determining which of these construction methods shall be followed. In the case of cofferdam work, for example, subsurface conditions will indicate fairly accurately the length of the piles required and will at least suggest what resistance to penetration will have to be overcome in driving. Further notes on cofferdams will be found in the next section of this chapter. Should varying strata be encountered, possibly a combination of two of the methods will be advisable, especially of the second and third, but such a possibility can be considered only if the details of the strata are accurately known.

If caissons are used, one of the first problems to be faced is the *skin friction* which must be overcome in sinking the caisson through the foundation bed material. In making preliminary estimates of the skin friction so to be overcome, knowledge of the strata to be passed through will be some guide, although the determining factors are so variable that only actual experience can be relied upon to give an accurate estimate of the resistance to movement. The overcoming of the resistance can sometimes be facilitated by the use of jetting pipes around the cutting edge of the caisson, discharging water under pressure, but this course can be followed only when it is known with certainty that it will not seriously disturb the surrounding foundation bed material. Table M gives some typical figures for skin friction which will serve as a general indication of its relation to the material being penetrated.

Trouble with caissons is often experienced because of uneven settlement. This may be due to purely mechanical causes, but often it is caused by variations in the underlying strata. In a general way, these may be foretold if a full soil survey has previously been carried out. So many factors are involved that it must be added that soil data can be only a contributory source of information relative to possible movements, to be considered in conjunction with other relevant information. Study of the tilting of bridge-pier caissons (several cases of which are fully on record) shows that generally such occurrences are due to a combination of causes, of which ground conditions are generally the most important, either because of uneven settlement due to varying strata or to the soil material not "breaking down" under the cutting edges of caissons, as had been anticipated. The tilting (through 42 deg.) of the 19,000-ton reinforced concrete open caisson for the east pier of the Mid-Hudson Bridge, at Poughkeepsie, New York, seemed to be due to this latter cause, a stiff stratum of clay and sand, in association with the design of the dredging pockets, resulting in excavation extending more than 10 ft. below the cutting edge. Sudden collapse of the unsupported wall of the bed material seems to have been the cause of the serious tilting that occurred, the righting of which constituted a construction operation of unusual interest and ingenuity.[12.19]

Knowledge of subsurface conditions in advance of the start of actual work will always be of assistance in the selection of construction methods and may even suggest new methods. This appears to have been the case with what is now known as the "sand island method," which was apparently originally evolved by Messrs. M. R. Hornibrook, Ltd., for their contract for the construction of the Grey Street Bridge, Brisbane, Australia, in November, 1927. The main part of the structure consists of three reinforced concrete arch spans, each 238 ft. center to center of piers. The river pier had to be founded in depths of water up to 50 ft. with rock varying from 83 to 107 ft. below high tide level; reinforced concrete cylinders were to be used for the two central piers, and two caissons for the main pier on the south bank. Preliminary information revealed mud as the surface stratum, and this, together with other data, led to the construction of artificial "islands" at each pier site, formed of cylinders of steel sheet piling filled up to above water level with sand. This enabled the cutting edges to be set up in the dry and the sinking of the cylinders and caissons to be carried out under constant control. The steel piling was salvaged, when once the foundation structure had been well founded in the river bed.[12.20] A similar method has been successfully applied to other leading bridge contracts in various parts of the world, notably the Suisan Bay Bridge of the Southern Pacific Railroad, in California (1930), and the Mississippi River Bridge at New Orleans, Louisiana (1934).

A further construction requirement of design is that there must usually be provided the means for thoroughly inspecting and, if necessary, cleaning the surface of the foundation bed on which the bridge pier or abutment is to be founded. In the case of bridges to be founded on unconsolidated materials using low unit-bearing pressures, such inspection is not so important as it is in the case of solid rock foundation beds. Personal inspection of rock surfaces is always a necessary supplement to even the best preliminary core drillings, to check up on the structure of the rock stratum and possible surface disintegration and also to ensure that it has been properly cleaned of all unconsolidated material so that good bond may be obtained between concrete and rock. There is on record at least one case, that of a bridge over the Colorado River of the Union Pacific Railroad, United States, the piers of which had to be rebuilt in 1925 only 15 years after the date of their original construction owing to the fact that the rock-bearing surface had not been properly cleaned before concreting was carried out.[12.21]

12.8. Cofferdam Construction.—Consideration of the essentially geological aspects of cofferdam construction has been left for discussion in this separate section as a matter of convenience, in view of the fact that cofferdams are used for much general foundation work in addition to providing facilities for the construction of bridge piers and abutments. They are now so common a feature of construction work that they are generally accepted without much thought being given to their development. It is sometimes believed that they are a relatively modern development, frequent reference being made to their "early" use on the river Thames in England for the construction of Waterloo Bridge (1809–1817), to which extended reference has already been made. The following extract from Vitruvius, dated probably about 20 B.C., is therefore of special interest:

Then, in the place previously determined, a cofferdam, with its sides formed of oaken stakes with ties between them, is to be driven down into the water and firmly propped there; then, the lower surface, inside, under the water, must be levelled off and dredged, working from beams laid across; and finally, concrete from the mortar trough—the stuff having been mixed as prescribed above—must be heaped up until the empty space which was within the cofferdam is filled up by the wall. . . . A cofferdam with double sides, composed of charred stakes fastened together with ties, should be constructed in the appointed place, and clay in wicker baskets made of swamp rushes should be packed in among the props. After this has been well packed down and filled in as closely as possible, set up your water screws, wheels, and drums, and let the space now bounded by the enclosure be emptied and dried. Then dig out the bottom within the enclosure.[12.22]

Cofferdams designed to be pumped dry (as distinct from those used merely to provide a still-water area) perform one main function, that of retaining the surrounding water and unconsolidated materials, thus providing an exposed dry area of river, lake, or sea bed on which construction operations can be carried out. The retention of sand, clay, or other foundation bed material, *e.g.*, after excavation has proceeded inside a cofferdam lower than the bed level outside, is not a difficult matter, being a special case of the retaining structures already considered. The retention of water above bed level is also a relatively simple matter provided that the piling used to form the cofferdam has watertight joints, calculations for the necessary bracing to support the water pressures being straightforward. One main problem remains, that of preventing the ingress of water around the lower edge of the piling—in other words, that of preventing *blows*.

The usual type of cofferdam structure consists of a single continuous wall of interlocking piling, driven into the foundation strata for a distance sometimes determined by the resistance offered to pile driving, sometimes merely on the basis of previous work, but almost always in an empirical manner. Experience must ever be a guide in such matters, and the actual evidence presented by the driving of the sheet piling into the strata to be met with will be a telling indication of the underground conditions encountered. Cofferdams have generally to be designed prior to construction; materials have to be ordered before they can be used. Estimates of piling penetration necessary are therefore an essential preliminary; and in this estimating work, accurate data on the strata to be encountered will be of value.

Permeability is naturally a soil characteristic of leading importance in such considerations, but it cannot be considered alone. Sands and gravels, when revealed by test borings, will naturally suggest the necessity of deep penetration and, if sufficiently permeable, may even dictate the use of a clay blanket all around the outer face of a cofferdam, as a means of sealing the direct path that would be taken by water when pumping inside began. Clays, on the other hand, will generally provide an impermeable barrier to the seepage of water, provided the piling is driven below any influence from surface disturbance. In cofferdams constructed in tidal water, or in any watercourse liable to fluctuation of water level, the greatest care must be exercised with all cofferdam structures in clay. Under certain maximum water-level conditions, the clay will be compressed by one face of the piling and may possibly deform permanently, leaving a gap between piling and clay when the water level changes to the other extreme. This type of water passage is a far easier course for the flow of water to follow than even the pores in sand and gravel; and if this condition is allowed to develop unobserved, serious damage may result, and a bad blow occur.

These notes indicate the importance that subsurface conditions occupy even in this detail of construction; they emphasize, above all else, the necessity for knowing as accurately as possible the nature of the material to be penetrated. Again, careful tests are necessary to distinguish various soil types. The author knows of at least one case in which a material that to the untrained eye appeared to be clay proved actually to be compacted limestone flour (with practically no clay content at all). It was assumed to be clay in cofferdam design, and yet it proved in practice to be "as porous as a sieve," leading to most serious trouble with blows.

Another type of cofferdam sometimes used for larger structures is that known as *cellular*, steel sheet piling being driven in the form of large cell structures which are then filled with sand, and which are stable as gravity structures.

One of the most extensive of recent cofferdam operations utilizing steel sheet pile cell units was that carried out during the construction of the Grand Coulee Dam, Washington. The Columbia River was diverted first into one half of its original bed and then into the other by cofferdams which were about 3,000 ft. long. The piles were driven generally into a consolidated glacial deposit which was difficult to penetrate, resulting in splitting and deformation of the steel piling. When excavation was carried out within the cofferdams, a sand seam 4 to 10 in. thick was encountered exposed across the excavation at an elevation just below the general elevation of the bottoms of the steel piles. In the west cofferdam, no trouble was met with due to this seam; but in the east cofferdam, owing to some complicating factors, including the deformation of the bottoms of steel piles, a leak developed through the seam which reached a maximum flow of 35,000 (U. S.) g.p.m. Much trouble resulted, including the wrecking of a small part of the cellular dam, some of the steel piles being split from top to bottom. The emergency was met by a combination of engineering skill and construction ingenuity the record of which is of great interest to all engineers, and the leak was finally stopped without serious interruption of the main dam-construction operations. The only feature of this emergency work that can here be mentioned is that 4-in. well

FIG. 12.9.—A general view of excavation proceeding behind the river section of the main cofferdam at the Grand Coulee Dam, on the Columbia River, Washington.
(Reproduced by permission of the Commissioner, U.S. Bureau of Reclamation.)

drills were used to locate the main passage through the sand seam that the leak was taking, cavities being found from 1 to 3 ft. deep. This passage was successfully plugged, despite the great flow through it, by the use of a special grout, ordinary cement grout washing out before it could set. The type of grout finally evolved for this unusual purpose is of interest in view of the frequent necessity for plugging leaks in cofferdam work, and its "prescription" is therefore given in Table N.

Four hundred batches of this grout were used to plug the leak described, the grout being pumped through a 7-in. concrete pipe line and forced into the passage under a pressure of 35 lb. per square inch.[12.23]

Another way in which geological advice can be of appreciable assistance in cofferdam work is in foretelling, with some degree of accuracy, the presence of boulders in the strata through which piles have to be driven. The presence of even one boulder can, on occasion, cause as much trouble in constructing and pumping out a cofferdam as all the rest of the work put together, as all engineers will know who have had such an experience. Test borings may possibly show up the existence of boulders; but unless the boulder formation is closely packed, it is quite probable that the few test borings put down on the

TABLE N.—GROUT "PRESCRIPTION" DEVELOPED AT GRAND COULEE DAM[1]

Sawdust	8 cu. ft.
Shavings (thin and wispy, from refinishing machines, measured loose)	10 to 16 cu. ft.
Portland cement (two sacks)	2 cu. ft.
Sand	6 cu. ft.
Lump bentonite	50 lb.

Mix the foregoing ingredients, dry, in a concrete mixer for about half a minute, and then add:

Water	8 to 10 cu. ft.
Pulverized bentonite (5 per cent by weight, of water)	30 lb.

After adding second part of the batch, mix 1½ min. more.

Product will be a 14-cu. ft. batch resembling wet cement plaster.

[1] Reproduced, by permission, from *Eng. News-Record* (*New York*), **119**: 15 (1937).

actual site of a bridge pier may miss any boulders that may be present. On the other hand, knowledge of the nature of the geological strata to be encountered in foundation work will at least suggest whether or not boulders are to be expected, *e.g.*, as in glacial deposits, in which case extra care may be exercised in the preliminary test-boring work. The value of foundation records for neighboring works can often be of great value in an application of geological considerations; their value has already been generally stressed but may be specially emphasized for this case.

Some cofferdam construction which was featured by unusual difficulties with boulders was an important part of the building of three new bridges across the Cape Cod Canal in Massachusetts, in 1934–1935. Wash borings had disclosed the existence of glacial deposits of sand and granitic gravel containing numerous granite boulders. Specifications were drawn up calling for foundations to be constructed inside steel sheet pile cofferdams driven to predetermined depths. What was not known and was revealed only by the foundation operations was the variable amount of boulders, their relatively large size, and their varying distribution. The contractor for the work estimated that 10 per cent of the piling might strike boulders that would have to be removed; actually, as much as 40 per cent of the piling in one cofferdam was so obstructed; and at all the main cofferdams for the three bridges, serious difficulties were encountered. It was found, for example, that the disturbance caused by pile driving and the operations for removing the boulders so disturbed the sand that it went *quick* and thus resulted in large and frequent boils. Ordinary pumping methods failed to keep pace with the flow of water into the cofferdams, to facilitate the removal of the boulders; and eventually an extensive well-point installation had to be used both inside and outside the sheeting. The presence of boulders had been revealed during the original construction of the canal, 20 years

FIG. 12.10.—Cofferdam for the Bourne south channel pier, Cape Cod Canal Highway Bridges, showing irregular driving of steel sheet piling due to presence of boulders, some of which are to be seen between tracks and cofferdam.

(Reproduced by permission of Messrs. Fay, Spofford and Thorndyke, Consulting Engineers, Boston, Massachusetts, U.S.A.)

before, but whether or not this earlier construction indicated the presence of boulders in the amount encountered at the bridge piers has been a matter of controversy.[12.24]

12.9. Inspection and Maintenance.—As a concluding note, mention may well be made of the importance of the inspection and maintenance of bridge foundations. The regular and thorough inspection of all civil engineering structures is a matter that can hardly be overemphasized. It is always important but nowhere more so than in the case of such features as bridge piers, underwater parts of which are not seen from day to day and yet are most prone to periodic damage. In the case of piers and abutments founded on dry land, inspection of the structures and regular checks on their positions and levels are the main requirements of inspection. Should any movement of the pier be detected, and no deterioration of the constituent material be evident, then inevitably some feature of the underlying geological strata is responsible for the trouble; and in consequence, foundation bed conditions must be investigated anew.

An outstanding example of the benefits to be derived from inspection work is given by the famous Lethbridge Viaduct of the Canadian

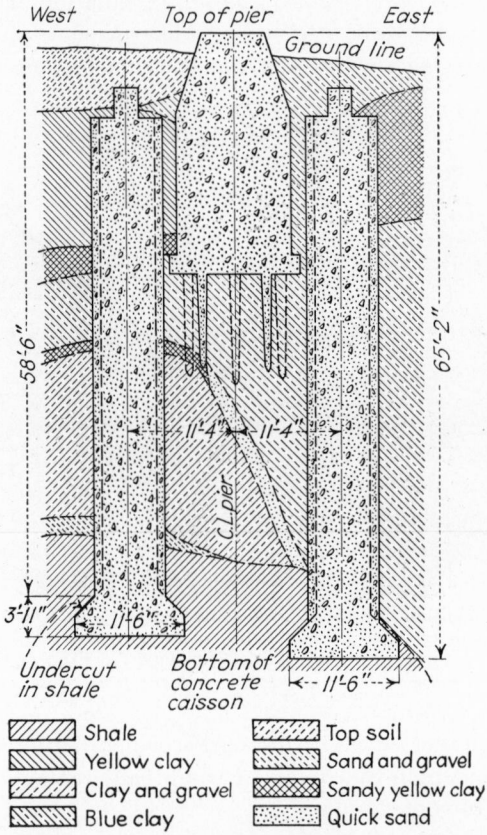

FIG. 12.11.—Pier 59 S., of the Lethbridge Viaduct, Canadian Pacific Railway—a geological section of foundation strata, showing pier as reconstructed.

(Reproduced by permission of the Canadian Pacific Railway Co., Mr. J. M. R. Fairbairn, Chief Engineer.)

Pacific Railway, a steel trestle structure 5,327 ft. long carrying track 312 ft. high above the bed of the Belly River of Alberta, Canada. The trestle bents were founded on concrete pedestals supported on concrete piles driven into the clay which overlies the whole site. Completed in 1909, the deck of the bridge showed sign of settlement soon afterward, a line of check levels run in 1913 disclosing a subsidence of 2½ in. at pedestal 53 south. Cracks and gopher holes in the clay below were plugged with clay, and drainage arrangements improved, but settlement continued; eventually, the pedestal had to be underpinned by caissons carried to shale rock 50 ft. below ground level, and a drainage tunnel constructed; as a result of these measures, subsidence has been arrested.[12.25]

The inspection of piers founded below normal water levels is a more extensive operation, involving underwater work and also the regular sounding of the river bed or watercourse between piers and for

FIG. 12.12.—Lethbridge Viaduct, Canadian Pacific Railway, from the N.W.
(Reproduced by permission of the Canadian Pacific Railway Co., Mr. J. M. R. Fairbairn, Chief Engineer.)

some distance on either side of the bridge site. Underwater inspection from a diving suit, although generally a matter for the field inspecting staff, is something that should regularly be undertaken also by the bridge engineering staff. It is not necessary to stress here the value of diving experience to engineers, but the importance to an engineer of being able to see the effect of his bridge structure on river bed stability is a special feature which may fittingly be mentioned. Examination of the pier structure is a part of underwater work, but equally so is the study of actual bed conditions, especially around the pier bases. This inspection, coupled with soundings over the whole water area adjacent to the bridge, if performed regularly will be a constant check on the dangerous possibility of river bed scouring. The regular nature of this inspection work must be stressed. Naturally, inspections should be timed to be in accord with periodic danger periods *e.g.*, those caused by regular spring flood discharges in rivers. They must also be supplemented by special surveys if any new structures are built in a waterway near enough to the existing bridge to affect it, even to a slight degree. Interesting records are available of the methods and programs adopted for bridge inspection work by leading railways. It may perhaps be mentioned that in order to speed up the inspection of bridge piers in shallow water, Canadian National Railways employs on one of its inspection crews an expert high diver who carries out underwater inspection without the cumbersome aid of a diving dress![12.26]

The importance of these and other features of bridge inspection work is well illustrated by two bridge failures that occurred in the United States in the late summer of 1933, both of which caused loss of life, injuries, and serious damage to stock and structures. The first occurred at the Anacostia River Bridge of the Pennsylvania Railroad, near Washington, District of Columbia, a four-span deck plate-girder structure, which, owing to the failure of the center pier, collapsed when a train was passing over it. Subsequent investigation showed that the gravel stratum on which the piers were founded had been seriously eroded, erosion probably culminating in that due to a tropical storm which was abating only when the collapse occurred. The railroad company had no record of soundings or underwater inspection from the time the bridge was built in 1904 until the date of failure, stating that as no settlements or cracks had developed, unsafe conditions under water had not been indicated. Some dredging had been done downstream, after the completion of the bridge, but at the investigation it was stated that even if this had affected the river bed at the bridge " . . . it was the duty of the railroad company to keep informed concerning changes that might have any effect on the safety of the bridge. . . . "

The second accident, which also occurred in August, 1933, was on the Southern Pacific Lines between Hargis and Tucumcari, New Mexico. It, too, was due to unusual floodwaters which undermined the east abutment of the deck plate-girder structure and thus caused the collapse of the bridge as a train was crossing. The Bureau of Safety of the Interstate Commerce Commission, in its report on the accident, attributed the failure to heavy rains resulting in excessive floodwater, the position of a highway bridge (constructed later, 150 ft. upstream) increasing the current velocity and probably diverting it against the railroad approach embankments; and to the fact that the embankment had not been protected against erosion.[12.27]

Mention of such illustrative examples prompts the thought that it is always easy to be wise after the event. This cannot be disputed; and yet at the same time, it must be recalled that one of the surest ways of learning constructive methods is by the study of past mistakes and errors. Bridge inspection can provide data of general value only by records of the discovery of features unsuspected and of serious consequence; and if some of these have been discovered too late, they may still serve as a reminder of the supreme importance of the regular examination by the engineer of the foundation strata in which he places such confidence as the sure and certain support of his bridge structures.

Suggestions for Further Reading

McCullough, C. B.: "Highway Bridge Surveys," *U. S. Dept. Agr. Tech. Bull.* 55, 1928 (reprinted 1938).

Standard textbooks on Foundation Engineering will naturally be found to contain useful data on the relation of Geology to bridge foundations.

CHAPTER XIII

THE FOUNDATION OF DAMS

> Formerly the richest countries were those in which Nature was most bountiful; now the richest countries are those in which man is most active. For in our age of the world, if Nature is parsimonious, we know how to compensate for her deficiencies; our engineers can correct the error and remedy the evil. For the powers of Nature, notwithstanding their apparent magnitude, are limited and stationary; but the powers of man, so far as experience and analogy can guide us, are unlimited.
>
> HENRY THOMAS BUCKLE.[13.1]

The construction of a dam to retain water causes more interference with natural conditions than does any other civil engineering operation. The validity of this assertion will be realized after even cursory consideration; the statement will later be amplified and explained, but of itself it constitutes a leading reason for the devotion of special attention to the foundation of dams. Equally striking is the critically important function that dams perform in storing water for domestic supply, for the generation of power, and for irrigation. The reliance that must be placed on structures carrying out these functions together with their existence in all parts of the world and in surprising numbers, ranging from the smallest timber check dam to such structures as the Lloyd Barrage or the Boulder Dam, combine to provide further reasons for the devotion of particular attention to dam-foundation design. Finally, although failures of civil engineering works are always of serious consequence, dam failures are possibly more serious than others, since they generally occur during periods of abnormal weather, often without warning, and almost always with disastrous results. Foundation bed defects are an unfortunate factor in many dam failures, and another telling argument is thus presented in favor of the necessity of neglecting no single feature of foundation beds that may possibly affect the dam structure that is to rest upon them.

Although the benefits to be derived from a critical study of engineering failures are duly stressed in other parts of this book, it may be noticed that constructive examples are offered, whenever possible, of works carried to successful completion with the aid of methods and investigations described. A slight departure from this course will be made in the case of dam structures, and a brief review will be presented of some dam failures of the past. This is no descent to the almost hysterical attention devoted to dam construction whenever a failure does occur, at which times the pages of the popular press would almost lead one to believe that dam failures are a regular occurrence. On these occasions, not a word is said of the thousands of successful dams performing their tasks without fear of failure, nor is the fact often mentioned that many other branches of civil engineering, *e.g.*, water purification, are far more liable to endanger the lives of many people. Dam failures are presumably "good news." It is not as news items that they will be considered here but rather as providing useful illustrations of some of the reasons for failure of engineering structures and of the significance of foundation bed conditions among these reasons.

13.2. Historical Notes.—Among dams of ancient times, that constructed by Joshua to assist the Israelites to cross the River Jordan is one of the earliest to be recorded, although the Marouk Dam across the river Tigris is of even older origin, its construction having been carried out in almost prehistoric times. Built for river regulation, it lasted until the end of the thirteenth century. The first masonry dam of which there are good records is that built by Menes, first King of the first Egyptian dynasty, some time before 4000 B.C. Located 12 miles south of the ancient city of Memphis, to provide the site for which it was constructed to divert the waters of the Nile, it was 1,500 ft. long and at least 50 ft. wide. It was maintained for 4,500 years and then neglected. Another famous old dam was that near Yemen in Arabia, of which little is known beyond its construction date (1700 B.C.), its great size (2 miles long, 120 ft. high, and 500 ft. wide), and the fact that it failed about A.D. 300 in a flood recorded in Arabian literature.[13.2]

These few examples are cited to illustrate the longevity of dam building as a major branch of civil engineering. Further examples could be described; but as information about foundation bed conditions is almost completely lacking in ancient records, they would have no particular reference to the purpose of this chapter. Coming to comparatively modern times, the Puentes Dam in Spain may usefully be mentioned. Constructed in the years 1785–1791, it was 925 ft. long, curved in plan, and 152 ft. thick at the widest part of the base, with a maximum height of 167 ft., a perfectly safe design from the structural point of view. Constructed of rubble masonry and faced with cut stone, it was finished off with ornamentation befitting its standing as one of the wonders of Spain. Although apparently founded in part on bedrock, a gravel pocket about 67 ft. wide, was encountered near the center of the dam site and the founding of the dam across this gap was solved by an ingenious piled design carrying a grillage and protected by an upstream apron. Although serving satisfactorily just after the dam was built, this method of overcoming geological difficulties was not reliable, with the result that on April 30, 1802, that part of the dam above the earth pocket failed, following the washing out of the underlying foundation beds. Still available is an eyewitness's account of the disastrous collapse of this "plug" of masonry, which went out like a cork, leaving what appeared to be two massive bridge abutments with a gap between them 56 ft. broad and 108 ft. high. This failure was somewhat exceptional, since other large dams, constructed in the same period, functioned successfully and thus inaugurated modern practice of dam design and construction.[13.3]

13.3. Dam Failures.—The record of dam failures in succeeding years provides a useful if a somewhat melancholy study. Analysis of the cause of failures indicates fairly definitely that the main reason has been the provision of inadequate spillway capacity, with defective foundation bed conditions as the second cause, these two accounting for the majority of all recorded failures. Spillway capacity is determined mainly from anticipated runoff from the catchment area above the dam, a subject on which geological conditions have a considerable indirect influence, as will be explained in Chap. XIV. The second group of failures noted are all dependent essentially on geological features, although the specific reason for the failure may vary from one case to another.

A number of reviews of dam failures have been prepared, and the number of failures so listed is remarkable. Dr. H. Lapworth mentions that over 100 failures of dam structures due to undermining of water-bearing beds below the dam foundation occurred between 1864 and 1876.[13.4] Mr. J. D. Justin lists details of over 60 failures of earth dams alone, between 1869 and 1919.[13.5] The study of failures in such numbers, although interesting and impressive, is not helpful with regard to avoidance of trouble in future work; a few special cases will therefore be referred to in detail. In his paper, Dr. Lapworth mentions several interesting cases including the Hauser Lake failure which bears a striking similiarity to the Puentes Dam failure, since a central portion 400 ft. long, founded on water-bearing gravel, was destroyed in 1908, the remaining sections being left intact.

Taking an example from relatively recent practice, there may be mentioned the Colorado Dam which was constructed in 1891–1892 to provide a power service for the city of Austin, Texas. Difficulties were encountered during construction, the Chief Engineer responsible for the work resigning owing to interference with his work. Eventually the work was finished, and the dam was placed in service, a structure 1,275 ft. long, of masonry, 68 ft. high and 66 ft. wide, resting on limestone, clays, and shales of the Cretaceous formation. The limestone strata were almost horizontal but alternated in texture, some being much harder than others. It is also said that the limestone was dissolved in places, giving rise to underground caverns, one of which was described by workmen but is not mentioned in engineering reports.

FIG. 13.1.—The site of the St. Francis Dam, California, U.S.A., as it was after the structure had failed.

(*Photograph, Wide World Photos, Inc.*)

The clays and shales were of a slippery nature and very broken, the site being on a fault zone. Flood flows over the dam caused some erosion of weaker strata and at least one serious washout which was repaired. This inevitably contributed to the sudden failure of the structure on April 7, 1900, when a central section collapsed completely, the dam then being overtopped by 11 ft. of water due to severe flood flow. The concentration of flow through the resulting gap caused two adjacent blocks, each about 250 ft. long, to be moved bodily downstream for a distance of about 60 ft. The causes of failure are complex, but they appear to include removal by erosion of rock supporting the heel of the dam, the slippery nature of at least part of the foundation, and the presence of percolating water under the base of the dam.[13.6]

The failure of the St. Francis Dam will be of more recent memory. This great gravity dam, 205 ft. high and 700 ft. long, was completed in 1926 by the Bureau of Water Works and Supply of the City of Los Angeles to assist in the storage of water for that well-known Californian community. It was located in the San Francisquito Canyon. Curved in plan and connecting with a long low wing wall, it created a reservoir with a capacity of 38,000 acre-ft. The dam was founded partially on schist and partially (for about one third of its length at the southwest end) on a reddish conglomerate with sandy and shaly layers. The contact between the two rocks is a fault, generally recognized to have been long inactive. Although the schist is a relatively sound rock, the conglomerate

> . . . is by no means a strong rock. A test . . . gave a crushing strength of 500 pounds to the square inch. . . . When wet, the rock shows a considerable change, [a sample starting] to flake and crumble when placed in a beaker of water and in about 15 minutes slumps to the bottom of the vessel as a loose gritty sediment that can be stirred about with the finger. . . . So far as can be ascertained, no geological examination was made of the dam-site before construction began and no crushing or immersion tests were made of the conglomerate.[13.7]

Seepage was noticed through the conglomerate when the reservoir first became full (in March, 1928); eventually, the dam failed with disastrous results which will still be generally familiar. Boards of inquiry were appointed by the state of California and by the city of Los Angeles; both agreed that the main cause of failure was the nature of the rocks under the dam. Quoting Dr. F. L. Ransome again:

> The plain lesson of the disaster is that engineers, no matter how extensive their experience in the building of dams or how skilful in the design of such structures, cannot safely dispense with the knowledge of the character and structure of the adjacent rocks, such as only an expert and thorough geological examination can provide.[13.7]

13.4. Review of Dam Construction.—A return may now be made to the statement with which this chapter started, and consideration be given to the effects of dam construction. A dam is an artificial structure erected to support a waterproof membrane designed to retain water above the level that it normally occupies at the site of the dam, suitable provision being made for passing a certain calculated flow of water past the dam, through it, over it, or around it depending on local circumstances. The membrane may be and generally is an integral part of the dam structure. It may be supported in several ways by such varying designs as an earth-fill dam, with the membrane either on the upstream face or in the center as a core wall; a rockfill dam, with similarly placed membranes; a gravity dam of masonry or mass concrete, the membrane being the upstream face of the dam itself; or a reinforced concrete dam of the arched type, multiple arch, multiple dome, or some other special design—all of these having an unbroken reinforced concrete skin serving as the waterproof membrane. It will be seen that dam structures can be generally grouped into two main divisions, earth and rock fill dams depending for their stability on the natural repose of unconsolidated material, and concrete or masonry dams depending for their stability on the structural performance of the material used for construction. It may usefully be mentioned that the choice as to which of these two types of dam shall be constructed at any location must be determined mainly from geological considerations, whereas the actual kind of dam, when once its general type has been decided, will be dependent to some extent on geological conditions affecting the supply of structural materials.

All dam structures retain water to a certain predetermined level. In this way, the flow of water in the watercourse being regulated is seriously affected below the dam site, the flow generally being regulated to a more uniform discharge than that given by the stream itself. In addition, the underground water conditions in the valley above the dam location are completely changed, the level of the underground water table being raised at least to the water level of the reservoir near the water line, with consequent changes of decreasing importance farther up the valley. Below the dam, the level of the water table may be lowered if normal stream flow is depleted. Between the two sides of the dam, there is thus set up a considerable difference in groundwater level. Although the waterproof membrane generally extends from side to side of the dam site, effectively isolating the two groundwater tables, this artificial condition will always exist while water is being retained. The dam structure will exert unusually high unit pressures on certain parts of the underlying foundation beds. These will be submerged well below water level in the reservoir area and so be subjected to hydrostatic pressure, which may be appreciable in the case of high dams. These are the main reasons underlying the assertion that a dam causes more interference with natural conditions than does any other type of civil engineering structure. In the special case of weirs, or dams founded on porous strata, in which the waterproof membrane merely deflects the flow of underground water but does not stop it completely, the serious change in the underground conditions caused by the dam will be obvious.

In the case of all dams designed to seal completely the flow of water, surface and underground, down the valley that they cross, their action in resisting the pressure of the retained water gives rise to four main geological problems: (1) determination of the soundness of the underlying foundations beds, and their ability to carry the designed loading; (2) determination of the degree of watertightness of the foundation beds at the dam location and of the necessary measures, if any are needed, to render the underlying geological strata quite watertight; (3) a study of the effect on the foundation bedrock of prolonged exposure to water; (4) an investigation of the possibility of earth movements occurring at the site of the dam, and the measures to be taken as a safeguard. Some of the geological problems affecting the reservoirs formed by dams may be encountered also during dam construction; but for convenience, they will be considered separately in the following chapter which is devoted specifically to reservoir problems.

Dams founded on porous foundations present their own special problems, mainly associated with the controlled flow of water beneath the dam structure. It may not be possible to make any dam foundation absolutely tight, but the slight leakage occurring with dams on so-called *impermeable* foundation beds, although possibly causing uplift pressure, is generally of no other consequence. In the case of those on admittedly permeable foundation beds, the exact nature of the flow of water through the underlying strata is a vital part of the dam design, and so accurate knowledge of the water-carrying properties of the unconsolidated materials encountered becomes a matter of importance. The accumulation of silt above dams, deposited as the water comes practically to rest in the reservoir formed by the dam, is another associated geological problem; it will be studied generally in Chap. XV. The scouring out of exposed strata immediately below dam structures, by reason of the excessive velocities of flood flows discharging over or around the dam, is also a matter of importance in dam design.

Many special problems naturally arise in connection with dam construction, so varied are the sites in location and nature, and so great are the possibilities for different designs. As special problems, interesting though they may be, have not usually a general application, they cannot receive attention here, apart possibly from one installation which is believed to be unique. This is at the Cheurfas Dam in Algeria, constructed in 1882, with a maximum height of 107 ft. Soon

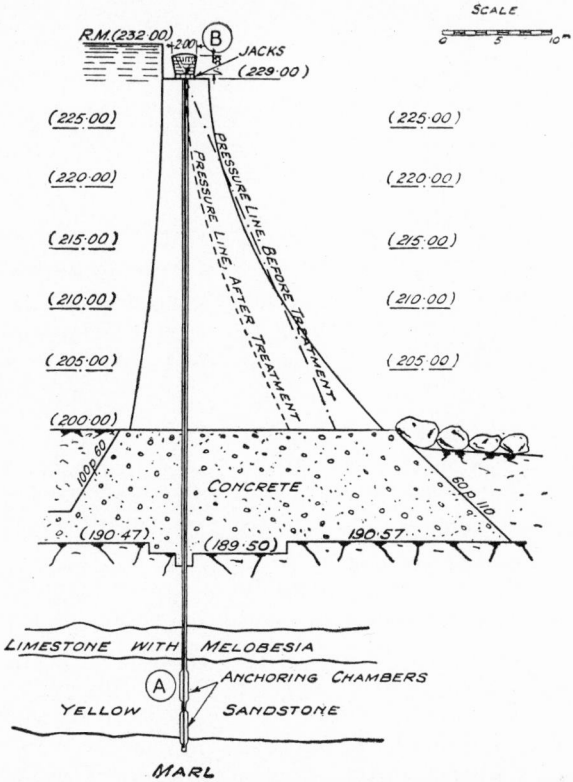

FIG. 13.2.—Cross section of the Cheurfas Dam, Algeria, showing foundation strata and method of "anchoring" the dam structure to them.

(Reproduced by permission of Mons. A. Coyne, Ingénieur en Chef des Ponts et Chausées, Paris, France, and the Institution of Structural Engineers, London, from the Structural Engineer, reference 13.8.)

FIG. 13.3.—General view of the Cheurfas Dam, Algeria, as reconstructed, and a close-up view of the head of one of the anchorage.

(Reproduced by permission of Mons. A. Coyne, Ingénieur en Chef des Ponts et Chausées, Paris, France.)

after construction, one abutment of the dam was washed away, but it was reconstructed and served well until recent years. An examination in 1927 disclosed serious weakness; and as the most economical means of strengthening the dam, 10-in. diameter holes 13 ft. apart were drilled from the crest of the dam to the foundation and then to a depth of 70 to 80 ft. into the calcareous sandstone, limestone, yellowish sandstone, and argillaceous marl on which the dam is founded. Specially prepared steel cables were then anchored in these holes to the foundation strata, secured in special caps on the dam crest, placed under great tension, and then anchored, their effect being to "tie down" the dam to the foundation strata.[13.8]

A final observation of a general nature which must be made before passing on to consider individual problems in detail is that as the material status of western civilization advances, the best dam sites will be progressively used up leaving for further development sites of increasing difficulty. Admittedly, this is a general tendency only, but there can be little doubt that it represents accurately the trend of dam construction. It will result or, more correctly, it is already resulting in increasing attention having to be paid to the geological conditions of the dam sites that have to be considered.

13.5. General Preliminary Work.—The location of a proposed dam will generally be confined between reasonably close limits by topographical, economic, and social conditions which usually result in the areas to be examined as possible sites being fairly well defined. Preliminary studies will follow generally the lines already suggested in Chap. VI. Accurate geological sections along possible lines for the dam will be a major requirement. The nature of valleys will sometimes result in bedrock being some distance below ground level, and therefore the determination of the rock surface across the valley will often be a first step in the preparation of the section. Geophysical methods can be of great assistance in this work when utilized in connection with strategically placed boreholes. The existence of buried valleys must be checked carefully in investigating, not only from test-bore results but also from general studies of the local geology, for there are on record cases of such valleys as much as 70 ft. deep being found between two boreholes as close together as 50 ft.

The possible presence of boulders is another danger to be guarded against, especially in glacial formations, the greatest care having to be taken with test drilling. The almost insuperable difficulties met with during the construction of the Silent Valley Dam for the Belfast City and District Water Commissioners, in Northern Ireland, were due very probably to the assumption that granite boulders encountered

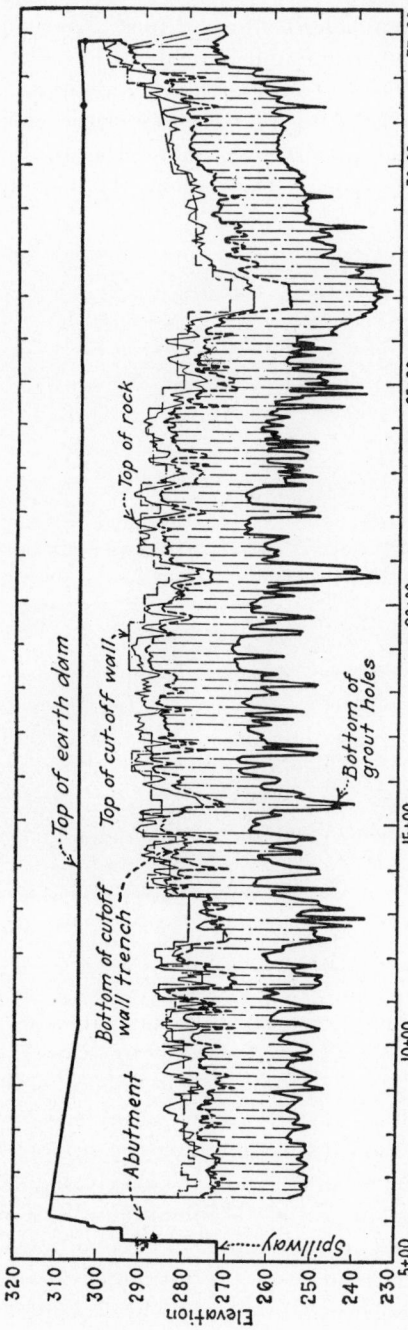

FIG. 13.4.—Profile along center line of the Ontelaunee Dam, Reading, Pennsylvania, showing irregularity of limestone foundation bed.

(Reproduced by permission of Mr. Farley Gannett, and the Editor, Engineering News-Record, from Reference 13.10.).

by the original test boreholes were solid rock. Actually, instead of a rock foundation bed being found at a depth of 50 ft. below ground level, the cutoff trench had to be carried to a depth of 180 ft. through running sand of such a nature that the use of the maximum working pressure of compressed air served to reduce the water level by only a few feet. The final solution of the problems thus encountered provides a fascinating study.[13.9]

FIG. 13.5.—Surface of eroded limestone as uncovered during the excavation for the cutoff wall trench for the Ontelaunee Dam, Reading, Pennsylvania.

(Reproduced by permission of Mr. Farley Gannett, President, Messrs. Gannett, Eastman and Fleming Inc., Engineers, Harrisburg, Pennsylvania, U.S.A.).

A similar hazard may exist when the dam site is located over limestone strata. The solution of limestone by water, after prolonged exposure, is a topic often mentioned in this book, and in the case of dam foundation beds it can often have most serious results. As an example may be mentioned the Ontelaunee Dam, a rolled earth structure, built for the city of Reading, Pennsylvania, in 1935. Even though extensive exploration of the underlying rock was made prior to designing the dam, utterly unforeseen conditions were disclosed during construction. Instead of the estimated 450 cu. yd. of rock, 4,719 cu. yd. had to be excavated; instead of 3,300 ft. of drill holes provided for in the contract, 10,450 ft. were drilled; and instead of

4,200 bags of cement expected to be used in grouting the drill holes, 61,368 bags were used—over fifteen times as much as forecast by the designing engineers. The whole job cost $209,018 (£42,000) as compared with the original bid of $82,992 (£16,600). After the job was completed and accepted, and the reservoir filled, leaks developed under the core wall and grouting, and another $20,000 (£4,000) was spent in plugging channels under the dam which were missed by the first grouting operation.[13.10]

FIG. 13.6.—Excavation in progress for the Vyrnwy Dam, N. Wales, for water supply of Liverpool England; S.W. end of site, as exposed in June 1882, showing glaciated surface of rock.

(Reproduced by permission of Col. F. Hibbert, Water Engineer, Liverpool.)

If it is possible to find another site that would be suitable, it is better not to build dams in limestone areas; large leaks may develop not only under the dam but from the reservoir where limestone is encountered. If no choice is possible, exploration should be very thorough in order to ensure, as nearly as possible, reasonable correspondence between engineering estimates and final costs as well as to ensure storage without leakage and the safety of the structure.

In the determination of the geological sections, local topographical detail and, more particularly, local geological features must always be given special consideration in valleys that have been subjected to glaciation. It was from surface observations at the site of the Vyrnwy

Dam for the water supply of the city of Liverpool, England, that Dr. G. F. Deacon (joint Engineer with Mr. Thomas Hawkesley) deduced that the valley was the site of an old glacial lake, held up at one time by a rock bar. The center line of the dam, which was the first large dam to be built in Great Britain, was located for the necessary Parliamentary plans along the inferred position of the rock bar. Subsequent detailed investigation by trial holes and shafts proved the inference to be practically correct, and the dam was built as

FIG. 13.7.—Excavation in progress for the Vyrnwy Dam, North Wales, for water supply of Liverpool, England; central area of excavation, looking northeast, as exposed in June, 1882.

(Reproduced by permission of Col. F. Hibbert, Water Engineer, Liverpool.)

originally located. Dr. Deacon estimated that a deviation of the center line up or down the valley of only ⅛ mile would have added £300,000 to £400,000 ($1,500,000 to $2,000,000) to the cost of the work.[13.11] As a contrast to this example may be mentioned the case of a small dam known to the writer, which was constructed without any such preliminary studies being made of the rock surface, as a result of which several thousand dollars were wasted owing to the center line being only 50 ft. upstream from what would have been a better location.

Whenever feasible, the rock surface should be traced right across the valley being investigated. This is not always possible, faulting sometimes providing clefts filled to great depths with unconsolidated material. The author has carried out some investigations in the northern part of Ontario, Canada, in a rock gorge barely 50 ft. wide through which passes the whole flow of a fairly large river. Solid diabase extends for a great distance on either side of the gorge, yet diamond-drill holes carried 180 ft. below water level in the river bed failed to reveal solid rock, passing through boulders, gravel, and finally over 100 ft. of compacted sand. This case is typical of valley faults, quite a number of which are to be found in Canadian practice; they always demand very careful study. When rock cannot be found at reasonable depths, the necessity for obtaining accurate data on all unconsolidated strata assumes increased importance.

The information thus made available will generally enable those in charge of the engineering work to decide on the general type of dam to be used and to begin their economic and design studies. If, for example, the rock floor cannot be reached by ordinary drilling methods, an earth- or rockfill dam or a "floating" concrete structure may be found to be necessary. If a rock bed is found to be available at reasonable depths, economic considerations will guide the designer in his choice between an earth- or rockfill dam and one of the gravity or some other structural type. The choice of the kind of dam to be used, although so closely related to geological conditions, must be left for study elsewhere, since it is essentially an engineering problem. Enough has been said to indicate the amount of geological survey work necessary before even a choice can be made. Once the choice has been made, geological work must continue, specially adapted to serve the particular requirements of the kind of dam now contemplated; and it must be carried to such a stage that from the information so obtained, assurance can be had that construction of the dam as proposed can be completed within the estimated cost.

In this connection, it may be fitting to quote a relevant extract from a paper by Dr. Charles P. Berkey to whose work on the relation of Geology to civil engineering tribute has already been paid. Professor Berkey has said:

> An impractical project must not be begun; all reasonable doubt of its feasibility must be removed. Serious questions as to the behavior of the ground must be settled before construction begins. Dams must stand. Not all of them do, and there are all degrees of uncertainty about them. Reservoirs must hold water. Not all of them do, and there are many ways by which water may be lost. The work must be done safely as a construction job. Not all of them are, and there are many sources of danger. The whole structure must be permanent and the work has a right to be done within the original estimates. Not all of them are, and there are many reasons for their failure or excess cost, most of them geologic or of geologic dependence.

If these requirements are not attained, it is generally either the fault of the geologist advisor, who ought to see the difficulties and dangers and impossibilities in advance, or it is the fault of the chief engineer who presumes too much on his own acquaintance with geologic matters and has failed to procure the necessary advice. The time has gone by for anyone to assume personal competence in every one of the fields that may require research ability on an engineering project; and the time has gone by for such work to be done by an untrained or unskilled man.[13.12]

13.6. Exploratory Work during Construction.—It has already been indicated that investigations of the geological conditions underlying a dam site do not cease when active construction work begins; on the contrary, the start of excavation means, in many cases, that geological study can be extended, actual rock surfaces examined, and previously formulated opinions checked up. In all cases of major dam construction, the services of a geological expert will be advisable, and his work will most certainly be continued until all excavation has been completed and the dam construction begun, the final inspection of the foundation bed as prepared for use being the most critical part of the investigation work.

The start of active construction operations will generally mean that equipment and power supply will be available for the use of aids to exploration other than those possible in ordinary preliminary work. As these additional methods are specially applicable to dam work, a word or two more about them may be usefully added. The digging of more extensive test pits is a first possibility but one that is merely an extension of work included in most preliminary investigations. A further extension is the excavation of shafts into rock and of exploratory tunnels. The purpose of such shafts and tunnels is to enable visual inspection to be made of the actual rock structure in the checking up on the geological sections across the site, previously prepared, and on the rock as being sound and strong enough to carry the designed loading. Again, in many deep gorges especially suitable for dam sites, borings cannot easily be taken in the river bed; and in such cases, underwater tunnels drifted from shafts sunk in the gorge sides can prove invaluable. Exploratory excavation work of this type can often perform a double purpose by providing access to rock well below the foundation bed surfaces into which cement grout may be injected, if this is necessary to render the foundation strata watertight. In this way, grouting can be made thoroughly effective, and more extensive in its influence than from surface operations.

As an interesting example of a dam at the site of which exploratory tunnels were widely used there may be mentioned the main dam of Le

Sautet water power development in the south of France on Le Drac, one of the headwaters of the Rhone. Located in a magnificent gorge 600 ft. deep and ½ mile long, the dam is 414 ft. high and yet has a crest length of only 263 ft. The dam proper is an arched structure, bearing on to the walls of the gorge, but it is backed with lean concrete designed to buttress the two sides of the gorge, the

FIG. 13.8.—Le Sautet Dam, powerhouse and gorge, Drac River, France. The dam is 414 ft. high. Exploration of limestone strata is described in text.

(Reproduced by permission of le Société Forces Motrices Bonne et Drac, Grenoble, France, Mons. M. Larond, Ingénieur Chef de l'Exploitation.)

resulting cross section being almost that of a gravity dam. The rock in which the gorge is located is a limestone formation, and this called for most careful exploratory work. Sixteen hundred and fifty feet of exploratory tunnels were driven, the preliminary investigations and surveys extending over a period of 10 years. As a result of this study, it was determined that the rock was as sound as could be desired, without faults or other imperfections. Watertightness of the gorge sides was assured by an extensive grouting program, about

20,000 ft. of grout holes being drilled and grouted, absorbing about 6,000,000 lb. of cement under pressures as high as 500 lb. per square inch. Much of the grouting was done from the tunnels, some of which have been left open so that further grouting can be done in the future if inspection shows this to be necessary.[13.13]

Another major aid to construction developed only within recent years is the use of special drilling machines capable of taking out drill

FIG. 13.9.—General view of excavation at site of the Prettyboy Dam, for the water supply of Baltimore, Maryland, U.S.A.
(Reproduced by permission of Mr. Frank K. Duncan, Chief Engineer, D. P. W., City of Baltimore.)

cores up to 36 in. in diameter, as described on page 103. The use of large-diameter drill holes at the Prettyboy Dam in Maryland was then mentioned. It is of special interest now to note that the highly foliated nature of the schist revealed in the foundation bed of this dam resulted in such overbreak and general rock movement after blasting operations that other methods had to be tried for the excavation of the necessary cutoff trench. After repeated trials, a novel expedient was adopted. Wire-rope saws were rigged up between pairs of the calyx-drill holes, and with these the necessary trenching work was successfully carried out, this being probably the first occasion on which wire saws were used on a highly quartose rock. Two of the wire-saw rigs are seen in the accompanying photograph.[13.14] The use of large-diameter holes at the Norris Dam, Tennessee, was also mentioned in Chap. VI; they proved most useful for checking up on the penetration of cement grout through the foundation strata.

The foundation of the world's highest dam calls for some mention in this chapter, even though it presents few unusual features.

FIG. 13.10.—Wire-rope saws in use for rock excavation at site of the Prettyboy Dam for the water supply of Baltimore, Maryland, U.S.A. (Holes also used for subsurface exploration by geologists.)

(Reproduced by permission of Mr. Frank K. Duncan, Chief Engineer, D. P. W., City of Baltimore.)

The continued study devoted to all the geological aspects of the Boulder Dam construction does, however, provide an unusually good example of the invariable requirements of dam construction work—close inspection of all rock surfaces as they are stripped and a final and thorough inspection before concreting is started. In 1928, a board of engineers and geologists recommended after much study that the dam be located at the Black Canyon site on the Colorado River.

The canyon walls were formidable, defying intimate inspection; the gorge floor was buried deep below the swift-flowing water of the river, heavy boulders, and river sediment. Predictions were made, however, explorations were renewed, and active construction soon started; by 1933, the canyon floor had been cleaned up (the river flow passing meanwhile through diversion tunnels), and the geological formation was thus available for direct inspection. Dr. Charles P. Berkey was a member of the original Board, and he made a final inspection

FIG. 13.11.—A general view of the site of the Boulder Dam, as finally excavated, showing nature of the canyon walls and the "inner gorge." (Photograph taken as the second bucket of concrete was being deposited, June 6, 1933.)

(Reproduced by permission of The Commissioner, U. S. Bureau of Reclamation, Washington, D. C.)

at this time, his official report to the Bureau of Reclamation of the United States Government being dated May 31, 1933. It revealed that every major contention and assumption made by the Board five years before had been confirmed by intimate examination of the gorge in the dry. The Report need not be referred to in detail, since many of its features are reflected in the publications describing this famous dam. All the rock formations of the gorge were found to be thoroughly capable of bearing the load and resisting the thrusts of the dam. Stability and general watertightness of the walls and floor as well as of the four great tunnels in the canyon walls exceeded all expectations, although grouting had to be carried out to obstruct leakage and otherwise guard against the great pressure of the water to be retained in this unusually deep reservoir. The rock of the foundation beds was found not to soften under prolonged submergence. The gorge did not follow a fault zone, but a most interesting feature revealed by excavation was the existence of an "inner gorge," forming a narrow and tortuous channel roughly along the center of the main gorge at a depth of 75 to 80 ft. below the rock benches on either side. The side rock benches were generally smooth and uniform (although

FIG. 13.12.—The Boulder Dam; a close-up view of the final excavation and clean-up work in the deepest section of the "inner gorge," showing some of the river gravel deposit.

(Reproduced by permission of The Commissioner, U. S. Bureau of Reclamation, Washington, D. C.)

showing some potholes), but the inner gorge was pitted and fluted, being generally very uneven in form and depth. This form suggested that the whole of the previously superincumbent mass of sediment had moved in great flood "tides," being subject to scour, whereas in the center there had been a considerable whirling action, setting up "pothole erosion" at greater depth. Although this would seem hard to believe, it was proved beyond doubt by the essential continuity of river fill from top to bottom and also by the finding of a sawn plank of wood embedded in the gravel at the edge of the inner gorge, a position that it is believed could have been reached only by burial during a recent flood. On this site now stands Boulder Dam, a tribute alike to those by whom it was planned, designed, and constructed.[13.15]

13.7. Soundness of Bedrock.—Turning now from general considerations to special geological problems likely to be encountered in dam construction, the essential soundness of the strata on which a dam is to be founded is a prime requirement that must be investigated. A question that the civil engineer desires to have answered well before his construction operations are due to start is the probable nature of the rock surface as it will be exposed on excavation and the possible presence of disintegrated rock, since on the answers will depend his estimates of cost. Test pits, if carried to rock surface, may give some indication of an answer to this question; test drilling likewise may give an indication, although with certain types of disintegration the rotten rock when still in its original position may give surprisingly good cores, showing up in its true nature only when exposed to the atmosphere. Thus it is that a careful geological study is a most necessary aid in determining whether or not disintegration is to be expected. A geologist may be able to deduce this from knowledge of the essential nature of the rock type to be encountered and of possible disintegrating influences, and from a detailed study of cores obtained from test drilling. As illustrating the possible extent of weathering, it may be noted that it has been found at Washington, D. C., that a bed of granite had decayed to a depth of 80 ft. to such an extent that it could be removed with pick and shovel. In the state of Georgia, limestone has been found decomposed to a depth of 200 ft.; in Brazil, shales have been found disintegrated to a depth of 394 ft. below surface level.[13.16] These figures will indicate how serious this problem can be. Taking actual examples from dam construction practice, before the start of construction of the great Assuan Dam, on the River Nile in Egypt, in 1898, it was assumed that in the deeper channels of the river bed, where drilling could not be carried out because of the swift currents, all decomposed granite would be found to have been eroded. When the site was unwatered, it was found repeatedly that there were depths of decomposed granite at the bottom of most of the channels the removal of which added greatly to the cost of the work. In one season, five times the estimated quantity of rock had to be removed; for the complete structure, excavation was 100 per cent in excess of that calculated. It is of some interest to note that Sir Benjamin Baker, Engineer for the work, consulted Sir Archibald Geikie, then Director of the Geological Survey of Great Britain, about possible fissures in the granite, but apparently in London and not at the site.[13.17] At another Egyptian dam, the Sennar Dam on the Blue Nile, the reverse experience was encountered, the excavation necessary being less than originally calculated, with consequent saving in cost. In a North American example, a cutoff wall resting on andesite had to be carried 40 ft. lower than the level originally suggested by core borings, in order to reach what was finally considered to be sound rock.

An added difficulty is that of determining exactly what is meant (or desired) by *sound rock;* the term may have quite different meanings for engineer and for geologist. Associated with structural soundness in the mind of the engineer is the strength of the rock and its impermeability, whereas the geologist may tend to consider the matter rather from the mineralogical standpoint. The subject is one of complexity, therefore, requiring experience for its consideration and a close cooperation of engineering and geological attention. A final note of warning must be stressed with regard to rock structure near fault zones. As indicated in Chap. III, and as exemplified by the rock encountered at the Prettyboy Dam, rock adjacent to fault planes may quite probably be brecciated and so useless as a sound foundation rock. Normally fractured rock can easily be detected from its appearance if not from its location, but under special circumstances this may not be easy. When work is proceeding in very cold weather, for example, the effect of frost may be to convert such breccia into an apparently solid mass. In the north of Ontario, Canada, the author has experienced this effect, diabase which had been fractured into fragments of about 1 cc. in size appearing to be quite solid when exposed at a temperature of −30°F., revealing its true nature (detected from its location) only when subjected for some time to the action of steam jets.

The structural strength of rock in proposed foundation beds is a main requirement of soundness and one of importance. Compressive strength is the property generally in question, dam structures (gravity, arched, or buttressed) being designed with a maximum unit pressure at the heel as one criterion for stability, this being equated to the maximum permissible compressive stress in the rock. Although it seems probable that no dam failure has been directly due to a failure of a rock foundation stratum in compression, this in no way minimizes the importance of this aspect of the subject. The compressive tests usually carried out deal with rock specimens in the dry, whereas part, at least, of the rock surface supporting a dam structure will be exposed continuously to water. A further consideration of rock soundness therefore must always be the effect of prolonged exposure to water. Not only is the bedrock to be exposed to water but to water under appreciable pressure. Any failure of the rock under the influence of water may therefore lead to serious structural weakness; and if the weathered rock is for any reason displaced as it weakens, as it may be by seeping water, serious damage may be done. The failure of the St. Francis Dam provides at once a most notable and most tragic example.

Preliminary testing can remove almost completely any doubts entertained on this matter, as has been instanced in many investigations. As an example may be mentioned tests made in connection with the Madden Dam site at Alahjuela, Panama Canal Zone. Among the rock formations investigated was a bluish-gray fine-grained sandstone, known as the *Gatun*(?) *formation*, samples of which were tested by the United States Bureau of Standards and found to have a compressive strength of 3,500 lb. per square inch when dry but only 850 lb. per square inch when wet. Similar figures for a light-gray medium-grained sandstone also tested [the *Caimito* (?) *formation*] were 2,300 and 550 lb. per square inch, respectively. The figures quoted are extremes. Petrographic examination disclosed the presence in the sandstones of a clay mineral occurring as films coating feldspar and other grains, abundant glaucanite contributing to the weakness.[13.18] This example serves to show how important are tests on thoroughly wetted rocks; incidentally, it illustrates yet one more example of the troubles that may be encountered due to the unwanted presence of clay.

Another possibility of gravity dam failure is to some extent dependent on geological conditions—the possibility of the dam sliding. Various values have been suggested for the coefficient of friction between either concrete or masonry and mortar and rock surfaces. Tests reveal that provided a rock floor is properly cleaned and prepared, the bond obtained between concrete and rock can be so efficient that failure will take place only by a shearing fracture of solid rock mass or of the concrete. Special preparation of rock surfaces may be necessary in the case of glaciated rock that may be worn so smooth that it must be roughened artificially in order to give a good bond. Perfect bond may be obtained between concrete and the rock surface, but the rock mass itself may slide forward under the influence of the pressure transmitted from the dam foundations. Movement of this kind will take place only along planes of weakness in the rock, such as bedding planes, and only when normal resistance to movement has been in some way removed. Naturally, geological structural arrangement will be the determining factor, movement being possible only in the case of bedding planes that are horizontal or that dip away from the dam. This arrangement can be determined only by adequate preliminary geological study. The Austin Dam already mentioned provides an illustrative example.

13.8. Possibility of Earth Movement.—Earth movement in its relation to dam foundations is of vital importance, the nature of dam structures being such that, unless due allowance has been made in design, any movement of the foundation beds may lead to serious structural damage. In areas subject to earthquake shocks, allowance

FIG. 13.13.—The Morris Dam, for the water supply of Pasadena, California, U.S.A., located in the San Gabriel Canyon, over a fault.

(Reproduced by permission of Mr. Morris S. Jones, Chief Engineer and General Manager, Water Dept., City of Pasadena.)

for seismic forces must always be made even for dams founded on the most solid of rock formations. A usual allowance is to consider an earthquake shock with a horizontal acceleration of one tenth that of gravity, vertical acceleration being neglected. In connection with the application of this design requirement to the Morris Dam built in the San Gabriel Canyon, near Azusa, California, to augment the municipal water supply of Pasadena, California, Professor H. M. Westergaard, of the University of Illinois, analyzed successfully the hydrodynamic effect due to the dam oscillating against the still water of the reservoir, and it was found that the natural period of vibration of the dam as a whole (328 ft. high,) was 0.16 sec. or less.[13.19] It was estimated that the volume of material in this dam had to be increased by 15 per cent to allow for the thickening necessary to resist seismic forces. Special dam designs have been evolved to be proof against earthquake shocks even if quite severe. As an example may be mentioned a special type of laminated concrete paving used for facing a considerable number of rock-fill dams in Chile, the rock-fills being designed so that they will not flatten under shock, the slopes

FIG. 13.14.—The Upper Crystal Springs earth-fill dam (of the San Francisco Water Department) showing displacement of dam on either side of the San Andreas fault, following the severe earthquake of 1906.

(Reproduced by permission of Mr. N. A. Eckart, General Manager and Chief Engineer, San Francisco Water Dept., San Francisco, California, U.S.A.)

adopted (1.6 and 1.8:1) having been arrived at after a long and careful study of rock slides in the district adjacent to the dam sites, many having been caused by earthquake shocks.[13.20]

Many dams, especially of old design, have failed as the result of earthquake shock, a fact that is not surprising in view of the relatively recent progress of seismic research work. What is perhaps more surprising is that dams built at right angles across faults along which movement has taken place owing to earth movement have *not* failed. Two good examples are provided by earth-fill puddle-core dams built across the San Andreas fault in California in 1870 and 1877, respectively. The San Andreas Dam is 95 ft. high above stream bed and 130 ft. above the bottom of its cutoff trench; and the Upper Crystal Springs Dam is 85 ft. high above stream bed and 190 ft. above the bottom of its cutoff trench. Both were constructed on clay foundation beds, the fault passing almost at right angles across their crest lines The disastrous San Francisco earthquake of 1906 reached a maximum intensity of 10 on the Rossi-Forrel scale, movement being concentrated along the fault line. At the two dams, permanent displacement up to a maximum of 12 ft. took place; and although outlet tunnels around the dams were badly fractured, the dam structures themselves remained stable. Despite the appreciable movement of one half of the dams, they remained watertight and so stand today as seen in an accompanying photograph. A large mass concrete gravity dam (the Crystal Springs Dam, constructed in 1877 and enlarged in 1888 and 1890) located only ¼ mile from the fault was undamaged.[13.21]

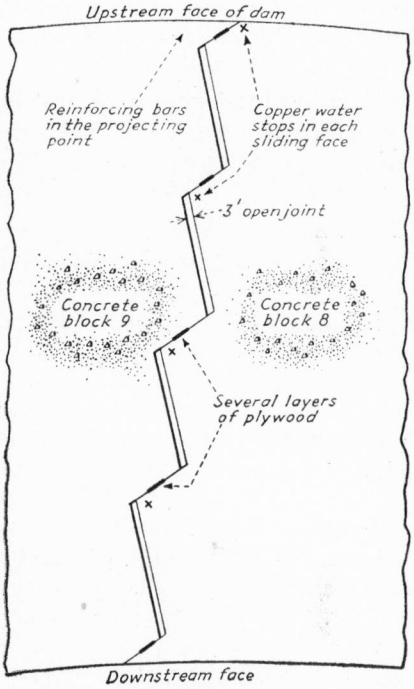

FIG. 13.15.—Sliding type of joint which divides the Morris Dam (for Pasadena water supply, California, U.S.A.) into two units over an old inactive fault.

(Reproduced by permission of the Editor, Engineering News-Record, from Vol. 121, p. 184, New York, August 11, 1938.)

This example of what can happen at a fault plane is a fitting reminder of the importance to be attached to the presence at a dam site of geological faults—certain signs of past earth movement. It will be increasingly difficult to find sites for dam structures without some such defect, and engineering ingenuity has already been displayed to advantage in devising means of overcoming the difficulties in design thus introduced. At the site of the Morris Dam, California, for example (see page 327) a minor fault intersects the dam foundations near the base of the right abutment in a direction almost normal to the axis of the dam. Study of the stratification of old stream-bed gravels revealed the fact that no appreciable movement along this and other fault planes had taken place at least for a period of approximately 10,000 years. The fault received special treatment in design, an open joint with vertical sliding planes being provided in the dam structure over the trace of the fault from top to bottom of the dam, between two of the blocks into which it was divided for construction purposes. The four sliding planes provided are at an angle of 45 deg. with the horizontal and so lie in the direction of past movement. Planes of contact are separated by a bituminous filler which can yield slightly if motion is not as anticipated.*[13.19]

Another interesting and illuminating example is provided by the foundation bed condition at the site of the Owyhee Dam built about 1930 by the United States Bureau of Reclamation to serve a large irrigation project. The dam is 355 ft. high above bed level but 530 ft. high above lowest concrete level; it provides 715,000 acre-ft. of storage on the Owyhee River in eastern Oregon.

The dam site was formed by the present Owyhee River cutting a narrow gorge 400 to 600 ft. deep through a flow of rhyolite which blocked the valley subsequent to the deposition of the original tuff underlying the district. The outer surface of this flow of rhyolite moving on the original tuff—probably in contact with water—chilled comparatively quickly, producing a layer of pitchstone agglomerate, roughly about 25 feet thick, extending as a contact between the two materials. Subsequent erosion has removed the upper levels of this rhyolitic flow, and the river had cut its gorge down into the remainder. This rock, which forms the abutment and foundation material for the structure, is an extremely hard crystalline mass due to the slow interior cooling of the flow. Beneath it lies the stratum of more elastic pitchstone agglomerate as a transition material blending into the softer tuff.

A fault in the center of the streambed parallel to the canyon subsequently has occurred with horizontal displacement. Exploration of this fault zone with diamond drill borings showed that faulted material was evident in the hard rhyolite in the form of small shattered pieces, but indicated that the disturbance lessened in the more yielding agglomerate and practically disappeared upon entering the tuff.

The original plan provided for the removal of this faulted material on the line of the upstream cutoff by means of a shaft to the undisturbed tuff. . . . Excavation work in this cutoff shaft and general foundation excavation along the upper portion of the fault revealed fractured and loose material, which

* It is of interest to note that the dam has been tested by the U. S. Coast and Geodetic Survey using a shaking machine consisting essentially of motor-driven eccentrically mounted flywheels, which produces minor "artificial" earthquakes. " . . . the tests were interpreted as verifying the computed periods of vibration of the dam and strengthen the confidence of the builders in its ability to resist earthquake shocks" [*Eng. News-Record* (*New York*), **121**: 184 (1938)].

made it advisable to remove the material in the fault zone throughout the entire width of the structure and refill the crevice with concrete to insure proper foundation conditions. This possibility had been foreseen as a result of the drilling and was provided for in the specifications. . . . Excavation proceeded through about 83 feet of faulted and broken rhyolite, 22 feet of the agglomerate and 10 feet into the practically undisturbed tuff to El. 2,145, at a depth of about 213 feet below low water level of the river. It is of interest to note that the level of actual excavation required to reach undisturbed material below the zone of fracture was within 2 feet of the estimated depth established by the exploratory borings in the zone.[13.22]

FIG. 13.16.—The Owyhee Dam, a general view of the unwatered site, showing nature of the canyon walls and the fault described in the text. (Scale may be determined from the size of the large crawler crane in the center foreground.)

(*Reproduced by permission of the Commissioner, U. S. Bureau of Reclamation, Washington, D. C.*)

This unusual foundation work is described in detail in view of its special interest and the close relation between construction procedure and geological structure that it reveals. Practically all recorded cases of major dam construction over faults include some features of interest and therefore call for some mention. Reference can be made, however, to one more example only, in view of its special interest, this being the arched foundation necessitated at the Rodriguez Dam on the Tijuana River near Tijuana, Mexico, constructed in the years 1928–1930. The dam is of the Ambursen buttress type, 240 ft. high above lowest foundation concrete, having a crest length of 2,200 ft. and providing a storage of 110,000 acre-ft.

Rock, which was exposed over much of the site, consists of rhyolite and granite with a fused contact, indicating watertightness on this plane. In the lower part of the gorge the rock surfaces were hard and fresh, although broken. . . . Cleavage planes are evident on either side, dipping toward streambed.

As preliminary investigation nine borings were made in streambed and about fifty test pits were excavated at higher elevations. Results of the borings in streambed indicated suitable foundation rock at five locations at depths not to exceed 50 feet. Cores could not be obtained in the other holes but two indicated sharp clean rock chips and the remaining two showed rather

FIG. 13.17.—Rodriguez Dam, Mexico; close-up view of the special arch incorporated in the foundation in view of the presence of a fault.
(Reproduced by permission of the Ambursen Dam Co., New York, Mr. S. W. Stewart, President.)

definite evidence of disintegration. Sidehill exploration indicated that suitable foundation rock could be secured at depths varying from 10 to 40 feet. . . . Foundation excavation work uncovered bedrock at depths of about 35 feet, as had been expected, but it proved considerably broken and unsuited for buttress foundations. A more serious difficulty was the finding of a pronounced fault about 20 feet wide parallel to the stream along the sound rock of the east bank. . . . A smaller, secondary fault about 4 feet wide near the center of the streambed parallels the major one and dips slightly toward it, indicating a junction at a considerable depth. In addition, deeper excavation along the line of the cutoff showed that a weak, inferior bedrock could be expected to continue below the existing elevation of foundation excavation.

Further and more extensive geological inspection of the site was made in the light of foundation excavation and a comprehensive report was made by

Dr. F. L. Ransome. . . . Excavation in streambed over the foundation area revealed sound rock walls about 90 feet apart at the upstream toe and 50 feet apart at the downstream edge of the structure. . . . Excavation in the cut-off indicated that conditions could not be bettered by going deeper. The alternative was to provide a foundation structure across streambed to equalize that portion of the load which could be carried by the weaker rock and transfer the remainder to the sound rock of the walls.

Finally, after much study and approval by a consulting board, a concrete arch with a span across the stream bed varying from 86 to 76 ft., the barrel being 225 ft. long, was designed, the space between

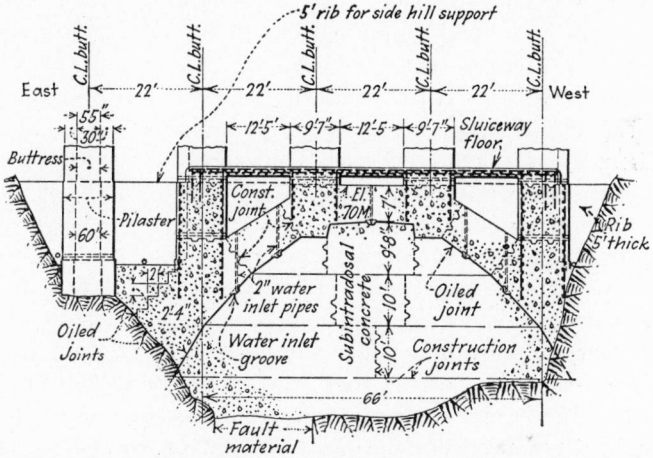

FIG. 13.18.—Rodriguez Dam, Mexico; cross section of the special arch incorporated in the foundation, near the downstream end.

(Reproduced by permission of the Ambursen Dam Co., and the Editor, Engineering News Record, from Reference 13.23.)

the intrados of the arch and the stream bed being built up with lean concrete to provide a support for constructing the arch and to transfer a portion of the load vertically, although no allowance for this was made in the arch design. The arch was successfully constructed, and the dam completed.[13.23]

13.9. Permeability of Bedrock.—The second essential requirement of a rock foundation bed for a dam is that the entire geological structure underlying the site of the dam, in addition to being strong enough to carry the designed loads, shall be sound also in that it provides a watertight barrier to the water impounded by the dam. Although the requirement is so obvious, it must be recorded that dams have failed because of its neglect, and an imposing and tragic list could be produced in confirmation of this statement. One example only will be mentioned, since it has been termed "the greatest object lesson that the history of engineering foundations has to offer." The Hales Bar Dam on the Tennessee River, United States, was founded on a pure but soluble limestone well known for its cavernous formation. Cavities were encountered to such an extent during the excavation of the dam site that completion was delayed several years, and the cost was increased far beyond the original estimate. Several hundred thousand barrels of cement were injected into the fissures and openings in the underlying rock. The dam site is located on a syncline, the lower side of which would be especially susceptible to solution water. Further trouble due to this cause was foretold, notably by L. C. Glenn, not long after the dam was put into use. About ten years after the dam had been completed (in 1926), leakage through this limestone had become really serious, and many unusual methods were used in attempts to provide an effective seal, including the dumping of rocks, gravel, clay, and bales of hay and also the use of various kinds of mattresses. Success was achieved only by drilling a large number of holes to an average depth of about 90 ft. and injecting into these hot liquid asphalt, the flow of which was assisted by ingenious devices; over 11,000 bbl. of asphalt was used. The process was later repeated, and leakage was eventually stopped. The dam stands today, but it is said to have cost about ten million dollars and more than ten years to stop the leakage through the foundation strata.[13.24]

The first reason for having rock strata underlying dams as watertight as possible is obviously to ensure that no water escapes. Not only is this necessary from the point of view of water conservation, but it is also essential because any steady flow through a solid rock formation is bound to have some erosive action, which in all probability will gradually but steadily intensify the defective conditions causing original leakage. The geological investigations already mentioned will indicate the underground structure to be encountered along the line of the dam, and from a study of this cross section a general idea, at least, as to the effectiveness of the strata in retaining water can be obtained. Naturally, limestone and all softer rock formations are suspect until proved to be sound, especially in the case of limestone formations known to be unusually soluble and so featured by underground caverns or open fissures.

In addition to study of all relevant geological data, including the detailed results of all boring, tests should be carried out whenever possible by means of the boreholes used for obtaining core samples. The use of carefully controlled and observed water pressure tests in these holes provides a fairly reliable confirmation of underground structural soundness. If the holes will hold water under the maximum test pressures, it is clear that the strata penetrated by the test holes are watertight in the neighborhood of the holes. This qualification is a necessary reminder that the use of water pressure tests in drill holes is not an infallible guide, although when considered in conjunction with ascertained geological structural data it can be a reasonably reliable indication of watertightness. Pressure tests of this nature were utilized in connection with the geological investigation of the Madden Dam site in the Panama Canal Zone, although in this case water was forced under pressure into flowing wells to see if they would provide underground connection to other wells or provide any other possible means of escape for the test water. For tests of this nature, the use of the color dyes dealt with in the following chapter in connection with leakage from reservoirs is often of value. Tests were also carried out, during the Madden investigation, on the permeability of test specimens of the rock types encountered, using core samples taken from drill holes, the apparatus used being not unlike the standard type in use for testing permeability of unconsolidated material, the specimens being packed into sections of steel pipe with lead wool and steam packing and subjected to a head of 80 lb. per square inch.[13.18]

In addition to the possibility of major leaks developing, an almost equally serious problem is presented by the danger of minor leaks, and especially on the site of the dam structure itself. Springs are included in this category; if noted either before or during construction, or if preliminary investigations suggest that they may tend to form at constant planes after construction is complete, the most careful precautions must be taken to box them out when the lower courses of the dam are being built so that they may be suitably connected with the drainage system.

This last possibility is probably the most difficult to investigate before construction and the most potentially dangerous, since the slight leakage of water that may occur between rock surface and dam structure after completion will probably give rise to what is generally known as *uplift pressure*. Although it is only in recent years that the matter has been studied in detail and by actual experiments on dam structures, uplift pressure was not, as is sometimes suggested, neglected in earlier dam designs, as the following quotation makes clear. Dr. Deacon, in describing the design of the Vyrnwy Dam, to which reference has already been made, stated (in 1896) that:

Although no visible springs of water issued from the beds of rock thus exposed, it was by no means certain that, when the reservoir was formed and the head on one side became 144 feet, springs subject to that pressure would not occur. Moreover, a mere moisture, rapidly evaporated when exposed to air might when sealed down acquire a pressure from the adjoining hills of far more than that due to the intended level of the lake . . . and the Author agreed with him [the late Mr. T. Hawkesley] in thinking it desirable to provide relief drains, which so far as he is aware, had not been done in connection with any former masonry dam.[13.11]

The predictions of Messrs. Hawkesley and Deacon have been generally confirmed at many dams on which tests of uplift pressure have been made. Many of these are founded on perfectly sound rock strata so that the uplift pressures may be said to be independent, in these cases at least, of geological structure. At other dams the reverse is the case, and therefore uplift pressures may rightly be classed generally as a problem associated with the geology of dam sites. As a record of interest, Fig. 13.19 is reproduced (by permission) showing a selection of test records conveniently plotted on a uniform scale for comparative purposes. The relation of pressures recorded to the rock type forming the foundation bed is indefinite, but the diagram shows that the usual assumption for uplift pressure, as varying from full head at the toe of the dam to tail-water head at the heel, is not unwarranted.

Drainage has been mentioned. It has long been a matter for keen debate, but the general consensus of opinion appears to be that, provided the drains are carefully located and are of such a size and character that they cannot possibly become blocked, their installation is a wise precaution. Drainage trenches filled with crushed rock and provided with suitable outlets constitute a usual expedient. Inspection galleries inside a dam structure are often connected with the drainage system so that regular checks on the drainage can be made; and in one case known to the author, one of the inspection galleries is located actually on the rock floor so that the rock surface itself can be periodically examined.

There remain to be considered possible means of increasing the impermeability of a dam foundation consisting of rock. In some cases, this work can change an unsuitable site into one that can be utilized with confidence. An obvious solution, often adopted to seal temporary cofferdams, is to use a blanket of impervious unconsolidated material such as clay upstream of the dam structure, so that any leakage passing under or around the dam will tend to draw the material into the cavities and in this way seal them tight. When large cavities are known to exist, bulky materials, *e.g.*, bales of hay, may first be used to block the main openings. Although these measures failed to stop leakage in the case of the Hales Bar Dam, they have been successfully utilized elsewhere. Of more general application is the practice of grouting or cementation of the underlying rock strata; it will be considered separately in the following section.

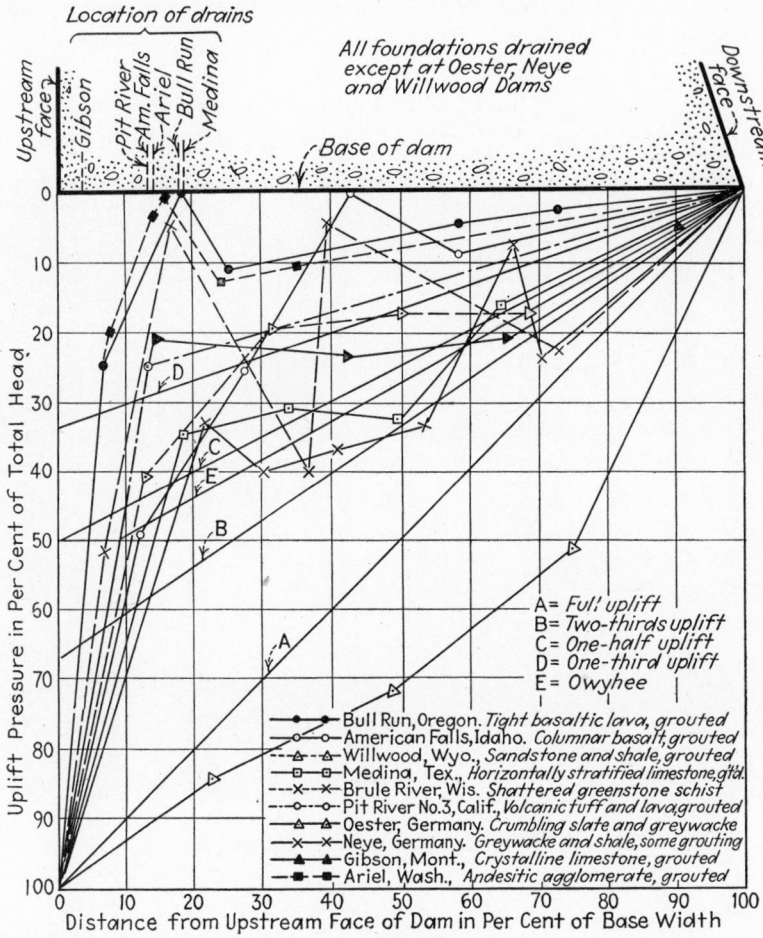

FIG. 13.19.—Maximum observed uplift pressures under existing gravity dams on rock foundations.

(This interesting record was prepared by Mr. Ivan E. Houk, Senior Engineer, U. S. Bureau of Reclamation, and first published in Civil Engineering, Vol. 2, p. 578, New York, September, 1932. It is now published by permission of Mr. Houk and the American Society of Civil Engineers.)

13.10. Grouting of Foundation Beds.—Grouting may be generally described as the injection of suitable material into certain sections of the earth's crust, under pressure and through the medium of specially drilled holes, in order to seal any open fissures, cracks, cavities, or other openings in the strata encountered. *Cementation* is a word applied to grouting, in some cases, when a slurry of portland cement (either alone or mixed with a proportion of sand) is used as the sealing material; it is also used in a proprietary connection. In addition to cement, clay, asphalt, and various combinations of chemical solutions are used to form the necessary seal; brief notes will be given on the use of each.

The use of asphalt has already been considered in connection with the sealing of the foundation beds of the Hales Bar Dam in Tennessee. Considering other special materials, the use of clay was

FIG. 13.20.—Typical view in limestone cavern, at Madden Dam site, Panama Canal, prior to sealing by clay grouting operations. Note reprecipitated calcite lining the walls.

(*Reproduced by permission of The Governor, Canal Zone.*)

apparently first featured on a major scale in the sealing of the limestone foundation beds underlying the Madden Reservoir in the Panama Canal Zone. The limestone formations encountered are badly weathered in certain localities and featured by underground caverns and fissures, rendering them far from the watertight barrier required to form the bed of this reservoir. Cement grouting was considered to be impracticable because of the presence of a loamy sand as a lining to caverns and crevices, with which the cement would not bond. Considerations of expense ruled out the possibility of using asphalt, and so the board of consultants for the project suggested the use of clay. A most valuable series of experiments were carried out, the results of which demonstrated that the best effects were obtained with mixtures of clay and water containing about 55 per cent of water. Grout of this consistency traveled up to 50 ft. underground and penetrated seams as narrow as half an inch. It was found that the mixture had to be prepared with the aid of blungers, as used in the ceramic industry; screens were also necessary. Three- and seven-inch diameter drill holes were used, the pressure varying with the consistency of the grout. Excavation of small pits at surface blowouts showed up several well-grouted seams, and other specific test results demonstrated the special utility of this method of grouting for the cavernous limestone encountered. Although clay does not give such a perfect seal as cement, for example, it can be employed in cases where cement could not be used and therefore is a useful tool of the foundation engineer.[13.18]

A more recent example of a somewhat similar kind was the use of a fine, red, sandy loam, having a high clay content, mixed with a small proportion of cement (varying up to a mixture of 1 part cement to 6 parts soil) as grouting material for sealing an old earth-fill dam near Huguenot, New York, supplying the Port Jervis community with water. The dam was raised without knowledge of subsurface strata, and serious leakage developed. To stop this, an extensive test-boring and grouting program had to be carried out, the grouting being used to seal boulder and gravel strata previously unsuspected. The case is a good example of the expense that can be caused by lack of adequate construction records.[13.25]

The use of chemical solutions, which enter into combination when mixed, as a sealing material for grouting operations is a relatively recent development. Available processes are largely associated with patent rights, notable among them being the Joosten process described on page 201. Although apparently not yet applied on any major dam-foundation work (except in France), chemical consolidation is here mentioned in order to complete this brief review of grouting operations considered in relation to geological conditions. Of comparatively recent introduction, the method has already proved of great value and is expected to develop into an important aid to construction.

Grouting with cement is the most usual method adopted, and civil engineering literature contains many references to its successful application to many different types of work. The method depends essentially on the fact that if portland cement is mixed in a slurry with water, it will, in course of time, "set up" hard, bonding satisfactorily with other materials such as rock with which it may be in contact. As originally used, the grout was simply poured into open cavities and allowed to set freely. Eventually, the idea of forcing the grout into normally inaccessible cavities was developed, and the use of pressure for forcing the grout into location was originated. The origin of grouting is a matter of conjecture, although it seems probable that the Romans were at least familiar with the idea, even though they had no means of using pressure for injection. It was certainly used to some extent in the nineteenth century, but not until James Greathead invented his grouting machine at the end of the nineteenth century for use in assisting the construction of tube railway tunnels in London did pressure grouting become a regular construction operation.

The first dam foundation for the sealing of which grouting was effectively used and of which the author has been able to trace written details was that of the Kinder Embankment, of the Stockport Corporation, England. During the construction of this work (1903–1905), the bedrock was found to be so fissured and faulted that it could not support the originally intended masonry dam, a satisfactory shale bed being eventually found only at a depth of 180 ft. below ground level. In the construction of the trench to this depth, much trouble was encountered with water. The drainpipes installed to tap this water, as concreting advanced, were carried up until water ceased to flow from them and then grouted up under pressure, a practice now frequently followed.[13.26] One of the first dam projects on which pressure grouting was used in the United States was the Estacada Dam in Oregon, constructed in 1912.[13.27]

Concurrently, the use of grouting in mining operations was progressing, especially for the sealing of water-bearing strata in such operations as shaft sinking. A pioneer in this work was Albert François, who in the early years of the century worked out and developed the ideas now embodied in what is known as the François cementation process, the special features of which are the use of a special direct-acting pump capable of developing 3,000 lb. per square inch pressure; flexible and reliable mechanical controls; the use of cementation pipes sealed into the ground; and the employment of special chemical solutions, lubricating in their action, to ensure the cementation of all fine cracks encountered. Development work has naturally taken place along general lines other than those associated with this particular process, so that it can now be said that cementation is an accepted part of foundation engineering practice, capable of accurate planning and control if considered with proper relation to the geological structure concerned.

British practice contains many interesting examples of cementation as applied to dam construction, including several cases in which the use of cementation has materially reduced the necessary depth of cutoff walls. Precementation has been another feature of British practice, the term being used to indicate the use of cementation of underlying strata before dam excavation is begun, the two objects in view being to consolidate effectively the material that has to be excavated (by means of the cement grout) and to obtain the benefit of the superincumbent weight of undisturbed strata over the fissures being sealed. This weight is necessary to prevent vertical lifting of rock strata above fissures which the pressure on the grout tends to cause, movement that has been readily observed in a number of cases. Precementation can be carried out even through surface strata of unconsolidated material if the drill holes used are suitably cased, but this introduces the risk of leakage occurring at the junction with solid rock strata; it is always advisable, therefore, to strip unconsolidated materials before cementation is undertaken.

Considerable savings can be effected by the careful application of this method, as was clearly brought out in a description of its use at the Scout Dyke Dam site in England. This is an interesting example, especially as vertical shafts were used from which horizontal drill holes were made, in addition to the usual vertical drill holes. Work on the

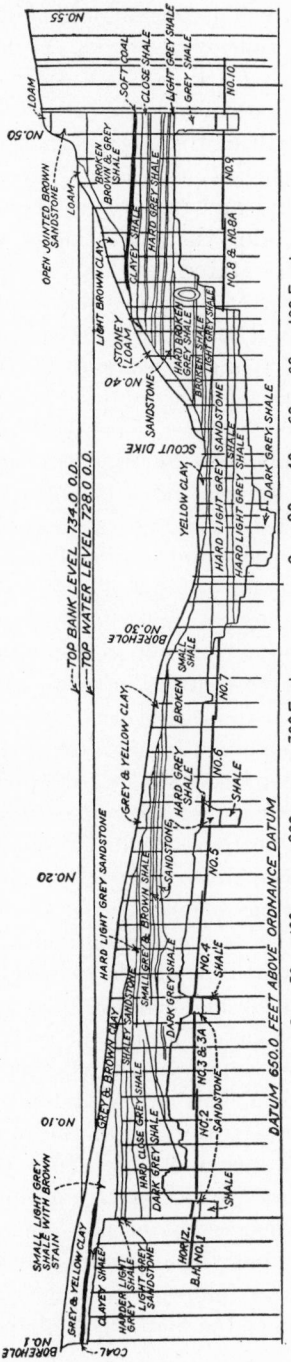

FIG. 13.21.—Geological section along trench of Scout Dyke Dam, Barnsley, England, showing boreholes used for precementation.
(*Reproduced by permission of Mr. J. R. Fox and the Institution of Water Engineers, London, from Reference* 13.28.)

dam, which is an earth-fill structure forming part of the Barnsley Corporation Waterworks, Yorkshire, England, was started in 1924. Underlying the site are alternating beds of sandstone and shale, a bed of permeable sandstone outcropping in the stream throughout the length of the reservoir; below this is a tight bed of shale. The sandstone bed was found, by means of trial borings, to be almost horizontal and uniform in thickness along the line of the dam. To avoid taking the trench down to the depths that the position of this bed necessitated (over 80 ft. at the ends of the dam), cementation was resorted to, and the trench afterward taken down only to the first sound shale or sandstone encountered. A series of vertical boreholes 25 ft. apart was put down into the impervious shale underlying the sandstone encountered at stream level, and cement was injected into them. Later, when the trench was excavated, shafts were sunk 200 ft. apart to the same level as the bottom of the boreholes, horizontal holes were drilled between them, and cement was again injected. In this way, any vertical joints possibly missed by the boreholes were sealed. All holes were drilled with a view to obtaining cores, and the evidence thus obtained was a valuable addition to preliminary data. It was found during the progress of the work that grouting pressures of not more than 400 lb. per square inch had undoubtedly lifted strata so that the limiting pressure was reduced to 100 lb. per square inch. Careful estimates of cost were kept which proved beyond doubt the great monetary saving achieved by this precementation program.[13.28]

Cementation under high pressure carried out for a very different purpose featured the construction of the Abitibi Canyon water power dam, in northern Ontario, Canada. This large structure, located on a river flowing into Hudson Bay, is founded on gabbro which preliminary test borings showed to be free from seams and fissures. When the site of the dam in the canyon was unwatered, it was found that along the stream bed there was a zone of rock which had been altered in composition and texture for a width of about 40 ft. by the passage of intrusive gases through the rock when molten and by possible faulting. Since it was not known whether or not this zone would be impervious to water, a shaft was sunk to a depth of over 100 ft. below river level at the upstream face of the dam. No water was encountered, and the rock appeared to be sound; but it was decided to carry on exploratory work by drilling from the bottom of the shaft, using cement grout when seams or porosity was found, at pressures not less than the full static head from the reservoir with water at top level. This course was followed, and the excavation concreted up so that construction of the main body of the dam could continue, a smaller shaft being boxed out, measuring 10 by 12 ft., with an enlargement at the bottom, as shown in Fig. 13.23. From this shaft, holes were drilled as shown, grouting being carried out at 10-ft. stages in each hole, using pressures up to 250 lb. per square inch. In addition to using the holes specially drilled, grouting was also carried out in old exploratory drill holes in the canyon walls and from inspection galleries left in the dam, a total of 1,527 bags of cement being used with satisfactory results.[13.29]

FIG. 13.22.—Abitibi Canyon power development, Northern Ontario, Canada; general view of excavation for dam site looking upstream and showing main cofferdam and diversion tunnel intakes. Shaft referred to in text is in the foreground, to the left of the center.

(Photograph by Dominion Construction Corp., Ltd. Reproduced by permission of Mr. G. F. Hardy, Consulting Engineer, New York, the Ontario Hydro Electric Power Commission, and the Dominion Construction Corp., Ltd., main contractors.)

Many of the major dam structures to be found in the United States have had their foundation beds sealed by pressure grouting, and some have presented features of special interest. Two only can be mentioned, despite the interest of the older structures; modern examples have been selected as demonstrating the latest technique, besides being also of unusual magnitude. The Norris Dam is a project of the T. V. A., located on the Clinch River about 80 miles above its confluence with the Tennessee River. It is a concrete gravity structure 285 ft. high, flooding an area of 34,200 acres at normal pool level. The entire valley above and below the dam site is underlain by Knox dolomite, featured in many locations by underground caves and porous beds, being described in official reports as " . . . a comparatively hard and substantial rock . . . quite massive, compact and sound. Some jointing has occurred . . . [with] occasional seams, of

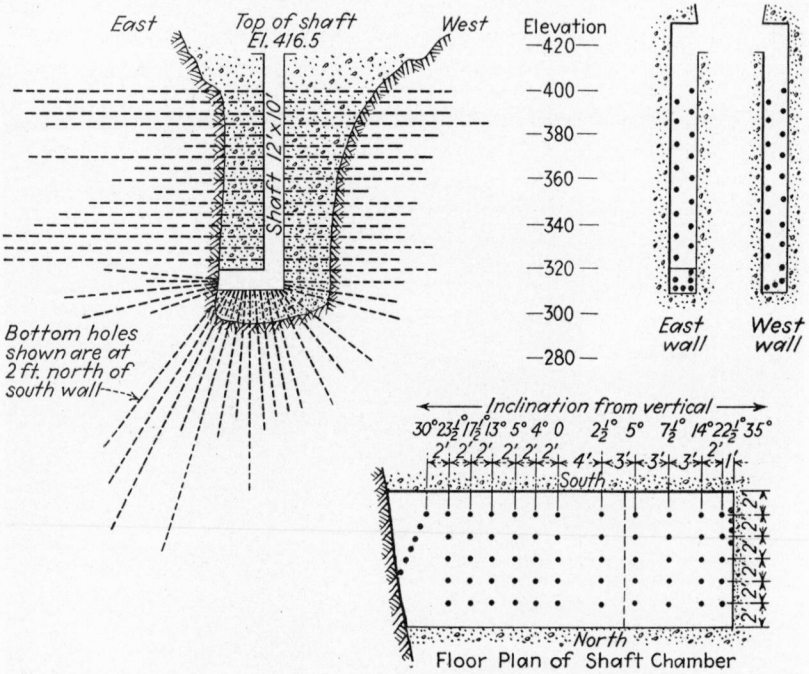

FIG. 13.23.—Diagrammatic sections showing arrangement of holes used for cementation in shaft at the Abitibi Canyon Dam, North Ontario, Canada.

(Reproduced by permission of Mr. M. Taylor and the Editor, Canadian Engineer, from Reference 13.29.)

importance in regard to possible leakage. . . . " Complete exploration of the dam site confirmed previously expressed opinions of consultative boards to the effect that the foundation rock itself is characterized by massive thick strata of excellent rock interrupted at definite intervals by horizontal seams which are partly open, partly clay filled, and for a portion of the area in close contact. A comprehensive grouting program was laid out covering the entire base area of the dam, spillway apron, and powerhouse. The grouting treatment was divided into two parts: shallow, low-pressure grouting covering the entire area of the foundation; and deep, high-pressure grouting to form an impermeable curtain under the heel of the structure. On the reservoir rim, the work involved the determination of the location of those portions in need of treatment to prevent excessive leakage, and the grouting of the parts found to be faulty.

Spacing of the drill holes through which grouting was done is shown in Fig. 13.24. A system of interlocking patterns was chosen for the shallow grouting. All seams were washed out by reversing air and water flow between adjacent holes prior to grouting. One complete pattern was first drilled and grouted to a depth of 20 ft. Following this primary grouting, a system of 40-ft. holes, evenly

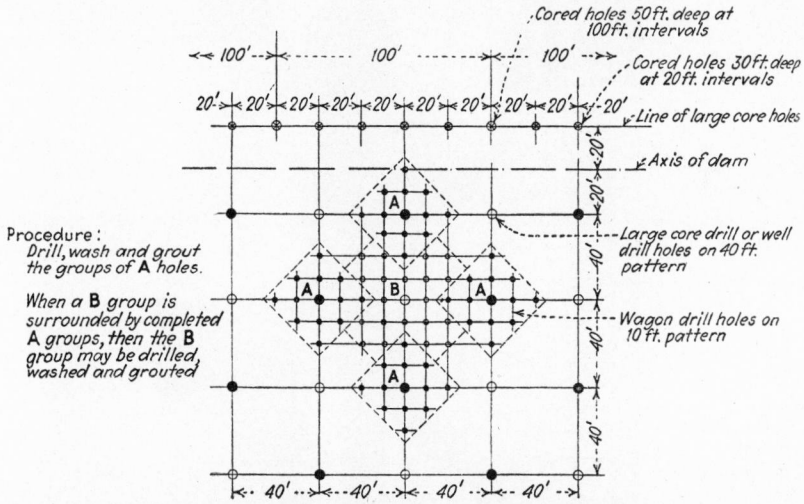

FIG. 13.24.—Diagrammatic plan of holes used for grouting foundation beds for the Norris Dam, Tennessee River, Tennessee, U.S.A.

(Reproduced by permission of the Tennessee Valley Authority, Mr. T. B. Parker, Chief Engineer, and the Editor, Engineering News-Record, from reference 13.30).

spaced between those first drilled, was superimposed upon the original pattern, and the same sequence of operations repeated. After the dam had been built to a height of 100 ft. or more, deep foundation grouting holes were drilled from the lower gallery to establish a deep grouting curtain. This curtain is located approximately 6 ft. downstream from the axis of the dam. The work of drilling and grouting the holes 10 ft. apart in the galleries was divided into three parts. First, groups of three holes on 100-ft. centers between groups were drilled, washed, and grouted, after which groups of three holes halfway between the first groups were similarly treated. This left space for two holes between the first and second groups, and these were drilled last. In general, a refusal pressure of 150 lb. per square inch was used in the curtain grouting.

A specially interesting feature of this work was the development of a special periscope for examining seams in exploration holes and the results of grouting in redrilled test holes. This was suspended from a portable A frame and arranged so that it could be raised or lowered. Illumination was provided by a 200-watt electric lamp, which permitted the inspection of holes to great depths. The use of this

FIG. 13.25.—Periscope for examination of drill holes, and the study of the results of grouting, as developed by T.V.A. engineers at the Norris Dam, Tennessee River, in use at this dam site.

(Reproduced by permission of the Tennessee Valley Authority, Mr. T. B. Parker, Chief Engineer, Knoxville, Tennessee, U.S.A.)

ingenious device, and the 36-in. diameter drill holes already mentioned, enabled a positive check to be obtained not only of the whole of the underlying rock structure but also on the results of grouting, as a result of which, despite uninformed predictions to the contrary, both dam and reservoir have proved to be watertight under full operating head.[13.30]

Finally, brief mention must be made of the grouting carried out in the foundation beds underlying Boulder Dam, the great structure on the Colorado River to which several references have already been made. All grouting work was specified in detail with regard to materials, pressures, and procedure by the engineers of the United States Bureau of Reclamation under whose direction the work was carried out. The contractors, Six Companies, Inc., developed special equipment for this, as for other parts of the work. Wherever concrete was to be placed against rock, advance provision was made for forcing grout into rock fissures and seams, if any existed. When concreting was in progress, special pipe connections were installed through which grout could later be forced when the dam was carrying full water load, in order to seal effectively the contact between concrete and rock. By means of carefully planned drill holes, the attempt was made to provide a grout cutoff "curtain" in the bedrock across the dam site; and to assist in attaining this objective, grout holes were drilled from the outlet tunnels up to 150 ft. deep. Grouting was also carried out from the inspection galleries left in the body of the dam, after the construction of the dam had advanced appreciably. Checks on the efficacy of the grouting carried out were made by taking out cores of grouted rock, and these showed satisfactory results. Pressures used varied up to 1,000 lb. per square inch, the average quantity of cement used (neat cement being used throughout) being 0.8 sack per foot of hole for all the longer holes.[13.31]

From relatively insignificant beginnings, pressure grouting, or cementation, has developed into an important feature of all major construction operations. The actual performance of grouting operations is essentially an engineering matter, but successful application depends on an accurate knowledge of underlying geological structure, just as does the need for the application of grouting. The subject is therefore one that must always be featured by cooperative endeavor of engineer and geologist, and the success of the operations described in the foregoing typical examples has been dependent to a large extent on such joint action.

13.11. Dams on Permeable Foundation Beds.—Discussion of dam foundations has been up to this point confined to dam structures constituting watertight barriers across a valley, watertightness being achieved either by the dam being founded directly on impermeable rock or through having some type of cutoff wall constructed below the dam structure proper, connecting it to a solid rock floor below. There are many sites at which dams have been required but at which solid rock exists only at very great depths below existing ground level. The unconsolidated deposits on which a dam structure must be founded at these sites present many unusual problems of design and construction. In permeable strata such as sands and gravels, the stability of *floating dams*, as they are sometimes called, depends on a predetermined safe rate of flow beneath the dam; in the case of structures on impermeable strata, for example, clay formations, reliance must be placed on the efficiency of the clay as a seal against uncontrolled flow beneath the dam. In both cases, accurate knowledge as to the nature of the unconsolidated deposits to be encountered is as essential for design as for construction.

Thus the importance of preliminary geological studies at dam sites again comes up for emphasis. Methods of investigation need not be detailed again, nor even an example given, since general procedure and the analysis of results are similar whatever the geological conditions at the dam site. Novelty enters in as a result of the fact that the unconsolidated deposits have now to be regarded as constituent parts of the structure, necessitating detailed testing prior to use in design, instead of as something to be got rid of by excavation or penetrated by cutoff trench or sheeting. Soil sampling must therefore be a feature of preliminary work, this being a part of the study of Soil Mechanics. The characteristics to be investigated in connection with dams on foundation beds of unconsolidated materials are common to other branches of civil engineering work. It will be convenient, therefore, to postpone discussion of these until Chap. XX when they can be reviewed against the appropriate background.

13.12. Materials of Construction.—The availability of suitable local materials for the construction of a dam will naturally be a determining factor in the economic studies of every dam project, and it may often affect the choice of the type of dam. The use of rockfill dams always calls for careful study of the material to be used. There may therefore be mentioned an interesting example of rockfill dam construction which emphasizes several matters already generally discussed.

The Los Angeles County Flood Control District was set up in 1915 to carry out flood protection work and water conservation for Los Angeles County in which live a population of about two and a half million. Twenty five million dollars of the funds raised by bond issue in 1924 were allocated for the construction of a concrete gravity dam of unprecedented size in the San Gabriel Canyon on which work was begun under contract in 1929, completion being called for within six years.

While excavation was still under way a serious earth slip at the west abutment resulted in holding up the work for a further study of foundations. Before the end of 1929 the contract for constructing the dam was cancelled.

Surveys for a new dam site developed what was thought to be a safe and satisfactory location just above high water level of the reservoir of the Morris Dam, constructed by the city of Pasadena lower down the San Gabriel Canyon. At this site it was proposed to build a rockfill dam . . . to have a height of 300 feet above streambed, a crest length of 1,670 feet, and a base thickness of 900 feet at streambed level. . . . A contract was awarded in the latter part of 1932. . . . After that contract was made, work in the quarry progressed satisfactorily, although the unexpectedly large percentage of material that

FIG. 13.26.—San Gabriel Flood Control Dam No. 1, in August 1934. Concrete toe wall in the foreground and Quarry No. 10 in the background.

(Reproduced by permission of the Los Angeles County Flood Control District, Mr. C. H. Howell, Chief Engineer, Los Angeles, California, U.S.A.)

had to be rejected resulted in low yardages delivered at the dam. . . . Up to October 1st, 1934, approximately 3,650,000 cubic yards of material, including stripping and waste, or the equivalent of about two thirds of the total volume of the dam, had been removed from the quarry. But little more than one tenth of this material actually went into the dam.[13.32] [Quotation amended and corrected].

After further conferences and reports, work on the new contract was temporarily held up, and the dam design revised again to take into consideration the rock available from the quarry. On this basis, the work was satisfactorily completed in July, 1937. Over ten million cubic yards of material was placed in the dam, as finally built, about one million cubic yards being a sand-clay mixture used as an impervious blanket near the upstream face; all other material was quarried rock. All loose rock fill was placed with the aid of sluicing, shrinkage approximating to 6 per cent. The dam is now functioning satisfactorily, having already been tested by severe floods. A smaller rock fill dam, known as San Gabriel Dam No. 2, was constructed about the same period. No trouble was experienced with rock supply in this case, the main difficulty encountered being the settlement of the rock fill when near completion, attributed to lack of sluicing for compaction of the fill. These brief details of a notable water conservation project (a full bibliography for which is noted) serve to illustrate the vital importance of material supply and, in particular, of rock fill. Geological advice had been repeatedly taken in connection with this work, some of the geological difficulties having been unpredictable.

13.13. Construction Problems.—It is fitting to include at the end of this list of special problems a note on those which arise during construction operations. To emphasize the fact that study of geological conditions at a dam site must be a *continuous* operation from the start of preliminary operations up to and beyond the final passing of the cleaned and prepared rock surface as the approved foundation bed, this paragraph is inserted. A problem that may be regarded as essentially part of the construction operations consists of the removal of all rotten and disintegrated rock, and equally important is the keeping accurate records of all rock that is so removed. When finally the approved foundation bed is ready for the undertaking of dam building, a complete and detailed survey of this must be made, this being one of the most critical duties of a resident engineer. Although it may possibly seem to be unimportant at the time, record surveys of this kind can often prove to be of inestimable benefit, their value being incalculable since if not accurately made when the foundation bed is ready, they can never be obtained again. The author writes on this point from personal experience of a rock surface survey so obtained in the depth of a Canadian winter which proved to be of great and unexpected assistance when an adjoining section of a dam was constructed in the following spring.

There are some detailed matters in connection with the preparation of a rock foundation bed which call for mention. One essential of construction is to get as good a bond as possible between concrete and rock. To obtain this, it is essential that the rock surface shall be clean, free from all loose particles of material (even though they be rock fragments), and having no standing pools of water, however small, when concreting operations start. Opinions differ as to the methods to be used when a glaciated rock surface is met with, but it is generally agreed that any polished surfaces so encountered should at least be roughened by "pop shots," even if excavation is not undertaken to break up the surface completely. Failure of the Gleno Dam in Italy has been attributed to sliding of the structure on a glaciated rock surface. Preliminary geological advice will generally be able to foretell the presence of such a feature, so that an appropriate construction schedule can be prepared. The exact location and regular discharge of all springs encountered must be most carefully recorded lest they prove to be possible sources of trouble. Any springs of unusual nature should be immediately brought to the attention of the geological adviser so that he may examine them and advise accordingly. The detailed inspection of all unusual features, in addition to general supervision of excavation operations, is just as essential a part of the duties of geological adviser as are the preliminary geological investigations.

13.14. Scouring below Spillways.—The erosive action of water discharging over a dam crest or spillway weir may be considerable, and on record are many cases of serious damage resulting from this action when uncontrolled. At the Sweetwater Dam in California, for example, a single flood carried away from below the spillway 10,000 cu. yd. of rock. The failure of the Colorado Dam, Austin, Texas (see page 306) was probably appreciably affected by the erosion caused below the heel of the dam. Several of the major dam structures in Egypt such as the Assuan Dam and the Sennar Dam have been the scene of serious erosion below their discharge sluiceways. This list could be extended to great length if it were necessary, but the examples quoted will illustrate clearly the importance of this problem of erosion. It is a problem that, although of special importance in the operation of dam overflow structures, affects other civil engineering structures also. As erosion is a phenomenon of general importance, a part of Chap. XV is especially devoted to it.

13.15. Inspection and Maintenance.—In conclusion, a brief note must be added with regard to the maintenance and inspection of dams. In some instances, notably in the case of Great Britain, this is a matter rigidly defined by law, annual inspections being obligatory for all structures above a certain size. The restriction of this legislative requirement to dams is a matter of some surprise to many engineers, in view of the numerous other types of civil engineering work at least equally hazardous to human life. With or without state regulations, however, regular inspection of all dam structures is an essential. Naturally, examination is restricted to those parts of the dam which can be seen—together with the inspection galleries, tunnels, and shafts that have been left open after the completion of construction. The practice in certain large dams of leaving open exploratory tunnels has already been mentioned; the use of this means of access is invaluable. One certain guide in inspection is the leakage of water through or around the dam. Records of this will naturally have been kept regularly, and any increase, either sudden or gradual, will be a call to remedial action. Study of the leakage, the sediment it carries, and its chemical content may reveal the reason for the increase and possibly suggest remedial measures. Such study of leakage in extreme cases when the damage is irreparable will at least enable due warning of impending trouble to be given. Failures should become more of a rarity as the years go on and especially so if, in the course of increasing the safety and reliability of dam foundations, due attention to the associated geological conditions is always regarded as a factor of the greatest importance.

Suggestions for Further Reading

Several notable engineering papers dealing with the geology of dam foundations have been prepared; the following are leading examples:

BRYAN, KIRK: Geology of Reservoir and Dam Sites, *U. S. Geol. Surv. Water Supply Paper* 597, 1929.

LAPWORTH, H.: The Geology of Dam Trenches, *Assoc. Water Eng.* (now *Inst. Water Eng.*), *London, Trans.*, **16**: 25 (1911).

and a symposium entitled Geology and Engineering for Dams and Reservoirs, *A. I. M. E. Tech. Pub.* 215, 1929.

Finally may be mentioned a valuable book in the French language dealing with the same subject and written by an eminent European geologist:

LUGEON M.: "Barrages et géologie," F. Rouge et Cie., Librairie de l'Université de Lausanne, Switzerland, 1933.

CHAPTER XIV

RESERVOIRS AND CATCHMENT AREAS

> . . . the question as to the possibility of retaining water in rocks for summer use is decided by the annual and periodical operations of nature. For the means of altering or improving some of these natural operations, so as to render the irregular supply of water which falls upon the earth more convenient to the general purposes of man, we must have resort to geology. . . .
>
> WILLIAM SMITH, 1827.[14.1]

The function of dams is generally to retain water above the elevation at which it normally stands. This impounding may vary from the control of water levels within a few feet of normal elevation by a regulating weir to the creation of the immense lake now behind the Boulder Dam, 145,000 acres in extent. Apart from specially restricted cases such as low-head regulating weirs, a reservoir will be formed by the dam to serve as an artificial storage basin for the water that is retained. The dam structure forms usually but a small part of the periphery of the reservoir, its bottom and the greater part of its sides (if this distinction may be used) consisting of the natural crust of the earth. A superficial assumption is that, provided the dam site is given due attention with regard to watertightness, all will be well. If this were the case, this chapter would not be necessary.

As explained at the outset of the last chapter, the construction of a dam and the subsequent impounding of water behind it cause more interference with natural conditions than do almost all other works of the civil engineer. The problems created at the dam site are common to the whole reservoir above this site although to varying degree. In brief, it may be noted that a difference will be set up in the level of the water table corresponding to the height of the dam between the two sides of the dam; all material in the bed of the reservoir and especially close to the dam will be subjected to considerable hydraulic pressure; all flooded areas will be in future subject to the action of water upon them; and finally the groundwater level up the valley in which the reservoir is located will be directly affected by the rise in water level, generally for a considerable distance away from the actual shore line of the impounded water. In consequence, there will be a tendency for the impounded water to find some means of escape through any weaknesses in the structure of the ground that may exist. Materials in the sphere of influence of the impounded water and liable to be affected by exposure to water may fail to retain their former stability, landslides being one possible direct result of failure. Features, such as wells, that come within the area of influence of the reservoir and are dependent on previous groundwater levels will be correspondingly affected. Cursory consideration will show that all these matters depend fundamentally on the geological structure of the reservoir basin. A study of this geology is therefore an essential complement of geological investigation at a dam site, since only by this means can possible dangers be foreseen and suitable precautionary measures be taken in advance.

Another important reason for the careful study of the geology of reservoir sites is to enable an accurate estimate of the capacity of the reservoir to be made, this requirement having often associated with it the determination of the necessary height of dam. This figure must often be determined purely from the point of view of elevation, when maximum working head is the criterion, as in the case of water power plants. In other cases, and especially with water supply reservoirs, effective storage capacity is the determining factor. For a reservoir formed by a dam and a basin of impervious strata, the calculation of this capacity is a simple matter, and reservoir capacities are often calculated on this assumption. If attention is given to the geological formations underlying the basin, it may be found that large masses of pervious rocks will be in contact with the impounded water and are so arranged that they can drain out (at least to some extent) as the reservoir water level is lowered. With the aid of reasonably careful preliminary geological investigations, an estimate of the additional storage thus available can be obtained, with consequent benefit to the economic aspect of the scheme.

14.2. Leakage from Reservoirs.—Leakage from reservoirs is obviously a likely source of trouble. Attention in the past has not always been given to the possibility of leakage developing. Of the many examples that could be cited in support to confirm this suggestion, there need be mentioned only the abandonment of the Jerome Reservoir in Idaho and of the Hondo Reservoir in New Mexico, in each case as the result of excessive and uncontrollable leakage which occurred after the impounding dams had been built. That structures of magnitude should have to be abandoned not because of defects of design or construction but because of the undetermined nature of local geological structure is surely most telling evidence of the need for adequate study of every feature connected with the geology of a reservoir site.

Failure by leakage may take place because of defects in the structural arrangement of the underlying strata, *e.g.*, a fault or excessive fissuring; through failure of one or more of the rock types encountered to stand up to exposure to water; as a result of the dip of pervious rock strata, such as to drain water away from a reservoir site; or through the absence of adequate impervious barriers at critical points in the topographic perimeter of the reservoir area. These different cases can best be explained by a discussion of typical cases.

As an instance of leakage from a reservoir most probably due to the first of the causes listed may be mentioned the experience met with at Ewden Valley works for the water supply of the city of Sheffield, Yorkshire, England. Construction of the two impounding earth dams in the Ewden Valley began in the year 1913, the respective reservoirs being known as the Broomhead and the More Hall, the former being for water supply, and the latter for compensation water purposes only. The Broomhead Dam is about 990 ft. long at the top of the embankment, overflow level being 91 ft. above lowest drawoff level. The configuration of the valley disclosed the presence of ancient slips near the dam site, and studies and test borings indicated a very disturbed condition of the underground strata consisting of shale, sandstones, and grits with overlying yellow clay. Excavation for the cutoff trench confirmed this deduction, the trench having to be carried to a depth of 120 ft. below stream level before the strata were found in undisturbed horizontal positions. Underground water flow in a band of grit was encountered and was tapped and drained as the trench was concreted up. Geological and specialist engineering advice were constantly taken as the works proceeded; but after the start of impounding water in May, 1928, increased flow below the dam was observed, indicating leakage from the reservoir. Further geological investigations were made (also in connection with a serious landslip) including many more borings, as a result of which a program of cementation was initiated. Fourteen million pounds of cement were injected in about six months (up to July, 1930), reducing the leakage from 2,478,000 to 1,577,000 (imperial) gal. per day. This latter figure has been still further reduced by additional grouting work carried out when the reservoir was drained and also by the sluicing of silt behind sandbag dams formed where rock outcrops were found in the bed of the reservoir. Leakage still persists through unknown underground passages, the works being a most striking example of the possibility of leakage occurring through rocks normally sound but locally distorted and fissured, and this despite most careful engineering work and geological investigation.[14.2]

As the dam site in a reservoir will usually be at the deepest part of the valley utilized to form the basin, the impounded head therefore being greatest at this location, structural geological defects and the existence of rocks that may be affected by prolonged exposure to water are of special consequence and greatest potential danger in the vicinity of the dam. Especially is this true of doubtful rocks such as those which are to any degree soluble in water. Rock salt is naturally the most dangerous of all to be encountered; but as it is not of frequent occurrence, it is not of great significance. Gypsum, although less soluble than rock salt, is yet seriously affected by exposure to water and so potentially dangerous. Of unusual significance in all reservoir investigations is the presence of limestone. Although the least permeable of the rocks mentioned, its relatively wide distribution renders it comparatively well known. This general familiarity, although not perhaps breeding proverbial contempt, does lead to the tendency to form incorrect judgments regarding its soundness, especially when considered in conjunction with the relatively solid appearance of many outcrops.

All limestones are at least partially soluble in ordinary water. If, through the movement of water, the resulting solutions can be removed from contact with the rock and replaced by fresh water, solution will continue. In all limestone formations subject to movement of underground water, therefore, special precautions must be taken to ensure that no potentially serious zones of weakness exist. The well-known caverns sometimes found in limestone districts are the result of water action. Sometimes these are the result of the action of flowing streams which for a part of their courses flow underground, parts of Derbyshire, England, being typical of similar well-known scenic areas to be found in many parts of the world. Some limestone caverns are the result of periodic movements of underground water tables. In consequence, a close study of existing underground water levels, although always of value in connection with the investigation of reservoir sites, is of special value in the case of those formed by limestone strata. Many examples from practice testify to the importance of close study of any limestone formation encountered in reservoir work. The troubles met with at the Hales Bar Dam (see page 334) were due to cavernous limestone. The pioneer work on grouting with clay at the site of the Madden Reservoir in the Panama Canal Zone (see page 338) was occasioned by the careful studies of the underlying limestone strata that had been made in the preliminary geological investigations, all points of weakness being effectively sealed up before the reservoir was filled. The Malad Reservoir, near Malad, Idaho, represents another example. A dam 440 ft. long and 70 ft. high was projected to impound water; construction was started in 1917. Work had to be stopped when the crest had reached a height of only 50 ft., since, although the level of the water being impounded had risen only 10 ft. above normal stream level, leakage was occurring at a relatively high rate. Investigation suggested that the leakage was occurring through a bed of soluble limestone 500 ft. long and 25 ft. wide in outcrop, on the side of the valley near the dam but effectively screened from view by a talus slope.[14.3] These particular examples include mention of cases previously noted in order to illustrate the close relationship existing between the geological conditions affecting dam sites and those germane to reservoirs and catchment areas.

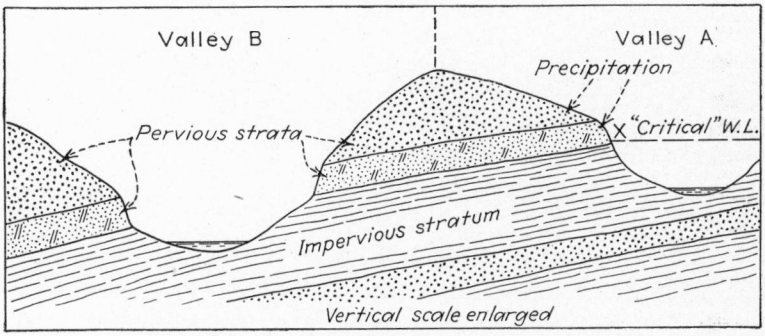

FIG. 14.1.—Simplified geological section, illustrating possible leakage from reservoir and catchment area.

The possibility of leakage through a pervious stratum dipping away from the reservoir valley is best explained by means of a simple diagram (Fig. 14.1). This is not an unusual geological section; and although the figure is diagrammatic only, many actual sections of a similar type can be found. The inclined structure of pervious and impervious strata will naturally result in any water in contact with the pervious beds in valley *A* (as, for example, water impounded in the valley by a dam with crest level above *X*) flowing underground until it is discharged into valley *B*. Discharge might take the form of springs. The structural arrangement as shown is as simple as possible; it is capable of much variation without the fundamental source of leakage being changed. For example, the lower pervious stratum may not outcrop in valley *B* but remain underground with no natural outlet for the entrapped water if the surface stratum were impervious. Artesian pressure will thus be set up, and a potential supply of underground water provided.

Finally, there always exists the possibility that leakage may take place through some permeable stratum at a low point in the ground encircling the reservoir. Rarely will the topography be such that there is no indication of any low spot in the perimeter of the reservoir site. When any depressions are observed, they will inevitably call for special geological study, since the fact that they are a departure from the regular topographical feature is an indication that some unusual process has been at work at some time in the past. Glacial action will frequently be found responsible for depressions—*saddles*, they may be called, if taking that particular form between two valleys. A section through a simple saddle in which the rock level has been

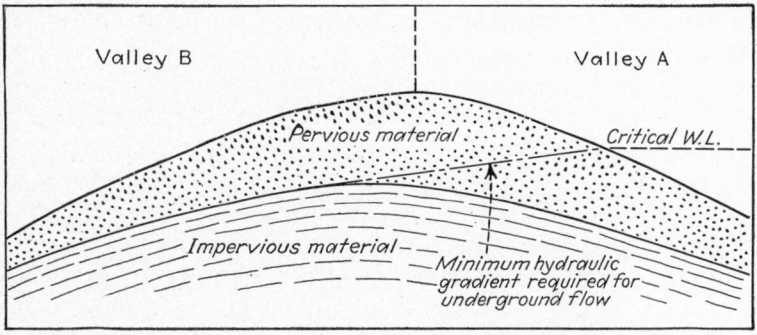

FIG. 14.2.—Simplified geological section, illustrating another possible means of leakage from a reservoir.

worn down well below surface level, pervious deposits making up to the observed level of the ridge at this location, is shown in Fig. 14.2. If water is impounded above "critical W.L.," and the broken line extending through "pervious material" is the necessary underground hydraulic gradient for flow through that material, it is clear that leakage will occur from valley *A* into valley *B*. In addition to causing loss of water from the reservoir, this underground flow of water may cause trouble in valley B, if there is any instability there of unconsolidated deposits which might possibly be intensified by the presence of excessive underground water.

There are available numerous records of leakage from reservoirs that appear to be generally of the type just indicated. Although few details of the occurrence have been published, leakage from a reservoir formed by a masonry dam across Cedar River, a tributary of the Snoqualmie River in the state of Washington, has been described by Dr. K. Bryan:

The north (right) end rested in glacial material. Some leakage through the rock at the south end and around the north end of the dam took place, but large and disastrous leakage occurred through the glacial material that formed the ridge between the two rivers. When the water in the reservoir stood at an altitude of 1,555 feet, springs broke out in Snoqualmie Valley at an altitude of 990 feet. The hydrostatic head was 565 feet, and the distance 6,500 feet, making a gradient of 1 to 11. The emergence of this leakage in Snoqualmie Valley caused slumping of the hillside. The total loss of water was 30,000,000 gallons daily (46.5 second-feet), and since 1914 the reservoir has not been allowed to fill. On account of continuous litigation full descriptions of the materials involved and of tests before and after construction are not available. The glacial materials of the ridge were, however, obviously so coarse grained and porous that it was evident before construction to both geologists and engineers, as set forth by Fowler, that the ridge could not resist the movement of water from the reservoir on a gradient of 1 to 11.[14.4]

In this interesting paper, Dr. Bryan also mentions leakage taking place through 2 miles of "landslide material" (in Colorado) and through 1,500 ft. of sand and gravel (at another site in the same state).

14.3. Detection of Leakage.—Leakage from a reservoir will generally be so obvious as to call for no special aids in detection. Sudden increases in stream flow below a dam site without a corresponding increase of rainfall will be a direct indication of some escape of water; and the appearance of springs, boils, or other surface indications of underground water in locations previously dry and stable will also indicate unsatisfactory conditions. There may sometimes arise cases of minor leakage which are more difficult to trace, especially in connection with the investigation of solution channels in limestone. The use of dyes in cases of this kind can prove to be invaluable, small quantities of the dye being added to the moving water at a convenient station, and samples taken from the suspected source of leakage. Various chemicals have been used for this purpose; permanganate of potash is perhaps the most common; in some cases, ordinary bluing can be used. Even the use of bacteria has been suggested; but for practically all special work, a chemical indicator is used. Many chemicals are thus employed, among them fluorescein and its sodium salt, uranin. These give water a characteristic color which can be detected even when diluted to the extent of one part in more than ten million parts of water.[14.5]

Even with the use of such sensitive dyes, however, leakage sometimes cannot be traced. As an interesting, although possibly unique, example may be mentioned the case of a small water power installation in Cephalonia, one of the islands of the Ionian archipelago (Greece), the undershot wheel of which is operated by the flow of sea water from the Mediterranean *into* the island. The strange stream so formed eventually disappears from view in the caverns and fissures of the limestone of which the island is there constituted. During British occupation of the island, large quantities of petroleum were poured into the sea water as it passed the mill, and the island was surrounded by boats, the crews of which were on the lookout for any signs of emerging petroleum. None appeared, and repeated tests have failed to reveal where the water goes! Although petroleum is a somewhat unusual indicator of leakage, it is sometimes so used where it can be with safety and convenience. The case is cited in view of its

FIG. 14.3.—Undershot water wheel in Cephalonia, Greece, operated by the flow of sea water *into* the island.
(Reproduced by permission of Messrs. John Blake, Ltd., Accrington, England.)

unusual interest and also because it presents yet another example of the vagaries of a limestone formation.[14.6]

14.4. Prevention and Elimination of Leakage.—If the preliminary geological investigations already described are applied to the study of reservoir and catchment areas when a dam construction project is planned, most of the possible sources of leakage from the reservoir should normally be discovered. The various possibilities of leakage having been noted, the nature of the means of leakage will be some indication as to whether or not it can be prevented by anticipatory remedial works. In the case of strata dipping across a valley into adjacent areas, little, if anything, can be done to interfere with the natural direction of underground water flow. When saddles occur with low spots in the bedrock elevation, as indicated in Fig. 14.2, cutoff dams can be constructed to seal effectively the open waterways thus available for underground flow. The simplest type of structure will suffice for this purpose, provided that a cutoff wall of some type is carried to the rock floor between the points at which it ceases to be above reservoir water level.

In other cases of possible leakage, and especially where localized fissuring of rock is encountered, grouting may effect a satisfactory seal if carefully carried out in close association with complementary geological study. Grouting operations may similarly be employed for remedial work if leakage is discovered to exist after a dam has been completed and the storage of water within a reservoir has begun. As it is seldom that a reservoir can be emptied completely, or even partially, for more than a short time when once it has been placed into active service, the carrying out of remedial works to eliminate leakage from a full or partially filled reservoir is often an extremely difficult matter. Another method of reducing and possibly eliminating leakage is to employ natural sealing material such as clay or the silt content of the watercourse feeding the reservoir, to block up the openings in the reservoir through which leakage is occurring, when these are definite and not too widespread. If the exact location of the area of weakness is known, sluicing or dumping suitable material into the reservoir immediately above this area may often be effective. Even mattresses of fascines, similar to those used in river-training work, may be used if they can be accurately located. The process may be a natural one. Mr. W. L. Strange has described how the Much Kundi Tank in Bijapur Protectorate, India, formed by a masonry dam on schistose rocks, was sealed and made tight after 12 years of natural silting.[14.7] In the discussion on Mr. Strange's paper, a Yorkshire (England) reservoir was mentioned, formed by an 80-ft. dam, which leaked as much as 440,000 (imperial) gal. per day but which silted up and sealed itself tight in two years. The permanence of any self-sealing operation will obviously depend on the nature of the leakage. If the sealing is complete in basalt and similar rocks, then it may be permanently effective; but if it is only partially successful in a limestone formation, then the constant flow of the small amount of leakage still persisting may be sufficient to enlarge existing leakage channels by further solution and cause a recurrence of serious trouble.

14.5. Secondary Effects of Reservoir Flooding.—The function of an impounding dam is to store a predetermined quantity of water and, in the case of a water power scheme, to raise the top level of the impounded water to a certain specified maximum. In achieving this result, areas of dry land must be flooded as the water rises, and the flooding will often entail extensive civil engineering construction of what may be termed a secondary nature. Buildings may have to be relocated, and sometimes whole settlements moved. Transportation routes, in view of the convenience of valley locations, will often be found traversing land that will be flooded when once the reservoir fills. Relocation work can sometimes attain major proportions. In the planning of all such work, the effect of the ultimate maximum water level must be considered. This type of work will often be

FIG. 14.4.—Bonneville Dam on the Columbia River, Oregon, U.S.A.; a view taken shortly after starting the impounding of water. The large area of land to the left of the dam is the ancient Cascade Slide, the old river channel having been located on the other side of this.

(*Photograph by Brubaker Aerial Surveys. Reproduced by permission of the Chief of Engineers, U. S. Army.*)

located on hillsides, and the possibility of landslides developing is one that must always be given special attention.

Unusual difficulties of this nature had to be faced in connection with the Bonneville Dam on the Columbia River, Oregon, construction of which started in 1934. The site of this dam is interesting, complicated as it is by the existence of the Cascade Slide, a great landslide which occurred about a thousand years ago after the river had cut a deep gorge close to one bank estimated to have been 200 ft. or more deeper than the modern channel. The slide deflected the river from its course, altering the topography all around the site selected for the dam, preliminary investigations having therefore to be carried out with unusual care. A railroad and a highway on each side of the river required relocation to bring them above reservoir level. The unusual part of this work was to obtain a stable location for the railroad on the Oregon shore some distance above the dam site, where an old landslide existed which had given the railroad company much trouble even before the dam construction started. The slide area is about 1½ mi. long and 3,000 ft. wide; its slow movement is attributed to the lubrication by groundwater of an underlying stratum of shale.

FIG. 14.5.—View of the Columbia River, immediately upstream from the Bonneville Dam, looking upstream and showing road and rail relocation work complete. The river here flows to the west; the work on the south bank was complicated by the Ruckel Slide area, which had long given trouble in the maintenance of the Union Pacific Railroad tracks at this location.

(*Reproduced by permission of the Chief of Engineers, U. S. Army.*)

Solution of the problem included extensive drainage tunnels and the construction of a retaining wall, measures mentioned in Chap. X.

Underground water levels will be materially affected in reservoirs the top levels of which are appreciably above normal river elevation, and this change in a natural ground condition may have widespread effects. Building foundations, and especially those on clay, must be carefully studied. Underground water supply systems will clearly be affected. If there is any possibility of even seepage, as apart from appreciable leakage, reaching catchment areas other than that in which the reservoir is situated, attention must be given to this contingency in case it should ever develop serious proportions. Peculiar problems may often arise due to special local circumstances. In a paper on dam construction in New South Wales, Australia, mention was made of the selection of a dam site for impounding water for the supply of the city of Sydney. Six dam sites were chosen:

The first selected for construction is situated on the Cataract River, having a catchment-area above it of 53 square miles. The geological formation of the Cataract catchment consists of Hawkesbury sandstone overlying coal-measures, and the area is bounded on its upper end by a range of high hills rising abruptly a short distance back from the ocean. Coal-seams outcrop on the side of the hills facing the ocean and dip with a fairly rapid inclination underneath the Cataract catchment. The seams have been worked from the face of the hills for some years, and coal-mining operations now extend under and past the divide. These operations, although so far well under the catchment-area, have not extended underneath that portion which will be covered by the stored water; when this does occur the minimum cover between the stored water and the coal-seams at the upper end of the storage will be 800 feet, increasing to 1,600 feet at the dam-site. The Water and Sewerage Board suggested to the Department of Mines the advisability of reserving from coal-mining operations all Crown land within the whole Sydney catchment-area that would be enclosed by boundary lines ¼ mile outside of and parallel to the edge of the stored water at each proposed reservoir. In accordance with the joint report made by Mr. E. J. Pittman, Government Geologist, and Mr. A. A. Atkinson, Chief Inspector of Coal and Shale Mines, however, the Government considered that there was not sufficient reason to prohibit mining under the site.

The following is an extract from the geologists' report:

We believe, however, that there will be no necessity to consider this alternative, for we are strongly of opinion that the coal can be extracted without detriment to the surface of the catchment area, and without affecting the holding capacity of the reservoir.

Accompanying this report is a geological section of a portion of this coalfield from the seam near Bulli to the Cataract Reservoir, showing approximately the thickness of cover over the Bulli seam. This cover is composed of what is known as the Hawkesbury series, consisting of considerable thicknesses of tuffaceous shales and massive beds of sandstone. The experience gained in the working of coal seams in other parts of the world convinces us that with this thickness of shales and sandstones intervening between the coal seam and the floor of the reservoir, the working of the coal can be accomplished without any danger of injury to or undue leakage from the reservoir. For while a slight subsidence of the surface may occur when the coal is entirely removed, there is no reason whatever to fear that this would be accompanied by any such leakage as would interfere with the water supply. With a view of guarding against the slightest possibility of damage to the stability of the retaining wall of the Cataract Dam through subsidence, a solid pillar of coal should, we think, be left beneath it, and this should be of such dimensions as to extend for a distance of 300 yards beyond the base of the wall in all directions.

If these precautions be observed, we feel convinced that the valuable stores of coal under the reserve can be extracted without any danger to the water supply of Sydney.[14.8]

Figure 14.6 is the geological section to which reference is made; it illustrates in an interesting way this unusual problem.

14.6. Catchment Areas.—Brief consideration will show that the water impounded by dams will be obtained in all normal cases from the rain that falls on the area of the valley, or *catchment area*, above the dam. In the ideal case, all the rain would flow down the watercourse to be stored in the reservoir; actually only a percentage of the total is obtained for use in this way, and the relation of this "runoff" (to use the convenient engineering expression) to the equivalent rainfall is influenced by the geological structure of the catchment area. Indeed, it can be said that the relation of runoff to rainfall can be fully studied only if due consideration is given to the local geology.

A catchment area that was formed of a continuous exposure of impermeable rock would yield at any gaging station in the watercourse a runoff equivalent to the rainfall on the total area above the station less the amount lost by

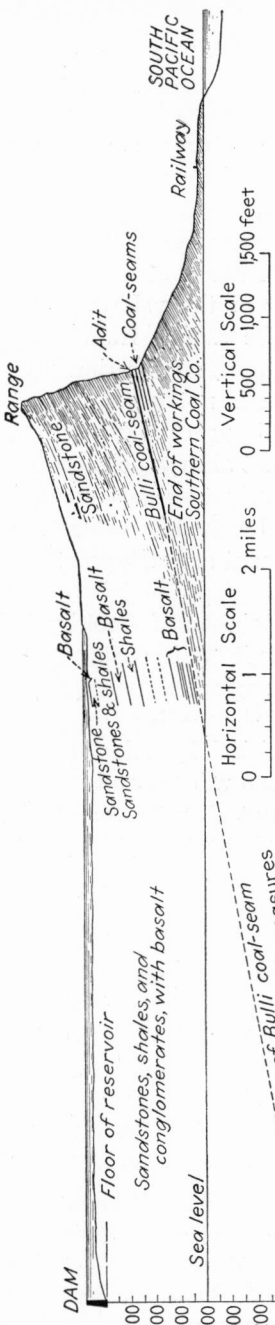

FIG. 14.6.—Geological section from the Pacific coast to the Cataract Dam, New South Wales, Australia, showing the probable depth of the Bulli coal seam and its relation to dam and reservoir.

(Reproduced by permission of the Council of the Institution of Civil Engineers London, from reference 14.8.)

evaporation. In the case of the catchment areas usually encountered, other losses must also be considered, of which the most important are those due to absorption by vegetation, the water that is temporarily absorbed into the underground water reservoir formed by any pervious rocks underlying the area, and the water that is deflected into adjacent catchment areas by means of peculiarities of geological structure.

Considering this last, a study of Fig. 14.1 will show that, in a way similar to leakage of water from the reservoir, rain falling on to the side slopes of the valleys depicted may not and in some cases will not join the runoffs from the catchment area on to which it falls. This diagrammatic section makes the source of loss so clear that elaboration is not necessary; suffice to say that the preparation of corresponding geological sections through typical parts of catchment areas should always be a component part of hydrological studies, so that, in evaluating anticipated runoffs, the true effective catchment area may be used in calculation. Somewhat naturally, the smaller the catchment area the greater the possibility of appreciable loss from this cause.

14.7. Relation of Geology to Runoff.—The calculation of rainfall losses is a vitally important part of hydrological study in connection with all engineering works dependent on assured water supply. Studies of the matter are to be found in many standard works, and many empirical formulas have been evolved to suggest a general relation between rainfall and runoff for catchment areas of different sizes and different climatic conditions. Of the factors already mentioned affecting rainfall losses, two are fundamentally dependent on geological structure. It will therefore be seen that although general formulas are undoubtedly useful in limited application, they must be interpreted in the light of information as to the local geology if they are to be effective and of use.

Loss of rainfall by percolation to underground pervious strata is the first of these two factors. Consideration of this possibility leads inevitably to a general study of groundwater, to which Chap. XVII is devoted. Fuller reference will therefore be found in this chapter, but there are some general features governing the relation of underground supplies to runoff that may well be mentioned here. The subject is one of great complexity, so that the following notes should be considered in conjunction with the chapter mentioned and not in any way as complete in themselves.

In all but the most exceptional watersheds (such as those used at Gibralter), some portion of the rain will fall on to pervious material and so sink below ground surface. There it will eventually join that large body of water which is permanently held in the interstices existing in the underground strata, taking its part in the slow but steady movements of the contents of this underground reservoir. It is water derived in one way or another from this source that constitutes the flow in the corresponding watercourse during periods of dry weather when the flow obviously cannot be the direct consequence of rainfall. This *dry-weather flow,* as it is termed in engineering practice, is the criterion on which fundamental calculations regarding water supply must be based. The determination of the minimum possible dry-weather flow at any gaging station is a critical hydrological investigation, since it will often be impossible to measure this by actual gaging before a water-impounding project is started. Records at the sites of works that have been constructed can usually be obtained, and these constitute a valuable guide to similar cases. It is in determining the similarity of catchment areas in this connection that geological structure must be considered.

In an interesting paper by Mr. D. Halton Thomson,[14.9] mention is made of a study of more than 30 records of various dry-weather flows for British streams, and these show discharges varying from 0.055 cu. ft. per second per 1,000 acres on the River Alwen in North Wales (drainage area, 6,313 acres) to 2.00 cu. ft. per second gaged on the River Avon, Scotland, with a drainage area of 5,000 acres, with various intermediate values. The ratio of these two records from these catchment areas, almost identical topographically and in respect to climate, is 1:36. This is a most surprising variation, and the explanation is wholly geological. Superficially, both areas might be regarded as impervious, the geology of the Alwen Valley in the part considered consisting of Silurian shales and grits covered with boulder clay; that of the Avon is wholly granite. Based on this general comparison, it would seem that the dry-weather flows from the two areas should be almost equal. Further study discloses the fact that although the Alwen catchment area is watertight, as might be expected, thus providing little or no underground storage, the granite over which the Avon flows is so fissured and decomposed that vast quantities of water are held in its interstices. The ample underground storage thus provided is drawn upon by the stream during persistent dry periods. Another interesting example is afforded by flow records at two different points on the River Exe in Devonshire, also quoted by Mr. Thomson, the lower one just above the range of tidal influence, and the other near the headwaters of the river, above the junction of two important tributaries with the main stream.

Table O shows that the unit dry-weather flow for the whole area is almost four times that of the upper part of the area alone, the reason for the variation again being solely geological. About one eighth only of the upper part of the area is formed of pervious rocks, the New Red Sandstone, the remainder being Devonian measures and igneous rocks, so that practically no underground storage exists. The basins of the two tributaries joining the main river below the upper gaging station are largely formed of the pervious New Red Sandstone, and they are both well served with underground storage, in consequence, to the extent that their unit dry-weather flows are so high that they raise the figure for the whole catchment area to that indicated.

TABLE O[1]

	Upper Exe basin	Whole of Exe basin
Drainage area.....	240 sq. miles	461 sq. miles
Average annual rainfall...	47 in.	42 in.
Dry-weather flow (aver. of September, 1911).	22 cu. ft. per second per 1,000 acres	87 cu. ft. per second per 1,000 acres

[1] Reproduced, by permission, from *Liverpool Eng. Soc. Trans.*, **44**: 118 (1923).

The examples quoted are by no means unusual, and they therefore illustrate vividly the effect that geological conditions have on dry-weather flows. This effect is felt to varying degrees for all stages of stream flow. Even during torrential rainfall, a pervious surface stratum will rarely be so completely saturated with water that rain falling on it will all run off, giving a flow equivalent to 100 per cent of the rainfall. This is confirmed by the unique conditions that combined to cause the catastrophic floods early in 1936 in the eastern part of North America. Unexpectedly early rains fell on catchment areas that were still frozen hard and ice covered, thus constituting catchment areas almost "ideal" if considered from the standpoint of impermeability. The resulting runoffs caused floods that broke all known records and in some cases exceeded the calculated "1,000-year maximum." That such floods do not normally occur is due to the regulating effect of geological formations, which determine to some degree the course of water from the time it falls as rain until the time it reaches the watercourse to join with the stream flow. The importance of this effect can hardly be over-emphasized. The figures quoted illustrate how geological conditions alone can explain some variation in runoff records. By means of similar studies, not only can records be compared, but the accuracy of recorded flows can be checked as possible limiting cases.

CHAPTER XV

EROSION AND SILTING

> Of these waters, six are drawn into covered catch-basins . . . in which, resting as it were from their run and taking a new breath, they deposit their sediment.
>
> SEXTUS JULIUS FRONTINUS (A.D. 40–103).[15.1]

The phenomenon of erosion and the complementary action of sedimentation can probably be regarded as the geological processes of most importance in civil engineering work, even when considered only in so far as they are operating at present. If the past is considered and even cursory thought is given to the difficulties introduced into construction work by the existence of underground eroded valleys and by the variation of unconsolidated sediments and glacial deposits, the significance of these processes will be considerably increased. They constitute, as was indicated in Part I of this book, a part of the geological cycle. By the glaciation now taking place in arctic regions, just as by the cutting of solid rock by the wind-driven sand of tropical deserts, parts of the earth's crust are steadily being eroded. Other parts are similarly being built up, the combined effect being to disturb the static equilibrium of the earth's crust, with the result that the erosion and sedimentation of today are probably contributing to earth movements of the future.

The civil engineer is repeatedly faced with problems caused by erosion—on the seacoast, where the action of the sea may be destroying valuable developed property; at bends in rivers, the current steadily encroaching into bankside property or adjacent transportation routes; and in areas where fertile topsoil is being eroded by the combined action of wind and rain. Similarly, the deposition of silt behind many artificially constructed barriers to river flow, the silting up of slow-flowing rivers, and the formation of bars in tidal estuaries each presents special problems in the solution of which the engineer is called upon to play a leading part. It will be noted that even in this brief introductory note civil engineering works of such important types as river-training work, coastal protection, and soil conservation have already been suggested. All have a common feature, in that in them the work of the engineer is concerned with an attempt to conserve or develop physical features that are constantly exposed to natural forces tending to alter them in some way. These forces will persist despite all the efforts of the engineer. Hence, in all civil engineering work connected with erosion or silting, not only are there structural problems to be faced, but consideration must also be given to the effect that the resulting structures will have on the future action of these natural forces. In all these considerations, a knowledge of the fundamental geological processes at work is essential; their neglect may result in failure of works structurally quite sound.

FIG. 15.1.—Dredging coal from the Susquehanna River, U.S.A., the product of erosion of anthracite strata exposed further upstream.

(Copyright National Geographic Society, Washington; reproduced with special permission from The National Geographic Magazine.)

Erosion and silting have been features associated with building work from the earliest days of civilization as it is known today. The valley of the Nile provides annually a vivid example of the silt-bearing function of rivers, as a result of which the fertility of that cradle of civilization has been long renowned. Records taken at the famous Roda gage near Cairo (which has been maintained for many hundreds of years) show that the average silt deposit over the valley is about 10 cm. per century. The subject is one that has many interesting and some unexpected features. For example, on the Susquehanna River, Pennsylvania, "coal mining" is carried out by dredging the coal dust that is transported down the river and which is gradually deposited after having been eroded to the northeast of Hanbury from outcrops of anthracite coal in the hills forming the watershed of the river! Perhaps no more telling example of the complementary nature of erosion and silting could be presented.

Erosion was not perhaps of great consequence in early days, as land values were then relatively unimportant. Under the Romans, shore protection works appear to have been constructed and also embankments for the reclamation of low-lying land from the sea, notably in the vicinity of the Wash on the east coast of England. These works were allowed to fall into disrepair until reconstructed and extended under the direction of Dutch engineers (notably Vermuyden) whose experience was based on their previous winning of coastal lands in Holland. Thereafter, protection works against erosion on river banks and coasts have occupied an increasingly important place in civil engineering work.

So diversified are the engineering works associated with erosion and silting that even to attempt to present a brief historical summary of their development is to essay the impossible. Similarly, they are so dependent on the details of local geological structure that considerations in this chapter must be confined to discussion of but a few leading and general cases. These will be indicative of the geological approach necessary to civil engineering design and construction of this character and will at least serve to demonstrate again the close relationship of Geology with the work of the civil engineer.

15.2. Coastal Erosion.—There must be few engineers who are not familiar with a stretch of coast on ocean, lake, or inland sea that is being eroded at a rate noticeable in the course of a few years or possibly months. Although sand spits and bars give some evidence of corresponding accretion, the balance always appears to be against the land in this constant battle with the sea. Rock-bound coasts and cliffs are similarly affected, although their increased resistance as compared with that of unconsolidated beach deposits usually renders erosion on rocky shore lines of small consequence. This relative unimportance is emphasized by the fact that coast lines featured by continuous rock outcrops are not so favorably placed for development, either for pleasure or for commerce, as are coast lines formed of unconsolidated material. Although the process of erosion is similar for all types of shore, and although the erosion of rock cliffs often presents geological features of unusual interest, attention will be devoted more particularly to the erosion of low-lying coastal lands.

The erosive action of pounding waves and swirling water is always obvious, so that the cause of coastal erosion need not be discussed in detail. The quantitative study of wave and eddy action is still a matter for debate; it is today being investigated by means of models as well as by mathematical analysis. There exists already an extensive literature on wave and wave action, over 200 technical references to this subject being listed in a standard publication on geological processes involved in shore formation.[15.2] Generally, it may be said that the essentials of the action are now understood, the direct hydrodynamic forces exerted by waves having been successfully measured, and the erosive effect of eddies being at least generally appreciated. When the nature of a beach is such that waves carry with them sand or gravel in suspension, the resulting intensified action in eroding the standing beach will readily be appreciated.

Wave action is only one factor in erosion, an equally important one being the geological formation of the coast line exposed to wave

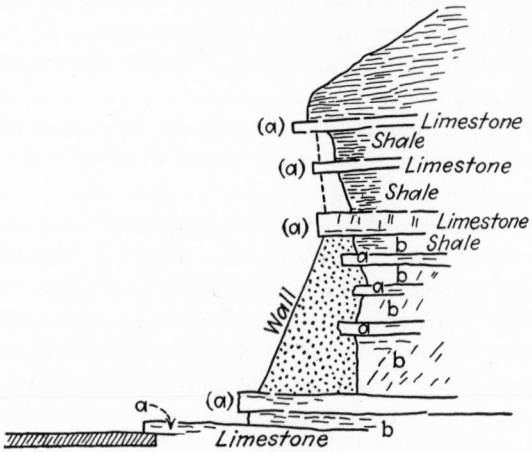

FIG. 15.2.—Geological section of sea cliff at Lyme Regis, England, showing protection wall constructed to halt erosion.

(Reproduced, by permission, from "Sixty Three Years of Engineering," by Sir Francis Fox, published by Mr. John Murray, London.)

action. On rockbound coasts characterized by regular strata of appreciable thickness, erosion will generally be uniform. When strata of varying types are exposed having differing resistance to abrasion and solution, differential erosion will result, often with somewhat fantastic effects such as the formation of "natural arches." It is only when erosion threatens a stretch of rocky coast that has been developed right up to the edge of cliffs that the civil engineer is called upon to exercise his skill and ingenuity. The cause of erosion will usually be obvious after even a preliminary survey but only provided that this has been made with due regard to the geological formation of the coast line.

A section similar to that shown in Fig. 15.2 might be found, for example, this representing a small cliff section close to the famous old church of Lyme Regis in Dorset, England. The sketch is based on one included in "Sixty Three Years of Engineering," a fascinating volume of reminiscences by Sir Francis Fox, who carried out the necessary remedial works. The alternating layers of shale, being much softer than the limestone beds, were easily eroded by the sea waves, leaving the alternate limestone beds overhanging. When the overhang became too great, a fracture of the limestone bed resulted, and renewed erosion of the shale took place. Other forces at work in completing the disintegration were the action of heavy seas in lifting the projecting layers of limestone and so loosening them from their beds, and slips in the upper layers of unconsolidated materials due to seepage of groundwater along the face of the receding cliff. The remedy successfully adopted was the construction of a continuous wall protecting the underlying layers of shale from further erosion and thus stabilizing the entire cliff face.

Considering shores consisting of unconsolidated material, they can rarely be regarded as stable, even prevailing winds sometimes affecting their stability. If any development work is to be carried out adjacent to the high water line on such beaches, consideration of general beach conditions is essential. Protection walls will be a frequent recourse, especially if the works are to project below the high water mark. Foundation bed conditions for these retaining structures must be given unusual attention. Scour at the foot of protection walls will always be a possibility to be seriously considered, making allowance for every change in beach regime that can be visualized. Cutoff walls of steel sheet piling below reinforced concrete superstructures provide a convenient and effective method of construction, ensuring protection against scouring. Provided that the wall structure does not encroach far below high water mark, the existence of this wall should not affect the stability of the beach. If a bulkhead wall has to be constructed either within the tidal range on the beach or below the low water line, then study must be made of the possible consequences of construction.

This leads to the study of beach formation—an extensive branch of Physical Geology and a perennial subject for debate among engineers responsible for coastal works. Of necessity, study must proceed empirically, local conditions being of unusual significance. Considering coast lines in general, two leading types may be recognized: the shore line of emergence and the shore line of submergence, the names being descriptive. The first is the result of a gradual uplift of the sea bed, the floor of the sea emerging to form a flat coastal plain consisting of unconsolidated deposits. Following the initial emergence, wave action will erode the edge of the emergent land mass, carrying this

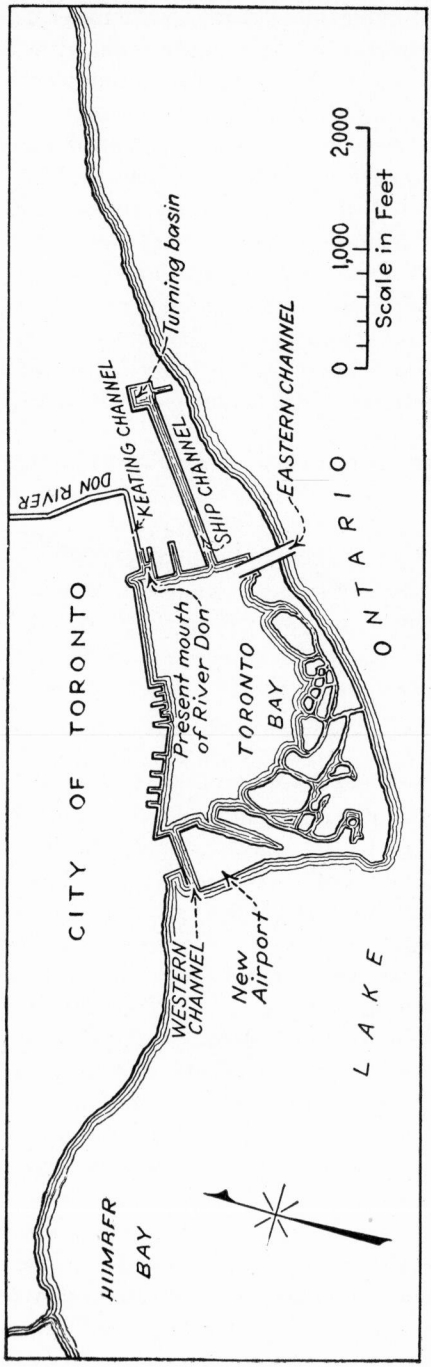

FIG. 15.3.—Sketch map of Toronto Harbor, Lake Ontario, Canada.

(Reproduced by permission of the Toronto Harbor Commissioners, Mr. E. L. Cousins, General Manager.)

back away from the water line to form an offshore barrier beach which will gradually work nearer in to the original shore line until ultimately the two become merged. Shore lines of submergence are the result of a gradual depression of a coast line, the original coastal area being submerged beneath the sea, which will thus stretch far inland and be in contact with an eroded land surface. The initial characteristic is extreme irregularity of outline, features such as fiords and deep bays being indicative of submerged valleys. Erosion by the sea will continue, its first effect being a tendency to straighten the shore line, a process in which strata that are harder than the surrounding beds will offer greater resistance and consequently give rise to special features such as stacks, arches, and small offshore islands. The products of erosion will be rapidly washed into deep water in initial stages of development,

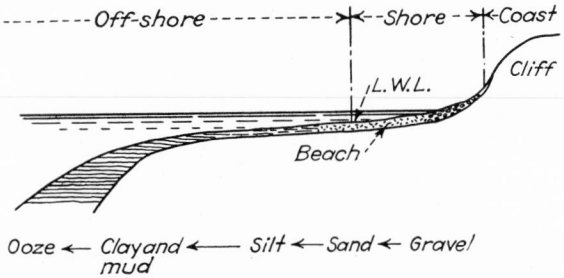

FIG. 15.4.—Diagrammatic section of developed shore.

but eventually a small beach will form at the foot of cliffs. If littoral currents exist, they will tend to sweep these beach deposits along the coast past headlands to be deposited in the deeper water of bays, forming what are, in effect, submarine embankments. With continued action, these formations will reach water level, when they become spits or hooks and in some cases are of so great an extent that they may almost completely block a whole bay. An excellent example of a *compound recurved spit* is to be found in the "island" that forms the great inland harbor of Toronto on Lake Ontario, Canada, the spit having been carefully developed and conserved as protection to the harbor. Eventually, a submergent shore line will lose its irregularities and acquire a beach formation. After this has reached equilibrium its development will be identical with that of a shoreline of emergence.

When a shore line has reached this state of stability, erosion of the coast or cliff will tend to proceed regularly. Movement of the eroded materials will be caused by currents and waves, the general tendency being for the finer material to be carried into deeper water, the coarser material forming the main beach deposit. The nature of beaches may vary considerably from place to place; but as a general rule, gravels will be found to be deposited within one mile from shore in depths of not more than 30 ft. of water, sand extending rarely beyond a depth of 300 ft. Figure 15.4 illustrates diagrammatically an ideal beach formation which has become reasonably stable. Studies have shown that the size of sand grains or gravel partially affects the slope of a beach—the larger the sand grain the steeper the slope. There may be some interrelation between these characteristics and the movement of sand along a coast line, according to the results of observations made on a part of the New Jersey coast. This latter phenomenon is known as *littoral drift* and is so important as to warrant individual attention.[15.2]

15.3. Littoral Drift.—The works of the civil engineer were not required to demonstrate the existence of that extensive movement of sand in regular and steady manner along many stretches of coast known as littoral drift. Its existence is obvious due to the regular accretion of products of erosion in not a few coastal locations. At river mouths, conditions are especially favorable for its observation, one peculiar result of the interaction of drift and estuarine conditions being the *barachois* (a French word meaning sand and gravel bars) which are a prominent feature of certain parts of the Atlantic coast of Canada. The records of Physical Geology contain many examples of great interest based on observations of coastal conditions and development. Similarly, studies undertaken in conjunction with engineering works present other valuable data, as an example of which may be mentioned the study, made in the course of investigations of the United States Beach Erosion Board, of Rockaway Point, Long Island, a sand spit near New York. The point of this spit has advanced at an average rate (at least between 1835 and 1927) of more than 200 ft. per year. When the progression of the point was halted by a rock-mound jetty, 300,000 cu. yd. of sand was found to have accumulated behind it in the course of little more than a year.[15.3]

With the advent of man-made structures projecting out into the sea, littoral drift has come to be of great practical importance, usually as a difficult problem to be faced but occasionally as a factor beneficial to coastal development. On any section of coast featured by appreciable drift (and few sand beaches are not), the projection of any solid structure below mean tide level will result in the building up of sand or other drift material on the one side of the structure and a corresponding although not necessarily equal erosion of material from the other side. If the drift is of any magnitude, the serious effects of the action will readily be apparent, especially in the case of works protecting harbor entrances. The possibility of trapping drifting sand by deliberately stopping littoral drift in order to build up a good beach section will be equally clear.

The coasts and major harbors of India have provided some of the most valuable data yet available on this important matter, much

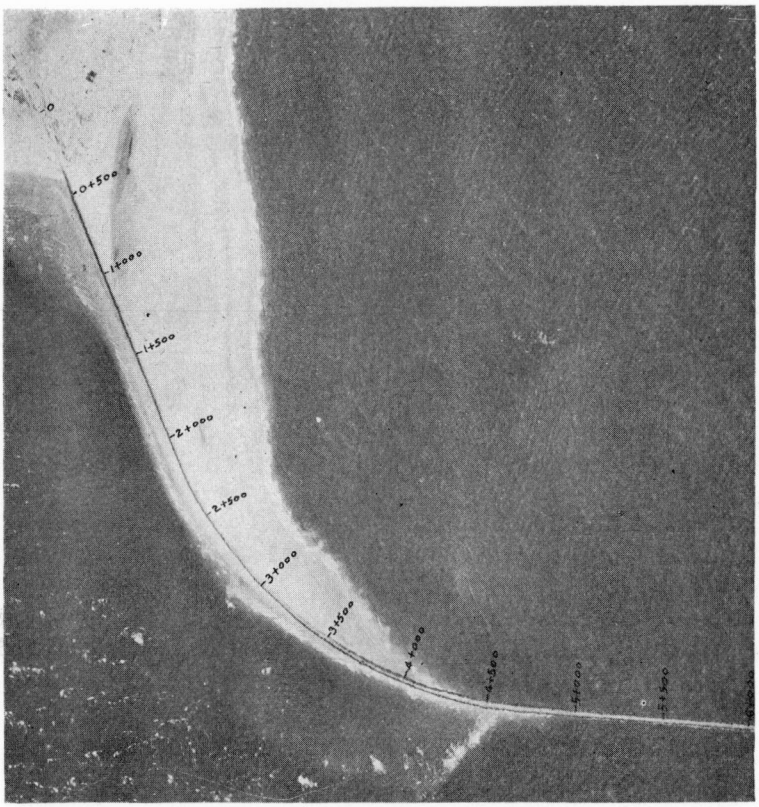

FIG. 15.5.—Accumulation of sand at the east jetty, Rockaway Point, Long Island, at the entrance to Jamaica Bay.

(Photograph by U. S. Army Air Corps. Reproduced by permission of the Chief of Engineers, U. S. Army, through the courtesy of the Beach Erosion Board.)

having been obtained as the result of bitter experience which has been faithfully recorded for the benefit of the profession generally. Thus it has been found that at Madras on the east coast of India, approximately one million tons of sand passes any given spot on the coast in a year, the direction of travel being from north to south. Similar conditions exist on the east coast of South Africa; and at Durban, in particular, great difficulties have been experienced in maintaining the harbor entrance. All this accumulated experience was applied to the design of the new harbor of Vizagapatam between Madras and Calcutta on the east coast of India. Initial studies suggested that no trouble would be encountered with drift, but the start of dredging operations at the harbor entrance in 1930–1931 soon demonstrated that conditions similar to those at Madras existed. Eventually, it was found that (1) the volume of sand traveling along the coast amounted to one million tons per year; (2) the travel took place largely

FIG. 15.6.—Protective breakwater (formed of two scuttled steamships) at Vizagapatam Harbor, India, the purpose of which is explained in the text.
(*Reproduced by permission of the Deputy Administrative Officer, Vizagapatam Port.*)

in shallow water, being generally confined to a zone 600 ft. wide; and (3) the travel was the result primarily of wave action although sometimes accentuated by shore currents. At Madras and Durban, erosion corresponding to the drift deflected by harbor works had occurred, and hundreds of acres of agricultural land had been lost in this way. At Durban, 700,000 cu. yd. of sand was required to restore an eroded foreshore to its original state. This aspect of littoral drift had also to be considered in connection with the Vizagapatam Harbor entrance. Eventually, a scale model was constructed by means of which the action of waves of varying degree on the coastal section was noted. This gave valuable indications of what protective works could be effectively carried out, the final solution being as shown in Fig. 15.7.

The breakwater (consisting, incidentally, of two scuttled ships loaded with rubble) ensures that the littoral drift is deflected between it and the coast line; it has therefore to fall into the dredged sand trap on the inner side of the entrance channel from which it is excavated by a dredge working under the lee of the protective breakwater. This discharges the sand on to the other side of the entrance channel, whence it is gradually moved away by the erosive drift action.[15.4]

This major example is cited since it introduces the main problems caused by littoral drift and demonstrates one method of circumventing these. Similar difficulties may be experienced all down the scale of marine structures; and although in the case of smaller works the same

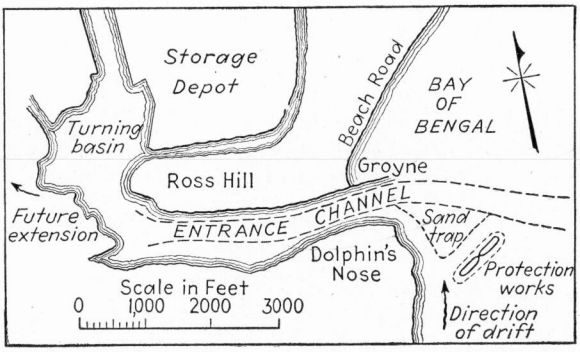

FIG. 15.7.—Sketch map of the entrance to Vizagapatam Harbour, India, showing arrangement made for dealing with littoral drift.

(Based, by permission of the Council of the Institution of Civil Engineers, London, on information given in reference 15.4.)

study cannot always be given to drift problems, they must always be carefully considered. Experiments alone can be relied upon to determine whether drift is taking place and if so what is the direction of sand travel. Experience with similar conditions will always be a guide in evolving remedial measures. In the case of channel entrances that must be kept open, the construction of guard jetties into water deep enough to ensure that the sand either crosses the channel below dredged depth or else does not build up so far out as the end of the jetty is often the only solution. In the case of small wharf structures projecting out from an open coast, the adoption of a *blockwork and bridge* type of structure or of an open pilework substructure will often enable the difficulty to be obviated.

Finally, there are many stretches of sand beach the progressive erosion of which will interfere with either recreational facilities or developed property and that must therefore be conserved as much as possible. When erosion is the result of continued littoral drift, the use of groynes is often the most satisfactory solution. Groynes are wall structures built out generally at right angles to the shore line and extending below low water line, their action being to interfere to such an extent with the travel of drift sand that some sand is trapped against the walls, gradually accumulating until a stable sand beach is formed. The shores of many parts of the British Isles have been protected in this way, as also has much of the low-lying eastern seaboard of the United States, erosion of the coasts of Florida and New

FIG. 15.8.—Groynes of interlocking steel sheet piling on the Atlantic Coast at Miami Beach, Florida, U.S.A., showing sand accumulation between them.
(*Reproduced by permission of Steel Union—Sheet Piling, Inc., New York.*)

Jersey in particular having been severe. In the latter case, records show an average recession of the coast line of about 2 ft. per year (153 ft. between 1840 and 1920); and local figures for Florida are similar. Problems arising on these coasts have been among those considered and investigated by the United States Beach Erosion Board; as a result, certain specific recommendations regarding groyne location were drawn up. It was found, for example, that groynes need not extend higher than the level of uprush at high water; construction at right angles to the coast line was found to be beneficial. A spacing tentatively suggested was 1½ to 4 times the length from the bulkhead or shore line, the longer spacing being successful when the volume of drift was large, and vice versa. The findings of the

Board emphasize the need for local study in coming to any final decisions.[15.3]

Practically all older installations of groynes were solid in construction, piled structures being common: interlocking steel sheet piling has provided a convenient modern method of economic construction, and many satisfactory designs are now available for the consideration of engineers. An interesting development of recent years has involved a departure from this standard type in the adoption of a permeable groyne; and as some notable examples have already been constructed, a

FIG. 15.9.—Installation of permeable reinforced concrete groynes at Sheridan Park, Wisconsin, on the shore of Lake Michigan.

(Reproduced by permission of the Regional Planning Dept., Milwaukee County, Wisconsin, U.S.A. Mc. E. A. Howard, Supervising Engineer.)

leading installation may be cited as an illustration. Much of the west shore of Lake Michigan, in the United States, is characterized by a steep bluff 50 to 100 ft. in height, formed of unconsolidated deposits. These range from an extremely fine sand to a heavy clay in the neighborhood of Sheridan Park, Cudahy, Wisconsin, where erosion was taking place at a rate of 2 ft. annually until protective measures were adopted in 1933. Surface erosion and seepage assisted wave action on the lake in furthering this erosion. The terrain at the top of the 100-ft. cliff was therefore reshaped so that the surface water drained *away from* rather than toward the cliff face, being led to catch basins and thence to a storm sewer. The face of the bluff was planted with vegetation in order to retard the surface velocity of water falling onto it. Subsurface drainage was installed in addition, designed to interrupt groundwater at levels such that it was kept away from the fine sand strata encountered. These were stable when dry but treacherous when wet. Means had finally to be found for reducing the effect of wave action at the foot of the cliff, since this was eroding the material that was periodically brought down by slides from the cliff face. It was felt that if this material could only be left in place, it would soon become stabilized, especially if vegetation could be induced to grow on it. An offshore breakwater was one obvious solution, but funds did not permit of this structure being built. Permeable groynes, an idea developed by Mr. Sidney M. Wood of Lake Bluff, Illinois, were finally decided upon; and in 1933, eleven were constructed of precast reinforced concrete units, 200 ft. apart and extending 200 ft. out from the shore line. Their arrangement is such that they become more permeable the farther they project from the shore, the shore ends being almost solid, whereas the lake ends permit a free flow of water through the structure. The groynes promote the accumulation of littoral drift, which is quite appreciable even in the Great Lakes, while serving to protect the coast by breaking up severe wave action. In the short time since their construction, these Sheridan Park groynes have proved successful, and erosion of this particular stretch of coast has been halted.[15.5]

15.4. Erosion by Stream Flow.—The flow of water in rivers and streams is another of the major causes of erosion. As the material eroded must be transported away by the stream flow, complementary problems of silt deposition arise whenever the natural flow of a silt-bearing river is interfered with in any way. Many and diverse engineering problems are thus created, for the correct solution of which an appreciation of the general geological characteristics of river flow is indispensable as a necessary background. The life history of rivers, although essentially a study in Physical Geology, is therefore a matter that should be familiar at least in outline to all engineers engaged on river works. Rainfall is the main source of all water reaching the surface of the earth. That portion which passes over the surface to become the runoff must naturally take a definite course and will at first gravitate into small gullies. These will lead, in turn, to larger gullies which will give place eventually to valleys down which a definite stream will flow on its way to lake or sea. Visualizing a complete profile of a river from source to mouth, it can be imagined as gradually decreasing in slope as it nears its mouth, increasing in volume and cross section on its way. This theoretical profile is not universal, although many of the larger rivers of the world, such as the Amazon and the Ganges, conform to it. Rivers rising in mountainous country near the sea may be steeply graded from source to mouth, whereas other rivers which have their sources in relatively low lying ground will have a flat profile throughout and corresponding characteristics.

The curved profile is one that a river will constantly tend to produce, and therefore it is a fundamental concept underlying all river studies, changes in gradient being achieved through the erosive power of running water and of the debris that it transports. Thus the steepest part of the theoretical profile may be regarded as the youthful stage of a river, the central section its middle-age period, and the flat gradient at the lower end its old age, the meandering of slow-flowing streams across surrounding plains being another characteristic of this last stage. All references to time apply to the geological time scale; but despite this, many examples are to be seen today of rivers in all stages of development, and many that show clearly their immediate past history. The canyons of the Colorado River in the United States, the Cheddar Gorge through which the River Avon flows in the south of England, and similar scenic river features present vivid evidence of progressive deepening due to erosion.

A photograph shows the River Bulkley in western Canada and illustrates well the differential erosion that may take place in strata of rock of varying characteristics and hardness. The main interest of the civil engineer in river history is in connection with the erosive processes that have been and still are at work in its development. Weathering of the exposed rock surfaces can be regarded as the starting point in the train of erosion that eventually finishes in the depths of the sea or of an inland lake. Gravity and the wind will have the effect of bringing the products of weathering down from the hillsides. After rainstorms, the flows in even the smallest gullies will start the work of transporting this debris downward with the stream. The erosive action of streams, although starting in this way, can be generally classified as three separate processes: (1) transportation, (2) corrasion, and (3) solution. The start of the transportation of material by a stream has just been described. This is a process easy to visualize and in some cases to watch, and it is one the general nature of which can be readily appreciated. It will be discussed in the next section of this chapter. Solution is also a type of action easy to understand.

The term *corrasion* is used to denote the combined erosive action, mechanical in nature, of the flow of water and of the material that it is transporting along the bed of a stream. The action has been likened to the passing of a continually moving file over the stream bottom. A direct analogy can be seen in the practice of cutting solid rock with wire saws, quartz grains being usually introduced beneath the moving wire rope, to be carried along by it as it passes over the rock. The wire rope is similar in action to the river flow, and the quartz grains to the material being transported by the stream. Corrasion will clearly depend on the velocity of the stream flow, on the amount and nature of the load that the stream is transporting, on the plan and profile of the stream course, and on the geological nature

FIG. 15.10.—Hawilget Canyon, Bulkley River, British Columbia, illustrating differential erosion.

(Reproduced by permission of the Geological Survey of Canada.)

of the stream bed. The direct influence of stream velocity on erosive action is a feature that needs no elaboration. The load of a stream is also of great importance, since the erosive action of water itself flowing in a straight and regular course over solid rock is small indeed. When it is carrying along sand and gravel, that portion of the load which travels on the bed of the stream will be constantly in irregular movement, bumping against the bottom and intensifying eddies. In this way, the resistance of the bed material to movement will be broken down, and erosion will gradually take place. A striking although rather localized example of this action is provided by the potholes which are often found in the beds of swift-flowing streams, the presence of pebbles or even boulders at the bottom of such holes being a telling reminder of their mode of formation. During the construction of the

Assuan Dam in Egypt, potholes were found in the unwatered river bed area, 4 to 5 ft. in diameter and up to 20 ft. deep, eroded out of the hardest granite.[15.6]

The plan and profile of a stream naturally affect its flow and so inevitably are related to its corrasive action. Bed profiles have been discussed in a general way; the increased erosion liable to take place in the more steeply graded section of a stream bed will be easily appreciated, especially since the erosive power of a stream depends not directly on the velocity but on some power of the velocity higher than the square. Another feature of a stream profile may be the

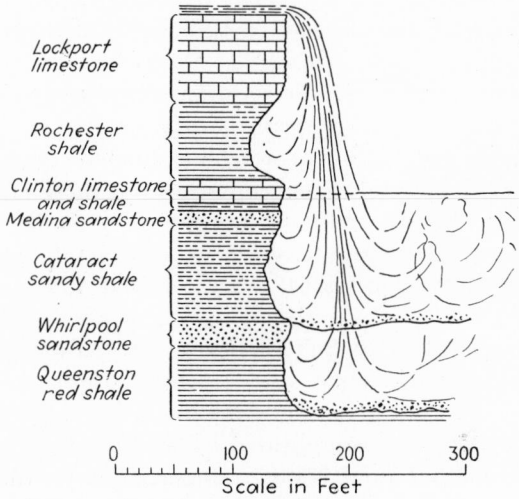

FIG. 15.11.—Geological section through Niagara Falls, illustrating alternation of strata which assists erosion.
(*Reproduced by permission of the Geological Survey of Canada.*)

existence of waterfalls. These will often be the result of geological conditions influencing bed erosion. No better example can be quoted than that provided by Niagara Falls, the Niagara River here dropping over the Niagara escarpment in two waterfalls, 167 ft. high, world famous for their majestic beauty and of peculiar interest to engineers in view of the water power developments associated with them. A typical cross section through the falls is shown in Fig. 15.11 from which it will be seen that above the falls the river flows over an outcropping bed of Niagara limestone, a relatively hard rock below which occur beds of shale with a thin bed of Clinton limestone interposed. The shale is affected by the action of water more quickly than is the limestone, with the result that even under the lip of the falls it is gradually eroded, this action leading eventually to an unstable projection of the limestone bed which then breaks off. Such rockfalls are periodically reported, and the recession of the lip is proceeding at a rate averaging 5 ft. per year. Originally, it is suggested by geological studies, the falls were near Lewiston, several miles below the present site. The erosion of this great gorge, which is known to have taken place in our present era, is therefore a vivid reminder of the potentiality of rivers as a cause of erosion.

The notes already presented have been confined to rivers that throughout their course were eroding material continually from their banks and to bed characteristics of swift-flowing streams or rivers in the young or mature stages. When rivers reach the state described as that of *old age,* they will have a bed profile that approaches the minimum possible if the discharge of the river is to reach the sea through the cross section of the natural waterway. This hydraulic criterion is related to the critical velocity for silt and scour, there being one particular condition at which a regulated stream will neither erode the exposed material in its bed nor deposit the silt that it carries in suspension. This silt and scour criterion is a condition of delicate hydraulic balance. It leads naturally to consideration of the transporting power of flowing water, a matter that will now be briefly reviewed.

15.5. Transportation of Solid Material by Stream Flow.—Many problems of river engineering, and especially those connected with river-development work involving some alteration of natural features, are liable to be seriously complicated by the continued presence in a river bed of solid material being transported down the river as a product of erosion. The matter has been the object of much engineering research work, records of detailed investigations by engineers going back at least as early as 1851 and recurring in increasing numbers in engineering publications since that time. Concurrently, the approach to a relatively simple geological process has changed from that of generalized observation to a detailed and sometimes complicated analytical mathematical investigation. The quantitative study of transportation by running water is now rightly regarded as an important branch of the science of hydraulics, with which this book is not directly concerned, but the basis of all analytical study is the fundamental geological process of erosion.

As an extreme example of how serious the matter may become may be mentioned an observation of the San Juan River, in Colorado, appearing when in flood as a moving stream of red mud and proving, on test, to be 75 per cent by volume composed of silt and red sand.[15.7] Admittedly, this is an extreme case, but the fact that careful estimates suggest that as much as 250,000,000 tons of silt annually pass the lower end of the canyon section of the Colorado River and that the Mississippi River carries almost twice this amount of silt into its delta each year will show that silt transportation is a process of magnitude. In many parts of the world, the regular deposition of silt so transported is a beneficial feature to fertile areas; the Nile Valley of Egypt is perhaps the best known example. In other parts of the world, deposition of silt transported by river flow during flood periods can be injurious to a like degree, and New Zealand, South Africa, and the Malay Peninsula are countries seriously affected in this way. As an example of what may result from this cause may be mentioned the town of Kuala Kubo in Malay, which had to be completely abandoned in 1929 and reestablished on a new site some miles away, owing to the original site being buried many feet deep in river-deposited silt.[15.10]

The source of the material transported by river flow is usually confined to the catchment area of the river in question. The quantity of material transported and the regularity of its movement are therefore dependent on the geological characteristics of the catchment basin. In preliminary studies of a river prior to the prosecution of development or remedial work, a rough survey of the geology of the drainage area is therefore an essential requirement. Coupled with the complementary topographical survey, this will indicate the stage of development that the river has reached above the point considered, the nature of initial causes of erosion (such as disintegration of steep rock slopes, wind action, surface soil erosion), and probably the main source of material transported by the river. This information is useful mainly as a guide to further studies, since the effect of transportation by water is to reduce practically all material, whatever its initial state, to particles fine enough to be classified as either fine sand or silt. For example, the vast quantity of silt transported by the Colorado River is found to be actually rock flour, mixed with some organic matter, ground so fine by water action that in grain size it is truly described as silt.[15.8]

A more detailed study and one of immediate value to the engineer is the securing of samples of river water together with the silt they carry and the quantitative analysis of these in order to obtain an estimate of the average quantity of material being transported. For this purpose, one of the several types of silt sampler now available may be used. Most of these consist essentially of a closed container to which river water can be admitted when desired, at some determinate point in the cross section of the river. Results given by sampling present only a partial picture, since transportation is not a simple matter of suspension but a threefold process. Described in the simplest way, the means of transportation may be said to be a rolling along the river bed of the heaviest particles, an intermittent bumping up and rolling of the next grade of particles, and finally the transportation in suspension of the finest particles of the river's load. The total load is therefore divisible into two parts, the bed load being that transported by the first two types of movement, and the suspended load, that moved by the third. The suspended load is capable of accurate measurement, by means of samples, but no really satisfactory method of measuring bed load has yet been devised for general application. As a result of experiments that have been carried out, however, it has been determined that (in the case of the Colorado River, at least), the bed load is about 20 per cent of the total load transported.[15.9]

The study of the several types of movement is what has naturally attracted the attention of investigators, and the correlation of the results of mathematical analysis with experimental data has provided engineering literature with many notable papers. The engineer is interested primarily in the results of theoretical investigations and especially as they may be applied to his specific problems. With the transport of material in normal river sections he will not usually be concerned; but whenever a river section has to be artificially enlarged or reduced, with consequent changes in velocity, he should be able to calculate in advance how these changes will affect the transportation of material, especially in relation to the scouring out of bed material if this is unconsolidated. An associated problem is that faced by engineers on irrigation work in the United States of America who have to design their canal systems so that the bulk of the transported load in the irrigation water shall be deposited before the water reaches the final feeder systems. The deposition of some river silt on to irrigated farm land has been found to have injurious effects on fertility. Elaborate precautions have therefore to be taken in order to remove silt before river water is admitted to a canal system. One installation of this type is at the entrance to the All American Canal which takes off from the Colorado River about 300 miles below Boulder Dam and serves the Imperial Irrigation District. Although Boulder Dam naturally holds up practically all the material transported by the Colorado River as far down as the dam site, the scouring action of the river is such that in the 12 miles immediately downstream of the dam, the river bed scoured out to an average depth of about 3 ft. in the first five months following the closure of the gates at the head of the diversion tunnels at the Boulder Dam site. This was not an unexpected result, preliminary design work for the canal intake works having been based on an assumed total load of 120 million pounds of silt per day in the maximum anticipated diversion of 15,000 cu. ft. of water per second. Estimates suggested that the cost of removing this amount of silt from the irrigation system if deposited in the canals would amount to $1,000,000 (£200,000) per year; a desilting installation at the canal intake proved therefore to be an economic possibility. The installation eventually provided consists essentially

FIG. 15.12.—Aerial view of the Imperial Dam on the Colorado River, and the intake of the All American Canal on the Californian bank—including the extensive desilting works (seen before admission of river water).

(*Reproduced by permission of The Commissioner, U. S. Bureau of Reclamation, Washington, D.C.*)

of six settling basins 269 by 769 ft. with a rated capacity of 2,000 cu. ft. per second, the maximum velocity in the basins being 0.24 ft. per second, and the deposited silt being removed by power-operated rotary scrapers. The efficiency of desilting at normal flow is calculated at 80 per cent, based on the researches of the engineering staff of the United States Bureau of Reclamation, which is responsible for the work.[15.11]

In India, the main problem faced by irrigation engineers is sometimes the reverse of that indicated by the American example just cited, as the silt transported by Indian rivers is not always injurious to agriculture. It is present in such quantities that the general practice has developed of designing Indian irrigation works so that the silt and sand are deposited in the channels on individual farms where they can easily be cleared out by the peasant farmers. Thus a main canal will be so "designed as to its section and slope that it will be able to carry forward its full quota of sand silt without either deposition or erosion."[15.12] These words are quoted from a publication of the well-known irrigation engineer, Mr. R. G. Kennedy, who, for the conditions of flow described, developed, as a result of prolonged investigation, a formula that now bears his name and is widely used. Mr. Kennedy's work was but one step in a large series of studies carried out in many parts of India and all devoted to the problem of so-called *stable flow*. In this work, hydraulic studies come into close contact with the geological background, grain size of the silt being an all-important factor. An appreciation of the geological origin of the silt considered will go far toward assisting in the correct application of the results of such hydraulic research.

15.6. Silting Up of Reservoirs.—In all but exceptional cases, the construction of a dam or weir across a stream bed will raise the water level above the line of the dam, thus increasing the cross-sectional area of the stream and so decreasing the average velocity. This will automatically eliminate a part of the suspended load of the stream, which will fall to the bottom of the stream or reservoir. Although all of the bed load may possibly move down toward the dam, it will eventually be caught in the basin formed by the dam structure. There exist many records of large dams rendered either ineffective or even useless in this way, by the unexpected and rapid filling of the reservoir behind them with transported material. To mention but a few of these, reference may be made to the so-called New Lake Austin, on the Colorado River in Texas, which lost 95.6 per cent of its capacity in 13 years; La Grange Reservoir on the Tuolumne River in California, which lost 83 per cent in 36 years; and the Habra Reservoir on the Habra River in Algeria, which lost 58 per cent in 22 years. These figures are quoted from an extensive compilation of valuable records contained in a paper by Mr. J. C. Stevens.[15.13] Reference is also made in this paper to the unusual siltation record for the Lake Macmillan Reservoir of the Pecos Irrigation District, United States, which lost 55.5 per cent of its capacity between 1894 and 1920. Since the latter year, no further accumulation of silt has taken place, because an extensive growth of tamarisk (salt cedar) in the flatlands immediately above the reservoir has had the effect of so reducing the stream velocity that the main part of the silt load is deposited before the reservoir is reached.

An essential part of preliminary work in connection with the construction of all dam structures is consequently an investigation of the amount of silt normally transported by the stream to be dammed. It may often be found that the quantity of material carried is small enough to be negligible; but in other cases, further and more detailed studies will be necessary. These will inevitably include a careful correlation of data as to the quantity of debris transported with its geological origin, so that the nature of the resulting deposit may be anticipated. This information is necessary in view of the engineering problems involved which may be summarized into two groups: (1) possible methods of removal of the transported material either before it is deposited in the reservoir or after it has come to rest behind the dam; and (2) the effect that deposits of material, if they cannot be avoided, will have on the stability and on the utility of the structure being planned.

Detailed discussion of these problems is mainly an engineering matter, but their essential dependence on geological conditions will be apparent. It is not always possible to provide underwater sluices through a dam by means of which silt deposits can be discharged. Designs must therefore often include other methods of dealing with the silt, possibly by allowing it to deposit behind the dam and then removing it by a floating excavating plant. An alternative method is to arrange small check dams in the upper reaches of a reservoir which will reduce the entrance velocity and so keep silt deposits away from the main dam. The use of check dams is valuable in connection with several phases of erosive action. In districts subject to heavy rains which are liable to wash down large volumes of debris, the construction of this type of dam can often reduce consequent damage by so slowing up the speed of debris-laden streams that much of the load is deposited above the dams instead of in locations such as bridge openings where its presence will cause trouble. An interesting example of this kind of work was cited in Chap. XI (page 260). The effect of silt on structures will be serious only in connection with mechanical moving parts, since it has been stated that under normal circumstances the deposition of silt against the face of a dam does not increase the normal pressure against the dam.[15.14]

Brief reference must finally be made to the radical changes that the construction of a dam and the formation of a reservoir have on the watershed above the site. Natural conditions are seriously interfered with; and although no immediate changes may be observed, an adjustment of the natural balance must inevitably take place eventually. The observation of these effects is a long-term study, and the time that has elapsed since the initiation of the modern period of major dam building is scarcely yet far enough advanced for any data to be available. The matter is one of some interest, and it may in the future become important. It is linked with general conservation problems, and the increased attention now being devoted to the husbanding of natural resources will fortunately allow its significance to be more widely appreciated and therefore the better observed in the future.[15.15]

FIG. 15.13.—Erosion of well-stratified limestone below spillway apron of the Back River power plant, Riviere des Prairies, Montreal. An illustration of severe erosion of rock below a relatively low-head structure (18 to 23 ft.); dentated sills installed as part of reconstruction.

(Photograph by D. Hutchison. Reproduced by permission of the Power Corporation of Canada, Ltd., Mr. L. C. Jacobs, Construction Manager.)

15.7. Erosion below Dams.—Another problem often associated with the construction of dams is the damage that may be done as a result of erosion of rock or other material below the spillway. The concentration of water flow and the high head under which it may take place render the hydraulic design of spillway channels always a delicate matter, and the proportioning of the artificial structure must be most carefully done. As in similar matters, unfortunate experiences have led engineers to undertake comprehensive research work in connection with studies, and there now exists in engineering literature a series of most valuable papers dealing with the hydraulic solution of the problems thus raised, many of them based on the use of model structures.

A typical example is a paper by Messrs. A. D. D. Butcher and J. D. Atkinson[15.16] describing extensive research work on models, undertaken as a result of certain problems of erosion that had arisen in the Egyptian Sudan at the Sennar Dam controlling the Blue Nile, located about 200 miles south of Khartoum. The dam is founded on a ridge of solid rock, but the river bed immediately below the dam consists of boulders and rock of doubtful quality. No masonry aprons were originally provided; and after the first year's operation, it was found that considerable erosion had taken place in this part of the river bed. A glass-sided observation tank was employed with models of dam cross sections in a similar way to that used by Dr. Rehbock in his original experiments on this problem, carried out in 1905 and 1909 at the Technical University of Karlsruhe. The necessity for the use of some type of sill at the foot of the outfall apron of a dam was demonstrated in these Egyptian experiments, as it had been previously during Dr. Rehbock's experiments when the patented dentated sill that now bears his name was evolved. This detail of spillway design is mentioned because of its importance when geological studies show that the rock immediately below the overflow apron is liable to be affected by the intensity of direct flow, the use of some type of sill often reducing the eroding effect of the discharge flow. The experiments described in the paper mentioned also led to the conclusion that symmetry of flow and therefore of design is essential if scour is to be minimized, a feature that can usefully be applied to any cuttings in natural rock that have to be made immediately below an overflow structure.

In the discussion of this paper, some interesting examples were cited of erosion in New Red Sandstone, and in clay and shale beds, adjacent to sluices at some of the locks of the Manchester Ship Canal. These openings did not have to pass large volumes of water and so illustrate how this problem of erosion may be of extreme importance even in the case of relatively small structures. At least one major structure has failed from this cause, this being the Islam Weir of the Sutlej Valley project. The weir is one of four similar structures on the Sutlej River in the Punjab, India, which constitute diversion dams for a large irrigation project. Consisting of 24 openings each of 60-ft. span, it is of the floating type of dam resting on a sand foundation and protected by upstream and downstream aprons. Although the head on the dam was only 18 ft., in two years downstream scour had amounted to 6 ft. 6 in.; and in 1929, six bays of the structure collapsed. The official report of this accident is a most valuable publication.[15.17]

FIG. 15.14.—The Islam Weir, Sutlej River, Sutlej Valley irrigation project, Punjab, India, as seen from upstream following the failure of six bays on September 19, 1929.

(Reproduced by special permission of the P.W.D., Punjab, through the courtesy of Mr. M. T. Gibling, Secretary, Central Board of Irrigation, Simla, India.)

FIG. 15.15.—The Calderwood Dam, Little Tennessee River, Tennessee, showing the special overflow pool which "cushions" the fall of spillway discharge and thus prevents erosion. Before the pool was completed, a flood which occurred during construction eroded a hole 30 by 50 ft. and 50 ft. deep (see *Eng. News-Record*, Vol. 104, p. 1064, June 16, 1930).

(Reproduced by permission of the Aluminum Company of America, Mr. J. P. Growdon, Chief Hydraulic Engineer.)

Numerous examples of the erosion of rock below dam spillways could be cited, most of which have developed only after the dam structure has been in use for some time. In some cases, although admittedly not in all, trouble of this kind can be foreseen if a detailed geological survey of the area immediately downstream of a spillway is made before construction begins. Careful study must be made not only of the effect of water and of concentrated flow on the rock but also of its structure. In addition to the more usual erosion processes which will operate with increased intensity below a spillway discharge, there always exists the possibility of concentrated flow loosening blocks of rock at its more definite joints and bedding planes, especially if these contain any disintegrated material or product such as clay. This loosening would not normally be serious, but the velocities that occur in a spillway discharge may be so high that blocks of rock of great size may thus be transported downstream bodily. They provide a dangerous "tool" for the stream flow to use in further erosion and leave exposed a further area of bedrock on which the concentrated flow will impinge. Such a condition existed on the Susquehanna River in Maryland, in the case of the fissured granite below the Conowingo Dam, the joints of which were filled with clay. Remedial measures include aprons and protective works, as already mentioned, and also the provision of deep stilling pools, as in the case of the Calderwood Dam.

15.8. River-training Problems.—The reference to river-control work which can here be made, despite the important place that the subject occupies in the field of civil engineering, must be brief. Decisions regarding design must necessarily be empirical; and as no two rivers are alike in all particulars, the local application of the results of experience elsewhere is often fraught with uncertainty and is liable to misinterpretation. River-control work is essentially the control of a natural river flow within a certain well-defined course—usually that which the stream normally occupies. Deviations from this course will be most liable to occur during periods of flood flow, so that control work is often a matter of flood regulation. In addition to this, the prevention of erosion of river banks during periods of normal flow constitutes another important branch of control work. Finally, the improvement of rivers in connection with civic or industrial development, such as the straightening of a river bed in connection with town planning and the canalization of a stretch of river to facilitate navigation, is another type of river regulation that is often of importance.

In undertaking work of this nature, an engineer is dealing with a fundamental geological process, and all his designs must be based on an acceptance of the natural characteristics of river flow. In the past, a rigid restriction of this work to assumed natural conditions has led to major controversy, especially with regard to the control of the Mississippi River, one of the outstanding river-control projects of the world. The control of some of the tributary rivers to this great waterway has also presented great problems, the popular name for the Missouri River, *Big Muddy*, being indicative of its silt-carrying nature and a clue to some of the difficulties encountered in its regulation. Attention may be concentrated on the Mississippi, since, despite its magnitude, it typifies problems encountered on many of the greater rivers of the world. The river is fed by a drainage area of 1,243,000 sq. miles and has a total length of about 2,400 miles. Its average discharge at Vicksburg is about 500,000 cu. ft. per second, flood discharge ranging as high as 2,000,000 cu. ft. per second. In general profile, it does not conform to the theoretical shape already described, being generally flat and slow moving for the greater part of its length. The valley of the Mississippi follows the general form for such river valleys. The flood waters, before confinement by levees was effected, covered a much greater area than the normal stream bed, at times as much as 30,000 sq. miles.

The first attempts were made in the early nineteenth century to confine the river during flood periods by raising the natural levees with artificial embankments, thus raising the next flood water level and promoting deepening of the channel through this restriction. In consequence, no local addition to a levee could be made without all levees being raised similarly. The first levee, built at New Orleans in 1717, was 4 ft. high. In 1932, the average height was more than 13 ft., and 2,500 miles of artificial levees were being maintained on the river. All this work, up to 1932, was based on the idea of maintaining the existing channel of the river. In June, 1932, after years of discussion, a new policy was inaugurated under the direction of General H. B. Ferguson, newly appointed President of the Mississippi River Commission. Another characteristic of the river was adopted for use, the natural process whereby the river makes cutoffs at narrow necks between bends, at long intervals of time, being expedited by making artificial cutoffs with the aid of powerful dredges. These cutoffs have materially reduced the length of the river and have lowered its flood stages. Supplementary dredging is done in the reaches between cutoffs to assist the river in restoring the steepened gradients to normal. A famous example of a natural cutoff is at Vicksburg, Mississippi, where in 1863 during the Civil War, General Grant tried to isolate the city from the river by cutting a canal through a narrow neck of land. This attempt was not successful; but in 1876, the river made the change for itself. Fortunately, the town of Vicksburg was able to maintain contact with the main river through the diversion of another stream; it is now well known to engineers as the location of the large-scale model of the river on which most valuable experiments have been made. This shortening of the river is being carried out, initially, at 12 sharp bends. When all these cutoffs have become

FIG. 15.16.—The site of the Sarah Island cutoff. Mississippi River, on October 25, 1935, when the right-of-way had been cleared and the first dredge cut started. Main river flows from right to left at top, and from left to right at bottom of photograph.

(Photograph by U. S. Army Air Corps. Reproduced by permission of the Mississippi River Commission.)

fully developed, the total shortening effected by them will amount to nearly 116 miles in a stretch of 330 miles. Associated with this work, relief floodways are being provided, designed to allow flood waters to spill over the natural banks but only under control and in specially selected locations. Reservoirs are being created on tributary streams by the construction of some important dams; and although this work is of relatively minor importance on a river as large as the Mississippi, regulation by this means on smaller streams is often extremely valuable. Finally, continued bank protection work is being carried out, involving mattress designs of much engineering ingenuity.[15.18]

The importance of bank-protection work must be emphasized, since it enters into so many phases of river-control work. Methods of protection range from the construction of permanent retaining walls in front of natural soil slopes, for preventing the initial formation of soil gullies, to the use of expedients such as mattresses of brushwood, dumped rock, and even sandbags in cases of emergency. In the evolution of the many successful types of protection, due consider-

FIG. 15.17.—The Sarah Island cutoff, Mississippi River, on September 20, 1938, after it had developed considerably, although it was not yet able to take all the lower water flow. The Worthington cutoff can be seen in the background, and the silting-up of the old river bed.

(Photograph by U. S. Army Air Corps. Reproduced by permission of the Mississippi River Commission.)

ation has inevitably had to be given to the nature of the river action at the location. For example, on many sharp bends, the use of a wall of deep-driven interlocking steel sheet piling has often proved successful where other expedients have failed, by enabling additional scour of the river bed to take place in front of the wall until a condition of equilibrium was attained. In the case of levee protection, many ingenious designs of flexible mattresses built of different materials, including reinforced concrete, take up initial variations of the river bed and also any alteration caused by further minor erosion at the edge of the mattress. Flexibility is also the keynote of the "wire sausage" type of protection adopted so successfully in India, cages of wire being filled with loose stones and so placed that the failure of one sausage will automatically result in the superimposed cages moving down to take its place. Another idea has been the planting of low-growing willow trees at or about normal water line, so that the foliage may impede the flow at flood periods and therefore reduce its eroding power.

The most spectacular type of bank protection work is that which occurs when breaches occur in raised river banks or levels. Then, indeed, quick action is required, the laws governing erosion giving ample evidence of their validity as soon as a breach is made. Ordi-

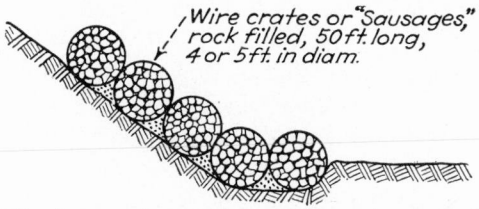

FIG. 15.18.—Wire "sausage" protection work as used on some Indian rivers. (*After Nolan; see reference* 10.19.)

narily, there are no special geological features to be mentioned in connection with breaches. Occasionally, a major catastrophe of great geological importance may occur in this way, as was the case in China in 1852 when the Hwang Ho River, which for at least 700 years had discharged eastward into the Yellow Sea, broke its banks at a point more than 300 miles above its mouth, formed a new channel across the province of Shantung and finally a new mouth in the Gulf of Chihli, almost 300 miles north of its old mouth, a position that it still occupies today. Probably no better example of the vagaries of river flow could be cited.[15.19]

15.9. Silting of Deltas and Maintenance of Estuaries.—The formation of deltas at the mouths of several of the major rivers of the world, for example the Nile, the Ganges, and the Mississippi, is one of the most obvious and generally appreciated demonstrations of the silt-transporting power of rivers. The somewhat similar phenomenon of the existence of bars at the entrances to tidal estuaries into which may discharge quite small rivers, such as the Mersey, is a telling reminder of the great transporting power of the sea. Both natural features introduce engineering problems of the greatest magnitude which, although simple in essence, are extremely complicated in detail, having provided a fertile field for discussion for at least 3,000 years. The building up of large areas of new land in deltas provides land for development work of various kinds, and this involves structural foundation problems often of unusual complexity. On the other hand, in the case of all major rivers, the maintenance of a stable navigable channel through the delta or estuary is a matter of prime importance and often of inordinate difficulty. There need be mentioned only Calcutta on the Hoogli branch of the river Ganges in India, New Orleans on the Mississippi River, and Liverpool on the River Mersey in England, with its famous approach channel through the sand banks of its estuary, to indicate the importance of the problem.

The rate of growth of the Nile Delta has already been mentioned; corresponding extension of other deltas has also been observed and studied. It is estimated, for example, that the delta of the Mississippi River is extending into the Gulf of Mexico at the rate of 250 ft. each year, although the river Rhône in France (under observation since 400 B.C.) has been growing at the rate of only 36 ft. yearly, the corresponding rate for both the Danube and the Nile being 10 ft. per year.[15.20] It may be noted that the nature of the discharge of the river is a factor of great importance in determining the existence of deltaic conditions, for rivers such as the Severn in England and the Amazon in South America, although carrying large quantities of silt, exhibit no special deltaic features, the funnel-shaped estuaries promoting self-maintenance of the main channels.

The geological conditions affecting the formation of deltas are well recognized: the load of the stream is deposited on the river bed as the velocity, and therefore the carrying capacity is reduced when the stream flow merges with the water of the sea or lake. Part of the load which might still be held in suspension may be deposited by flocculation of the colloidal material by the salt solution provided by the sea water. As a result, deposits are formed in regular beds, and they constitute some of the newest of sedimentary deposits. These may reach to surprising depths; for example, a boring put down in Venice on the delta of the River Po in Italy reached a depth of 500 ft. without penetrating the bottom of the deltaic beds.[15.21] It will be appreciated that, although a delta is popularly imagined to be above water level, a deltaic formation is essentially an underwater geological structure, a fact influencing its engineering significance. It may also be emphasized that although a delta may be the formation of a single stream, breakages of the natural levees above the river mouth may possibly result in the main stream branching out into numerous distributaries through which the river finally discharges into the sea.

The Nile Delta and that of the Ganges are two excellent examples of this special feature.

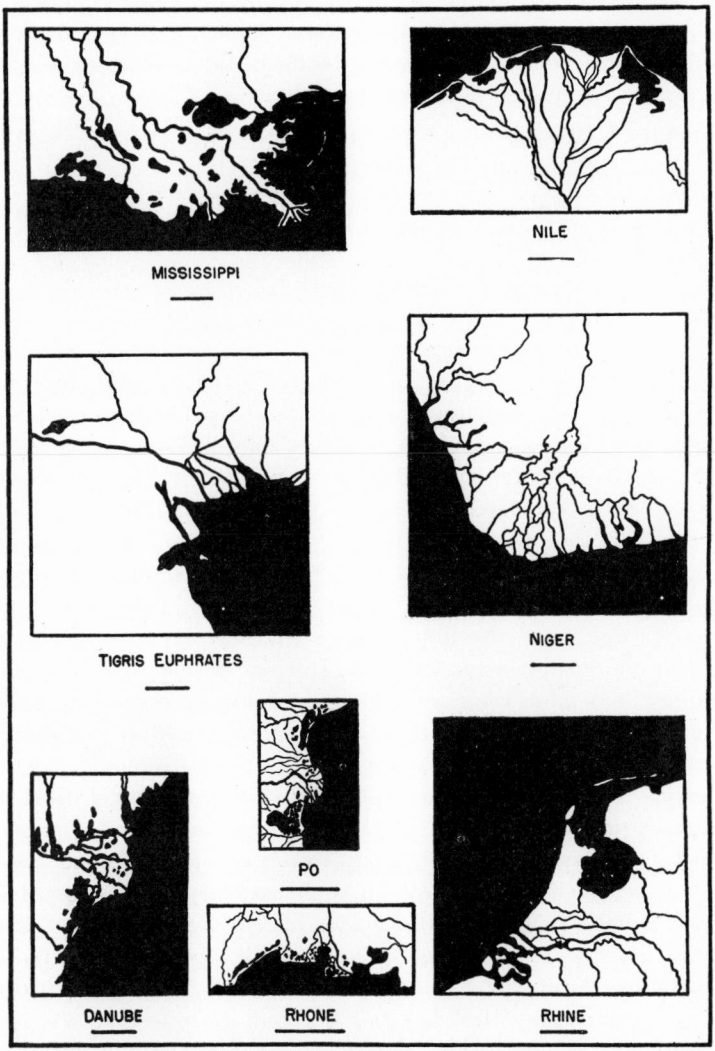

FIG. 15.19.—Some of the main river deltas of the world. (The bar under each name represents 50 kilometers, about 31 miles.)

(Reproduced, by permission, from Geological Bulletin No. 8, Louisiana Geological Survey, New Orleans, Louisiana, U.S.A., Mr. Cyril K. Moresi, State Geologist.)

The formation and persistence of bars at the entrance to many tidal river estuaries has long been a matter of keen debate and investigation. Bars may be said to have been the object of some of the earliest hydraulic research of which record exists; the Abbot Castelli included them in the subjects that he investigated in the years around 1660. Other early Italian and French engineers devoted considerable attention to them, and many explanations of their existence have been advanced in the intervening years. Bars in tidal estuaries and river mouths may be divided into two general classes:

1. Those consisting of hard material not affected by the scouring action of normal currents passing over them.
2. Those of sand, shingle, or other unconsolidated material which, while retaining their general features, are constantly subject to alteration from effects caused by winds, waves, and varying currents.

In addition, there are occasionally formed temporary bars of sand or shingle heaped up by the action of waves in heavy gales and afterwards displaced by currents. The famous Chesil Bank in the south of England on the east side of Lyme Bay may be cited as an extreme example of this variable type of obstruction. It is 15 miles long and varies in width from 500 to 600 ft. and in height from 20 to 40 ft. depending on the tides and waves. Sir John Coode has recorded that in one night a quantity of 63,000 tons of shingle was moved along the bank.[15.22]

As an example of the first type of bar may be cited that which existed at the entrance to Aberdeen Harbor, Scotland. It consisted of boulder clay and when removed by dredging was formed again by the piling up of sea sand from the neighboring shore. Continued dredging finally gave a permanent channel by the removal of all the sand exposed in the vicinity as it filled into the bar channel, thus leaving exposed only boulder clay which was not eroded to any appreciable degree by tidal currents.[15.23] Many examples of the second type of bar could be given, since details of such bars have been presented in connection with the many discussions of the general subject before technical societies. These discussions would appear to have reduced to a certainty the cause of the formation of such bars from out of the wealth of controversy that has raged about the question. It seems almost certain that sand and shingle bars across the mouths of tidal estuaries are formed by the interaction of tidal and other shore currents, local topographical detail having a considerable indirect effect. The influence of the discharge of river water into these estuaries appears to be small. An interesting fact is that although the exact location of a river discharge channel in the sandbanks of its estuary may change appreciably, a sand bar will always exist at the entrance to any new channel if such was the case in the original channel.

It may be fitting to describe what is at once one of the most remarkable and one of the best known of these natural impediments to navigation, that at the entrance to the River Mersey in England, on which stands Liverpool, one of the world's greatest ports. The river is relatively small—only 56 miles long and draining only 1,724 sq. miles. It has a high tidal range (up to 32 ft.), and the lower part of the river is therefore swept by large volumes of water on every tide. The actual

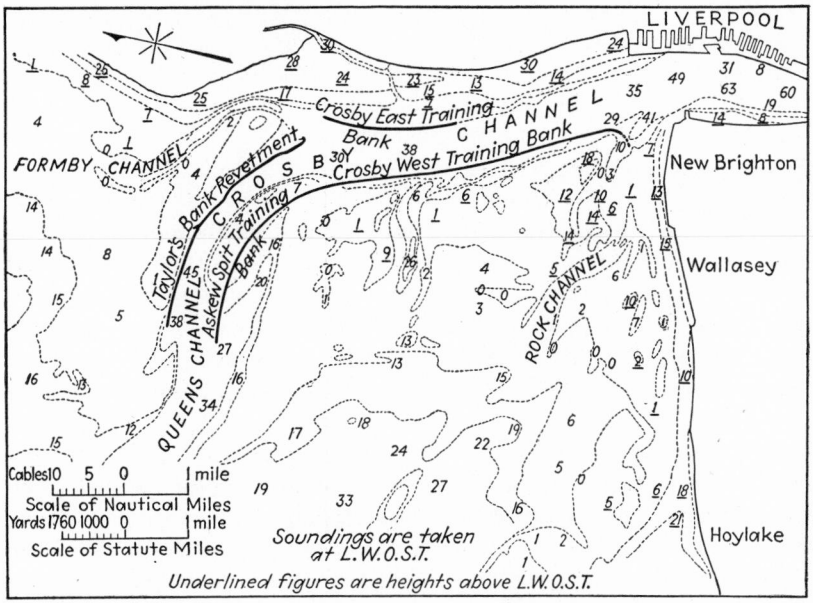

FIG. 15.20.—Liverpool Bay and the estuary of the River Mersey, England, showing ship channel and training walls.

(Reproduced by permission of the Mersey Docks and Harbour Board, Mr. T. L. Norfolk, Engineer-in-Chief.)

mouth is located in what may be called the "corner" of Liverpool Bay, in which there are certain well-defined sandbanks, some of which are high and dry at low water. Through these, the river maintains a regular channel with a depth of between 30 and 50 ft. below low water level. The lower section of the channel has undergone considerable change of position in the last hundred years, limiting positions being 3 miles apart. In every case, the channel finished in a large sand bar, that which existed before dredging started rising directly out of 53 and 46 ft. of water at low tide in the landward and seaward directions, respectively, giving a depth at the bar itself of between 7 and 17 ft. Flood conditions in the upper reaches of the river appear to have the effect of increasing the available depth of water. The interference which this obstruction meant to navigation can well be appreciated; but it was only in 1890, after exhaustive studies, that dredging across the bar was started experimentally. The success obtained in deepening the bar brought into prominence other shoals in the main approach channel, and it was foreseen that if the depth of this channel throughout its length was to be commensurate with the deepening that had been effected on the bar, considerable further dredging would be necessary. For some years, therefore, extensive and systematic dredging was practiced; and since the desired depth in the main channel became achieved, the natural tendency to deposit on the former site of the bar has been easily countered by occasional dredging. Inside the bar, the channel has been stabilized by the construction of large rock revetments or training walls.[15.24]

Although dredging is often an essential part of improvement work, it is dealing only with an effect, whereas the use of training walls is an attempt to modify a cause of silting or shoal formation, the general function of training walls being to train a river in a permanent course in which its own velocity will have the necessary scouring action. The use of training works is a more hazardous operation than the initiation of dredging, since it involves large expenditures and is liable to make matters worse instead of better if not successful. As an illustration, the development of the entrance to New York Harbor may be mentioned. Four long shoals obstructed the main ship channel and Gedney Channel in providing access into the lower bay from the ocean. One proposed form of improvement was by dredging, maintaining this channel, should it be necessary, either by periodical dredging or by contracting the entrance by the construction of a dike running across the shoals from the Coney Island side. The estimated cost for obtaining the dredged channel was $1,490,000, with a total cost of the improvement, should the contraction works be necessary, estimated at between $5,000,000 and $6,000,000. Between 1884 and 1890, after much discussion and some trials of dredging, the proposal for the dike was discarded and the channel dredged to 30 ft., saving approximately $4,000,000 (£800,000). Since that time a new channel 2,000 ft. wide and 45 ft. deep has been dredged from the main entrance to New York Harbor. Practically no maintenance is required on this channel due to the normal scour action of the lower section of the Hudson River.[15.25]

A sidelight on the second point mentioned, and an incident of general interest in connection with tidal rivers, is provided by some observations made in the river Garonne, France, at the end of the last century. A steamer was sunk by collision at the mouth of the river opposite Verdon and rested with her keel at the bottom of the channel, masts and funnel alone showing at low water. On examination, it was found that the vessel was completely buried in sand, the sandbank extending 300 ft. fore and aft and 150 ft. on either side. Subsequently, a second examination showed that the sandbank had completely disappeared, the hull being unsupported at the ends, resting on sand only in the central section. Further study showed that the sandbank formed and disappeared with every tide; the first examination had been made at the end of the ebb tide, and the second at the end of the flood, the ebb depositing the sand, and the flood removing it. No more striking evidence of sand travel with tidal currents could be imagined; it serves also to illustrate what troubles can be invoked by a training wall not correctly located.[15.26]

Many rivers and river entrance channels have been successfully developed and deepened by the judicious use of training works. Liverpool is a case in point, but of more importance (since regular dredging is not involved) is the mouth of the Mississippi River, improved in 1875–1879 by General James B. Eads at his own expense until the improvement had definitely satisfied governmental requirements. The method adopted by this distinguished American engineer was to concentrate the scour of the current in one of the "passes" of the river mouth by two parallel jetties of fascine work, thus making the water the agent for effecting the improvement of its own course. A full account of this important example of river-training work is available in a volume that has long been of value and interest to engineers.[15.27]

There may also be mentioned the case of the Fraser River in British Columbia which, although not perhaps so important commercially as some of the rivers mentioned, is yet one of the great rivers of the Pacific Coast of North America. It has a length of 790 miles and drains an area of 91,700 sq. miles, the average yearly flood discharge being 380,000 cu. ft. per second. The river passes through drift deposits and is therefore fairly heavily charged with sediment. This has resulted in the formation of an unusual delta the outbuilding of which occurs because of the dominance of the river currents over the tidal currents and because the wave action is not great. The tides and tidal currents have had the effect of giving an unusual form to the seaward part of the delta and of introducing unusual and complex problems into the successful training of the estuary of the river. Borings have disclosed the depth of the deltaic formation as at least 800 ft., and evidence has also been found of the

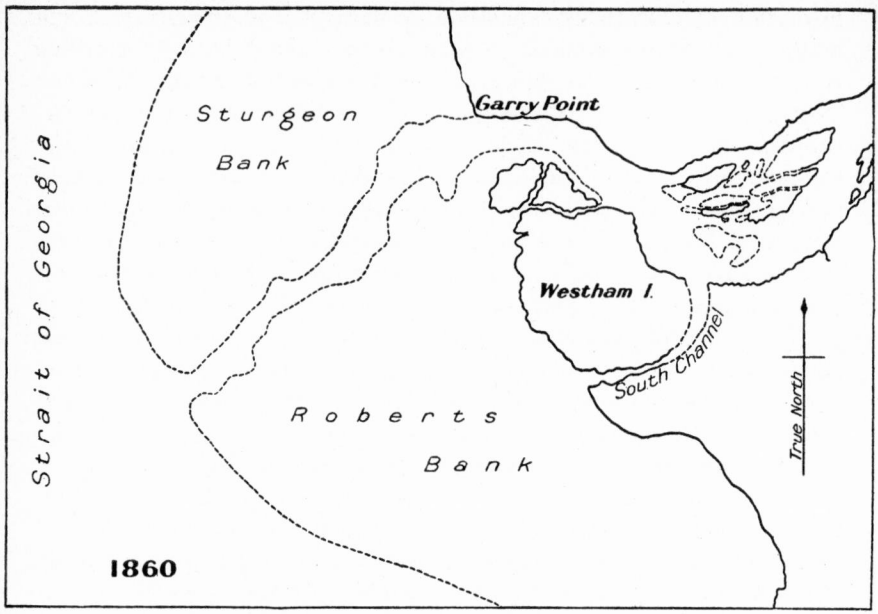

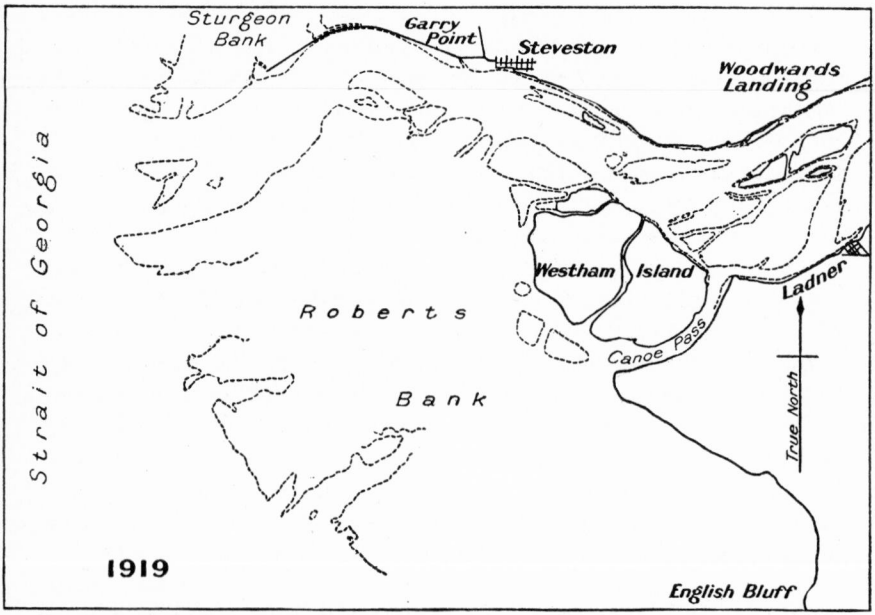

FIG. 15.21.—Sketch maps showing the development of the Fraser River Delta, British Columbia, and the changes in river channels from 1860 to 1919.

(Reproduced by permission of the Geological Survey of Canada from reference 15.28.)

formation of calcareous sandstone in relatively recent years in the delta. The main navigation channel is maintained by extensive training walls and by dredging; in dredging a bar at the mouth of one of the arms of the delta, a deposit of this sandstone had to be cut through by the dredge. The training works are in charge of the engineers of the Department of Public Works, Canada; and in connection with their work they have cooperated with the Geological Survey of Canada, a most interesting report (from which the foregoing data have been obtained) being one result of this cooperation.[15.28]

Local conditions affecting tidal estuaries are so variable that detailed study of each case is usually required. In recent years, this has been done in many cases by means of a specially constructed scale model, one of the most interesting and important branches of civil engineering to which the use of models has been applied. By the use of the principles of hydraulic similarity, it is possible to construct a model of an estuary to such a distorted linear scale that observation in a vertical direction can accurately be made without the use of a model too large in plan. By extension of the similarity to the scale of time, tidal effects occurring over long periods of years can be reproduced in a matter of days. The nature of the material composing the bed of the estuary or river is that most difficult to fit into the revised scale system, since it is clearly impossible to reduce the grain sizes of sand, for example, in the same ratio of reduction as is used for the linear dimensions. Even if this reduction could be approached, the resulting fineness of the particles would inevitably mean that they would not obey the same laws as the larger grains in the actual estuary and would certainly not reproduce accurately the erosion and deposition that actually takes place.

From the time of the earliest experiments, the problem has been approached experimentally. Mr. L. F. Vernon-Harcourt in his model of the Seine used, in turn, silver sand, powdered sulphur, coal dust, pumice, and bath brick, finally relying on a very fine sand having a small admixture of peat from the Bagshot beds of the south of England. Professor Osborn Reynolds, after much experimenting, finally used fine Calais sand in his classical experiments on the Mersey estuary; and although the grain size of sand used in his various experiments varied, he did not remark on any discrepancy in his results. Some of the most recent experiments in this direction are those of Professor A. H. Gibson in connection with his work on an important model of the Severn estuary with reference to the proposed tidal barrage scheme. He tested 13 different bed materials consisting of powdered pumice, silica sands, and emeries of varying grain sizes and found that the best agreement with actual results obtained from surveys was obtained not with the finest material tested but with a sand having a mean diameter of 0.007 in., which is about 25 per cent less than the average mean diameter of the sand in the estuary itself.

This work on the Severn estuary model could be compared with variations in the bed of the estuary over a period of 78 years, known

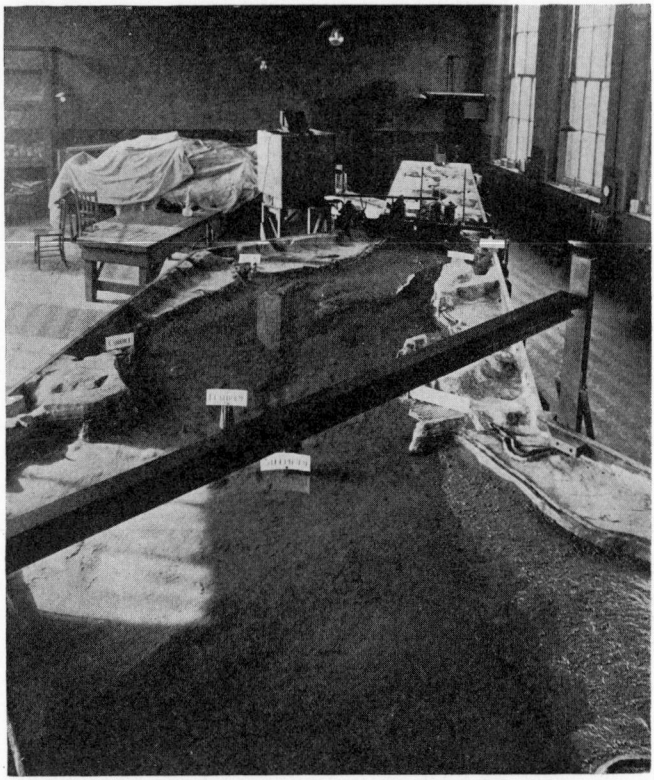

FIG. 15.22.—Scale model of the estuary of the River Severn, England (looking upstream) at the University of Manchester.
(Reproduced by permission of Professor A. H. Gibson, University of Manchester.)

from the results of accurate and regular hydrographic surveys. The agreement was surprisingly close and confirmed the great value of this branch of model work, which appears to have been first used in France in 1875 when a model was constructed to study the action of certain proposals for improving the river Garonne between Bordeaux and the sea. The first real attempt to apply scientific principles in the design of models was made in 1885 by Professor Osborne Reynolds at the University of Manchester, and his work is there being continued today by Professor A. H. Gibson.[15.29]

15.10. Soil Erosion and Wind Action.—Consideration must finally be given to the runoff of rainfall from the ground on which it falls to the nearest defined watercourse. The possible erosive action of surface flow over rock is negligible when considering a period of time of relatively few years; but in the case of unconsolidated materials, erosion may be far from negligible. When vegetation covers the ground on to which rain falls, the binding effect of plant root systems, in addition to the protection afforded by foliage, will generally result in the rain reaching the actual ground strata gradually. Alternatively, after heavy rains the rain may flow over the surface vegetation at a speed and in a manner such that little erosion of the ground takes place. It is only when vegetation is lacking that erosion can become serious.

Intensive cultivation and the wholesale cutting down of trees are factors that in places have contributed to the creation of serious surface soil erosion which naturally occurs also in those barren areas of land which are subject to intensive rainfall. Erosion of barren lands may not give rise to local problems, since the areas will probably be undeveloped, but erosion of cultivated land and potential forest sites will usually be of some consequence. The occurrence of droughts followed by winds and sudden rains is naturally conducive to erosion; and in the years culminating in 1936, extensive damage of this type was done in North America. Estimates suggest that in a period of 50 years, 50,000,000 acres of good crop land has been ruined outright for further cultivation in the United States alone. Another 50,000,000 acres is in a condition almost as serious. On another 120,000,000 acres, the topsoil is mostly gone. Erosion is actively under way on another 100,000,000 acres. In Canada, similar damage has been done in the prairie provinces; and the Magdalen Islands located in the Gulf of St. Lawrence, once a fertile group, have had all their topsoil removed, leaving only sand, not a stick of standing timber now remaining on the islands. The problem is not confined to North America. It is estimated that in South Australia alone, over 1,000 sq. miles that were once pastoral country are now desert. Sandstorms are an inevitable result of erosion by wind action; and when extensive, they serve only to increase the area of wastage. The damage done by the great dust storm of May 9–11, 1934, when a cloud of dust measuring 1,000 miles from north to south and 1,500 miles from east to west extended over the central part of the United States, was incalculable.[15.30]

These few illustrative figures will demonstrate how serious this problem has already become. A complete study of the causes of soil erosion inevitably leads to complicated discussion, because so many factors are involved. The combined effect of wind action and the runoff of rain water on cleared land is generally admitted to be largely responsible for the damage that is being done, excessive cultivation having been perhaps the original troublemaker. The problem is now being tackled on a large scale. In the United States, a federal Soil Conservation Service has been initiated, and through cooperation

FIG. 15.23.—An early stage in soil erosion—"fingered" erosion of idle bean land in Ventura County, California.
(*Reproduced by permission of the U. S. Soil Conservation Service.*)

with local forces it has already achieved real progress. The attack must be a cooperative effort if it is to be effective, and therefore engineers will be found working with agriculturalists, geologists, and soil experts, the services of the meteorologist being also utilized. Conservation methods adopted vary greatly, but a fundamental feature is that designs shall aid natural processes. Thus it is that the labor of the engineer on such work is closely allied with geological considerations and that the construction of small check dams across gullies in order to check flood flows and reduce flood velocities, and therefore erosion, is a leading type and one demanding this cooperative endeavor.

The action of the wind does not often affect the work of the civil engineer apart from the part that it plays in soil erosion. Trouble is generally present when wind does have any appreciable effect, one of the main evidences of its work being drifting sand dunes. Although these are generally confined to desert regions, they occasionally present troublesome local problems either on seacoasts with sandy beaches or in relatively arid regions as in Utah and Arizona. The travel of coastal sand has wrought great damage in Europe, especially on the

FIG. 15.24.—Barrier and spillway structure of check dam at Farmington, Utah, showing the accumulation of detritus, including boulders, above the barrier following a period of spring high water.

(Photograph by L. M. Winsor. Reproduced through the courtesy of Mr. L. M. Winsor, District Engineer, Salt Lake City, Utah, Bureau of Agricultural Engineering, U. S. Dept. of Agriculture.)

shores of the Bay of Biscay. At Liège in France, for example, sand dunes have been moving for two centuries at the rate of 81 ft. per year, necessitating two successive reconstructions of the local church. The geologist can but observe this phenomenon. The botanist, however, is here of assistance, and much has been done in the development of plants, grasses, and shrubs that will take root in sand and provide a matting of root material which will effectively bind the surface together. Among the flora used may be mentioned:

Sueda fruticosa, used to prevent movement of coarse sand and shingle.

Ammophila arenaria, planted by the quadrant method given in Gerhart's "Deutcher Dunenbas" for dune reclamation in Palestine.

Pinus halepensis; Acacia cyanophylla, trees found to be successful in Palestine, the latter growing as dense thicket and giving good charcoal.

In a few places, these wind-formed dunes have an economic value where they serve as a storage ground for rain water; in western Texas, for example, rangers get their water supplies from wells put down at the summits of large dunes.

Suggestions for Further Reading

Coastal Erosion and Marine Works

Johnson, D. W.: "Shore Processes and Shoreline Development," John Wiley & Sons, Inc., New York, 1919; contains an extensive bibliography.

Matthews, E. R.: "Coast Erosion and Protection," Charles Griffin & Company, Ltd., London, 3d ed., 1935.

These two books provide a background for further study, which can be prosecuted with the aid of the extensive technical literature now available on practically all branches of marine erosion and protective works. Fortunately, a carefully prepared bibliography is available:

"Sand Movement, Beaches, and Kindred Subjects," a bibliography (over 1,000 items), The Engineer School, Fort Humphreys, Virginia, 1930.

Action of Stream Flow

Twenhofel, W. H.: "Treatise on Sedimentation," Williams & Wilkins Company, Baltimore, Maryland, 2d ed., 1932.

Silting of Reservoirs

Eakin, H. M.: "Silting of Reservoirs," *U. S. Dept. Agr. Tech. Bull.* 524, 1936; contains a bibliography.

Edgecombe, A. R. B.: "Report on the Silting of Reservoirs and Measures Taken or Projected for Its Prevention or Control," Superintendent, Printing and Stationery, United Provinces, Allahabad, India, 1934.

Stevens, J. C.: The Silt Problem, *A. S. Civil Eng. Trans.*, **101**: 207 (1936); contains a bibliography.

River Training Problems

In addition to the references listed in Appendix D, there are available many publications—mainly in the records of Indian engineering—dealing with river-training works. A notable example is

Spring, F.: "River Training and Control on the Guide Bank System," *Govt. India Tech. Paper* 155, 1903.

As this book goes to press, the following paper, containing as an appendix Mr. J. R. Bell's original paper on his "Bund" system, has appeared:

Gales, Sir R. G.: The Principles of River Training for Railway Bridges, and Their Application to the Case of the Hardinge Bridge over the Lower Ganges at Sara, *Jour. Inst. Civil Eng.* (*London*), **10** (1938).

For a complete survey of the training of the Mississippi River, see:

ELLIOTT, D. O.: "The Improvement of the Lower Mississippi River for Flood Control and Navigation" (prepared under the direction of Brigadier General T. H. Jackson), U. S. Waterways Experiment Station, Vicksburg, Mississippi, 3 vols., 1932.

SOIL EROSION

AYRES, Q. C.: "Soil Erosion and Its Control," McGraw-Hill Book Company, Inc., New York, 1936.

GENERAL

See also WILLIAMS, G. R., and others: "Selected Bibliography on Erosion and Silt Movement," *U. S. Geol. Survey Water Supply Paper* 797, 1937.

CHAPTER XVI

WATER SUPPLY

> But if anyone will note the abundance of water skilfully brought into the city, for public uses, for baths, for public basins, for houses, runnels, suburban gardens, and villas; if he will note the high aqueducts required for maintaining the proper elevation; the mountains which had to be pierced for the same reason and the valleys it was necessary to fill up; he will conclude that the whole terrestrial orb offers nothing more marvellous.
>
> PLINY THE ELDER.[16.1]

The appearance of a short chapter dealing with water supply may seem strange at first sight. The subject is presented in this way in order that its importance may be duly emphasized and that the relation of geological considerations to water supply may be clearly defined. All too often in presentations dealing with Geology and engineering work, the "Geology of Water Supply" is featured in a manner not in keeping with the importance of the engineering side of water supply work, to which geological study and data can be no more than a serviceable aid. That geological conditions have a profound effect on sources of water and on the means by which it is supplied for public use cannot be denied. It is to be observed, however, that geological conditions are a part of the natural order and cannot well be altered, whereas the task of the civil engineer in designing a water supply system is the utilization for public good of the particular natural conditions with which he has to deal, often a task of great difficulty.

Much can be written even in summary form about the importance of water supply; this can hardly be over-emphasized. It is a prime requirement of public health, an essential to cleanliness, and a fundamental requirement of almost all modern manufacturing processes. Indeed, a satisfactory public supply of pure water is one of the engineer's greatest gifts to his fellow men and at the same time one of his greatest responsibilities. So vital is the maintenance of a public water supply that no possible risks can be taken in connection with the necessary engineering design. It is but seldom that a city is in the fortunate position of having a duplicate water supply system, thus being immune from serious trouble should anything interfere accidentally with the functioning of the main supply. Unusual precautions have therefore to be taken in providing for temporary storage facilities; for this reason, conservatism in water engineering design and practice is so often justified.

16.2. Historical Note.—This importance is reflected in the attention that was paid to water supply systems by some of the earliest engineers. Much has been written on this branch of ancient engineering work and its beginnings. The inscription on the Moabite stone, dating from the tenth century B.C., contains references to water conduits and cisterns. There is evidence to suggest that reservoirs for water supply existed in Babylon, the earliest dated 4000 B.C. The Bible contains not a few references to water supply systems some of which were of considerable magnitude, and parts of which have remained in use up to the present day. Notable among these were the waterworks for the city of Jerusalem installed in Solomon's newly conquered city about 900 B.C. and showing highly developed technical skill. They were extended by King Hiskia in 700 B.C.; an inscription records that his engineers successfully bored a tunnel almost a mile long by working from the two ends. The Romans also augmented the supply, and parts of the original installation are still in use today supplying the Mosque of Omar.[16.2] Works in Asia Minor at Priene (dating about 350 B.C.), a town of only 5,000 inhabitants, are hardly excelled by those of any modern town, since every house was connected by earthenware pipes to mains supplied from a large and steady spring on the hillside above the town.[16.3]

The waterworks of ancient Rome are widely known and appreciated, but the fact that the supply to the city is estimated to have exceeded eighty million gallons per day is not perhaps so generally realized. The principal sources of the Roman water supply were the great beds of limestone feeding springs in the valley of the river Amio. Many of the springs fed into some of the 11 aqueducts that were constructed to carry the supply to the Imperial City, those monumental undertakings even the ruins of which testify to the engineering skill of their Roman builders. Fortunately, a contemporary description of these works exists, written by Sextus Julius Frontinus, Water Commissioner of Rome in A.D. 97, and also translations of his work from the Latin. This earliest of engineering authors was an enthusiast, commenting in one place thus: "Will anybody compare the idle Pyramids, or those other useless though much renowned works of the Greeks, with these aqueducts, with these many indispensable structures?" This has a modern touch, as indeed has much else of what Frontinus wrote; his descriptions are worthy of study by all engineers interested in the history of their art. It will be found that he does not neglect geological features, although he does not, of course, describe them in modern terms.[16.4]

The Middle Ages were not devoid of their water supply engineers. The modern emancipation of women is possibly reflected in the fact that the first post-Roman artificial water supply system in England appears to have been that installed by St. Eanswide, of whom it is reported "that she haled and drew water over the hills and rocks against nature from Swecton a mile off to her Oratoria at the sea-side." The good lady in question was the daughter of Eadbald, King of Kent (A.D. 616–640), and was the first Prioress of St. Peter's Priory near Folkestone.[16.5] So the fascinating story can be unfolded; it will be found related in detail in more appropriate places, but sufficient has been presented to show that the water engineer of today is following in a great tradition and that, in some respects at least, modern problems were reflected in these ancient works.

16.3. Sources of Water Supply.—This plural subheading is almost a misnomer, since the *source* of practically all water supply is the fall of rain, assuming that the natural cycle may be broken into in this way for classification purposes. The estimation of the quantity of rain that will fall upon an area is always of importance in water supply studies, whether the supplies are to be drawn from catchment areas, lakes, rivers, or wells; these are the features that are usually referred to as sources of water supply. The quantity appears to depend on many local factors, including the physical configuration of the district; its elevation above sea level; and its relation to mountain masses, prevailing winds, and the proximity of the sea. For a full discussion of rainfall prediction and evaluation, reference may be made to standard treatises on water supply engineering. It is a matter only indirectly and problematically related to Geology. There appears to be some reason for believing that the presence of forests has an influence on rainfall. If this is the case, it will readily be seen that Geology, in so far as it has been favorable or not to vegetable growth, will indirectly influence rainfall characteristics.

It is in the study of what happens to rainfall when it reaches the ground that geological conditions assume importance. Part of the rain is evaporated; part will run off the surface; part will be absorbed by the vegetation on to which it falls; and part will eventually find its way into the subsoil. Cursory consideration will show that all four processes are dependent to some extent on the geological nature of the ground on which the rain falls. Only a portion of the rainfall on any area is therefore available for use. That which runs off the surface to form the flow of streams and rivers is clearly of special interest to water engineers. The available runoff naturally includes also that part of the water which eventually finds its way into watercourses by seepage through the local ground formation. Geology has therefore an appreciable effect on the relation of runoff to rainfall, and attention was consequently devoted to this matter when catchment areas were considered in connection with reservoir construction (Chap. XIV).

The most satisfactory method of evaluating probable runoffs is by means of stream gaging in conjunction with the study of rainfall records. By thus obtaining a measure of actual stream flow, the various reductions in quantity of water from the amount falling as rain are automatically taken into account. This does not mean that study of the geology of the catchment area can be neglected, unless long-term stream flow records are available, since, after varying intensities and durations of rainfall, geological peculiarities may affect the relation of runoff to rainfall to a varying degree and consequently interfere with the regularity of stream discharge. The geological structure of a catchment area is therefore always of importance and must be considered in association with the rainfall and stream flow records that are available in estimating probable runoff. The loss of water due to evaporation into the atmosphere and absorption by plant life is of interest to the water engineer only in so far as it represents a quantity of rainfall that cannot be used. That which percolates into the ground, in addition to representing an amount that has to be determined as a loss in the rainfall-runoff relationship, goes to replenish the great volumes of underground water which are of vital importance to life on the earth's surface. Clearly, the amount that thus enters the ground is directly dependent on the geological nature of the exposed surface and on the type of vegetation growing thereon.

This brief recapitulation of the main features of rainfall and runoff will show that there are many aspects of the application of Geology to water supply engineering. The application is of unusual importance in connection with groundwater; and for this reason, as also because groundwater is of importance in civil engineering in several ways other than as providing a source of supply, this particular aspect will be treated separately in the next chapter. There remain other important sources of water supply to a consideration of which the following notes are devoted.

16.4. Water Supply from River Flow.—Probably the simplest and most straightforward method of obtaining a supply of potable water is to abstract it directly from the flow of a river or from a fresh-water lake; it is also probably the first method ever used, if imagination may be allowed to range back to the first conservation activities of early man. Today, some of the greatest cities in the world secure their water supplies in this way—London, from the river Thames; Montreal, Canada, from the St. Lawrence; Chicago and other American and Canadian lake cities from the Great Lakes.

What relation does Geology have to this branch of waterworks engineering? First, the indirect influence that geological conditions have on the rate of river flow. The local geological structure at the site of the water intake will affect to a considerable degree the type of structure adopted to provide the necessary intake. At Toronto, Canada, for example, the intake to the city's duplicate water supply, constructed in 1932, was designed as a tunnel in the local shale for a distance of 3,300 ft. out from the lake shore, finishing in a vertical shaft leading to a series of precast concrete and steel pipes laid on the lake bed in a trench, at the ends of which are located the actual intakes, suitably screened. The location of the necessary control buildings for intake conduits from rivers or lakes, adjacent to a watercourse or shore line, will often introduce unusual geological conditions in the foundation beds that have to be used. As in other respects, however, this is not a feature peculiar to waterworks engineering.

The abstraction of water from a river may be undertaken in a manner similar to that followed in Toronto; Montreal and London obtain their supplies in this way. In some instances, however, when a river is bordered by porous beds of sand or gravel, advantage is taken of this geological feature, and the water is abstracted through the porous stratum and not directly. This method has many advantages, the most important of which probably is that the water is naturally filtered by its passage through the porous stratum, the use of artificial slow sand filter beds in other water supply schemes being a special adaptation of this process. Many such installations are to be found on the continent of Europe on some of the highly industrialized rivers. The cities of Berlin and Frankfurt in Germany and Gottenburg in Sweden may be mentioned as obtaining at least a part of their water supplies in this unusual way, which is also followed largely in the great Ruhr industrial area of Germany. Hamburg, Germany, in addition to obtaining water that has filtered through from a river bed, uses two irrigating channels into which river water is pumped as additional "distributors," located to the north and south of the boreholes from which the city supply is finally obtained.[16.6] The city of Breslau, Germany, has adopted a similar practice. The devices used to abstract the water from the porous strata vary, but in some cases they are similar to the well points used on construction projects to dry up water-bearing ground (see page 196). Thus, Düsseldorf, Germany, uses borings 70 ft. apart with gravel-packed suction pipes 12 in. in diameter, the yield of each being about 24,000 (imperial) gal. per hour.[16.7]

This interesting method is not confined to Europe. The city of Des Moines, Iowa, obtains its water supply in this way from the Racoon River which flows through a valley about a mile wide consisting of impervious material with a superficial deposit of water-bearing sand. Collecting galleries consisting of reinforced concrete pipes 4 ft.

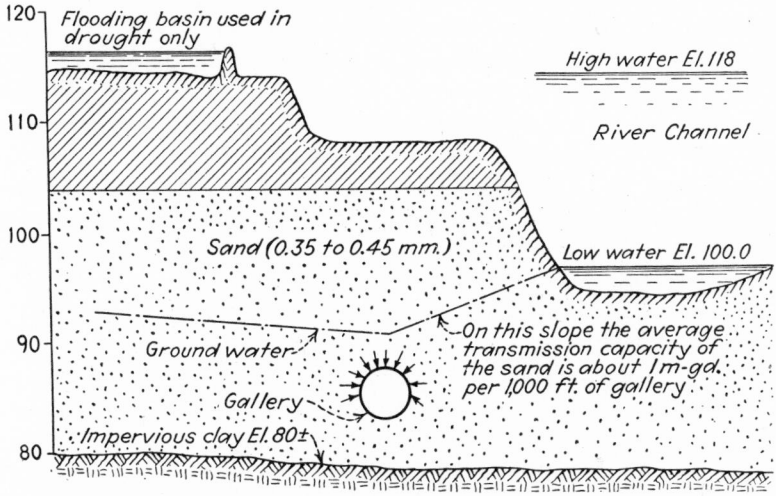

FIG. 16.1.—Diagrammatic section illustrating intake arrangement used for the water supply of Des Moines, Iowa, U.S.A.

(Reproduced by permission of Mr. Dale L. Maffitt, Secretary and General Manager, Des Moines Water Works.)

in diameter are located parallel to the river bank and about 200 ft. away from it. They are located in this way in order to obtain fairly uniform percolation; a minimum good-quality supply of 300,000 (U. S.) gal. per day is obtained.[16.8]

A somewhat more unusual installation of this kind is that which supplies the city of Kano in northern Nigeria, Africa, now an important center with a population of over 100,000. Prior to the construction of the new waterworks, the necessary supply had been drawn from wells dug in the local laterite. As this was unsatisfactory, a new installation was completed in 1931 at a cost of about $1,000,000, (£200,000). Five intake "wells" in the form of reinforced concrete cylinders with internal diameters of 9 ft. (three) and 15 ft. (two), respectively, were sunk by grabbing from the interior to depths of between

35 and 52 ft. The wells rest on rock and penetrate the sand which constitutes the local river bed; they are located in the river bed over a stretch of about 2 miles. Four-inch intake pipes are located in two, and porous concrete blocks in three, of the wells, near the rock surface through which water is drawn from the surrounding sand which acts as a filter and also as a storage reservoir during dry seasons; a daily flow of over 500,000 (imperial) gal. has been obtained.[16.9]

FIG. 16.2.—Sinking intake well No. 3 for the water supply of Kano, Nigeria, Africa (1929).

(Reproduced by permission of Messrs. Binnie, Deacon, and Gourlay, Consulting Engineers, London.)

16.5. Water Supply from Impounding Reservoirs.—When water for public use cannot be obtained directly by abstraction from river or lake, a first alternative source is the reservoir of groundwater found under many parts of the earth's surface, the geological aspects of which are considered in the following chapter. The second main alternative is to utilize a distant source of supply such as the streams found in mountain areas, impounding these artificially and conveying the water thus obtained by means of aqueducts to the cities for which it is required. Although these schemes are always bold engineering conceptions, they are not a development of modern engineering, as a glance at the records of the Roman aqueducts will make clear. These ancient water supply systems have now been eclipsed in size and conception by those of modern times, but they will always retain their interest. Water is conveyed today for city use in aqueducts hundreds of miles long; that for the water supply of Los Angeles, California, brings water 250 miles to the city from the Owens Valley, and 43 miles of this length is made up of 142 tunnels of varying length. This installation is but one of a distinguished group of major engineering enterprises of this type which include the Hetch-Hetchy scheme supplying San Francisco and the Catskill system for New York; and in Great Britain, the Vrynwy scheme supplying Liverpool, the Rhyader supplying Birmingham, the Haweswater for Manchester, and others that supply major inland cities of England.

A typical water supply scheme of this type consists of a catchment area, part of the rainfall on which is impounded behind a *storage dam.* From the reservoir formed by the dam, water is led away through an *intake structure* into the main *aqueduct* in which it is conveyed to a point on the city distributing system, generally a *storage reservoir.* It will be seen that the engineering problems introduced by the works mentioned are not peculiar to water supply engineering; all of them are important parts of many water power schemes for example. In so far as the relation of Geology to the works indicated is concerned, no unusual features are introduced by their use for water supply purposes. The problems that they introduce can therefore be treated generally. In the case of catchment areas, dam foundations, and tunnel construction, this has already been done; general foundations are considered later.

Although the engineering works for water supply derived from distant sources introduce no unique problems in relation to Geology, individual examples of this type of project have provided the engineer with most valuable data on the general geological questions then introduced. Especially notable was the work of this nature during the construction of the Vyrnwy Dam for the water supply of Liverpool, England. The Ewden Valley works, also in England, have similarly been mentioned as providing most valuable data to engineers by reason of the geological conditions encountered. These and other English and European water supply systems have involved works of great magnitude, but some of the North American water supply schemes are even larger and may well be briefly mentioned.

The Catskill system for the city of New York, to which extended reference has already been made, is one such example. It includes two great tunnels, one going under the Hudson River 1,200 ft. below water level; and two great dams; the close study of Geology made in connection with this and other associated works was to a large extent pioneer work in this field. The Board of Water Supply for the City of New York has continued to correlate geological study with their contract work and has applied this particularly to the initial contracts for the new Delaware River system which has been inaugurated (1939) to supplement still further the water supply of the great metropolitan area of New York. The Delaware River aqueduct will consist of 85 miles of tunnel, in six sections, the longest of which will be 45 miles long, the predominating rock types being shales, slates, sandstones, gneisses, and schists. Initial contracts covered the construction of the first 13 of the 30 shafts for which the tunnels will be bored and about 200 ft. of tunnel in each direction at the shaft bottoms. Tunnel contracts were awarded subsequently. Full and carefully recorded geological data were available for all those submitting tenders on these contracts.[16.10]

San Francisco now receives its water supply from an impounding reservoir system of great magnitude; the Hetch-Hetchy water supply and power scheme brought water to the city for the first time in 1934. Water is obtained in a catchment area in the Yosemite National Park and conveyed to the city through an aqueduct 155 miles long, 82 miles of which consists of tunnels (the remainder being pipe lines). Several dams are also included, the largest being the O'Shaughnessy (impounding) dam which will be 430 ft. high when completed to full height. It is founded on granite which was seen to contain many large and deep potholes when uncovered by excavation. The tunnels were driven through varying rock strata, including granite, slate, and recent sedimentary formations. In these last, much trouble was encountered due to "quicksand" and to excessive pressure from chlorites, methane gas, and sulphureted hydrogen; an explosion of methane caused a serious fatal accident. Despite all difficulties, the tunnels were completed, the granite rock in some being so satisfactory that they are in use without a concrete lining.[16.11]

Reference may finally be made to one of the largest of all these projects, the Colorado River Aqueduct of the Metropolitan District of Southern California, which is to convey water a distance of 240.5 miles from the Colorado River to the Los Angeles district for the use of the thirteen cities which constitute the Metropolitan District. The aqueduct is designed for a peak flow of 1,605 cu. ft. per second, or approximately 1,000 million (U. S.) gallons per day. There are to be five pumping stations having a total lift of 1,618 ft. The aqueduct will consist of 91.9 miles of tunnels, 63.4 miles of open canal, 54.7 miles of conduit, and 28.9 miles of siphons. Excavation in the aqueduct will total about 25 million cubic yards; and concrete, 3½ million cubic yards. Of the total tunnel footage, 58 per cent is supported by structural lining; 14 per cent is coated with gunite; and the remainder is unlined. These figures will give some idea of the magnitude of the project, but its unique nature will fully be realized only when it is considered that the eastern part of the aqueduct is located in an arid desert region, construction operations in which could begin only after the building of 148 miles of surfaced highway, 471 miles of power lines, and 180 miles of water supply lines. The aqueduct passes through the desert as a concrete-lined open canal, and the lining is watertight, as the surface of the ground consists generally of alluvial deposits. It crosses three earthquake faults which are classed as *active;* flexible pressure pipe lines across these locations have been considered desirable. Excavation for the Parker Dam, impounding the reservoir from which the supply of water is obtained, and the construction of the tunnels have featured unusual geological conditions. It has not been possible to mention these works in detail but—as is the case so fortunately with many leading civil engineering projects—full descriptions are available in technical journals to which reference can be made by those interested. In these descriptions, repeated references will be found to geological features that affected the work.[16.12]

16.6. Unusual Sources of Supply.—Brief reference may be made to one or two unusual sources of water supply which have been adopted in some parts of the world. The supply to the community living at the foot of the Rock of Gibraltar is one such example. The great Rock consists of a hard and compact Jurassic limestone with the result that it provides no natural supply of spring water. For strategic reasons, the potable water supply has had to be obtained within the fortress area and so, until recently, the supply was limited to the rainfall on the surface of the Rock itself. This is steeply sloped on the east side, the town being located on the west, and a catchment area of 24 acres was there constructed using corrugated iron sheets laid on 40-deg. sand slopes. The sheets are secured by means of purlins and piles, originally wooden but now iron because of the ravages of white ants. An additional catchment area of 15 acres is secured by utilizing steep rock slopes on the west side. Rockfalls frequently damage the east catchment area, but careful watch enables trouble to be quickly located; washouts of considerable areas have nevertheless occurred. Recently, an additional supply of potable water was found by Mr. W. H. Pearce, City and Water Engineer, by means of test borings sunk in the sand of the flat tableland to the north of the Rock. Two wells were later dug and these now supply about half of the total consumption. The brackish water supply to the town for all purposes other than drinking is obtained from shallow wells in the same area; it is kept separate from the potable supply. All drinking water is stored in reservoirs tunneled out of the Rock, with a total capacity of 14 million (imperial) gallons; these are generally in the form of galleries, leading off from two main construction adits, driven from west to east at a level of roughly 375 ft. above sea level.

This interesting example shows how geological conditions distinctly unfavorable to the type of water supply system that might have been expected (springs, wells, or other methods of underground extraction) have been adapted to give an almost perfect although miniature example of a surface water supply. It is of interest to note that the water thus obtained at Gibraltar is used without treatment and, although rather "flat," is quite palatable and consistently safe.

Another unusual type of surface water supply is provided by the *dew ponds* that are still found in Sussex and other parts of southern England and probably also in other parts of the world of a like character. Some of these dew ponds seem to be natural formations, possibly finished off to some extent by the farmers who use them, but they are occasionally made artificially. They have for long been a subject for debate; but if Mr. P. à. M. Parker may be cited as an authority, the water deposited in them as dew is the result of condensation of the atmospheric moisture on the cold surface provided in the pond. Mr. Parker describes the construction of a pond as follows:

> A basin-shaped hollow is excavated in an open space, well exposed to damp sea winds. The hollow is covered by a layer of straw and twigs, or other non-conducting material, about 18 inches thick. On this is laid a continuous bed of puddled clay, about 2 feet thick, which in its turn is covered by a layer of broken stone. The object is to provide a surface of stone and clay which rapidly grows cold at night, and the dew thus collected is caught by the layer of puddle clay, and conducted to the central pond. Such a prepared area, about 200 feet in diameter, under favourable conditions, will keep the pond in the centre about 20 feet in diameter, and say 3 feet maximum depth. So far as can be gathered, in default of systematic observations, the yield is about 0.01 inch per night during the summer, over the prepared area.[16.14]

The geological aspect of this source of supply is concerned solely with the provision of an impermeable basin, the clay surface deposits found in the south of England being well suited to this. The ponds illustrate vividly the direct relation of water supply to atmospheric moisture; the water thus obtained so directly is of admirable quality.

In parts of Australia, "ironclad catchment" areas are sometimes used. These are extensions of the "100 per cent catchment areas" provided by such impermeable surfaces as the roofs of farm buildings. By the laying of corrugated iron sheets on ground specially graded to give a convenient fall to special storage tanks, a large number of ironclad catchments have been installed at relatively low cost. All of the scanty rainfall in these locations is thus obtained for use; an area of 26,000 sq. ft. has been found sufficient for the average farm.[16.15]

The use of corrugated iron can hardly be classed as a geological feature; but in other parts of Australia (notably Western Australia), a similar result is achieved by the use of rock catchment areas. Outcrops of solid granite are sometimes there found so shaped that, by the construction of small concrete walls, all rain falling upon them can be led to suitable storage tanks. After the first fall of rain, all that falls runs off to the tanks, the granite outcrops being remarkably solid. In some places, holes excavated in the granite are used as the storage tanks.[16.16]

Other examples could be cited, but those already mentioned clearly demonstrate the ingenuity adopted in unusual locations to obtain a satisfactory water supply and the dependence of even these unusual methods on geological features.

16.7. Quality of Surface Waters.—Rainfall absorbs gases and floating solid particles from the air before it reaches the ground. Rain water is therefore by no means pure, and it is often definitely acidic in nature. Especially is this the case during thunderstorms when nitric acid may be formed; small portions of carbon dioxide are always absorbed. Consequently, in flowing over rock formations of various types, rain water has a slight chemical effect on them, and the characteristics of lake and river water can therefore vary considerably. Although the detailed examination of water intended for use as domestic supplies is now a matter for specialist study, more particularly from the bacteriological aspect, it is one with which the civil engineer has to be generally familiar. Knowledge of the geological conditions causing the various impurities in water is a prerequisite for such familiarity. The subject is treated briefly here but more fully in so far as groundwater is concerned in the following chapter.

Many upland gathering grounds, by reason of their topographical arrangement and of the underlying geological structure, are covered with a layer of peat for much of their total area. Rain falling on peat deposits and running off to become stream flow inevitably takes with it traces of the peat and of the organic acids found in peat deposits. It thus becomes brackish, often slightly discolored, and usually slightly acidic. If the acidity is pronounced, the water may be slightly plumbo-solvent, a grave matter in England and other European countries where lead piping is still in use for water supply purposes. Similarly, such water may have a serious effect on concrete surfaces exposed to it. Investigations in Scotland have shown that the only normal types of open conduit lining that successfully resist the action of moorland water are Staffordshire blue brick (an engineering brick of high quality) and aluminous cement concrete. Ordinary portland cement of varying consistencies was seriously affected.[16.17]

The brackish taste and discoloration are objectionable from the point of view of the user of the water, although they are not usually of serious consequence. Both can be removed by modern filtration and treatment methods. In many cases, recourse has been had to stripping off all the peat from the area to be submerged by impounded water, often at great cost. Thus, a total volume of four million cubic yards of peat was removed from the site of No. 5 Reservoir of the Boston (United States) Waterworks. The bottom of the reservoir formed by the Silent Valley Dam in Northern Ireland (see page 312) was found to be covered with peat, in some places 30 ft. deep. Instead of this being removed, the valley floor was covered with a layer of clean sand 2 to 3 ft. in thickness. An area of 342,000 sq. yd. was covered in this way, the sand being obtained from local glacial deposits.[16.18] It will be seen, therefore, that thorough examination of the surface condition of any proposed reservoir site is a necessary part of preliminary investigation work.

Hardness, temporary and permanent, is perhaps the most widely recognized of possible impurities. In the case of surface waters, both are the result of contact with rocks containing calcium or magnesium carbonate or calcium or magnesium sulphate. If a river is known to flow over limestone or dolomitic deposits, the source of the contamination can easily be found. Cases do occur in which the strata in the bed of the river do not suggest the cause of the salt content. The River Derwent in Derbyshire, England, for example, flows over grit beds, and yet its water is hard. The hardness is found to be due to the stream flow being fed by tributary streams coming down from mountain limestone deposits. Study of geological features both in river bed and over catchment area will suggest the type of hardness when hard water is encountered and so the necessary treatment. Although generally troublesome, hardness of water is not always a disadvantage, as some beer drinkers may know. For it is usually the mineral content of water that gives special local flavor to pale and bitter ales, hard water being desirable for their manufacture. To take one well-known example, ales brewed at Burton-on-Trent, England, are made from hard water obtained from valley gravels and Keuper sandstones. It was once carefully calculated that in the year 1869 drinkers of Burton ales consumed 350,000 lb. of solid gypsum in the process of assuaging their thirst![16.19] Soft water is required, on the contrary, for the making of stout. Waters used in the north of Scotland and elsewhere for a similar, although stronger, purpose are also dependent on hardness for their peculiar efficiency.

Suggestions for Further Reading

Much useful information relative to the geological aspects of water supply projects will be found in standard textbooks on water supply. In addition, the following volumes may be noted as being of special interest in connection with Geology:

Dixey, F.: "A Practical Handbook of Water Supply," Thomas Murby & Co., London, 1931; based on African practice.

Lacey, J. M.: "Hydrology and Ground Water," The Technical Press, Ltd., London, rev. ed., 1934; based on Indian practice.

Walters, R. C. S.: "The Nation's Water Supply," Ivor Nicholson & Watson, Ltd., London, 1936; a popular description of British water supply.

Woodward, H. B.: "The Geology of Water Supply," Edward Arnold & Company, London, 1910; based on British practice.

See also references at end of next chapter, p. 473.

CHAPTER XVII

GROUNDWATER

And it came to pass the same day, that Isaac's servants came and told him concerning the well which they had digged, and said unto him, We have found water. And he called it Sheba: therefore the name of the city is Beersheba unto this day.

Genesis 26:32–33.

In certain mines of the Pittsburgh district, United States, 20 tons of groundwater have to be pumped out of the mines for every ton of coal that is won. Twenty million acres of land in India are irrigated by groundwater obtained from wells, an area equal to the total expanse irrigated in the United States. In 1931, an analysis of the water supply to 734 towns and cities in Germany was made, including the water service to a total population of 33,395,000, and it was found that 1 per cent was supplied by special or combined methods, 11.4 per cent from surface water, and 87.6 per cent from groundwater. It is estimated that two thirds of the waterworks in the United States, supplying at least twenty million people, are solely dependent on groundwater for their supplies. And although the city of London is popularly supposed to obtain all its supplies from the rivers Thames and Lea, actually it obtains no less than 18.8 per cent of its public supply (or about eight and a half million gallons daily) from wells and probably a considerably larger percentage of its total supply, since many industrial concerns in London have their own private well systems.[17.1]

These figures demonstrate clearly how important a place groundwater occupies in the life of the world today. Its importance will also be appreciated if it is considered that the quantity of water within that part of the earth's crust relatively close to the surface is believed to be equal to one third of the total volume of water in the sea. It is further estimated that water can exist to a depth of 6 miles below surface level, and it is generally admitted that water exists almost invariably under all parts of the earth's surface, although it may often be located at such depths as to be of no possible use to man. To all who have to work in the earth's crust, therefore, this vast reservoir of water is of great importance whatever their work may be. The miner comes first to mind; to him groundwater may be a matter of life or death, and it may be all that swings a venture from success to disaster. Water below ground surface is equally of vital importance to the civil engineer as a source of water supply and also as the controlling factor in all drainage operations, a hazard to be countered in tunnel driving and other underground operations, and a feature of foundation work of unusual complexity and occasionally of unique significance.

The use of groundwater as a source of potable water supply has today a generally accepted importance even greater than its statistical position might indicate. This is probably due to two main reasons. In the first place, although a large community not located on a river or lake can contemplate the idea of bringing its water supply from a distance, this course is practically impossible when an individual wishes to supply his own home or farm. In consequence, the possibility of utilizing as a source of supply water that lies in the ground below is always attractive. Individual well installations and spring supplies have been perhaps the leading factor in enabling primitive nomadic life to be gradually converted to the fixed agricultural existence that is the basis of our present civilization. The second reason follows directly. As the individualistic agriculturalist gradually gave way to modern industrial organization and city building, the new centers of population could not always be located in convenient relationship to a good water supply. Hence, in many cases in which industrial conditions determined the site of a town, reliance had to be placed on the possibility of obtaining a good supply of groundwater in the vicinity.

Naturally, there are reasons other than these two that have served to make the use of groundwater so widely appreciated. The two mentioned go far to explain why the development of subsurface waters is almost synonymous with written history and consequently why it is that the literature of the subject is surprisingly extensive. Only in comparatively recent years, however, has groundwater been scientifically investigated, more particularly by the staffs of geological surveys, one of whose main tasks has been to study and inventory underground water resources for the general benefit of pioneer settlements.

17.2. Historical Note.—The existence of groundwater has probably been realized by man from the dawn of history, for the coyote and possibly other animals will dig down to water if it is close enough to the surface to be reached by an animal's scratchings. This interesting observation is made by Dr. O. E. Meinzer in a paper[17.2] on the history of hydrology, which covers the subject so thoroughly that a brief note such as this must inevitably be gratefully indebted to it. Considering some of the earliest documents known to man, the twenty-sixth chapter of the Book of Genesis discloses a thorough familiarity with groundwater conditions; similar references could be traced throughout the Bible. It is well known that the Romans were familiar with the use of wells; in England they had one well 188 ft. deep, and they supplemented their well supplies by means of adits driven in the chalk, supposedly a modern development.

Throughout the Middle Ages, groundwater continued to be widely used without being understood. Almost until the end of the seventeenth century, it was generally conceded that spring water, as observed on hillsides, could not possibly be derived from rain. Many explanations were advanced as to its origin, one of the most interesting being that, owing to the curvature of the earth, the water in the middle of the ocean is actually at a higher altitude than the springs, thus furnishing the necessary head. The idea that springs have their origin in the sea was often based on passages from the Bible such as that to be found in the Book of Ecclesiastes 1:7—"All rivers run into the sea, yet the sea is not full; unto the place from whence the rivers come thither they return again."

Not until the sixteenth century did these views even begin to be questioned, Bernard Palissy (1509–1589) being one of the first to develop the theory of the infiltration of groundwater. Pierre Perrault (1608–1680) was one of the first to put Hydrology on a quantitative basis relating the total rainfall and runoff for the basin of the river Seine, France. Although roughly done, this work may be regarded as the starting point of Hydrology as it is known today, especially as it was just about this time that Edmund Halley (1656–1742) was conducting the first known quantitative experiments on evaporation, thus proving definitely the origin of rainfall. Thereafter, hydrological studies developed steadily into the scientific study of today although hampered to some extent by the strange ideas that persisted with regard to water witching, or divining.

A scientific approach to groundwater problems was thus established when modern engineering began in the early years of the last century. It was an approach adopted and developed by several early civil engineers, and at least two of them made notable contributions to the subject. One of these was William Smith; the other was Robert Stephenson whose name is well known in connection with the early development of British and other railways. William Smith was one of the first to relate the study of groundwater to the study of Geology; he did even more than this in applying the results of such combined studies to the solution of engineering problems. Although he appears to have left no written work on the subject of groundwater, the works that he carried out show clearly that he had a very vivid conception of the main principles of groundwater location and movement. As examples may be mentioned the water supply system that he developed for Scarborough and the drainage of ground at Combegrove, near Bath, in the south of England.

17.3. Groundwater Characteristics.—That part of the rainfall which is absorbed into the ground is important in many ways. Sometimes this absorption accounts for almost the entire rainfall, as in certain parts of Australia, in the drier sections of which are many rivers that disappear and others that flow only for certain limited portions of each year. Even more striking is the elevated area of 2,000,000 acres to the north of Casterton and Colerdine in the state of Victoria, which acts as the "intake" for the great Murray River artesian basin; it has an average rainfall of 25 in. and no runoff.[17.3] This is an extreme case. If an average case is considered, that part of the rainfall which does seep into the ground may be subdivided as follows, after Dr. B. Smith:[17.4]

1. That absorbed directly by plants.
2. That drawn by capillary action to the surface and there evaporated.
3. That which unites with molecules of mineral substances and so becomes chemically fixed.
4. That which flows directly into the sea (in coastal districts) through springs and underground channels.
5. That which escapes at the surface of the earth through springs or by feeding rivers.
6. That which is retained in the earth.

The first two divisions are important in agriculture, the last two to engineers, both civil and mining; the fourth is sometimes of importance in engineering work.

Civil engineering considerations of groundwater refer almost wholly to that part of the water which is retained in the ground; but in order to make clear the general relationship of all subsurface waters, Fig. 17.1 is here included. The term *vadose* water has been used in more than one way, but the diagram indicates its correct use as describing that water which is still in the zone of aeration, *i.e.*, in the part of the crust which is not *saturated* with water. General vadose water is not of great moment in engineering; it does affect the drainage of shallow excavations and is naturally related to surface drainage and irrigation. Below the zone of saturation will be seen indicated what is termed *internal water*, this being located in that deep part of the earth's crust (estimated by some as a depth of 6 miles) where rock pressure is so great that no interstices occur in which free water can exist. This water and that associated with rock magma are of geological interest

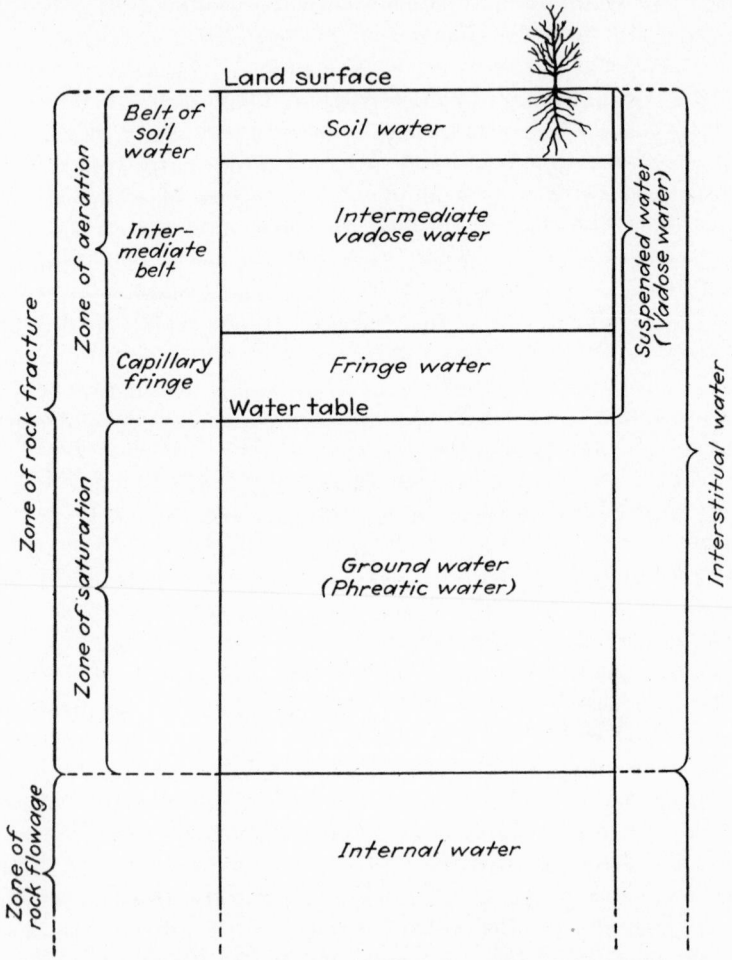

FIG. 17.1.—Diagram showing the subdivisions of underground water.
(Reproduced by permission of the Director, U. S. Geological Survey, from U.S.G.S. Water Supply Paper 494, by Dr. O. E. Meinzer, Outline of Ground-Water Hydrology, Washington, 1923.)

only. They are mentioned since, in connection with certain artesian basins and the drainage of some mines, the question may arise as to whether or not water is encountered from plutonic sources, *i.e.*, water that has not previously existed as atmospheric, or surface, water. It is known as *juvenile*, or plutonic, water and may be of two kinds:

(1) that which has been imprisoned in the earth's interior since its formation or (2) that created by chemical combination of primitive hydrogen with oxygen of external origin.

One of the most striking examples mentioned by those who support the idea of plutonic water is that which was met with during the driving of the Simplon Tunnel. This was so hot that workmen were scalded to death. It was found to contain no chlorine, whereas all surface waters contain at least a small percentage of this element.[17.5] Then there are many notable hot springs such as those near Kivu in Central Africa, at Karlsbad in Germany, and at Bath in England, which are thought by some to be of internal origin. Similarly, there are the *fumoroles suffaroli* of northern Italy, the steam of which may have a similar origin. Some students of the matter have suggested that probably the sea is fed by plutonic water emerging in the sea bed. At Bendigo, Australia, there are mines that have a zone of workings in which meteoric (surface) water occurs; below this zone is a perfectly dry one; but below it, at a depth of 4,500 ft. from the ground, springs appear.[17.6] Controversy regarding plutonic water has centered chiefly on the Great Australian Artesian Basin. In this basin, there are wells from which water is obtained so hot that it can be used for making tea or cooking. The origin of this water is an interesting topic and is mentioned here in order to emphasize the fact that in the construction of very long tunnels such as the Simplon, an unusual factor of considerable importance may arise. *Connate*, or *fossil, water* is water trapped between impervious beds which have been raised and distorted at some time in the past. The trapped water may be encountered in subsurface engineering work.

Reverting to Fig. 17.1, it will be seen that the zone of saturation is topped by what is there called a *water table*. This is a most useful concept, apart from the fact that the use of the word table suggests a level surface, and a water table is rarely if ever level. It can readily be imagined that the distance to a water table may be great or small depending on local geological conditions. Seepage has been actually observed through depths of 90 ft. of glacial drift, into a tunnel on the Pietermaritzburg Railway, South Africa.[17.7] If the earth's crust were of uniform composition, all that is essential for a summary would have now been said; but underground conditions are far from uniform, and so it is that the disposition of groundwater is one of the most complicated problems in Geology.

Essentially, the disposition of water below the saturation level depends on two main factors: (1) the texture of the local rocks, their composition and relative stability; and (2) their structural arrangement and relationship to other neighboring rocks. The variations possible in both these factors will make clear the complexity of the distribution of groundwater. Although underground conditions may be so variable, the water there retained is still governed by the natural laws governing surface water. Its movement is in accordance with the law of gravity, apart only from that motion of minor degree which is due to capillary action. It may be well to emphasize the fact that groundwater does move. The vertical oscillations of the upper surface of saturation, or water table, will perhaps be readily appreciated since so many direct results of this movement are to be observed, such as the alteration of well water levels and the intermittent flow of springs. That groundwater is often moving in a horizontal direction corresponding to the slope of the water table is perhaps not quite so clear. No more vivid explanation of this underground travel has been given, perhaps, than in this statement of Dr. Herbert Lapworth:[17.8]

> It was well recognized that underground water could travel for vast distances. For instance, it was known that in the artesian basins of Australia underground water travelled for hundreds of miles; and there were cases in the United States where it travelled 200 miles, with pressures which ran up to 500 feet head at the surface and something like 2,000 feet head at the bottom. In Algerian Sahara it was known, from borings, that water came a distance of 300 miles from the Atlas Mountains, passing along definite channels, some of which had actually been traced for a length of 70 miles.

These cases quoted by Dr. Lapworth, are admittedly extreme, but they indicate clearly how extensive underground travel can be. As a more utilitarian example there may be cited the water supply of the city of Leipzig, Germany, obtained from an underground source that is almost a "stream," being 2 miles wide and 40 ft. deep. Again, it may be noted that a well at Grenelle, Paris, 1,798 ft. deep, was found to be drawing its water from Champagne, 100 miles away. With these examples of underground travel as reminders of the dynamic nature of the problem being considered, attention will now be turned to the relation of rock texture and composition to groundwater distribution.

17.4. Influence of the Nature of Rocks.—The variation between the distribution of water in ground consisting in one case of uniform coarse sand and in another of solid igneous rock is an obvious illustration of the effect of the composition and texture of rocks on groundwater distribution. Between these extreme cases occurs a wide range of conditions of varying complexity. Before considering the influence that the nature of rocks can have on groundwater, some definitions are necessary. *Perviousness* is a word commonly used in this connection; it may be defined as the capacity of a rock to allow water to pass through it. A pervious rock has communicating interstices of capillary or supra-capillary size. Its degree of perviousness cannot be correlated with any standard scale, but it is usually expressed as a volume of flow per unit of cross-sectional area per unit of time (such as gallons per square foot per day). *Porosity,* on the other hand, is a measure of the interstices contained in any particular volume of the rock; it is generally expressed as a percentage indicating the aggregate volume of interstices to total volume. Strange as it may appear at first sight, a high degree of porosity is no assurance of perviousness. Clay, for example, has a high porosity; examples have been found of newly deposited Mississippi clay with a porosity of between 80 and 90 per cent. When saturated, however, it becomes impervious, the water it contains being held firmly by molecular attraction.

Despite the fact that porosity and perviousness are not always synonymous, the convenience of the property of porosity as an index to the water-bearing capacity of solid rocks and as one that can easily be investigated in the laboratory has resulted in much attention being devoted to it. Table P presents some typical values for well-known types of rock.

TABLE P.—THE POROSITY OF ROCKS[1]

Type of Rock	Maximum Porosity, Per Cent
Soil and loam	40 to 60
Chalk	Up to 50
Sand and gravel	25 to 30
Sandstone	10 to 15
Oölitic limestone	10
Limestone and marble	5
Slate and shale	4
Granite	1.5
Crystalline rocks generally	Up to 0.5

[1] These figures are presented as illustrative only of the general trend of porosity in relation to various rock types. Tests on actual samples should be performed if a value for porosity has to be used.

It must be emphasized that these figures are approximate only. In the case of the chalk of England, for example, porosity varies from 26 to 43 per cent for that found in southern England and is about 18 per cent for that found in Yorkshire, whereas the chalk of Antrim, Ireland, has sometimes a porosity of less than 10 per cent. The following data given by Professor W. S. Boulton[17,9] are of interest. Seven sample cores of Bunter sandstone (usually pervious and permeable and without joints and fissures) taken from depths varying from

300 to 750 ft. gave porosity values from 13.2 to 30.2 per cent and proved to have a permeability ranging from 0.05 to 17.4 (imperial) gal. per square foot per 24 hr. These figures serve to demonstrate the wide variation possible in a fairly regular type of rock. This is even more strongly emphasized by samples taken from a 40-ft. bed of this same rock at approximately 100-ft. centers, the porosity even under these circumstances varying from 3.58 to 20.3 per cent. Despite this appreciable porosity, the permeability in the direction of bedding, *i.e.*, perpendicular to the cores, was very small indeed, and checks showed that leakage through this rock was practically all taking place along joints and fissures. That this should be the case for such a relatively good "water-bearing rock" as sandstone is indeed surprising, so that the fact that the same feature has been demonstrated to be the explanation of the water-bearing properties of some chalk deposits will the more easily be appreciated.

It is sometimes observed that, in addition to flowing along joints and fissures, water traverses chalk along the bands of flints and rubble that quite often distinguish chalk strata. Another feature to be noted in connection with both sandstones and limestones is the significance of the material that acts as a natural cement joining the individual grains to form the solid rock. An unusual but striking example of this is given by Mr. R. Schaffer, who compared Monk's Park and Ketton stones.[17.10] Both are British oölitic limestones, and in each case the oölitic grains are microporous and show a similarity of size and structure. In the Monk's Park stone, however, the intergranular spaces are filled with a nonporous matrix of crystalline calcite; whereas in the Ketton stone, they form interconnected pores. The Ketton stone is therefore pervious and yields water freely, whereas the Monk's Park stone is impervious. It is sometimes thought that porosity decreases with increase of depth as a result of the pressure of the superincumbent rock. Professor P. F. Kendall has pointed out that the Norfolk chalk of England, usually regarded as the most porous British rock, can be found giving a very small yield because, it is suggested, of the pressure of overlying rock on this relatively soft material.[17.11] Few determinations of porosity at great depths have been made.

Confirmation of the importance of fissures and joints in connection with the movement and storage of groundwater is available from many sources. In all but pervious rocks such as sandstones and some limestones, fissures are usually the leading factor in determining the water-bearing capacity of a rock. The distribution of joints is a matter of some complexity. It seems clear that beyond a certain depth the distribution and the size of joints diminish appreciably. The critical depth may possibly be about 500 ft. Below this, a regular diminution of supply from joints may be expected, and therefore an economic pumping limit will soon be reached. In any case, pumping is not generally economic below depths of 700 or 800 ft. It follows that although water may be generally available below these levels, it cannot usually be regarded as commercially available unless under artesian pressure, a special feature to be considered later.

Distribution of groundwater is therefore the prime queston in underground-water surveys, rather than the mere location of water; and for this reason, some general notes on the water-bearing properties of the more common rock groups will next be presented.

Sands and Gravels.—These materials, being at once porous and pervious, may be classed as almost the ideal water-bearing strata. Their wide use as artificial filtering and drainage mediums renders unnecessary any detailed comments. Extremely fine sands, however, are of little use.

Clays and Shales.—As a general rule, these will be useless as sources of groundwater, since, although clays are often wet, the water present is not readily available. Hard shales may yield water at joints.

Sandstones.—Sandstones are so variable in texture and composition that they may vary from an almost impermeable rock to one so pervious that water actually squirts out from it when under pressure, as in some of the Bunter sandstone of England. In these cases, the diameters of boreholes are of importance, some being made as large as 30 in. When pervious, sandstones form an admirable source of groundwater, since in addition to giving a high yield they constitute an effective filtering medium. Especially is this true of the New Red Sandstone of England; water from deep wells in this stratum is almost invariably clear, sparkling, and palatable.

Limestones.—This extensive group of rocks is second only to sandstones as a source of groundwater. Its importance is well recognized by underground solution channels and associated caverns and other familiar features, as also by the wide dependence on chalk strata as underground reservoirs. The water-bearing characteristics of chalk have already been noted. It may be further observed that as water in contact with limestones dissolves a small quantity of the rock, the use of limestones as sources of water tends always to increase the yield obtained from them. Thus, it is calculated that with every million gallons of water pumped out of the chalk of the south of England, about 3,300 lb. of chalk is also removed, pore spaces and any existing underground channels being correspondingly enlarged. The interesting suggestion has been advanced in France that the effective rock volume available at any well or borehole in limestone might be increased by the judicious injection of hydrochloric acid. No records have yet been noted of a practical trial of this idea (except in oil-field work), but it suggests interesting possibilities.

Crystalline Rocks.—In general, these rocks are not classed as water bearing, although few are absolutely dry when encountered in excavation. When decomposed, as the result of weathering or some other cause, and also when fractured and fissured, they may yield appreciable quantities of good potable water. Fresh crystalline rocks in massive formation will not generally yield any useful quantity. Granites, when decomposed, may have a relatively fair yield; quartzites, slates, and marbles may yield a useful supply because of jointing; gneisses and schists, unless they have decomposed badly, cannot be relied upon for any appreciable quantity. In Cornwall, England, several important towns obtain their water from wells in local granite; it is probable that the water collects in the blanket of disintegrated rock at the surface. In Maine, United States, many similar wells are found in other igneous rocks, but their yield is usually small unless jointing is very marked.

17.5. Quality of Groundwater.—The foregoing notes lead directly to consideration of the quality of groundwater, a matter of vital importance whether the water is to be used for industrial or for domestic purposes. In general, groundwater is free from bacteria, the passage of the water through the ground strata constituting a natural filtering process. This possibility does not remove the vital necessity for the bacteriological examination of all groundwater that is to be used for domestic purposes, especially when the arrangement and nature of the strata are such as would permit contamination of the water stored in them from surface sources.

Groundwater is usually found at a fairly constant temperature. This temperature is usually moderate (between 50 and 60°F.), so that the water is cool in summer and warm in winter—one feature that commends it as a source of domestic water. In recent years, this even temperature of groundwater has taken on an added significance in connection with its use as cooling water. The development of air conditioning systems in many of the larger buildings of cities calls for steady supplies of cooling water, not in themselves large but appreciable when many buildings are dependent on the same source of supply. Consequently, in New York, for example, several of the larger theaters are allowed to use well water for cooling air conditioning systems provided that (1) a closed circuit is used for the water, to obviate any possibility of contamination; and (2) all water used is returned to the ground by means of wells. The rise in temperature averages about 10°F. In some American cities, the extensive use of groundwater for this purpose is already giving rise to serious problems.

There is at least one location at which groundwater was used for domestic supply where the reverse process has had to be used. Riverside, California, obtained a supply of 2,000 (U. S.) g.p.m. from an

FIG. 17.2.—Cooling plant for groundwater supply to the city of Riverside, California, U.S.A. Unfortunately this unique plant was destroyed in March, 1938, by severe floods on the Santa Anna River, which changed its course and now flows over the site of the plant.

(Reproduced by permission of the Municipal Water Dept., City of Riverside, Mr. R. L. Boulden, Superintendent.)

artesian well 965 ft. deep installed in 1931. The temperature of the water so obtained was 110°F., and in consequence a special and elaborate cooling system had to be installed by the local water authority in 1936. Two adjacent wells, each 1,100 ft. deep, provided cool water. Clearly, some peculiar underground structural arrangement must be held responsible for this unusual condition.[17.12]

Groundwater may contain dissolved solids and gases. Of the gaseous impurities, methane may occasionally be encountered in water that has not traveled far from surface deposits or that has been in contact with deep-buried strata of organic origin. The gas is usually due to the decomposition of organic matter in the absence of free oxygen. It is dangerous; and therefore if it is present in such quantities that it is liberated from the water on reaching atmospheric pressure, it can be a source of trouble. Hydrogen bisulphide is occasionally met with and is easily detectable if present in any appreciable quantity; its origin may vary, but a possible source is the interaction of organic acids from surface deposits with underground sulphates. The gas is easily removed by aeration; in the case of medicinal spring waters, its presence is often their chief feature. Carbon dioxide is the most important gaseous impurity of groundwater; its origin is generally the atmosphere. Its presence gives to water a sparkle which is not unpleasant. It is not therefore objectionable; but in water, the gas makes a weak solution of hydrocarbonic acid. This acts as a solvent for several different rock constituents, and for this reason the presence of the gas is often significant.

Free carbon dioxide gas is occasionally encountered in civil engineering work in such quantities as to be troublesome. On the great aqueduct conveying Owens Valley water to the Los Angeles area, California, severe corrosion was discovered in the steel pipe lines forming a part of this aqueduct; tubercles formed very quickly, and constant cleaning and painting were necessary for adequate maintenance. Careful tests were made of the water at the entrance to the aqueduct, but no peculiarities were found. Further checks were made; and it was eventually found that between the two ends of a tunnel 7,000 ft. long, the carbon dioxide content increased from a negligible amount to 4.2 p.p.m. The tunnel was therefore unwatered; and on examination, the gas could actually be heard hissing as it escaped through cracks in the concrete lining. The tunnel penetrates what is locally known as *Soda Hill*, a granitic rock formation containing many badly fractured fault zones and near to which hot limestone beds are thought to exist. The consequent acidity of the water was clearly the cause of the corrosion troubles. The remedy adopted (in 1936) was to lead the gas through special chases into a pipe beneath the tunnel invert which, in turn, was connected with a new adit, 2,000 ft. long, through which the accumulated gas could escape.[17.13]

Pure water will dissolve only 20 p.p.m. of calcium carbonate and 28 p.p.m. of magnesium carbonate, but water containing carbon dioxide will dissolve many hundreds of p.p.m. of these solids. It is these and similar solid impurities in groundwater that may give rise to the main purification problems of the water works engineer. It follows that water obtained from limestone strata is always suspect, even without test. The dissolved carbonates give to the water what is termed *temporary hardness*, since the dissolved solids can be removed by simple chemical processes. If the dissolved solids are the corresponding sulphates, the corresponding hardness of the water is termed *permanent*, in that it cannot be removed by simple processes used for eliminating the carbonates. Chemical analyses are necessary to determine the degree of hardness of the water; if this exceeds about 200 p.p.m. of calcium carbonate, the water requires *softening*. Clark's lime process, a chemical method, is used to remove temporary hardness. For permanent or mineral hardness, various processes are available, but the use of zeolites is usually a fundamental part of all such methods. Zeolites have the property of exchanging non-hardening constituents for the hardening constituents present in water; when exhausted, they can be regenerated by treatment with a brine solution. Artificial zeolites are now available, formed of complex silicates, but the original zeolites were from a natural source, being made from glauconite (a complex silicate of alumina, iron, and potassium).

Iron salts are another impurity in groundwater that can be of serious consequences in industrial work, *e.g.*, dyeing; and they are even a nuisance in connection with domestic supplies. Soft water is liable to contain less iron in solution than hard water, the maximum iron content for which is generally about 1 p.p.m. depending on the other salts present. Manganese salts may be present in a similar way, being often associated with iron salts although always in small quantities. They affect the taste of water and give black stains, whereas iron salts produce red or brown stains. Deferrization of such water is a simple matter in principle but difficult in actual practice. It is essentially a chemical process, aiming at the production of ferrous hydroxide as a coagulant which can be precipitated or removed by filtration. Sodium chloride remains for mention as the final important mineral impurity liable to be encountered in underground water. It may have its origin in the sea, as explained later in this chapter, but may also be of mineral origin. Rock salt deposits are an obvious source of supply; some are "mined" by means of brine solutions.

The quality of groundwater has so far been treated by considering the nature of possible impurities; it may be useful to add a brief summary of what may be expected from typical water-bearing strata. The only type of *clay* from which water supplies are regularly obtained is boulder clay, which may sometimes have a pervious structure. Water thus obtained may be quite hard; traces of hydrogen sulphide are often found. *Sands* usually give a fairly pure water, but occasionally they will yield water with a small ferruginous content. The same may be said of water from *sandstones;* if the cementing material is calcium sulphate or calcite, the water can be quite hard. Water from *limestone, chalk,* or similar rocks will almost certainly be hard, but occasionally other impurities will also be present. Water from some of the East Anglian chalk in England, for example, contains so much ferrous carbonate that it must be treated.

To deal with the possible impurities in groundwater in this brief space necessarily means that only a sketch of the subject can be presented. This is done with one main purpose in view—that of demonstrating the relation of such impurities to the geological conditions of the ground from which the water is being obtained. This correlation is useful for two reasons. In the first place, preliminary study of the local geology will at least suggest the possible impurities to be expected in any prospective groundwater supply; and, second, accurate knowledge of the relevant geology will demonstrate the course of any known impurities and indicate the probability of any increase or decrease of the amount of contamination.*

17.6. Influence of Geological Structure.—Underground conditions affecting groundwater differ from the ideal case not only by the wide variation of materials in contact with the water but also in respect to the way in which the various rock strata are arranged, in general and in relation to each other. Maintaining the concept of a level ground surface, the first variation from the ideal case that may be considered is the alternation of pervious and impervious beds. If the impervious stratum constitutes the surface layer, no water would normally penetrate to the potential underground reservoir provided by the pervious bed. Possibly a more usual case is that in which the pervious bed is at the surface; rain will soak through until it reaches the main body of groundwater retained above the impervious stratum from which supplies can be drawn by means of boreholes or wells. The arrangement is capable of duplication, and a "sandwich" structure can be imagined. Such structures actually occur; the middle Lias formation in Northampton, England, only 93 ft. thick, contains no less than five water-bearing horizons separated by impervious clays, any four of which may be present at any one location.[17.4] The condition illustrated in sketch *A* will be interfered with only if the impervious stratum is pierced in any way. This would seem hard to imagine, but cases have occurred in which wells have been drilled too deep. At Kissingland, near Lowestoft in England, the bottom of an old well penetrating 50 ft. of gravel was pierced, all the water tapped by the

* Professor John Wyllie of Queen's University, Kingston, Canada, has encountered unusual cases of chronic arsenical poisoning due to the presence of arsenic in farm well water. This, in turn, was traced to minute quantities of ferrous arsenate naturally occurring in the limestone stratum through which the well was excavated. The details of the investigation are given in *Canadian Public Health Jour.*, p. 128, March, 1937.

well running away to the lower stratum; a similar instance happened in the drilling of a 70-ft. well at Nipigon, in Ontario, Canada.

A further variation will be obtained when the strata are inclined instead of being level, as indicated in sketch *B*. In this case, the inclined impervious stratum will constitute a barrier between the two pervious beds, so that the elevations of the respective water tables need not and probably will not be the same. A similar effect is to be

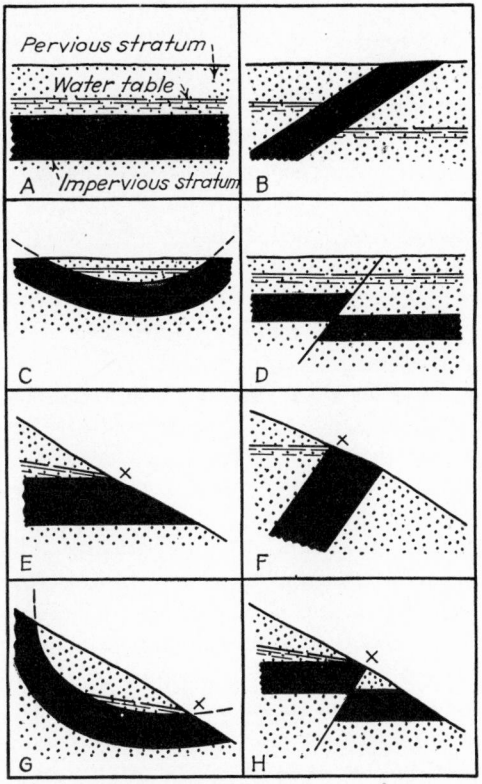

FIG. 17.3.—Diagrammatic sketches illustrating effect of geological structure on groundwater distribution.

noted when alternating layers constitute part of a fold; many arrangements are naturally possible for this case of which one only is illustrated in sketch *C*. This is a general case only, but it shows how water will collect in such a distorted stratum. It may be noted that the geological structure underlying the city of London is of this type. Finally, sketch *D* demonstrates the effect that a fault may have on the distribution of groundwater in alternating strata. The variations possible in this simple case are dependent on the relative thicknesses of the strata, the nature of the fault, and the throw of the fault. In the case illustrated, the barrier provided by the impervious stratum remains, but a slight increase in the throw of the fault would leave a gap in this barrier, with consequent alteration of the groundwater conditions. The shatter zone generally found in rock bordering a fault plane may also alter the water-bearing character of a rock by providing water passage along the planes of fracture.

The four simple cases already considered are illustrated in sketches *E*, *F*, *G*, and *H* with the surface of the ground inclined: the altered groundwater conditions are indicated diagrammatically and need no elaboration. It will be seen that two peculiarities are immediately introduced. At the points marked *X*, bodies of groundwater will

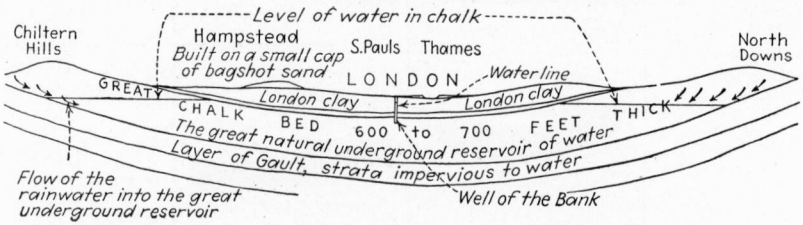

FIG. 17.4.—Simplified geological section through the London basin.

come into contact with the atmosphere. This leads directly to the second feature, *viz.*, that the surface of the groundwater will not be level but will be inclined at the hydraulic gradient necessary for flow to take place through the material of the stratum. This is a general condition, as has already been explained. An important consequence is that, as a rule, the "water table follows the ground surface"—a broad statement but useful as a guide.

Of particular significance and importance is the variation in the yield of water obtained from the same rock at various positions along the range of a fold in which the rock has been distorted. If the folding has been anticlinal and is of relatively short radius, there will be a distinct tendency for fissures to develop. Similarly, if the folding has been synclinal, there will be a tendency for the rock to "tighten up" and have its usual perviousness reduced, the tendency being most marked in the case of the softer rocks. This has been demonstrated in connection with the water supply of the municipality of Richmond, London, where the construction of a well and over 10,000 ft. of adits in the chalk have failed to develop an appreciable supply, owing to the selected location being on a syncline in the local chalk formation.[17.4]

17.7. Springs.—Springs may sometimes provide a useful and pure water supply, in which case their reliability will often have to be investigated to ensure continuity of supply. Alternatively, an engineer may be called upon to study the source of springs in connection with possible water-lowering operations such as pumping for water adjacent to springs already in use. Again, as springs often accompany instability of ground and especially of steep natural slopes, their course may often have to be traced in drainage operations. Sketch *E* (Fig. 17.3) represents in simple fashion the geological arrangement

FIG. 17.5.—Thousand Springs, north side of Snake River Canyon, Idaho. The water issues from the open-textured basal portion of a lava sheet and flows over compact clay and volcanic lapilli.

(*Photograph by I. C. Russell. Reproduced by permission of the Director, U. S. Geological Survey.*)

necessary for the existence of steadily flowing springs; if the principle therein presented is applied to local conditions, the nature of many springs will readily be made clear. The flow may vary; but except in very dry spells, a spring will not dry up because of the relatively slow movement of groundwater. This slow movement when related to rainfall explains the lag between major movements of the water table and heavy rainfall, a lag that averages between two and three months in temperate climates and on which depends the real efficiency of groundwater supplies.

Although not numerous, springs that flow intermittently have an interest all their own which may excuse a brief reference to them here.

Sketch *F* shows the basis of one type of intermittent spring. If the surface level of the groundwater drops below *X*, as it might through depletion elsewhere during dry weather, the spring previously existing will disappear, returning only when the groundwater has been so replenished that the level again rises above *X*. The principle illustrated may again be applied to local geological conditions and thus explain natural phenomena such as the "breaking out of the bournes" in early spring in England after spring rains, *bourne* being a name applied in southern England to this type of spring. They are a feature of the chalk uplands and also of the oölitic limestones of the Cotswold hills, in England; they occur frequently also in northern France. Names of villages in England betray their existence; *Winter-*

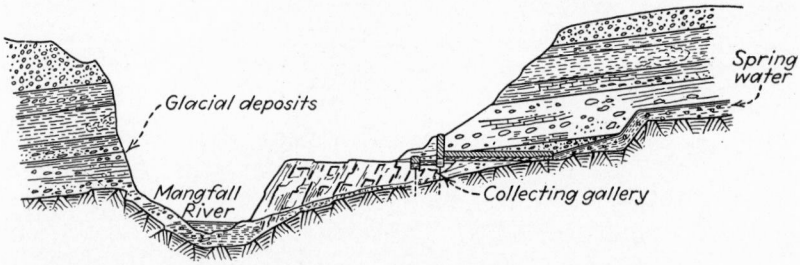

FIG. 17.6.—Typical geological section showing method of collecting spring water in the Mangfall River valley for the water supply of Munich, Germany.
(Reproduced by permission of Dr. Ing. E. Marquardt, Munich, Germany.)

bourne Zelston and *Waterbourne Whitchurch* are two delightful examples from the county of Dorset. Many springs discharge into the sea owing to local geological conditions, as could well happen in the case given in sketch *G*. Dr. F. Dixey reports that in the Persian Gulf, natives regularly dive to the sea bottom with leather bags which they fill with drinking water from spring discharges.[17.14]

Many small communities obtain their domestic water supplies from springs, but so also do some larger centers of population, notably in the south of Scotland and in the Whitehaven district of the northwest of England. One of the most important water supply installations dependent on springs is also reported to be one of the most perfect in the world—that of Munich, Germany. The works lie in the Mangfall River valley, the river flowing out of the Gegemsee, a large lake at the foot of the Bavarian Alps. The water flowing from the lake does not follow the river bed but finds its way in great quantities through deposits of gravel, joining the river on its flow farther down the valley. As it leaves the valley deposits, it is interrupted and collected for diversion to the city aqueducts.

17.8. Groundwater Surveys.—The effective use of the vast quantity of water stored underground is naturally dependent, in the first instance, on accurate knowledge as to where groundwater may be encountered by excavating or boring operations from ground level and, secondly, on what quantity of the water present in the strata can be economically obtained for the use of man. Provided that an adequate supply is found, data are also necessary as to the quality of the water and thus as to the possibility of treatment being necessary. The determination of these facts is the object of groundwater surveys.

FIG. 17.7.—Attendant's house at the portal of one of the adits for collecting spring water for the supply of Munich, Germany.
(*Reproduced by permission of Dr. Ing. E. Marquardt, Munich, Germany.*)

They cannot be considered as directly comparable to topographical surveys, as they are most complex and generally call for trained and expert hydrological advice. Local geological conditions are always involved, and so a working knowledge of local geology is a prerequisite for any groundwater investigation. In addition, familiarity with the hydrological characteristics of the rocks met with is essential before even general principles can be applied to the particular problem of the locality. Thus it is that for the carrying out of a complete groundwater survey, the services of an expert are always desirable and often essential. The following note is an outline only of what is involved in this survey work, presented so that engineers engaged in groundwater problems may know how to regard them and may appreciate the advisability of expert assistance when the necessity arises.

The problem is not merely that of "finding water," as is often and popularly imagined, for there are probably few parts of the earth's surface under which water could not be found at some depth. The engineer's problem is to find water at such a depth, in such quantities, and of such quality that it can be utilized economically in the public service. Despite this economic limitation, finding water is an important part of groundwater investigations. It seems probable that plants and other vegetation were the first indicators of groundwater to be used regularly; they still constitute a fairly reliable guide. Vitruvius lists plants known to him that indicate the presence of groundwater near the surface of the ground and even endeavors to suggest when they are reliable. The subject is one that has received little concerted attention until comparatively recent years. Scattered references to water-indicating plants, shrubs, and trees occur in geological and botanical literature, but it has been left to the United States Geological Survey to publish a general survey of existing knowledge, prepared by Dr. O. E. Meinzer, who has suggested the term *phreatophytes* to denote those plants which habitually grow in arid regions only when they can send their roots down to the water table.[17.15] This publication shows that the nature of vegetation encountered will be some indication of the depth to the water table, and it shows that there is at least some basis for the idea that plants will also indicate the quality of the groundwater. In other parts of the world, local information can generally be obtained from observant inhabitants with regard to plants that may serve as good guides.

The search for water can be likened to prospecting for minerals, water being, in a sense, a mineral, although of a special kind. The two branches of investigation have sometimes been closely related, as when Vogt in his water investigations discovered one of the world's richest potassium salt deposits at Mulhouse, Alsace. They have also been associated throughout a long period of time, in that both have been the object of attention by "diviners," those gifted individuals who are said to have the power of being influenced by a divining rod when they carry this over a deposit of mineral or groundwater. Divining rods have been used not only for the alleged locating of water and minerals but also for attempting to find hidden treasure and even criminals. They appear to have been used by the Scythians, the Medes, and the Persians; Marco Polo found them in use in the Orient; and there are those who believe that the rod of Moses was none other than a divining rod. Incantations were originally always used in conjunction with the rods, and it is said that in Cornwall, England, the belief still exists that the rod is guided to the deposit by pixies. For about a century (1669–1760), religious authorities debated as to whether water divining was an invention of the devil or not.[17.16]

Despite these mystical associations, the use of the *virgula divina*, as it was known to the ancients, has now been treated more realistically for some considerable time, probably the first scientific account being that in "De Re Metallica" by Georgius Agricola (1490–1555). Since that time, a constant stream of literary efforts has emerged from those who feel strongly about the matter; a recent publication lists almost 600 works published between 1532 and 1916.[17.16] There are now a special journal and a society in Germany devoted to the study and encouragement of the *Wünschelruthe* (wishing rod). There is a British Society of Dowsers, and there has even been held an international conference on *Rhabdomancy*, the modern name for the ancient belief. This conference was held at Verona on March 15, 1932; it was honored by the presence of a delegate of the Italian Ministry of Agriculture. One delegate is reported to have suggested that divining could aid in climatology and in "hydromineral therapeutics." The conference agreed that the principles of rhabdomancy are established but that there is some charlatanism; it was proposed to form a union with the idea of keeping out the charlatans.[17.17]

Few will deny that water diviners sometimes find water, but engineers generally will need much to convince them that this has not been a combination of the ordinary laws of chance, coupled with a good working knowledge of the principles of hydrology. To both engineer and geologist, actual field tests alone are of value; the evidence that they supply seems to be conclusive. In 1920, the Commission for Water Conservation and Irrigation of New South Wales, Australia, reported that in an official test of 56 bores at sites selected by diviners, only 70 per cent were successful; the tests then were stopped, as the divining rod was held to be discredited. Tests conducted at Sunbury, England, in 1925 by the (then) Anglo-Persian Oil Company with regard to "oil divining" were even more conclusive. In a sports field, where no oil had been buried, eight definite locations of oil were divined. When taken to within a few feet of the Company's oil store, the divining rod gave no indication at all of the presence of the several hundred gallons of light oils in the store.[17.18]

If it is remembered that, whether groundwater is located by a diviner or in some other way, its presence is due to the local geological structure in association with surface water conditions and the character of the local strata, all of which are capable of determination in a scientific manner, the attraction of rhabdomancy will be reduced. This possibility is mentioned quite seriously; even today, one may meet senior engineers who will employ water diviners while caring little for all that Geology can do for them. As recently as 1925, an important contract on a leading building of the city of London was "assisted by the services of a diviner." Naturally, it is not suggested that a divining rod is not affected in some way at some time when carried by some persons, but it is suggested that, as nobody can find water when the geological conditions are not favorable, recourse should be had to Geology for finding it and not to a method that, in 1883, was described by Mr. R. W. Raymond[17.16] as follows:

> Its claims to virtues derived from Deity, from Satan, from affinities and sympathies, from corpuscular effluvia, from electrical currents, from passive perturbatory qualities of organo-electric force are hopelessly collapsed and discarded. A whole library of learned rubbish about it which remains to us furnishes jargon for charlatans, marvellous tales for fools, and amusement for antiquarians; otherwise it is only fit to constitute part of Mr. Caxton's "History of human error."

To enumerate the geological studies by which groundwater may be investigated would be but to repeat much of what has already been outlined with regard to geological mapping in general, since, although groundwater is being investigated, accurate knowledge of the geological structure is first necessary. In addition to the usual investigations, however, close attention must be paid to all evidences of groundwater. Records of all existing wells and boreholes must be examined and correlated. By this means, when records are available, it is possible to prepare contour maps for groundwater levels over large areas of country. In England, this has been done in Dorset by Mr. Baldwin-Latham, and over a considerable part of the Midlands by Mr. R. G. S. Walters.[17.19]

Contour maps provide one means of estimating the quantity of groundwater available, this quantitative work being the other main branch of groundwater survey work. Various methods are available; their relative suitability is dependent on local conditions and the data available. One method is to attempt to prepare a water balance sheet for the area considered, measuring the total runoff and rain, estimating the losses due to evaporation and absorption by plants, and determining the remaining balance which may normally be assumed to be the addition to the groundwater supply. Measurements may be also made with regard to actual discharge of groundwater into watercourses from areas in which the water table is known to be reasonably steady; a time lag will always be found between the falling of rain and increase in groundwater flow; but with this determined, estimates can be made as to the percentage of the rainfall available as groundwater. A good example of a survey somewhat on these lines is available in connection with the valley of the Pomperaug River, Connecticut, United States, having a catchment area of 89 sq. miles.[17.20] The most reliable method and that most widely employed is to determine and use the *specific yield* factor for the rock in question, the one serious obstacle to the application of this method being possible variations in the nature of the local rocks, which may invalidate the calculations to some extent.

By specific yield is meant the "interstitial space that is emptied when the water table declines, expressed as a percentage of the total volume of material that is unwatered."[17.21] For the measurement of specific yield, various methods have been suggested some of which can be utilized in a laboratory using samples of the materials encountered in the area under investigation, others necessitating field observations. Laboratory methods are convenient and interesting, but the possible variations not only of the materials tested but also of their structural arrangement in the area being studied make these methods of doubtful value unless unusual precautions are taken. On the other hand, they can often provide a useful guide to probable yield of groundwater before major operations can be undertaken in the field, since the most effective field methods for estimating specific yield depend on the recording of water levels underground during pumping or recharging. These observations can be made only after boreholes have been sunk to required levels. Groundwater levels will always be the most useful guide to groundwater supplies, since inevitably they take into consideration all the subsurface variations that may possibly be disregarded in the application of any experimental test results. Even in the use of recorded water levels, however, it must always be remembered that observations from adjoining boreholes and wells should be carefully correlated with each other and with what is known of geological conditions in the vicinity, so that any underground irregularity may be detected if this is possible from the data available.

To select any one example to illustrate the wide field covered by this branch of groundwater study involves a difficult choice. Possibly a record of one of William Smith's many engineering activities in England may be cited, in view of its historic interest and of its intrinsic illustrative value. In a letter dated February 24, 1808, from John Farey to Sir Joseph Banks the following descriptive note has been found. This is quoted in "William Smith: His Maps and Memoirs," a most interesting volume compiled by Mr. T. Sheppard, Director of the Municipal Museums, Hull, England, as a result of many years of research.

. . . we met the Reverend Mr. Le Mesurer, Rector of Newton-Longville near Fenny-Stratford, who related his having undertook to sink a well, at his parsonage house, within a mile or two of which, no good and plentiful springs of water were known, but finding clay only at the depth of more than 100 feet, was about to abandon the design: Mr. Smith, on looking into his map of the strata, pointed out to us, that Newton-Longville stood upon some part of the clunch clay strata, and that the Bedford limestone appeared in the Ouse river below Buckingham, distant about eight miles in a north-west direction, and he assured Mr. L. that if he would but persevere, to which no serious obstacles would present themselves, because all his sinkings would be in dry clay, he would certainly reach this limestone, and have plenty of good water, rising very near to the surface; Mr. L. accordingly did persevere in sinking and bricking his well, and at 235 feet beneath the surface (the first 80 feet of which were in alluvial clay with chalk and flints, etc., similar exactly to what I have uniformly found on your estate at Revesby, and in the bottoms of many of your fen drains) the upper limestone rock (8 feet thick) was reached and found to be so closely enveloped in strong blue clay as to produce not more than 9 feet of water in the well in the course of a night; from hence, an augur hole was bored in blue clay, for some distance, to the second limestone rock, which produced a plentiful jet of water, which filled and has ever since maintained the water, I believe almost up to the surface of the ground. . . .

17.9. Underground Dams.—When the presence of adequate groundwater at an economical depth has been ascertained with certainty, there will arise the problem of devising means for drawing upon it for the intended supply. This may be done in several ways, the use of wells and boreholes being the most common. It is sometimes possible to adopt an even simpler means when the groundwater is known to be immediately under the surface and moving in a definite direction. It is then possible to construct an underground storage dam provided that under the pervious water-bearing stratum is an impervious bed into which the dam may be keyed to provide a watertight joint. It may be thought that such a combination of fortuitous circumstances would be so rare as to be of negligible importance. This is not the case, probably because the geological characteristics of buried river valleys often coincide with the conditions requisite for this type of water storage scheme. Thus it is that in several countries, including Germany, South-West Africa, and Brazil, are to be found numerous examples of underground dams and that from the underground reservoirs thus formed public water supplies are drawn.

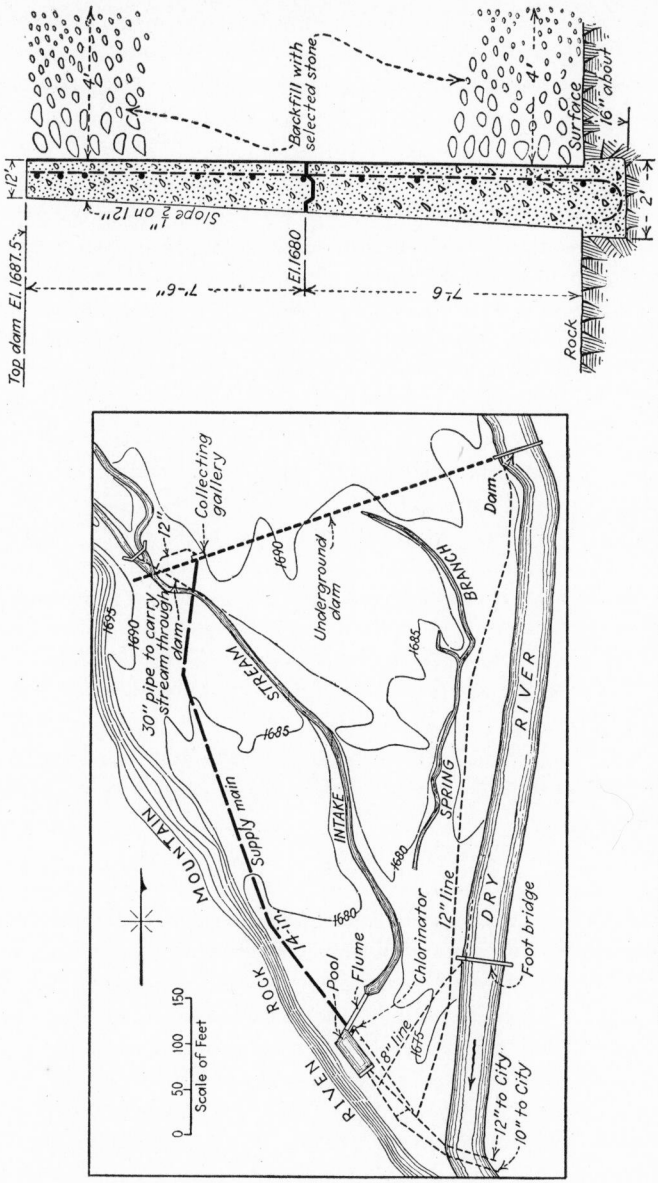

FIG. 17.8—Details of underground dam for the water supply of Harrisonburg, Virginia.
(*Reproduced by permission of Mr. A. B. McDaniel, and the Editor, Engineering News-Record, from Reference* 17.22.)

As an example may be mentioned an interesting recent installation in Virginia. The city of Harrisonburg was led to the consideration of augmenting its water supply system on account of a water shortage during the great drought of 1930. Surveys and investigations finally led to the conclusion that this could be obtained from groundwater flow in the valley of the Dry River from the normal stream flow of which the city already drew its main supply. It was found possible to extend an existing concrete diversion dam across the river bed for a distance of about 900 ft. through a relatively level part of the bed of the valley as far as a steep rock cliff. This necessitated a thin reinforced concrete wall, varying from 12 to 24 in. in thickness and keyed into the underlying fine-grained closely cemented Pocono sandstone rock for a distance of 16 in. The entire wall was backfilled on the upstream side with selected stone, obtained to some extent from excavation. At the location of the lowest level of the rock, a suitable concrete collecting gallery was constructed upstream of the dam, from which a 14-in. supply main leads to the city. The valley bed consists of products of disintegration and erosion varying in size from fine sand to large boulders; and in this deposit, 12 underground streams were encountered in definite courses, carrying such an amount of water that it caused difficulties during excavation. This flow is now permanently collected in a perforated concrete pipe laid under the stone backfilling and connecting with the collecting gallery. An average flow of about 700 (U. S.) g.p.m. is obtained from this unique installation, which has performed satisfactorily since its completion in 1934.[17.22]

17.10. Wells and Boreholes.—Wells and boreholes constitute the principal means of obtaining for use a supply of water from underground sources. They may be considered together, since they are generally alike in principle and action. The nature of the excavation is a matter of convenience; and in consequence, wells are generally used for small depths, and boreholes for greater depths. Unless passing through very stable strata, *wells* (using this term to denote both types of hole, as will be done generally throughout this chapter) will be cased in some way throughout the depth of all strata apart from that from which the supply is to be obtained. Details of construction cannot here be given, but occasionaly wire screens are used to protect that part of the well into which the water is being drawn. Another common precaution is to fill the excavated hole with specially selected coarse gravel through which the discharge pipe is led. This may decrease the yield of the well by about 5 per cent, but it will prevent plugging.

In the case of wells penetrating strata having a low specific yield, it may be necessary to attempt to secure a greater exposed surface in the pervious strata than is given around the perimeter of the well shaft. This is easily effected by the driving of adits into the strata from suitable locations in the well. This will generally be a wet construction operation; but it has the advantage that as soon as water problems become too troublesome, the object of the adit is achieved. The Romans are known to have used adits in wells that they excavated in Great Britain; today, adits are quite common in wells in the chalk of southern England and in the Bunter sandstones. In the Canary Islands and Hawaii, adits alone are sometimes used as water-collecting galleries, driven into steeply sloping lava beds in order to obtain the water that seeps through this porous rock. The *Karez* of northern India and the *Keghriz* of Iran (Persia) are primitive types of a special kind of adit of considerable interest but restricted application. The planning and building of these unusual collecting galleries appears to be a matter of intuition rather than calculation, and the work is generally hereditary. Russian engineers working in the Caucasus have tried to construct *keghriz* scientifically but without success.[17.23]

Fortunately, the ordinary type of well installation does not call for anything beyond scientific skill. When once the excavation has tapped the water-yielding stratum, a body of relatively free water will be available which can be drawn upon as required. The level of this water may vary from time to time quite apart from the variation due to pumping. When the water table is very near the surface, barometric pressure may affect the level, and even temperature has been found to have some effect due to the capillary properties of the strata above being affected by a change in temperature. Water levels in adjacent wells may not coincide if there is a marked change in geological conditions between them. A striking example occurred at St. Andrews, Scotland. A farm drew its supply from a borehole 28 ft. deep in sands and gravels. An architect sank a well to supply some new cottages, the site of the well being 60 ft. from the borehole; but although the well was sunk to a depth of 30 ft. in the gravel, only a trace of water was found, owing in all probability to variation in level of the underlying impervious boulder clay floor.[17.24]

When pumping is started in a well that has been satisfactorily completed to a free water surface, the water level will naturally fall and continue to do so until the hydraulic gradient in the surrounding material is such that flow will take place in the well to equalize the amount withdrawn. It can be shown theoretically that for uniform material, the curve of the gradient is a parabola. As this will be the same all around the well, the phenomenon known as the *cone of depression* is obtained. This phase of hydrodynamics has been studied extensively in theory, although it is not easy to verify in practice. The intersection of a cone of depression with the water table will trace out a circle which will mark the range of influence of the pumping operation. It will be clear that any other well located within this range will be affected, its effective yield being interfered with to some extent. Such interference is a critical matter in built-up city areas where groundwater resources are being taxed to the limit. It is surprising to note how wide these ranges of influence may be in porous strata. In Liverpool, England, the effect of pumping has been noticed as far as 2 miles away from the well in which pumping was taking place, the intervening ground being generally pervious red sandstone.

The wells so far considered have been those which are usually encountered, having no unusual geological features. In addition, not only artesian wells but also a few special types of wells may be noted briefly, since, although they are not of common occurrence, they are often of great importance in some localities. There are, in the first place, shallow wells which depend for their supply on a *perched water table, i.e.,* a body of groundwater perched well above the zone of saturation by means of an impermeable stratum overlying the main body of porous material. Some striking examples are to be found on Long Island, New York, where a thin layer of almost impermeable till separates a pervious surface deposit from the main body of pervious strata, and the main water table is as much as 250 ft. below this perched water.[17.21]

Australia provides numerous examples of two other types of unusual wells. *Soak wells* is the name that has been given to the shallow wells used for obtaining groundwater that collects, after running off exposed sloping surfaces of impervious weathered granite, in the accumulation of decomposed or disintegrated granite at the foot of these slopes. The products of decomposition vary and often yield quartz grains in a matrix of kaolin; from such a deposit no water supply can be obtained. But if the borehole is continued, a water supply will often be obtained from the body of water trapped between the bottom of this deposit and the top of the completely fresh granite; the transitional space is often open enough to yield water itself and to serve as a drain for the overlying kaolin.

The second type of unusual well encountered in parts of Australia (and elsewhere) is that known as a *tray well.* This is really a case of tapping a localized body of perched groundwater, a body of water

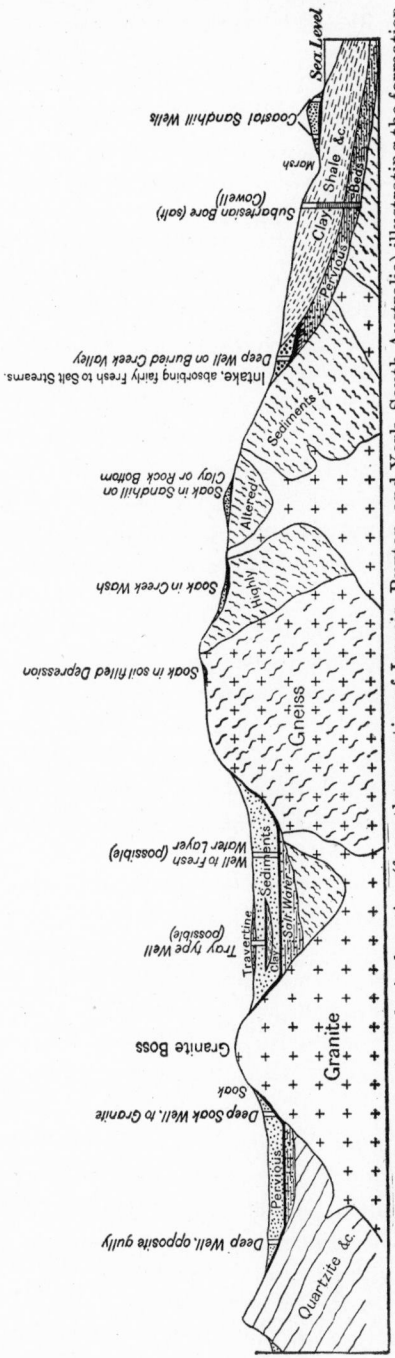

FIG. 17.9.—Ideal diagrammatic geological section (from the counties of Jervois, Buxton, and York, South Australia) illustrating the formation of tray wells, and the dependence of these and other groundwater features on local geological structure.

(Reproduced by permission of Dr. L. Keith Ward, Director of Mines, S. Australia, from reference 17.3.)

perched not on a continuous impermeable stratum but on a lenticular bed of impervious material often of limited extent. A good example is furnished by a well at Challner, where a saucer-shaped bed of impervious clay with some ironstone nodules about it, located halfway down through a deep stratum of sand which overlies solid granite, collects a small quantity of fresh water as it percolates through from the surface. This supply is tapped by the Challner Well; endeavors were made to increase the supply by boring further, with the result that the bed of clay was pierced, the fresh water starting to empty into the sand below. Fortunately, the well was saved by the plugging of the hole. Ten chains away from this well is another, known as Nugent's Well, the bore of which encountered no water above the thin clay layer, so that it had to be extended to the main body of groundwater overlying the granite directly.[17.25] This second well is of an unusual nature, also, since it cannot be extended right down to rock surface because the water immediately above the rock has a strong saline content, which makes it quite unsuitable for domestic use. The thin layer of fresh water that tops this salt water layer can alone be tapped for use.

17.11. Salt Water in Wells.—The presence of salt water in wells is so important that it calls for some detailed attention. Salt water may be present for two general reasons: (1) encroachment of sea water into wells located near the seacoast and (2) contamination of inland well waters with excessive mineral content. In both cases, as soon as the salt content passes a certain limit, the water ceases to be fit for human use. The following general limits have been suggested, based mainly on Australian practice:[17.26]

Ounces per Imperial Gallon	
½ (3,125 p.p.m.)	Can be used domestically
1 (6,250 p.p.m.)	Horses can live on this and keep in good condition
1¼ (7,812 p.p.m.)	Horses can live on this
1½ (9,375 p.p.m.)	Cattle can live on this
2½ (15,625 p.p.m.)	Sheep can live on this

The contamination of groundwater supplies with excessive salt content, such as is found in parts of Australia and in the Sudan, has in the past been thought to be due to the leaching out of decomposition products of overlying or adjacent rocks. Consideration of the chlorine content of samples investigated in South Australia disproved this supposition, pointing to the fact that saline content is due to insufficient drainage for carrying off the salts brought down by rain. Continued evaporation of water gradually concentrates these dilute salt solutions, with the results already indicated. (Decomposition products probably contribute to the actual salt content found.) As evidence of what rain water may contain may be mentioned the fact that for England the average chlorine content is about 2.2 p.p.m., equivalent to 3.62 parts of salt. Rain falling at Land's End during strong southwest winds, blowing in from the sea, has been found to contain one hundred times this amount.[17.4] As can be imagined, the problem is a complex one, further complicated by varying local conditions. In all such cases, the greatest care must be taken to prevent wells from being driven into the salt water layer and, in the case of correctly driven wells, to prevent the possibility of over-pumping.

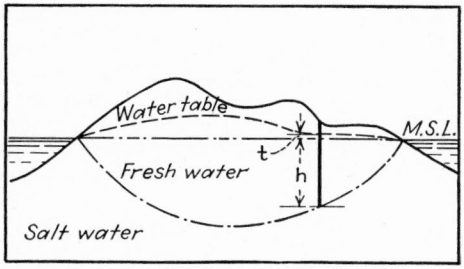

FIG. 17.10.—Section through Island of Norderney, Germany, (from Herzberg) illustrating the theory that $h = \frac{t}{g - 1}$, where g is specific gravity of salt water and h and t are as shown.

(Based, by permission of the Director, U. S. Geological Survey, on Fig. 2 of U.S.G.S. Water Supply Paper 537, A Study of Coastal Ground Water, by J. S. Brown, Washington, 1925.)

Overpumping is often responsible for the contamination by salt water of wells adjacent to the seacoast. Wells along the shores of the Thames Estuary and in the Liverpool district of England have been made brackish by this means, salt water becoming mixed to varying degrees with the fresh water because of excessive drawdown of the latter. The matter is one of widespread importance, and it has therefore been widely investigated, particularly in England, Holland, Japan, and North America. The existence of wells is indicative of the existence of water-bearing strata; and if these are adjacent to the sea, then it will normally follow that at some point the fresh groundwater will be in contact with infiltrated sea water. It has been found that the zone of diffusion between the two is relatively small, so that any appreciable change from the normal position of the groundwater (such as might be caused by excessive pumping) will result in a corresponding change in the position of the salt water. As the average salt content of sea water is about 35,000 p.p.m., it will be clear that it takes but a small contamination of the fresh water being pumped to render it useless for domestic use.

In the ideal case of an area of land consisting of uniformly pervious material surrounded by sea water, a state of equilibrium between fresh and salt water can readily be visualized. This has been investigated, theoretically and practically, by Herzberg in Germany (about 1900) and by other investigators; the law governing the equilibrium is often referred to as *Herzberg's theory*. Although the theory treats of an ideal condition, it has been verified to a remarkable degree in the sandy coastal lands of Holland. The principle that it embodies can be applied in some form to all geological conditions actually encountered in coastal lands, at least in a general way, and so assist in the prevention of serious trouble from this cause.[17.27]

17.12. Artesian Wells.—This type of well has been left for special comment, in view of its unusual geological interest and of its importance in water supply. Its name is one of the few in connection with subsurface conditions that is not self-descriptive; it originated from that of the town of Artois, France. A well is said to be *artesian* when, because of local geological structure, the groundwater rises either up to or above ground level as soon as the water-bearing bed is pierced, the water being stored in the bed under hydrostatic pressure. The term *subartesian* is used to denote wells in which the groundwater is under hydrostatic pressure but not to such an extent that it is forced to ground level, coming to rest as soon as it is freed at some point below ground level but above the top level of the water-bearing bed. The usual cause of artesian pressure is the hydrostatic head of that part of the body of groundwater which is confined above the level at which it is tapped by means of impervious strata. This condition may arise in the case of a synclinal structure of alternating pervious and impervious beds, as illustrated in Fig. 17.11; two alternative structural arrangements are illustrated in the same figure, and other similar conditions can be imagined.

There are many well-known examples of artesian waters, but naturally the water encountered at Artois may be said to be the leading example. The groundwater in the well-known chalk deposits there does not find its natural level until the overlying Tertiary clay is pierced. Despite the general adoption of the name, the use of artesian water was revived rather than initiated at Artois, since it had been extensively used in earlier periods, particularly in northern Africa. At Thebes, shafts were sunk, and excavation continued as bores 6 to 8 in. in diameter to water-bearing sands, in many cases 400 ft. from the surface.[17.4]

Artesian conditions may exist in the form of artesian slopes or artesian basins. In both cases, but especially in the former, artesian water may be present at depths to which it is uneconomical to drill in order to tap the water. Local conditions and demands will determine the relative economy of using such supplies, and the inevitable variation of conditions will explain the great variation in the depths to which artesian wells have been sunk. In Berlin, Germany, and St. Louis, Missouri, depths up to 4,000 ft. have been necessary and have actually been drilled; in England, a well 1,585 ft. deep at Ottershaw appears to be about the maximum depth utilized; in Australia, depths of over 4,000 ft. are used, a well at Boronga, New South Wales,

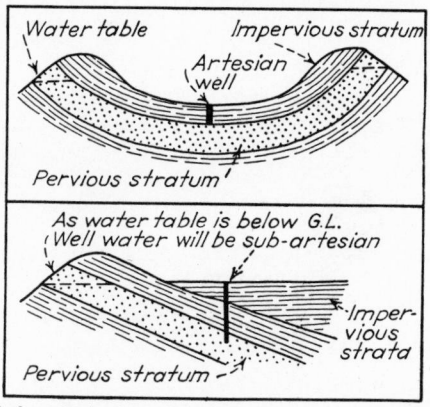

FIG. 17.11.—Simplified geological sections illustrating artesian groundwater conditions.

being 4,338 ft. deep; whereas along the Atlantic seaboard of the United States, artesian supplies at depths of a few hundred feet are relatively common. One famous well on this coast is the 12-in. well at St. Augustine, Florida, which originally supplied 10,000,000 (U. S.) gal. a day from a depth of 1,400 ft.

It will be appreciated that the recognition of the existence of artesian conditions is essentially a geological problem, to be solved from knowledge of the local geological formation. Deductions from such data need not be discussed again, but reference may be made to the example quoted on page 452 from the records of William Smith's work which deals with an interesting example of subartesian conditions.

Although the areas in which artesian water can be tapped are often relatively small, there are, on the other hand, some extensive areas under which groundwater is stored under pressure. One of the best known of these is the Great Dakota Artesian Basin, located in North and South Dakota, with an area of about 15,000 sq. miles. It is widely used as a source of water supply, the water-bearing bed being Dakota sandstone, and it has been the subject of detailed groundwater investigation. Particular attention has been paid to the possibility that some, if not all, of the artesian pressure may be due to the weight of the superincumbent rock strata. This is still a subject of debate in geological circles. It may, perhaps, be noted that there are records of observed changes in water pressure due to the passage of railway trains close to a well under observation. It is of interest also to note that the pressure due to the weight of rock was discussed by

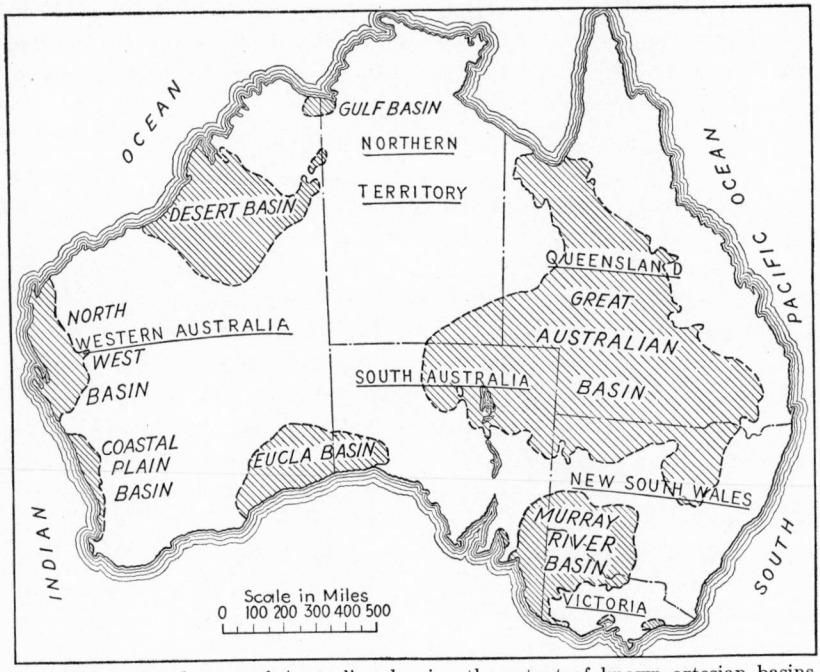

FIG. 17.12.—Sketch map of Australia, showing the extent of known artesian basins. (*For information on these basins, see reference* 17.28.)

Thales about 650 B.C. and also by Pliny, and was suggested as the agent elevating sea water to the level of springs.

Australia is a continent in which artesian water is of major importance, the Great Australian Artesian Basin being almost worthy of classification as one of the wonders of the world. Its total area is over 660,000 sq. miles, and it is located in the states of South Australia, Queensland, and New South Wales. Over 3,000 artesian wells draw on the supply that it provides, yielding a daily flow of over 500,000,000 (Imperial) gal. In form, the basin approximates the ideal type of a complete artesian basin; the eastern rim is tilted up to a greater altitude than the western. The main aquifer extends as a practically continuous body under the whole of the basin and consists, in general, of Jurassic sands, occurring as very soft sandstones. Artesian conditions proper extend over the greater part of the basin; in the western section, the head is sufficient to give rise only to subartesian conditions. The water obtained is of good quality, but its temperature is notably high, being extremely high in the deeper parts which reach a depth of between 5,000 and 6,000 ft. Experiments in New South Wales have shown that the temperature varies almost directly with depth, records ranging from a temperature of 86°F. at 700 ft. deep to 122°F. at 3,000 ft. deep. From among the many notable records obtained from this great basin may be mentioned the drilling of an experimental well at Careunga, near Moree, New South Wales, to a depth of 4,014 ft. In this depth, three separate artesian flows were encountered, at different pressures, with different compositions, and with different emissions of methane gas; records of all three flows were kept separately. There are several other large and important artesian basins in Australia, the most notable of which is the Murray River Basin, lying in the states of South Australia, New South Wales, and Victoria and having a total area of 107,250 sq. miles. It is of special interest to engineers to note that the conservation of these great natural reservoirs of groundwater has been recognized in Australia as overriding ordinary state boundaries. In consequence, a series of regular interstate conferences has been held, attended by geological and engineering delegates from most of the states concerned, beginning in 1912, at which records have been exchanged, experiences discussed, and general conclusions arrived at with regard to many problems involved. The printed records of the conferences, together with the relevant state geological bulletins, constitute a most valuable part of groundwater literature.[17.28]

17.13. Some Outstanding Groundwater Supply Systems.—The following notes describe three outstanding water supply systems dependent on groundwater. The examples have been selected because of their unusual features from the numerous major water supply systems of this type throughout the world. Some of these, such as that provided by the great London Basin in England, are so well known that they have been passed over in favor of the following lesser known installations.

1. *Long Island, New York.*—Long Island constitutes the coastal plain of the states of New York and Connecticut, separated from the mainland by Long Island Sound and isolated geologically from it with respect to water-bearing beds. The island supports a population of over four million persons, and in recent years a large part of its

water supply has been obtained from groundwater. Crystalline bedrock outcrops in some of the western parts of the island, the rock surface dipping to the east so that in places it is at least 2,000 ft. below the surface. Overlying the rock is a great deposit of stratified Cretaceous sands, clays, and gravels, the dip of the strata decreasing toward the surface. At the surface, stratified sand and gravel drift deposits and two terminal moraines give rise to a fairly regular surface topography; the highest ground elevation is about 420 ft. above the sea. These unconsolidated deposits constitute the great underground reservoir from which has been drawn probably the greatest concentrated groundwater supply utilized in North America. As the unconsolidated

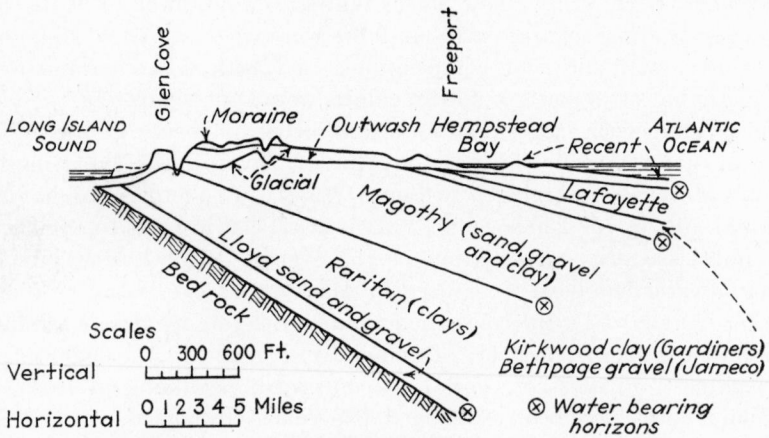

FIG. 17.13.—Geological section through Long Island, New York, U.S.A., indicating water-bearing horizons.

(Based on work of Prof. W. O. Crosby; from Mr. R. Suter. Reproduced by permission of the Division of Water Power and Control, Conservation Dept., State of New York, Mr. R. Suter Executive Engineer.)

strata do not cross the Sound to the mainland, the groundwater underlying the Island is obtained only from the rain falling on it and consequently the relation of this body of fresh water with the surrounding sea water is a delicate one, providing an unusually good example of the state of balance described on page 460. The rainfall averages about 42 in. per year, of which possibly one half reaches the zone of saturation. Opinions differ as to how much of this water can be obtained for use; but from the records available, it is known that over 200,000,000 (U. S.) gal. per day have been pumped for domestic and industrial use. Unfortunately, this pumping has been concentrated in developed areas, with the result that a serious situation with regard to saline contamination now exists with regard to many of the main wells. Under the older part of Brooklyn, the water table has dropped to El. 40.0, and most of the water under Brooklyn is now unfit for drinking purposes. The situation is complicated by the fact that as the area has been developed, the construction of paved areas and sewerage systems to carry off the rain water to the sea has been reducing the groundwater recharge. As is sometimes the case, the present problem is more a legislative one than an engineering matter, but the development and use of this great source of water supply constitutes one of the best examples of groundwater utilization on a large scale.[17.29]

FIG. 17.14.—A part of the area providing Southern California's great underground storage basin, the photo showing the apex of the detrital cone of the San Gabriel River as it emerges into the South Coastal Plain.
(*Photograph by Spence Air Photos.*)

2. *South Coastal Basin of California.*—Southern California is well known in many parts of the world for reasons other than engineering. What is not so well known is that a large part of the water supply to this area is obtained in a singularly interesting manner from underground sources. The South Coastal Basin of California comprises the Los Angeles metropolitan area and territory to the east, mainly agricultural but including 26 incorporated cities, the total population being about 2,500,000. The area consists essentially of a coastal plain, separated from the desert hinterland by three ranges of mountains

generally parallel with the coast. In these mountains rise three major streams, the Los Angeles, the San Gabriel, and the Santa Anna, the valleys of each of which are well defined. Geologically, the area is most complex, with the result that there are 37 fairly distinct basins, although the boundaries of these are not always evident at the surface, many being fault planes. The slope of the valleys is quite steep, but that of the coastal plain is relatively flat; the junctions of valleys and plain provide almost perfect examples of discharge cones, as the accompanying photograph well shows. Rainfall in the mountains is high but irregular, with the result that the discharge of the streams is "flashy." Their courses through the coastal plain are often dry, the dry-weather flow percolating into their beds to replenish the groundwater stored in the unconsolidated deposits forming the plain. These are varied in the extreme although generally coarse. Faulting has complicated their distribution, but the area as a whole constitutes a great underground reservoir. It has been calculated that its capacity, in the 100-ft. layer immediately below outflow level, is no less than 7,000,000 acre-ft., of which only about 60 per cent has so far been used. The total surface area of the basins is 1,305 sq. miles, and the average specific yield of the unconsolidated strata is 8.4 per cent by volume. Despite the high cost of pumping, the groundwater is used not only for domestic purposes but also for agricultural irrigation. So great has been the demand that between 1916 and 1935 the average drop in water level in the entire San Gabriel Valley (which is said to be fairly representative) has been 60 ft. In other locations in the basin, the drop has been considerably greater. It is to augment this supply that some of the great aqueducts already mentioned have been constructed.[17,30]

3. *Nassau, Bahamas, British West Indies.*—The Bahamas are very interesting islands, geologically and in other ways; they are flat and low lying, composed almost entirely of debris derived from corals and calcareous organisms, and resting on a shallow submerged platform, which is separated by deep submarine troughs from the neighboring land masses of North America and the West Indies. The sand derived from the corals, together with infiltrated calcium carbonate, has formed as the local rock a very soft and porous limestone which weathers quickly. Many of the low-lying areas have large brackish ponds in them, the water level of which rises and falls with the tides, being connected to the sea by means of underground passages, the salt and fresh water being in a state of balance. New Providence, the island on which is located Nassau, the capital city of the Bahamas, is only 21 miles across from west to east and 7 miles from north to

south. Its topography is featured by two ranges of low hills which nowhere rise more than 100 ft. above sea level. Water supply was obtained solely from shallow wells and the flow off roofs until the year 1933. A serious outbreak of typhoid fever in 1926–1927 caused reconsideration of the whole question of the island water supply. After exhaustive surveys, and despite the unfavorable local geological conditions, an area was found that would yield fresh water in quantities sufficient for the desired supply without danger of overpumping, provided that the wells installed were not carried too low, as had often been the case in the past. This area measures about 3,300 by 2,000 ft. Sixty-five wells were sunk to the necessary elevation to reach the

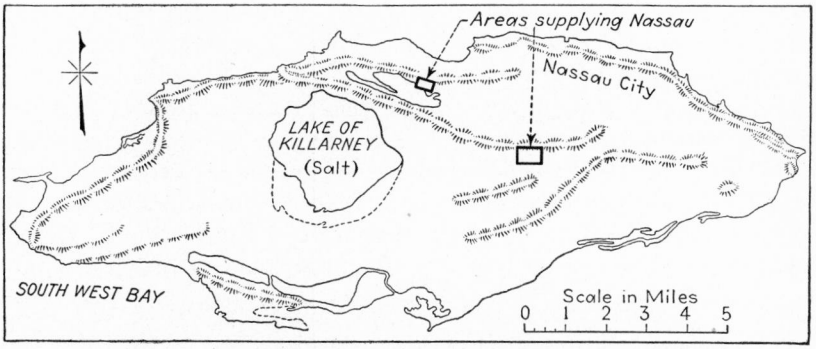

FIG. 17.15.—Sketch map of the island of New Providence, Bahamas, showing areas supplying Nassau with fresh water.
(*Reproduced by permission of Mr. F. C. Van Zeylen, Director of Public Works, Bahamas.*)

fresh water layer, generally at intervals of about 300 ft. and staggered in plan. The depth to fresh water level varied up to 40 ft., but many of the wells were much shallower than this, pumping arrangements being designed in accordance with the various lifts. The water obtained from this ingenious installation is calculated to be enough to supply a population of 15,000; it is hard (230 p.p.m.) and so has to be softened before use; it is also chlorinated as a safeguard to public health.[17.31]

17.14. Underground and Surface Drainage.—Underground drainage is often a reciprocal operation, for the water drained from one area must be disposed of elsewhere and may constitute a replenishment of the water supply for the latter area. Thus, what is believed to be meteoric water which appears as flow from coal seams in the Rishton Colliery, Lancashire, England, is pumped out of the mine and, by means of a suitable connecting system, augments the public water supply of the town of Accrington.[17.32]

The drainage of underground workings is always dependent on groundwater conditions and the position of the water table in relation to the workings and to water-bearing beds. With these essential data available, methods can be determined for the abstraction of water from that part of the ground in which work has to proceed. There are many instances of successful drainage operations of this type. One famous case—that which developed during the construction of the Kilsby Tunnel in England—has already been described. Another example may also be mentioned. The lead and zinc mines of northeast Wales, bordering on the estuary of the river Dee, which have been worked for many years, have always been difficult to work because of the presence of water in the surrounding strata, which is a "block

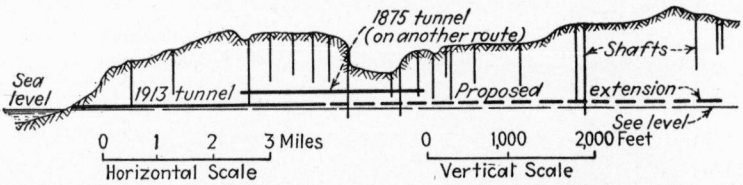

FIG. 17.16.—Cross section through the drainage tunnels to the Halkyn Mines, North Wales.

(*Based, by permission of the Institute of Mining and Metallurgy, London, on a drawing in reference* 17.33.)

and checkerboard" arrangement of limestone, with many cross veins providing a free flow of groundwater. As early as the year 1818, the mines were drained by a tunnel known as the Halkyn (deep level) Tunnel, which started in a hillside at Nant Flint, at an elevation of 200 ft. above Ordnance Datum, being driven thence in a southerly direction eventually for a distance of 5 miles. After most of the inaccessible ore had been removed from the space thus drained, a new tunnel, the Milwr, or Holywell-Halkyn, Tunnel, was started in 1896 from the banks of the Dee estuary, 190 ft. below the level of the older tunnel. In a similar way, it drained a large volume of the limestone, thus permitting a great extension of mining operations. Figure 17.16 illustrates this interesting scheme.[17.33]

The dependence of all surface drainage on the groundwater conditions at the locality in question has been made evident by examples already quoted. Clearly, if ground is to be effectively drained, the water table must be at such a level that gravitational flow will automatically lead excess water away from ground surface. Where this is not the case, artificial measures to achieve the result must be adopted. An interesting example is on record from experience in Arizona. In the Salt River Valley of that state, 203,000 acres of improved farm lands were threatened about the year 1923 with serious water-logging due to excessive irrigation. The surface materials consisted of sandy clay, loam, and loess, underlain by clay and caliche sloping gently toward a river channel. Drainage by means of open channels proved to be ineffective owing to the nature of the surface materials, with the result that the water table rose 1.4 ft. per year after irrigation started. Sixty-four thousand acres soon had groundwater within 10 ft. of the surface. After study of all local conditions, especially the geological formation, pumping was decided upon; this was an economic possibility due to the availability of cheap offpeak power. It was started in 1923, and the area of water-logged ground was quickly reduced.[17.34]

Surface drainage is sometimes dealt with by allowing it to flow down specially excavated bores (sometimes called *dumb wells*) into pervious strata located below impervious beds near the surface. This method has been used, with varying degrees of success, in many districts. In South Australia, a swamp (suitably named *Dismal*) was drained, with partial success, by boring dumb wells from the surface through to the underlying limestone strata. Dumb wells have been used in Lincolnshire, England, for draining from the upper estuarine clays into the Lincolnshire limestone. During the construction of the tube railways of London, through the London clay, at least one contractor bored a dumb well and drained the water collected during construction into the underlying chalk, the water table being well below the level of the tube tunnel.

17.15. Replenishment of Groundwater.—In addition to their use as drainage outlets, dumb wells are also used in some places as a means of replenishing groundwater supplies that are already in use but that are taxed unduly. This expedient was tried in the Lea Valley, England, in connection with the water supply of London, filtered water being used for the replenishment, and some success was attained, although the process does not appear to have been continued. Naturally, for such a course to be successful, the well must be in a region where the water table is at some distance below ground level so as to obtain the necessary head for the disposal of the water, and the water used must be reasonably clear so that it will not block up pores of the material through which it flows as it enters the ground.

Another method of groundwater replenishment that has been adopted on a larger scale, with much success, is akin to those river supply systems in which the river water is pumped from galleries paralleling the river course so as to use natural filtration. In Southern California, for example, in the San Fernando Valley, spreading grounds have been created by means of which water is fed into the ground, thus artificially replenishing and safeguarding one third of the groundwater supply pumped for the city of Los Angeles. The work was begun in 1931, as an experiment; but after 5 years' successful operation, the system took on a definite character. At the end of this period, 47 out of a possible 180 acres were in use for this purpose, through which 106 cu. ft. of water per second were fed to the valley bed. Basins 100 by 400 ft. are separated by dykes 4 ft. high and 15 ft. wide. To each basin a 6-in. pipe supplies water to a depth of 6 in. during filling operations. The water is then left to stand and seep into the ground; this takes between 8 and 24 hr., depending on the state of the beds

FIG. 17.17.—Replenishing groundwater for Los Angeles: one of the spreading basins of the Tujunga spreading grounds.

(Reproduced by permission of the Dept. of Water and Power, City of Los Angeles, California, U.S.A., Mr. H. A. Van Norman, Chief Engineer and General Manager.)

which have to be harrowed and cleaned of silt at intervals of 10 days. Percolation goes on at the rate of between 3 and 10 ft. per day, depending on the state of the beds. The water eventually reaches the North Hollywood Pumping Station, some distance down the valley, the following year and is here pumped out for use through eighteen 20-in. wells 400 ft. deep.[17.35]

17.16. Conservation of Groundwater.—The apparently unlimited supply that is popularly supposed to be available whenever a body of groundwater is tapped, either by a well adapted for pumping or by an artesian bore, has led in many cases to serious abuse of the great natural resource thus provided. In other instances, even though the limitations of the source of supply have been generally realized, lack of regulation of private operations has led to serious depletion of groundwater resources. Overpumping has already been mentioned in dealing with the contamination of groundwater supplies and also in the brief description of the water supply of Long Island. The drilling of an excessive number of bores into an artesian basin is a corresponding abuse. The following summary of some actual records will show how serious and widespread the effects of this excessive use of groundwater have become:

1. Records obtained from the area of the Great Australian Artesian Basin have disclosed that the discharge of 233 bores in New South Wales (taken as examples) decreased by 39.92 per cent between 1914 and 1928, *i.e.*, from 83,543,586 to 50,192,429 (imperial) gal. per day.

2. In the case of the Great Dakota Artesian Basin, United States, the artesian pressure at Woonsocket, South Dakota dropped from 250 lb. per square inch in 1892 to 35 lb. per square inch in 1923.

3. In the Paris syncline, France, the level of the water table has dropped about 250 ft. in 93 yrs.

4. In the Chicago district, United States, the drop in the level of the groundwater surface averages about 4 ft. per year.

5. In the London Basin, England, the underground water level has been dropping since about 1878 at the rate of about 18 in. per year.[17.36]

The importance of this depletion of groundwater supplies in populated areas can hardly be overemphasized. The reduction in the quantity of water available in storage is the most obvious and probably the most serious result. Associated with it is the increase in the lift for all such water which has to be pumped and the corresponding drop in pressure of artesian waters. As the level of the water table drops, flow conditions from the surface will be altered, and the possibility may develop that contaminated waters, previously discharging safely into streams, may now reach the main body of groundwater, with serious results as a consequence. This has actually been found to take place in some areas in England. Flow of groundwater to rivers may be interfered with, and the vitally important dry-weather flow of such streams seriously affected. Finally—and in some cases most important of all—a lowering of the groundwater level in unconsolidated strata that are being utilized as the foundation beds for large structures may have most serious and disastrous results, if in this way the bearing value of the strata is reduced.

In addition to the problems presented by steadily falling groundwater levels, caused by overpumping, there exists in areas dependent for their supply on artesian water another possible condition of waste which may assume serious proportions. Reference is made to leakage of artesian water into water-bearing beds above the artesian beds.

Records exist of several instances of serious losses from this cause and of the remedial measures taken to combat the problem, of which perhaps the best known are the examples of the supplies of the Roswell Basin, New Mexico, and of Honolulu, Territory of Hawaii. Leaks were found to be due in some cases to a general disintegration of the casing used for the bore; in others, one or more large leaks occurred at joints; improper sealing of the casings was sometimes responsible, as was also insufficient casing to seal off the bore effectively. Detection of this type of leak has been facilitated by means of special deep-well current meters. Once discovered, leaks can readily be cured; the wells generally have to be cleaned out before recasing, the entire operation being naturally the work of specialists. Small wells that can be dispensed with and wells no longer wanted can in similar manner be sealed up, with consequent conservation of the water that they were previously wasting. This can be done, in the case of slow-flowing wells, by means of concrete placed from an underwater dump bucket so as to form an effective plug at the correct horizon in the well. In the case of swift-flowing wells, the current must be reduced before concrete can be effectively placed; this has been done by means of crushed rock dumped into the well and, in the case of very fast-flowing wells, by the use of iron turnings and filings placed by means of a dump bucket. Precast concrete plugs have also been used in well-formed bores, being concreted firmly into place when once set into position. By these means, a flow of ten million (U. S.) gallons per day has been conserved in Honolulu, equivalent to about one half the total water supply previously going to waste.[17.37]

Leakage from bores, however, will be found much less frequently than overpumping, with its attendant unfortunate results. Control of the latter necessarily involves legislative action. Always a difficult matter to correlate with general engineering work, legislation with particular respect to underground water is unusually complex. One country at least has dealt with this problem in a courageous manner. The Burma Underground Water Act of 1930 restricts the making of all new wells except under license, licensing old wells and closing harmful wells.[17.38]

Despite the intricacies of the subject, however, it is one of great importance to all engineers having to do with works affecting or affected by groundwater. The matter must inevitably be considered in close relation to study of all local geological conditions. And it is one that cannot be neglected, despite all the difficulties. In connection with a comparable matter, Thomas Carlyle sounded a warning which may usefully and appropriately be quoted:

"Nature keeps silently a most exact Savings Bank and official register correct to the most evanescent item . . . and at the end of the account, you will have it all to pay, my friend, there's the rub."

Suggestions for Further Reading

The books mentioned at the end of the last chapter contain much useful information on groundwater. In addition, practically all geological surveys issue publications dealing with the distribution of groundwater in their respective areas, and these may usefully be consulted for local particulars. Some survey publications, especially those of the United States Geological Survey, contain general treatments of the subject. Two of these are outstanding:

MEINZER, O. E.: The Occurrence of Ground Water in the United States, with a Discussion of Principles, *U. S. Geol. Survey Water Supply Paper* 489, 1923.

MEINZER, O. E.: Outline of Ground-water Hydrology, with Definitions, *U. S. Geol. Survey Water Supply Paper* 494, 1923.

Special attention may also be directed to:

The bulletins of the Geological Survey of South Australia, particularly Nos. 1 (1912), 3 (1914), 11 (1925), 14 (Geological Structure and Other Factors in Relation to Underground Water Supply in Portions of South Australia, by R. Lockhart Jack, 1930), and 17 (1938).

The reports of the (Australian) Interstate Conferences on Artesian Water (issued by various Australian state printers).

After the completion of the foregoing chapter, the following comprehensive volume was published:

TOLMAN, C. F.: "Ground Water," McGraw-Hill Book Company, Inc., New York, 1937.

CHAPTER XVIII

BUILDING FOUNDATIONS

> If a builder build a house for a man and do not make its construction fair, and the house which he has built collapses and causes the death of the owner of the house, that builder shall be put to death.
>
> The Code of HAMMURABI (2130–2088 B.C.)

All engineering structures must be supported, in some way, by the materials that form the upper part of the earth's crust. There thus exists an inevitable connection between geological conditions and foundation design and construction. This connection has already been discussed with regard to dams, bridges, and other structures associated with transportation routes. These are usually of considerable size and so achieve in the popular mind an importance not strictly in accord with fact—for the major part of engineering construction is made up of a large number of smaller projects. It is to this great group of miscellaneous structures that many civil engineers have to devote their attention. They know the trials and difficulties that small works may cause no less than larger projects; among these, foundation problems are far from being unimportant.

To describe the supporting elements of this great miscellaneous group of structures as building foundations is to use a general term in a singularly wide way. But so varied are the types of work involved that generalization is inevitable. On the other hand, the main types of foundation design are so well recognized that a general discussion of them can safely be made without fear of over-simplification. In addition, the founding of building structures represents a relatively large part of this miscellaneous foundation work, and many of the associated problems are common to general foundation practice.

For many years, building foundation work was largely a matter of rule of thumb, restricted only by such building regulations as were relevant. All too often, construction methods, which are of vital importance in many foundation operations, and sometimes even the final details of foundation design were left entirely to building contractors. In the hands of a competent construction company, this practice may not have been objectionable; but in other cases, it led almost certainly to unsatisfactory foundation work. Today, the foundation of structures is coming to be adequately recognized as an important part of building construction. On many projects, large and sometimes small, architects are cooperating with engineers in this branch of building work which is so essentially in the domain of the engineer. Preliminary investigations are more thoroughly made, and building plans are frequently prepared in association with a definite scheme of construction, the contractor being thus protected against the uncertainties that he might otherwise have to face.

Simultaneously, the science of foundation design is developing. Theoretical design methods are being correlated with the results of their application in practice; and by the improvement of field and laboratory soil-testing methods, better advantage is being taken of the possibilities that many building sites present. The design of building foundations, as is the case with all foundations, consists of three essential parts: (1) calculating the loads to be transmitted by the foundation structure to the strata supporting it, (2) determining exactly the nature of the foundation beds that are to act as a support, and (3) designing a foundation structure to fit the conditions ascertained as the result of operations (1) and (2). Building loads may be affected in a general way by the local geology; but as a rule, for any particular region, they will be related only to structural design. Design of the foundation, however, is completely dependent on the nature of the ground underlying the building site. The determination of ground conditions is essentially a geological problem. When once the foundation bed conditions are known, selection of the type of foundation to be used is a matter of engineering judgment. In the actual preparation of design features, due account must be taken of all geological details liable to affect both the construction and the successful performance of the foundation. Soil Mechanics studies are today offering a major contribution to design work of this nature, but even the accurate designs made possible in this way may be invalidated if a full study of the geology of the underlying strata has not also been made.

With this broad outline providing the necessary background, consideration will be given to leading aspects of the application of Geology to building foundation work. The fact must be emphasized that geological conditions alone are to be discussed—a warning necessary in view of the incomplete picture of foundation engineering that the following notes display. Adequate determination of geological conditions at a building site is but one of the three essential components of complete foundation design, and its isolation from the other two parts and particularly from the final detailed correlation of loading and ground conditions necessarily distorts its significance in foundation work. Foundation failures, however, are rarely due to faulty structural design; it is safe to say that, in general, they are related in some way to failure of the foundation beds to carry the loads put upon them. It is, therefore, an extremely important part of foundation work that is here discussed and one that, either consciously of unconsciously, is associated with the design and construction of every civil engineering structure.

18.2. The Influence of Geological Conditions on Foundation Design. Ground conditions at a building site may be one of three general types:

1. Solid rock may exist either at ground surface or so close to it as to render possible the founding of the building directly upon it.

2. Bedrock may exist below ground surface but at a distance that may economically be reached by a practical form of foundation so that the building load can be transmitted to it.

3. The nearest rock strata may be so far below the surface that it cannot be used as a foundation bed, the structure having to be founded upon the unconsolidated material forming the surface strata.

The influence of Geology is evidenced directly by this broad classification, the three types of ground condition being the result of geological processes of the past. The city of New York typifies the first general type, in certain parts; some of the buildings of Manhattan are founded directly on the Manhattan schist which outcrops over part of that famous island. Much of the area covered by the city of Montreal, Canada, is of the second type, Palaeozoic bedrock being overlain by unconsolidated materials of the Pleistocene and Recent periods of thickness varying to more than 100 ft. The city of London may be taken as an example of the third type, the great London Basin with its deposits of London clay, sands, and gravel overlying the chalk, rendering impracticable the foundation of buildings on rock. It would be of interest to study the influence that ground conditions, of these broad types, have had on architecture, but engineers will readily appreciate the significance of the groups without the aid of this comparison.

Determination of a ground condition of the first type will be a relatively simple matter, even as it is also a rare one in urban areas (possibly because cities are usually to be found on river banks where unconsolidated deposits are the rule rather than the exception.) The only additional services of Geology will be in connection with the estimation of the soundness of the rock and the significance of any structural features revealed and possibly with a determination as to the allowance to be made for seismic disturbances. It is but rarely that the bearing power of rock will be in question as being sufficient to withstand building loads; but if compression tests of the rock are made, great care should be taken to see that they are made with specimens loaded in a direction corresponding to that at the site. The difference in the bearing power of sedimentary rocks when tested parallel to the stratification and at right angles to it may be as much as 50 per cent. As sedimentary rocks include those rock types which have low bearing capacities, this point is sometimes of importance. (Typical figures are the following for slate: 7,220 lb. per square inch at right angles to bedding and 5,180 lb. per square inch in the direction of bedding; these are average figures obtained in connection with the foundation of the Boston Parcel Post Building, United States.[18.1]) Values for the usual bearing stress utilized for various rock types will be found in engineering handbooks and treatises on foundation design; independent tests for all but the smallest projects are usually advisable. Provided that no serious structural defects exist in bedrock, this first general type of ground condition provides probably the most satisfactory of all foundation beds.

If rock does not outcrop at the surface in the vicinity of a building site, underground exploratory work will be essential. If this discloses the existence of the surface of bedrock within a reasonable depth below the surface, thus suggesting the possibility of the second general type of ground condition, test boring and drilling should be carried out as extensively as possible all over the site to determine accurately the contours of this rock surface. Simultaneously, groundwater observations should be made. The word *reasonable*, although frequently so objectionable, is here used to cover the many variable local conditions that may affect the depth to which it proves economical to carry foundations to rock. The maximum depth so far utilized appears to be 250 ft., having been reached by the foundations of the Cleveland Union Terminal Tower, Cleveland, Ohio.[18.2]

Geological conditions will be a major factor in the determination of the best type of design to use. The practicability of using piles will depend upon depth to rock and the nature of the unconsolidated materials above the rock. The nature of the soil will also determine the feasibility of using what is known as the Chicago well method of pier construction or a variant. The method was apparently originated by General W. Sooy Smith about the year 1891. The Chicago subsoil consists generally of a soft blue clay; the bedrock is a hard limestone formation occurring at a depth of about 130 ft. The clay will stand vertically for a height of about 6 ft. when exposed and so permits of the use of the dry, sheeted, circular shafts that characterize the method.

As it is essential to the method that the excavation be made exactly the size required by the braced sheeting, it can be adopted with safety only when excavation is to be through a suitable stable clay. When groundwater is met with underground, construction difficulties will abound; and if solid piers are to be used as the foundation, the use of caissons under compressed air will be advisable unless open dredging through the water is permissible.

FIG. 18.1.—Foundation shaft excavated through clay by the "Chicago well" method, for foundations of Stop and Shop Building, W. Washington Street, Chicago.
(Reproduced by permission of Mr. W. J. Newman, President, W. J. Newman Wrecking Co., Chicago, Illinois, U.S.A.)

It is of importance to note that if the surface of the bedrock slopes at an appreciable angle, special precautions may have to be taken in designing and constructing a foundation. This restriction is obviously of importance in construction with buildings founded directly on rock. It is also of some significance in cases of the third type of ground condition when, although not used as the foundation bed, solid rock is within a moderate distance of the ground surface. The action of gravity naturally affects all loose material lying above a solid rock surface; and if the latter is lubricated (as by groundwater), slipping of the whole mass may occur down the slope of the rock. This was the explanation generally recognized by engineers familiar with the serious movement of the Cahuenga Pass multiple-arch retaining wall constructed in Los Angeles in 1925. Four hundred and fifty feet long and varying up to 60 ft. high, the centers of buttresses being 30 ft. apart, the wall was founded on bedrock at its ends but not in the central portion. Rock was here 40 ft. below ground level; spread footings were therefore used and carried to a depth of 20 ft. Movement of this central section was observed before all the fill had been placed; it continued even after the loading on the wall had been reduced and in July, 1927, amounted at one place to 17 in. outward and 15 in. downward. The load on the spread footings was limited to 8,000 lb. per square foot, but the underlying rock sloped steeply away from the wall, so the conclusion was that soil strata and wall were moving down on the rock surface as a unit.[18.3]

It can be seen, therefore, that exploration of bedrock contours over an area larger than the actual building site is advisable in order to check against any unsuspected slope of the rock surface. An allied problem is to have stratified bedrock available as a foundation stratum but dipping at such an angle to the horizontal as to render doubtful the stability of the upper layers. In parts of New York City, the Manhattan schist is so inclined; it is a feature that is encountered elsewhere not infrequently. Local details will determine the best solution of the problems thus presented, but a usual method is to drill the surface rock and anchor it by dowels to layers that are so far below the surface as to be beyond the range of possible movement.

The last general type of ground condition, when rock cannot be used as the foundation bed, is probably the most common of all three, particularly if due consideration is given to the vast number of small structures that could be founded on rock (by methods already described) but that need not be, because of their small weight. Adequate subsurface investigation is again imperative if the foundation design is to be satisfactory. There was a time when it was generally thought that to carry test borings down as far as the proposed length of bearing piles was enough; today, this limitation of test boring would often properly be regarded as tantamount to waste of the cost of what exploratory work was done. Even before the type of foundation to be used has been selected, accurate and extensive knowledge of subsurface conditions is essential. As a general rule, exploratory work should give definite results for a depth *at least* equal to twice the width of the structure and greater than this if possible, especially if the presence of relatively soft strata is suspected. The extra cost of a few feet more penetration of test borings, when once the boring equipment is set up, is negligible compared to the value of the results so obtained.

The vital necessity for obtaining subsurface data to such depths can be illustrated only by a brief glance at one of the fundamental concepts of soil action. It may first be observed that foundation loads may be transmitted to unconsolidated strata in two general ways: (1) by constructing a continuous raft or spread footing which rests directly on the surface stratum and (2) by forcing piles into the soil or by constructing piers therein, their tops being structurally connected in a suitable way to the bearing columns of the structure. A variation of the first type is to have the load from a single structure distributed through a number of isolated footings; the second method may depend for its stability on the bearing of the ends of the piles or cylinders resting on a reasonably hard stratum, or the bearing may be the result of skin friction. Other minor variations might be listed, but these four main methods include generally all leading classes of foundation structure for unconsolidated strata.

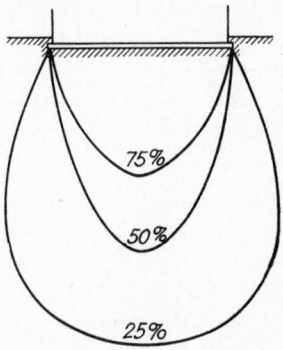

FIG. 18.2.—Stress distribution in soil under a foundation slab, percentages being in relation to the load per unit of area.

(*After Terzaghi.*)

How do these transmit the load from the building to the soil on which they rest? Figure 18.2 is a diagrammatic answer, illustrating the formation of what are often called *bulbs of pressure*. These represent the way in which the initial foundation loads are gradually dispersed in the strata that carry them. Bulbs of pressure have been recorded from many experiments on actual foundations; their investigation and theoretical calculation are an important part of Soil Mechanics. They have also been studied by an interesting application of photoelastic methods. They cannot be further discussed in this place but are illustrated in order to show why subsurface conditions have to be so carefully investigated to considerable depths. The bulbs as drawn in Fig. 18.2 are for a homogeneous soil. They are drawn proportionately in scale to the foundation dimensions. For building work, a general rule is that the depth of the zone of appreciable compression in the bulb is from one to one and a half times the width of the building, if a raft foundation is used. Hence it will be seen that with all these methods of constructing foundations in soil, materials well below the surface are appreciably stressed, the actual amount being calculable if loading and soil data are known. It follows that if, instead of a homogeneous soil, the strata include a layer of softer material—even at a considerable depth—this material may fail under load, leading to serious settlement of the structure, despite no evidence of ground failure at the surface.[18.4]

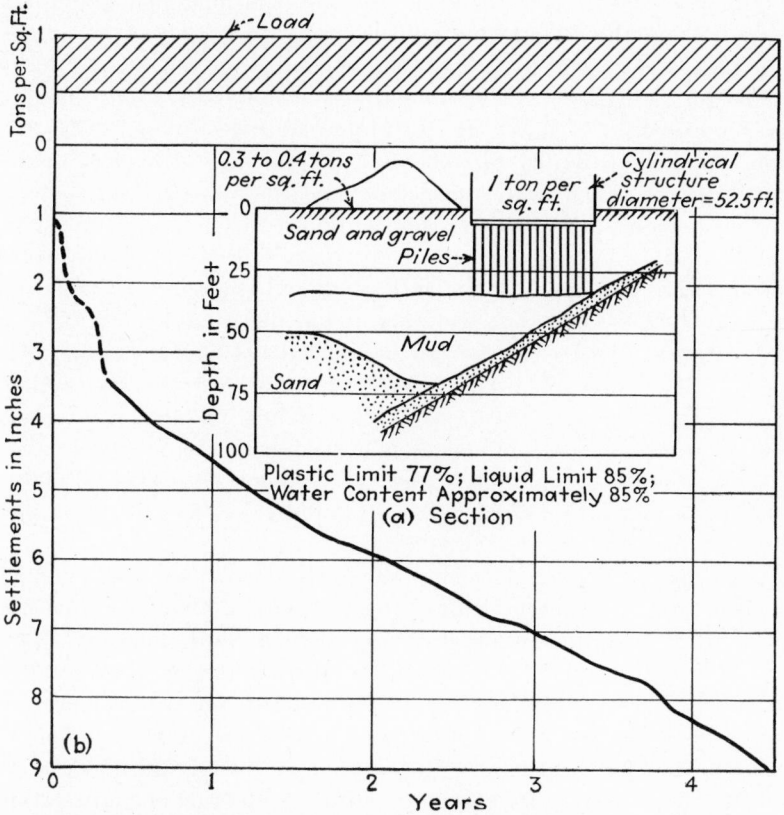

FIG. 18.3.—Settlement record for the cylindrical structure shown, due to mud stratum. (*Reproduced, by permission of the American Society of Civil Engineers, from reference* 18.6.)

This is no impracticable theory but a summarized explanation of many building failures and unusual subsidences in the past. Dr. Terzaghi has stated that in one of the cases he had under observation, "a building settled more than one foot owing to the compression of a layer of clay located between a depth of 100 and 130 feet below the surface of the ground."[18.5] The Special Committee on Earths and Foundations of the American Society of Civil Engineers, in their *Report* of January 18, 1933, mention an interesting example (Committee Case D) of an oil tank, 52.5 ft. in diameter, supported on wooden bearing piles driven into a layer of sand and gravel. Driving was hard, and the usual penetration formulas gave satisfactory results, the distributed loading being only 2,000 lb. per square foot. Unfortunately, the sand and gravel layer was superimposed on a thick pocket of wet mud (water content approximately 85 per cent) at about 35 ft. below ground level, the piles finishing just above this level. When the Committee reported, the tank had settled almost 1 ft. in less than 4 years despite the apparently excellent foundation conditions, settlement being due to the consolidation of this mud layer under load.[18.6] Mr. H. A. Mohr mentions (as his Case 3) a large building, carefully planned and designed, for which good test borings were taken before construction, the results being closely studied. Before completion, this building was discovered to have moved out by 4 in. and settled 4 in. at one place in its front wall. New borings were put down, to a depth of 87 ft., original borings having stopped at 42 and 31 ft., respectively; it was then found that a stratum of "very soft black clay and fine sand" existed at a depth of 55 ft. below the surface, the upper strata being clay, with some sand of varying consistency. The front of the building was about 90 ft wide. The necessary underpinning and pile driving to correct the unsatisfactory condition thus discovered cost about $18,000 (£3,600)—"probably four times what the cost would have been had the piles been driven before the wall was built."[18.7]

This aspect of foundation design on soil strata is so important that one further example will be cited—again not an exceptional case but one from ordinary practice having general significance. The Westinghouse Electric and Manufacturing Company's Building in Philadelphia, Pennsylvania, was constructed on a building site, boring records for which were supplied by the previous owners. These showed 8 to 10 ft. of loose fill and then uniform clay and sand to rock level, 45 ft. below the surface. Bearing pile foundations were therefore decided upon for the 10-story building. The concrete pile contractors drove test rods to determine the length of piles, which was fixed at between 25 and 30 ft., a penetration shown by the test rods to be satisfactory. When 8 of the 10 stories had been constructed, the building was found to be settling noticeably. It had already sunk 4 in. on one side. New core borings and an open test caisson were therefore put down to rock, a mica schist which was found 57 ft. down, the strata passed through being 27 ft. of fill, 4 ft. of peat, 1 ft. of gravel, 15 ft. of silted peat, 8 ft. of silt, and 2 ft. of coarse sand and gravel instead of the fill, clay, and sand expected. The weakness that permitted the excessive settlement was at once obvious;

an elaborate and ingenious underpinning operation had immediately to be undertaken, and the footings to be supported on 15- and 16-in. pipe piles, concrete filled and carried to rock.[18.8]

Other examples could be cited; all would confirm the necessity for adequate subsurface exploration and study of the soil types encountered. The latter is now essentially a laboratory process; and on the basis of the results of such study, the foundation design is prepared. It is determined primarily by the nature of the loading and the soil strata available as foundation beds; each of these may vary widely. Frequently it will be found that a raft or spread foundation will be more suitable than the conventional bearing piles, especially with some types of clay soils the disturbance of which may appreciably affect their physical properties. These are some of the matters now receiving attention in Soil Mechanics studies; they can be adequately investigated only if complete and accurate subsurface exploration work has preceded them.

18.3. Subsidence of Buildings.—It will have been noticed that the foundation troubles to which reference has so far been made have been generally in the nature of excessive subsidence. Structures that are supported on bedrock, directly or through piles or piers, will settle by extremely small amounts only, provided that the rock foundation strata are sound. The displacement of such structures will generally be of negligible importance. If a foundation has to be supported by unconsolidated strata, appreciable settlement is always to be expected. Modern practice takes this settlement into account; foundation structures are now usually so designed that uniform settlement may be obtained throughout the building to obviate serious structural distortion; some of the methods used to achieve this result are ingenious.

The extent and nature of foundation subsidence will clearly be dependent on the nature of the soil strata. When these have been thoroughly explored and tested, it may be found that major settlement is unavoidable with the loads to be carried unless special measures can be taken. Especially in this true in connection with the continuous settlement caused by the plastic deformation of clay. With buildings this may not be very serious, but with bridges and similar structures it may; jacking devices are therefore sometimes incorporated into foundation designs subject to this disadvantage.

Geological study may suggest other occasions when the incorporation of provision for jacking in an original design may be a useful and economical safeguard. An interesting example is afforded in the special form of support and foundations for four circular reinforced concrete water tanks, each 36 ft. in diameter with its top about 32 ft.

above ground level, designed by Mr. H. C. Ritchie for the Urban District Council of Heanor, Derbyshire, England. The tanks were built in 1937 to replace a reservoir that failed as a result of subsidence of supporting ground owing to colliery workings beneath. The same site had to be used; and although the foundations could be carried to a stratum of hard shale, future subsidences of the substrata had to be anticipated. Each tank is therefore carried on *three* points of support (instead of the usual four or more) so as to obviate indeterminate stresses at any stage of irregular settlement. The feet are not structurally continuous with the supporting foundations but rest

FIG. 18.4.—Reinforced concrete water storage tanks at Heanor, Derbyshire, England, showing unusual three-point foundation arrangement.
(Reproduced by permission of Messrs. Ritchie and Partners, Consulting Engineers, London.)

freely on them, and the load is transmitted through special steel bearing plates and shoes which will permit of the jacking up of the structure at each individual support so that the tanks may be raised and releveled as subsidence takes place.[18.9]

Water bearing sand which, by reason of its water content, may be what is popularly described as *quicksand* may cause trouble in the way of foundation settlements as well as adding difficulty in construction. An old but effective device in the case of small works is to "pack" the sand with dry coarse gravel; many a water-logged trench has been salvaged simply by the treading in of gravel to a depth of 2 or 3 ft. On a larger scale, the same idea can be applied by the driving of hollow pile shells, filling these with dry gravel, and then withdrawing the shells. Dr. Terzaghi has referred to an interesting case in which loose sand 50 ft. thick overlaid an inclined rock surface. Concrete piles were to be used; but as the groundwater contained

40 mg. of carbon dioxide per liter, regarded as dangerous for concrete, piles were formed of "masses of sand, gravel, and trass." These compacted the sand stratum, the surface of which actually settled 1 ft.[18.10] Another type of treatment for similar ground conditions was adopted on the construction of the St. Louis Telephone Building, St. Louis, Missouri, United States, which is founded on concrete piers carried to solid limestone 74 ft. below surface level. The pier excavations were made as open wells through the overlying strata of clay, fine sand, and boulders. The fine sand stratum was water bearing, and so two wells were first sunk to rock near the center of the site and used as pumping sumps, until all other piers had been concreted.[18.11] Open pumping such as this is not, of course, always advisable because of the possible subsidence that it may cause of adjacent ground; but if subsurface strata are known, it can sometimes be safely adopted.

Another means of improving the bearing capacity of weak soil strata is by the injection into them of either cement grout or chemical consolidation solutions, which when set will at least partially consolidate the strata. Clearly, such methods can be applied with certainty only if accurate data are available as to the nature of the subsurface conditions. European practice includes several notable examples of this class of work, one of the most interesting of which has been the safeguarding of the Leaning Tower of Pisa; Fig. 17.5 is a familiar view of this tower. It was started in A.D. 1174 but not completed until the year 1350. It has continued to tilt since then, the present displacement being over 16 ft. in its total height of 179 ft. Its foundation consists of a circular slab 64 ft. in diameter, with a central hole 15 ft. in diameter. Whether or not there are piles under the slab is still uncertain. Foundation strata consist of a bed of clayey sand 13 ft. thick, underlain by 21 ft. of sand resting on a bed of brackish clay which is of unknown thickness. Recent investigations and laboratory tests suggested that the cause of the tilting is attributable to the clay layer 28 ft. below the foundation slab. After careful study, it was finally decided to strengthen only the masonry of the foundation slab and ring wall; this was done with cement grout under low pressure. To complete the work, the sand stratum surrounding the foundation was consolidated by the use of a chemical method employing a single gel medium, injected at a low viscosity. The method is the invention of Dr. Rodio under whose direction the work was carried out.[18.12]

All records of building subsidence are reminders of the fact that the earth's crust is not the solid, immovable mass so popularly imagined. When solid rock outcrops at the surface, it provides a foundation stratum of material that is as susceptible to stress and strain as any other solid matter. If unconsolidated material constitutes the surface layer, it provides a foundation bed even more liable to evidence movement under load—possibly continuous settlement in the case of some clays if loaded beyond a certain limit. Modern foundation design aims not at eliminating settlements—the impossible—but at so

FIG. 18.5.—The Leaning Tower of Pisa.

controlling them that the structure supported will exhibit no undesirable effects. Should there be any reader of these words to whom the idea of all buildings settling is strange, let him reflect on the quite typical fact that a part of the Tower of London—that symbol of stability—is now known to rise and fall as the tide in the river Thames alongside ebbs and flows.[18.13]

18.4. The Influence of Groundwater on Foundation Design.—Groundwater has already been mentioned several times in this discussion of foundations; its significance is even greater than even these references might suggest. Naturally, its importance is most evident in connection with structures founded on unconsolidated material; although when foundations have to be taken through soil strata to rock, it is frequently a vital factor. As indicated in the discussion of foundation methods for this latter type of ground condition, if groundwater is present, the use of piers excavated in the dry to rock is automatically eliminated. In certain very stable kinds of soil, open dredging of small piers may be possible (as is sometimes the case in bridgework), but frequently the presence of groundwater means the adoption of a more elaborate construction method—usually the employment of a compressed air plant but including also such special methods as the use of well points and freezing. As these methods have already been discussed in another connection (see page 196) further reference to them will not be made here. The possibilities that groundwater thus presents in the selection of a suitable kind of foundation design will make very evident the necessity for obtaining accurate groundwater data as an essential preliminary.

More than this is involved, however, by groundwater conditions. A large percentage of building foundations, more particularly those of the past, utilized timber-bearing piles as an essential supporting element. It is well known that most kinds of timber have a long and useful life if kept either always wet or always dry (alternate wetting and drying being the exposure condition liable to cause rotting). Many timber bearing piles have been driven into ground in which the water table is near the tops of the piles. If, therefore, this water table varies in elevation, the timber piles will be alternately wet and dry and consequently may deteriorate. How frequently this condition occurs is only now being realized, as older buildings are gradually renovated and their foundations investigated.

An interesting and historic example is that of the great tower of Strasbourg Cathedral, designed in 1439; its stone footings were originally supported on timber piles. The installation of a new drainage system in 1750 lowered the water table appreciably and in time caused the tops of the piles to decay. Serious settlement of the tower resulted. In recent years, the tower was jacked up after its columns had been encased in concrete, and new reinforced concrete foundations installed.[18,14] Several other European buildings of historic interest have been similarly affected; but as a second example, there may be mentioned a famous building of the New World, the Boston (Massachusetts) Public Library. About the year 1929, cracks were noticed in the building; and after investigation by the city authorities and their consulting engineers, it was found that the tops of many of the timber piles supporting the structure had decayed, some having rotted

FIG. 18.6.—Upper sections of timber piles under the Public Library, Boston, Massachussetts, U.S.A., showing deterioration attributed to varying groundwater level.

(Reproduced by permission of the Boston Public Library, Dr. Milton E. Lord, Director, through the courtesy of Mr. Bayard F. Snow, Boston.)

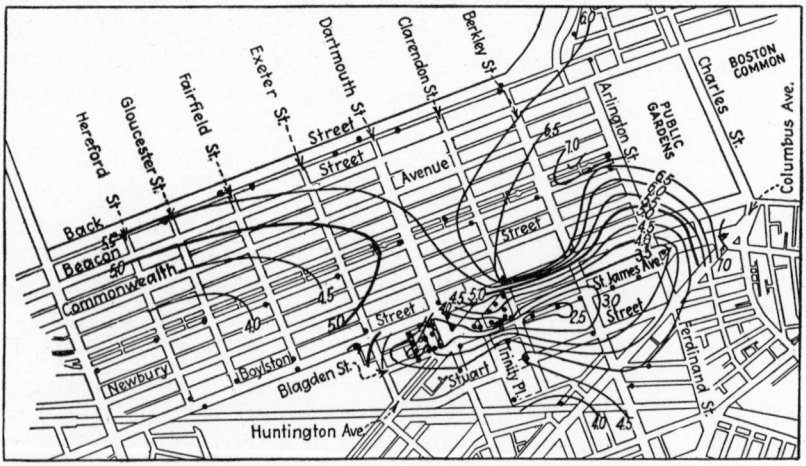

FIG. 18.7.—Typical groundwater contour map, for part of Boston, Massachussets U.S.A., showing relationship between area of low groundwater and low-level sewer in St. James Avenue—as prepared from groundwater survey described in text.

(Reproduced by permission of Mr. Bayard F. Snow, President, Messrs. X. Henry Goodnough, Inc., consulting engineers, Boston, and the Editor, Engineering News-Record, from reference 18.15.)

completely. About 40 per cent of the building had to be underpinned, affected pile heads being cut off and replaced with concrete. The foundation strata in this part of Boston consist of boulder clay, blue clay, and silty sands overlying the local bedrock; pile foundations are therefore usual in this city area. The trustees of Trinity Church, adjacent to the Library building, alarmed at the possibility of similar trouble, had an investigation made in their behalf by Messrs. X. Henry Goodenough, Inc., consulting engineers of their city. Their detection of the cause of low groundwater in this vicinity is an interesting example of groundwater exploration carried out at relatively low cost, and it finally revealed, in association with city authorities, that groundwater was leaking away into a low-level sewer constructed about 1912 through the area in question. The sewer was partially blocked off by a suitable dam, and immediately groundwater levels in the area rose; the test dam was therefore left as a permanent feature to maintain a satisfactory groundwater level. Figure 18.7 shows groundwater contours before the leakage was checked; as a telling piece of evidence showing the possible danger of groundwater leakage in porous soil strata, it would be hard to equal.[18.15]

How can such foundation troubles be prevented and remedied? One method, followed in some building regulations, is to ensure that all piles are either cut off or embedded in concrete at a level well below the assumed minimum elevation possible for the local groundwater. An alternative method is the use of "composite" piles in areas of fluctuating groundwater level, protected sections being used for the range of water level variation and timber below this. Another practice was followed in connection with the foundation of the Northwestern Mutual Life Insurance Building in Milwaukee, Wisconsin. Foundation beds consist generally of glacial strata, with bedrock 250 ft. below ground surface; the water table at the time of construction (1930) was 17 ft. below the surface. Timber-bearing piles were used, and lengths of 4-in. pipe were built into the foundation slabs, fitted with screw caps and strainer bottoms. Their use is considered a remote possibility; but if the groundwater should, in the future, drop to such an extent that the piles are endangered, water can be pumped into the ground through these pipes, and the pile tops kept submerged.[18.16]

Timber-pile foundations are not alone in being affected by lowering of the local water table. The bearing power of soils, and particularly of clays, may be appreciably changed by an alteration in their moisture content. Allied troubles are the shrinkage of clays as they dry out and the possibility of the finer particles being washed out of water-logged sand as groundwater recedes or is drawn from it. All these conditions have been responsible for notable foundation troubles. To take another Old World cathedral as an example, the troubles experienced in connection with the foundations of St. Paul's, London, are related to local groundwater conditions, apparently having begun with the start of excavation for a deep sewer close to the Cathedral site by the Corporation of the City of London in 1831. "Hundreds of tons" of quicksand or silt were removed by steam pumps set to work at a construction shaft; but after protests by many eminent engineers and architects (including Thomas Telford and John Rennie), the work was abandoned, and the shaft filled up; damage, however, had already been done. Following recent investigations, all foundation work in the neighborhood of the Cathedral is now subject to severe restrictions.[18.17]

Foundation troubles consequent upon the drying up of clay strata used as foundation beds are unfortunately so frequent (especially in the case of small building construction) that it is hardly necessary to cite examples. A walk along the streets of any city the smaller buildings of which have been founded directly on clay, from which water has drained either into leaky sewers or because of a general drop in groundwater level, will show all too well the results of clay shrinkage. Parts of the city of Montreal, Canada, provide typical examples, but the cracked buildings to be seen there can be duplicated in many other places.

Just as the shrinkage of clay can cause foundation troubles, so may its swelling when its moisture content is increased. An interesting case of this has been reported by Professor F. E. Giesecke.[18.18] The foundations of the University Baptist Church in Austin, Texas, were built as concrete walls carried to rock, thus surrounding a 6-ft. layer of clay (decomposed shale) under the building proper. Owing to an increase of the moisture content of the clay, it expanded vertically by as much as 4 in., lifting the church building by breaking concrete footing slabs, the expansive force of the clay overcoming the tensile strength of the concrete footing wall.

Attention to the possible action of groundwater, whether absent or present in foundation strata, is thus a necessary part of subsurface exploration work for building foundations. The necessary studies can most usefully be made in association with test drilling. If the information then obtained is not sufficient, observation wells can be installed simply and at low cost. (Those used in the Boston studies already mentioned cost only $45 (£9) each.) Any such study will tend to confirm the necessity for strict control of groundwater resources, the depletion of which is of great importance in connection not only with public health but also, as will now be seen, with the stability of many building structures.

Finally, and although not strictly associated with building foundations, there may be recorded the most unusual groundwater trouble of which the author has either heard or seen any record. Details were carefully recorded by Mr. T. Chalkley Hatton, Chief Engineer of the Sewerage Commission of Milwaukee, Wisconsin, at the time of the occurrence for the benefit of his fellow engineers in view of its most serious and unusual nature.[18.19] The Sewerage Commission was constructing in 1922, by contract, a 36-in. concrete intercepting sewer in National Avenue, Milwaukee, 35 ft. below the existing impervious street pavement. Soil strata were 18 to 20 ft. of dense clay overlying fine gray clay and sand and gravel beds, all water bearing. A shaft (No. 1) 35 ft. deep was sunk but had to be abandoned owing to the water that accumulated in it. A second shaft (No. 2) 334 ft. to the west of No. 1 was sunk to a depth of 52 ft.; and a third (No. 3), 672 ft. farther to the west, to 49 ft. Shaft 2 was used as a sump, a million (U. S.) gallons of water per day being pumped from it. A compressed air plant and lock were set up at shaft 3; and on May 23, 1922, air was finally "put on," but a pressure of only 3½ lb. per square inch could be maintained owing to leakage that could not be detected. In the search for leakage, shaft 2 was entered several times that day. The air compressor was left running all night. Next morning, a pumpman descended shaft 2 and did not return; a workman went down to retrieve him. Later three of a rescue crew went down the shaft. All five were killed. It was found that the air at the bottom of the shaft contained only 5.9 per cent of oxygen but held 8.9 per cent of carbon dioxide, the cause of the tragedy. Investigation of the source of this sudden accumulation of gas traced it to the decay of organic matter in the subsoil. The pumping had lowered the water table by 12 ft.; and as the water level dropped, carbon dioxide was liberated and occupied the vacant pore space in the clay. From this position, the gas was displaced by the pressure of the compressed air and driven toward the convenient hole provided by the shaft. Stopping and starting the compressed air plant demonstrated later the accuracy of this explanation. This tragic example is a reminder alike of the chemical activity of some clays and of the subsurface disturbance that can be caused by a sudden change in groundwater conditions.

18.5. Preliminary and Exploratory Work.—The investigation of subsurface conditions at all open building sites follows generally accepted methods, geological studies and test boring operations being carried out as intensively as necessary. In urban areas, especially in the case of buildings that are to be constructed on sites already occupied, the matter becomes a difficult one. Frequently it will be impossible to make test borings until existing buildings are demolished, before which time foundation designs have to be prepared at least in outline. In many cases, direct geological investigation will be impossible.

For preliminary work in cities, therefore, those minor exploratory aids mentioned in Chap. VI assume a major importance. All old maps of the locality should be examined; they will often show the site of old watercourses, possibly now covered, and sometimes will suggest the existence of recently filled land at a building site. Messrs. Prentis and White, foundation engineers of wide experience, have testified to the value of the Viele map of New York in connection with foundation work in that city.[18.20] The map was prepared in 1860 when much of Manhattan Island was open land and shows old watercourses, swamps, and a pond. Naturally, the use of old maps, and similarly of old descriptive documents, is of greatest importance in Europe, owing to the greater age of Old World cities as compared with those of North America and other newly developed countries. Groundwater conditions can usefully be studied from available water level and pumping records of local wells and boreholes. If records are not available, observation wells can easily be put down. The best of all information, however, apart from actual test holes at the site, is that afforded by the foundation records of adjacent buildings and the borehole records obtained for their construction. When information of this nature, which can be relied upon with certainty, can be procured, it will often enable preliminary foundation designs to be prepared, although it never obviates the necessity for confirmatory tests at the actual site when these can be made.

As an interesting example of exploratory work at a building site may be mentioned that carried out in connection with the foundations of the unusually large Battersea (steam) Power Station of the London Power Company, Ltd., on the south bank of the river Thames, London, for which Mr. A. C. Dean was Consulting Engineer.[18.21]

A preliminary examination of the available records of borings and other works in the vicinity of the station showed that the level of the London Clay might be reached at a depth of 30 feet to 35 feet below ground level. It was noted that in some cases the London Clay was at a considerably greater depth, and that in sinking a well at the Nine Elms Brewery no clay had been encountered. From these indications, and from a study of the map accompanying the Memoirs of the Geological Survey for the London area, it was anticipated that depressions in the clay of considerable depth might be encountered, and this proved to be the case.

Figure 18.9 is a key plan of the site, the irregular spacing of the boreholes representing the efforts made to trace the depth and extent of the depressions in the clay. A total of 73 test holes were bored, and eight trial pits were sunk. These showed the alluvial gravels of the Thames flood plain overlying the London Clay. As water accumulated in the pits, it was decided to surround the entire building site

FIG. 18.8.—Site of the Battersea Power Station, London, at completion of excavation. (*Reproduced by permission of the London Power Co., Ltd., Sir Leonard Pearce, Engineer-in-Chief.*)

with a retaining wall, to exclude water, and to found the station on the London clay. The Geological Survey was given full information from the test borings and consulted as to the results. The two main depressions detected from the borings were later confirmed as excavation proceeded and found to be of the "pipe" type and not a continuous channel. Their origin is possibly due to subsidence of the clay, following solution of the underlying chalk.[18.22]

The cost of the necessary preliminary work is often a difficult matter to arrange, and building construction is no exception to this. It will be useful to point out, therefore, that this thorough investigation for the Battersea Power Station is that featured in the table on page 88, and which will there be seen to have cost only 0.2 per cent of the total cost of the station structure—a percentage that is not

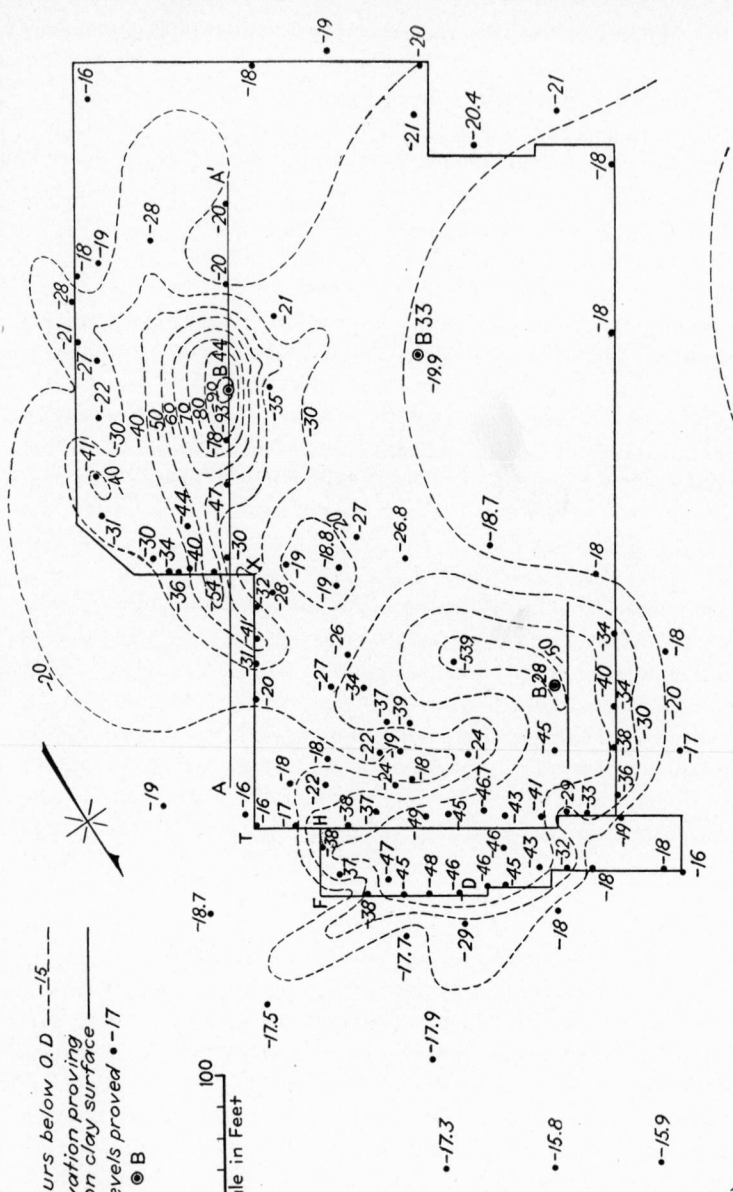

FIG. 18.9.—Plan of the site of the Battersea Power Station, London, showing location of boreholes, and the contours of the London clay surface thus revealed.

(Reproduced from "Summary of Progress; Geological Survey of Great Britain," 1930, Part Two. Crown Copyright, reproduced by permission of The Controller of H. M. Stationery Office, London.)

unusual but illustrates vividly the economy of adequate subsurface exploration.

18.6. Cooperative Subsurface Investigations.—In view of the fact that in urban areas, data on local geological conditions can be obtained only from a study of the test boring and drilling records for numerous small building sites, it might naturally be supposed that there would be some generally recognized method of correlating these exploratory results. Unfortunately, and surprisingly, this is not the case. Judging by printed and published records, there are very few cities in the English speaking world in which any cooperative venture in connection with foundation bed data has been carried out. The situation in many cities is, indeed, almost the reverse of this. Private institutions of one sort and another regard information as to foundation strata beneath buildings with which they are, or have been, associated as of a confidential nature and their own particular property. This attitude is hard to understand. Were minerals of value involved, the case would be different; but in the general practice of foundation work, such secrecy appears to be singularly out of place. It persists, however, particularly as legislation on this matter is not yet available.

Fortunately, there is another side to the picture. In a number of cities, engineers have cooperated with ready will and made available for public benefit accumulations of foundation bed data which are of unusual value in that they give composite pictures of foundation conditions for complete city areas. The following brief notes outline the most interesting of these ventures of which the author has knowledge; the record is probably far from complete and so is to be taken as suggestive only. It may be useful to mention that, in some cases, data of this nature have been assembled by geologists in geological publications from which they can of course be obtained by engineers. As an example may be cited Memoir No. 53 of the Geological Survey of Canada, "The Pleistocene and Recent Deposits of the Island of Montreal," by Dr. J. Stansfield. Despite its geological title, this useful publication gives a summary of many records, obtained from foundation records, of strata underlying the city of Montreal. The utilization of these results as the basis of a geological study is another reminder of the reciprocal services of the science and the art.

It will be noticed that local engineering societies and local sections of national engineering societies have been prominently associated with these cooperative ventures. In the field of civil engineering, there seem to be few more useful tasks that such local groups of engineers can perform. The functions of the local engineering group are often in question. Here is an essentially local task requiring concerted action, which takes years to complete, and which results in lasting benefit to the community. It is one that might usefully be adopted by many more engineering groups than appear yet to have entered this field.

Boston, Massachusetts.—In the first published copy of the *Journal of the Boston Society of Civil Engineers* (1914) appeared a paper by Mr. J. R. Worcester entitled Boston Foundations in which an assembly of boring records was featured. As a subsequent development of the interest aroused by this paper, the Society set up a Committee on Boston Subsoils, the purposes of which were stated as follows:

> The purpose of this Committee is to gather data regarding the character of the subsoils in Boston and adjacent areas, and to present it to the Society in such form as to add to the general knowledge and to make it available for reference by any who may wish to get a clear idea of the geological construction under this City.

This Committee made an extensive report in 1931 and a second one in 1934, which has not yet been published.[18,23] With the hearty cooperation of those making borings in the city area, the Committee has assembled over 9,000 records, locations of which are shown on 16 sectional maps. The borings are so well located that it has been possible to prepare six geological sections through the city at many points and to make maps of the rock and boulder clay contours and to locate old shore lines of Boston and Cambridge. The Committee gave special consideration to the complex groundwater situation under the city area and made tests of local soils as well as compiling a classification of these. The rock and boulder clay maps were made by the geologist member of the Committee (Mr. I. B. Crosby) who had started work of this nature as early as 1923. The work of the Committee aided in large measure in preparing the proposed revision of the Boston Building Laws, as mentioned later. As a result of this cooperative work, there are now available to the public sufficient data on the geology of the city area to give a reasonably clear picture of all local foundation bed conditions.

New York City.—A project known as the Rock Line Map of New York was initiated in 1935 by the W.P.A. of New York, under the direction of Dr. C. P. Berkey of Columbia University and the sponsorship of the Board of Estimate of the City of New York. The project covered the five boroughs making up the city of New York and was concerned with the depth to the rock floor under the city area and the character of the bedrock and overburden. Every available bit of information was painstakingly searched out and recorded on large-scale section maps of the city, and from this basis the topography of the rock floor has been drawn in with as much detail as possible. About 20 years ago, a similar project was sponsored by the Municipal Engineers' Society of New York. All borings then available were recorded on standard sheets and on a large-scale map of the city. A collection of standard types of rock met with, properly named, was installed in the Municipal Building, and structural drawings were made of the local geology. All this information was available for the use of anyone interested who made application. The new Rock Line Map project served to bring this original and most interesting record up to date.[18.24]

New Orleans, Louisiana.—Initiated in 1934, at the request of the Louisiana Engineering Society, a survey of foundation data was made as a project of the W.P.A. of Louisiana. Serving on the Advisory Committee were representatives of Tulane University, the local section of the American Society of Civil Engineers, and the two organizations already mentioned, the project being sponsored by the Board of State Engineers of Louisiana. The results of the Survey were published in a notable volume in 1937 ("Some Data in Regard to Foundations in New Orleans, and Vicinity"). In this are assembled over 200 boring records and details of over 100 examples of foundation design in the New Orleans area through the cooperation of local engineers, architects, and officials. As the city is located near the mouth of the Mississippi River, the local geological formation is typically deltaic, consisting of sand and clay strata extending to depths greater than 350 ft. Interesting features of these beds include the existence of two layers of cypress tree stumps, the trunks having been cut off; one layer is a few feet below the surface, but the other is about 25 ft. below mean Gulf level. A human skull was found in undisturbed clay at a depth of 37 ft. below ground surface, and layers of oyster shells occur about 50 ft. down.

Winnipeg, Manitoba, Canada.—At the instigation of the Winnipeg Branch of the Engineering Institute of Canada, a Committee on Foundations in Winnipeg, 1937, was set up in May, 1937. Professor A. E. Macdonald was Chairman, membership being made up of representatives of the University of Manitoba, the Engineering Institute of Canada, the Association of Professional Engineers of Manitoba, the Manitoba Association of Architects, and the Canadian Institute of Chemistry. The function of the Committee was to investigate and report on (1) why difficulties are experienced with foundations, (2) how best to repair faulty construction, (3) proper design for new foundations. In August, 1937, the Committee submitted an interim report, discussing the formation of Winnipeg soils and subsoils, the foundation of new buildings, and the cause of settlement of existing buildings. It discussed the complexity of building foundation design in Winnipeg and suggested that research work in both field and laboratory should be carried out.[18.25]

The report includes the following summary account of the local geology:

The Winnipeg area is part of the bed of old Lake Agassiz, the great glacial lake which many thousand years ago covered the greater part of Manitoba. There is evidence that this great sheet of water at one time stood at a depth

FIG. 18.10.—Typical settlement cracks in a single-story building, Winnipeg, probably due to dehydration of clay subsoil. Building had to be underpinned. (*Photograph by A. E. Macdonald.*)

of over five hundred feet above the present city of Winnipeg. Great masses of clay or powdered rock settled to the bottom of this lake and with the melting of the ice barrier and consequent release of the impounded water the site for the future city of Winnipeg was provided.

Limestone bedrock lies between 60 and 70 ft. below surface level, the intervening strata being a succession of clay and silt beds superimposed on a layer of boulder clay, or till; occasionally, beneath this impervious layer sand and gravel are found carrying water under high pressure. The serious settlement that has developed in many Winnipeg buildings is due to a variety of causes, but an important one has been the reduction in moisture content of the clay strata close to the surface and consequent shrinkage of the material supporting shallow footings. Heavy buildings are now founded on the bedrock through the medium of concrete piers.

Philadelphia, Pennsylvania.—In January, 1938, the Philadelphia Section of the American Society of Civil Engineers appointed a Committee on Foundations, with Professor G. P. Tschebotareff as Chairman, for the purpose of obtaining data concerning the engineering properties of local soil deposits, by means of full-scale observations on new structures, and of comparing the results of its observations to theoretical values.

The Committee has adopted the following statement covering the purpose of the investigations proposed:

> The purposes of the Foundation Committee shall be: first, to secure suitable projects for investigation by obtaining the permission of the owner or owners of new structures in the Philadelphia Area; second, to arrange and cause to be performed, certain investigations relative or pertinent to the problems of Foundation Settlement and Earth Pressures.

In its first *Report*, dated October 10, 1938, the Committee was able to state that cooperation with the architectural profession had been obtained, the Philadelphia Chapter of the American Institute of Architects having appointed a representative to the Committee. Permission to observe the behavior of four new buildings had been granted, and the "Committee proposes to engage in the collection of boring and other records [and] subsequently to prepare maps for publication showing the location of old creeks, the elevation of rock, and of the original natural soil surfaces."[18.26]

Shanghai, China.—The engineering Society of China has received many communications relative to subsurface conditions in Shanghai; reports of the Whangpoo Conservancy Board have supplemented these. A correlated bibliography is available. The local strata consist of the alluvial deposits of the Yangtze delta, extending to a depth of about 900 ft. below ground level, no rock having yet been located at a depth less than 700 ft. Strata consist of varying thicknesses of sand and clay.[18.27]

18.7. Building Regulations.—Even a discussion of the geological aspect of building foundations would not be complete without at least a brief reference to building regulations. These are ordinances, usually enforceable by law, issued by responsible local authorities such as city councils to regulate and safeguard the design and construction of all buildings within their jurisdiction. Foundation design has naturally to receive attention in these regulations. Although the details may vary according to locality, a usual form of regulation for the bearing values of rock and soils is as follows:

The load imposed on rock or soil in its natural bed shall not exceed its safe bearing value and, in the absence of a load test, the load shall not exceed the value prescribed for such material in the following table:

	Tons per sq. ft.
Solid rock	10 to 100
Gravel	6
Compact coarse sand	4
Fine dry sand	3
Dry clay	3
Sand and clay mixed	2

NOTE.—The foregoing are typical values for North American practice, given in short tons of 2,000 lb.

Soil tests of a simple character are usually specified and reference is generally to be found to borings and test pits although this is often brief and not always mandatory.

Geological references in building regulations are usually conspicuous by their absence. In many cases, there is no necessity for any reference to local geological conditions, apart from mention of the various soil and rock types. This is not always the case, and a note of warning may therefore be sounded against accepting the requirements of building regulations as a certain guard against all foundation troubles, including excessive settlement of completed structures. If there exists a weak stratum beneath the surface at a building site, the use of the lowest allowable bearing stress for the surface stratum will not automatically ensure that this material will not cause trouble. If this suggestion is considered in conjunction with the examples of excessive settlement mentioned in Sec. 18.3, the importance of carrying out adequate subsurface investigation at *all* major building sites, irrespective of what local building regulations may allow, will be clear. Engineers are not, as a rule, among those who believe all that they read in printed form, but they may, on occasion, be led away from salutary uncertainty as to subsurface conditions by accepting without question the foundation design data of official building codes.

This comment is in no way meant to disparage any set of building regulations or to detract in any way from the acknowledged and essential value of these codes in general. In order to be kept down to a reasonable size, their contents must be limited to the most important aspects of building—of which foundation conditions constitute only one division. It is as a reminder of the unusual conditions that may occasionally be encountered that this suggestion of caution is advanced, and also in connection with the possibility of inserting special provisions into the foundation chapters of codes for such districts, if and when these are revised.

The last few years have seen a great increase in the attention devoted to building regulations; opportunities for revision may therefore occur more frequently than in the past. One notable suggested revision relates to the proposed Building Code for the City of Boston, Massachusetts. Local engineers cooperated closely in all aspects of the work of revision, including that of the sections referring to foundations. The whole of Chap. 29 of this proposed Code is worthy of wide study; only four sections will here be quoted (by permission) as being especially relevant to the preceding discussion[18.27]:

Section 2903; *Borings and Test Pits.*

(*a*) Before issuing a permit for erection or alteration of a permanent structure, the Commissioner, in the absence of satisfactory data, shall require the owner to dig pits or make borings at such locations and carried to such depths as will disclose the character of the materials underlying the site of the proposed structure. When it is proposed to support a structure directly on bedrock, loading the rock to more than ten tons per square foot, the Commissioner may require a drill hole or a core boring to be made at each pier location, carried into the rock a sufficient depth to prove that sound bedrock has actually been reached.

(*b*) Copies, in duplicate, of the results of all borings and pits made or started, together with samples properly protected from evaporation representing the various classes of soil as nearly as possible in their natural state in the ground, including samples of the materials selected for the support of the foundation, shall be filed with the Commissioner for his information and classification. Washed or bucket samples will not be accepted.

Section 2904: *Classification & Allowable Loads of Foundation Materials.*

(*f*) Where the bearing materials directly under a foundation overlie a stratum having smaller allowable bearing values, these smaller values shall not be exceeded at the level of such stratum. Computation of the vertical pressure in the bearing materials at any depth below a foundation shall be made on the assumption that the load is spread uniformly at an angle of 60° with the horizontal; but the area considered as supporting the load shall not extend beyond the intersection of 60° planes of adjacent foundations.

(*h*) Whenever, in an excavation, an inward or upward flow of water develops in an otherwise satisfactory bearing material, special methods satisfactory to the Commissioner shall be immediately adopted to stop or control the flow to prevent disturbance of the bearing material. If such flow of water seriously impairs the structure of the bearing material, the allowable bearing value shall be reduced to that of the material in loose condition.

18.8. Some Unusual Foundation Bed Conditions.—In conclusion, brief notes are here presented on two unusual foundation bed conditions of geological interest. Although they may not be exactly duplicated elsewhere, the examples cited have special value in demonstrating something of the difficulties that some engineers have to face, and to overcome, and also in showing what strange conditions may be met with beneath the surface of the earth.

Detroit, Michigan.—Detroit is located in the glaciated area of North America; and the foundation bed conditions there encoun-

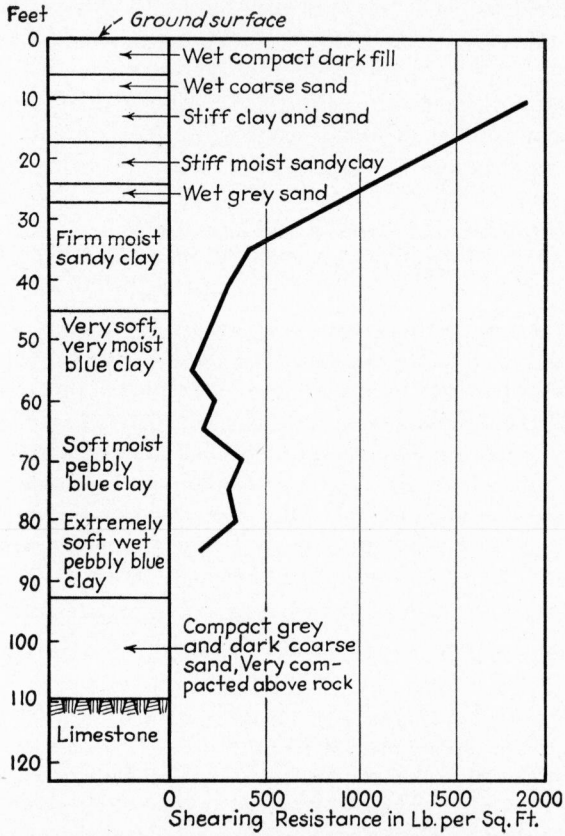

FIG. 18.11.—Typical test boring log, and foundation-strata shear characteristics, for the subsoil of Detroit, Michigan, U.S.A.

(Reproduced by permission of Mr. G. R. Thompson, City Engineer, Detroit.)

tered, although unusual, are not unique and may be found in other cities similarly located, *e.g.*, Winnipeg, Canada. The local bedrock is fissured limestone, overlain by clay strata of varying types with a layer of very hard glacial till ("boulder clay," or "hardpan") immediately above the rock. Depths to this impervious layer may vary up to 120 ft. If it is not pierced, foundation design and construction is not unusual; shallow foundations may bear on to clay strata, and deep foundations on to the impervious layer through the medium of piles or concrete piers. If the glacial till is pierced at all, or if a crack is encountered in it, subartesian water is met with which may rise as high as 100 ft. above the rock. During the construction of the Greater Penobscot Building, 565 ft. high, giving a total foundation load of 200 million pounds, these conditions were encountered, and bell-mouth open concrete caissons which were to have been founded on the impervious layer had to have compressed air applied to them and be carried right to rock. The water has an unusually high sulphur content, making it injurious to concrete. Efforts made to counteract this additional trouble included grouting of the bedrock around the base area of concrete caissons *before* the impervious layer was pierced.[18.28]

Mexico City, Mexico.—One of the most unusual geological formations on which a city has ever been built is the great basin at the south end of which is located Mexico City. The basin measures about 68 by 44 miles and is completely surrounded by mountains of which one is still an active volcano. Its surface is almost horizontal, the result of the fill that overlies the irregular rock surface. Below the topsoil are found horizontal layers of brightly colored material resembling marl or clay, separated by more or less pervious strata of sand and clay. The colored material is now known to be made up of fine vitreous particles, smaller than the particles of clay found in the bottom of the basin, originally formed as volcanic dust during volcanic activity which has characterized the region in the geological past. It does not possess the usual properties of ordinary soils, having a spongelike gel structure. Its weight decreases with increasing depth, reaching as low as 38 lb. per cubic foot; its porosity is indicated by a void ratio as high as 3.25, but when saturated it is not very permeable; it is, somewhat naturally, highly compressible. The difficulty of founding heavy loads on this material may be imagined. The building subsidences that have occurred in Mexico City in consequence of this condition are now widely known. Modern technique is being applied to the peculiar problems involved, and already technical publications dealing with local foundations are available. Notable among these are papers presented at the First International Conference on Soil Mechanics and Foundation Engineering, to which reference is made, and from one of which, by Dr. Jose Cuevas, the foregoing note has been prepared.[18.29]

Suggestions for Further Reading

Standard textbooks on foundations will usually be found to contain data relevant to some of the special aspects of foundation work considered in this chapter.

Special reference may usefully be made to:

PRENTIS, E. A. and L. WHITE: "Underpinning," Columbia University Press, New York, rev. ed., 1931.

Proceedings of the First International Conference on Soil Mechanics and Foundation Engineering, Harvard Univ., Cambridge Massachusetts, 3 vols., 1936.

TERZAGHI, K.: The Science of Foundations—Its Present and Future; *Am. Soc. Civil Eng. Trans.*, **93**: 270 (1929).

CHAPTER XIX

MATERIALS OF CONSTRUCTION

And they said one to another, Go to, let us make brick, and burn them thoroughly. And they had brick for stone, and slime had they for mortar.

Genesis 11:3.

Most of the materials used by civil engineers in the construction of the projects they design are obtained, directly or indirectly, from constituents of the earth's crust. Probably little thought is given to this fact in connection with the use of steel and other metals, for example, but the discovery and working of metallic ores are branches of economic geological work of great importance to civil engineering. With such indirect contacts the civil engineer is rarely concerned, but there are other materials that he frequently uses, the Geology of which is of direct significance. The construction of an earth dam, using local deposits of unconsolidated materials, provides an obvious instance of this nature. The use of clay for the manufacture of bricks, tiles, and similar products is clearly related to the geology of the clay thus utilized. The peculiar properties of the natural materials used for the production of limes and plasters are to a large extent determined by geological features. Finally, the extensive use of stone for building construction and the steadily growing use of artificial building "stone" —if concrete may be so described—are intimately connected with the geological history and mineralogical nature of the rocks thus put to use.

All the materials mentioned are regularly used in ordinary civil engineering work. Despite the fact that they may be supplied for use after having been inspected and approved by independent testing organizations, the civil engineer should be at least generally familiar with their origin and nature, particularly as he may one day be faced with the problem of selecting and approving such materials without aid, when engaged on pioneer work. No treatment of Geology as applied in civil engineering would therefore be complete without at least a brief reference to the Geology of these natural materials of construction. Inevitably, in the space available, only an introduction can be presented to consideration of the Geology of each of the divisions discussed. For convenience, sources of more detailed information are noted at the end of each section instead of at the end of the chapter. It will be seen that specialized literature available is of wide scope, in general, this being an indication of the attention that has been devoted to this contact of Geology with civil engineering. Similarly, textbooks dealing with the general subject of *Economic Geology* will usually be found to include sections devoted to so-called *structural materials.* The following are examples of such treatments, in which useful data are to be found:

GREGORY, J. W., "The Elements of Economic Geology," Methuen & Company, Ltd., London, 1927; see Chaps. 13 to 16.

TARR, W. A., "Introductory Economic Geology," McGraw-Hill Book Company, Inc., New York, 1930; see Chap. 14, Structural Materials.

19.2. Earth Dams.—Despite the usually restricted meaning of the word *earth,* its use to describe dams constructed of natural unconsolidated materials has now been generally accepted. Earth dams are essentially simple structures. Frequently, they are the most economical of all types possible at any locality. They are probably the most common type of small dam; but in recent years, many earth dams have been constructed that may truly be classed as major engineering structures, some being over 200 ft. in height. Their simplicity has resulted in their use in primitive communities, exemplified by some of the notable ancient dams of India. Their economy is often due to the fact that they can be constructed of material found adjacent to the dam site, a feature well demonstrated by some of the great natural earth dams that have been observed. One of the most recent of these was the Gros Ventre Dam, which was formed on June 23, 1925, by a great landslide in the Gros Ventre Canyon, forty miles south of Yellowstone Park, Wyoming. Eyewitnesses stated that the major slide occurred in less than 5 min. The resulting dam was 8,000 ft. long and 1,000 ft. wide on the top. River flow in the valley was cut off by the dam for 12 days, at the end of which a part of the dam was overtopped. Although considered safe, the dam partially failed on May 18, 1927, and six people lost their lives in the resulting flood. Figure 10.7 is a cross section of the valley at the site of the slide (see page 227).

Foundation requirements for earth dams are no less rigid than for dams of masonry or concrete, so that the preliminary investigations at the site of an earth dam must include due attention to the local geology. Studies will not be restricted to the dam site, however, if an earth dam can be considered as a possibility for the final design; subsurface studies must be continued around the site until uniform beds of suitable soils have been located that are sufficiently extensive to serve as sources of material for the dam. This work will ordinarily be carried out along standard lines, and test borings, test pits, and possibly geophysical methods used to supplement and confirm the results of surface observations. Samples of material in sufficient quantity to permit of detailed study will be required; they need not be undisturbed, since the soils sampled will have to be moved if they are to be used for dam construction. If preliminary tests, of grain size, permeability, and porosity, show that the material is suitable, more detailed investigations of the sources of supply will be necessary. These may be carried out by means of more frequent test borings and sampling; in some cases, test cuts with a steam shovel are made. Close study of the local geology, and especially that relating to the origin of the soils, can obviously be of great service in this work.

Up to relatively recent years, the design of earth dams, involving selection of the material, proportioning of the cross section, and a decision as to the necessity for a concrete or clay core wall, was based to a large extent on previous experience and general engineering judgment. The utilization of scientific design methods initiated a notable change of procedure which the use of the more recent results of Soil Mechanics studies has developed still further. Today, none but the smallest earth dams should be designed on any basis other than a complete and detailed analysis of all the factors involved, utilizing modern soil testing methods, scientific stability calculations, and accurate field control. These belong to the sphere of Soil Mechanics, the relation of Geology to which is outlined in Chap. XX. There are few types of soil containing a reasonable amount of fine material that cannot successfully be used for earth dam construction, so that once a suitable source of material is located, its geology ceases to have any special significance as regards design. Geology is here used in a general sense and in distinction to the detailed soil characteristics investigated in the laboratory, on which the main features of design are naturally dependent. Interesting though design methods are, especially those relating to the investigation of subsurface flow of water, they cannot here be considered.

Construction methods will be dependent to some extent on the type of soil being used. Dams built in rolled layers are frequently composed of material high in sand and gravel, at least for the downstream side of the core wall if one is included in the design. The ease with which material can be spread and rolled is a determining factor which automatically eliminates many soils with an appreciable clay content. Soils of this nature are more suitable for the hydraulic fill method, in which the soil forming the central part of the dam is pumped into place. The semihydraulic fill method is similar, material being transported to the dam in the dry state and then sluiced into position in the core. Both methods result in the formation of a core of fine particles, but the amount of fine material must be rigidly controlled if the core is to be stable and effective. Soils with a high clay content are therefore the least suitable of all natural formations for use in earth dam construction.

Sources of Further Information

The bibliography of the design and construction of earth dams is extensive, but geological features are usually mentioned only incidentally. General reference may usefully be made to the following volume, which includes a large number of descriptions of successful earth dams, details of some earth dam failures, and lists of further references:

JUSTIN, J. A.: "Earth Dam Projects," John Wiley & Sons, Inc., New York, 1932.

19.3. Puddle Clay.—It has been a common practice in Great Britain and other European countries to have a core wall in the center of earth dams constructed of "puddle clay," natural clay mixed up with water to form a plastic mass which can easily be handled and which can be "worked" into position so as to form a continuous wall of material. The puddling is usually done in a pugmill of special design. Puddle clay has also frequently been used to seal leaks in canal banks and similar water-retaining structures. The term is one that is somewhat loosely applied, since many types of clay can successfully be used for the purposes indicated. In North America, for example, it has a somewhat different connotation. The three main requirements of the puddle clay used in dam construction are its tenacity, its imperviousness to water, and the property of drying without serious cracking, shrinkage on drying being a minimum. These properties have been tested in the past by means of simple field tests. Although these are now liable to be superseded by the more detailed tests evolved in Soil Mechanics laboratory practice, they may be given at least this passing attention, since they have been successfully applied to the puddle clay requirements of many notable earth dams. Tenacity was often tested by making up a roll of the puddle clay about 1 in. in diameter and 12 in. long; this had to remain as one piece when suspended by one end. Permeability was tested by making a bowl of the clay to hold several gallons of water and comparing the loss therefrom with that from a similarly placed impervious basin. Rough and ready as they may now seem to be, tests such as these have proved of great service in the past.

The design of earth dams is changing its status rapidly, as has just been indicated, and puddle clay may not in the future be regarded with the same veneration as has been shown to it in the past. Despite this, it may be of use to present some data on puddle clay recently assembled in connection with the use of a core wall of this material in the dam across the Silent Valley in Northern Ireland, constructed to impound a supply of water for the city of Belfast. This is presented, by permission, as Table Q, reproduced from Mr. McIldowie's interesting paper on this work.

TABLE Q.—MECHANICAL ANALYSIS OF PUDDLE CLAYS[1]

Name of reservoir	Proportion of clay, per cent	Proportion of sand, per cent	Proportion of water when pugged to max. plasticity, per cent	Contraction on drying in bar 100 mm. long, per cent	Percolation of water through a layer of clay 2 in. thick under a head of 7 ft., cc.			Tenacity of molded briquettes 1 sq. in. in section when dry, pounds	Date when impounding began	Height of puddle core, feet
					After 24 hr.	Total in 432 hr.	During last 24 hr.			
1. Pontian Ketchil (Singapore Municipality)	94.6	5.4	29.8	7.2	Nil	Nil	Nil	65	1931	45
2. Blaen-y-cwm (Ebbw Vale Steel Company)	79.0	21.0	21.0	5.0	0.8	11.6	0.6	67	Completion, deferred	80
3. Carno (Ebbw Vale U.D.C.)	71.5	28.5	20.0	8.0	1.4	11.4	0.5	73	1911	95
4. Taf Fechan (Merthyr Tydfil Corporation)	71.5	28.5	17.4	2.9	0.2	2.2	0.1	...	1926	100
5. Silent Valley (Belfast Water Commissioners)	60.0	40.0	20.9	4.0	Nil	Nil	Nil	156	January, 1932	85
6. Burnhope (Durham County Water Board)	57.3	42.7	20.1	5.9	Nil	Nil	Nil	107	Under construction 1916	130
7. Alwen (Birkenhead Corporation)	50.6	49.4	15.0	2.5	2.3	17.6	0.6	30	.....	20
Average	69.2	30.8	20.6	5.1	0.7	6.1	0.2	83	.....	79

[1] Reproduced, by permission, from G. McIldowie, The Construction of the Silent Valley Reservoir, Belfast Water-supply, *Inst. Civil Eng. Min. Proc.*, **239**: 504 (*London*) (1935).

19.4. Bricks.—Civil engineers are rarely required to become brick-makers; but as is the case with other materials of construction, the source of bricks and the leading features of their manufacture should be familiar to those who have frequently to use them; civil engineers are numbered in this general company. Bricks are of two main kinds: sand lime bricks and those made by "firing" clay. The former are not met with so frequently as are the latter and have little direct relation to Geology. They are manufactured by specialized processes from mixtures of sand and lime; limiting requirements for sand and lime, to be mentioned later in this chapter, apply generally to their use in brickmaking.

The peculiar properties of many clays that make them so useful to man have been known from very early years in human history, as recent archaeological work has demonstrated. Working rules based on long established practice had to suffice until comparatively recent years for the guidance of brickmakers. Today a vast amount of scientific data is available, obtained not only in connection with brickmaking but also in connection with the ceramic industry generally, which has made brickmaking a specialized and highly technical industry. (It is of more than passing interest to contrast this well-established study of clays in connection with their use in manufacturing with their relative neglect until recent years—so far as scientific study goes—in general civil engineering practice.)

The two properties that have made clays so important a source of useful articles are their plasticity, by reason of which they can be molded into desired shapes when wet, and the way in which they will harden when subjected to the action of heat, thus establishing the molded shape permanently. The term *plasticity* is here used in relation to molding processes; if too plastic, a clay will be hard and sticky to mix; and if deficient in plasticity, may break up as it is being molded. Mixtures of clay and sand are therefore sometimes used in order to obtain a satisfactory material for the molding process. Shrinkage of clays is another important property in connection with brickmaking, uniformity of shrinkage being desirable. The actions that take place when clays are heated are complicated. Generally, however, two recognizable stages are of importance when bricks are made: that at which the clay vitrifies or becomes hard by a fluxing together of the whole mass which still retains its shape; and that at which the whole mass becomes soft or viscous, as a result of excessive fluxing, and loses its shape. Clearly, in the manufacture of bricks in large kilns, an appreciable temperature interval between the two stages is desirable.

A combination of these properties is what determines the suitability of a clay for brickmaking, provided that the clay is uniform, clean, and free from gravel and boulder intrusions. Thus it will be apparent that although most clays can be baked to give some sort of brick, by no means all will give a brick that can satisfactorily be used in construction. The presence of gravel in clays is an objectionable although not an insuperable difficulty; but if pebbles of limestone are present, or if the clay is of a calcareous nature, it will be advisable to continue the search for a suitable brick clay elsewhere. It is in connection with the location of possible sources of clay supplies that the contact between Geology and brickmaking is made, since when once a uniform clay supply has been located, its suitability for brickmaking depends primarily upon its physical and chemical properties, which have to be determined by suitable laboratory tests.

FIG. 19.1.—Excavating shale for brickmaking, by means of a mechanical shale planer, at the Acme Brick Company plant, Oklahoma City, Oklahoma.
(Reproduced by permission of the Eagle Iron Works, Des Moines, Iowa.)

Adobe clays, containing sand, are often used for making sun-dried bricks in districts having hot climates. Common bricks are made from a wide variety of suitable clays. Special "engineering," or high-quality pressed bricks are made from specially pure high-quality clays or shales and sometimes from inferior fire clays. The latter are clays that will stand a high degree of heat without fusing; their use for making fire bricks is well known. They are frequently found in association with coal deposits, in adjacent strata from 1 to 5 feet thick, but not all clays found with coal are fire clays. Paving bricks are made from clays that can be accurately machine molded and that vitrify at a moderate temperature.

It is sometimes thought that the color of bricks is an indication of their quality. This is so in the case of the famous Staffordshire blue bricks of England, made from clays and marls of the county of Stafford, which contain from 7 to 10 per cent of iron oxide; the bricks are burned at unusually high temperatures. But this case is a rare exception, for, in general, the conception is erroneous. Pure clay is white, but the usual color found in most clay deposits is due to oxides and carbonaceous matter. When clays are burned, those which are white usually remain so; but colored bricks result from colored clays, although the tint may change. The effect of iron oxides is generally to give a red or buff color; if lime is present, its effect may counteract that of the iron oxide, giving a cream- or yellow-colored brick. Terms such as *red burning* and *buff burning* are used to describe the final color of brick clays after they have been vitrified.[19.1]

In the course of the long-established business of brickmaking there has gradually developed, especially in Europe, a definite correlation between clay deposits of well-defined geological position and their suitability for brickmaking. Data as to this correlation and the areal distribution of the clays so described are usually obtainable from local geological surveys. Finally may be mentioned the fact that in certain parts of the United States in which there are no outcrops of solid rock for use as railway track ballast, burned clay ballast has been extensively and successfully used. It has been suggested that the same material could be used in similar areas even for such a purpose as the riprap protection for the slopes of earth dams. Both of these applications suggest the great use to which clays are put in civil engineering practice.[19.2]

Sources of Further Information

Dobson, E., and A. B. Searle: "Bricks and Tiles," The Technical Press, Ltd., London, 14th ed., 1937.

Ries, H.: "Clays: Their Occurrence, Properties and Uses," John Wiley & Sons, Inc., New York, 3d ed., 1927.

Wilson, Hewitt: "Ceramics—Clay Technology," McGraw-Hill Book Company, Inc., New York, 1927.

Publications of local geological surveys for areal distribution of brickmaking clays.

19.5. Tile and Other Structural Clay Products.—An extension of the use of clay for the provision of building materials has taken place during recent years in the development of *structural clay products* (as they are commonly called) such as hollow tiles and architectural terra cotta. These products are now widely used in connection with buildings constructed as steel or reinforced concrete frames. As showing how important they have become may be mentioned the fact that in the year 1935 there were produced in the United States of America about 900,000 tons of structural clay tile alone, this figure not

FIG. 19.2.—Starring Potteries, Littleborough, Lincolnshire, England. The fire clay here used for the manufacture of drain pipes, etc., is about 4 ft. thick and immediately underlies an extensive coal seam, one inch thick (which is worked as the "Inch Mine").

(Geological Survey of Great Britain photograph. Crown Copyright, reproduced by permission of The Controller of H. M. Stationery Office, London.)

including ceramic tiles, roofing tiles, sewer pipe, or any other of the miscellaneous structural clay products now available.[19.1]

Common to the manufacture of all these articles are processes similar to those used in brickmaking. Fine-grade clays are used, processed under more rigid control than in the case of common brick in view of the nature of the finished shapes. All the clays used are regularly submitted to detailed testing and analysis; excess of impurities such as iron, magnesia, or lime has to be carefully guarded against. Stoneware is usually made from refractory clays which burn to a dense mass at relatively low temperatures; clay mixtures, sometimes with the addition of sand, are frequently used. Sewer pipes have to be glazed during manufacture, and this is done with salt; it is said that a high iron content is of assistance, but the main requirement of clays for pipe making is that they shall be high in fluxes so that they will vitrify readily. The materials most commonly used for structural tiles are fire clay of first and second grades, high-grade brick clays, and suitable clay shales. Some of the latter have to be obtained with the aid of blasting so that regular quarrying operations may have to be planned and directed in connection with tile manufacture, in addition to the mining that has to be carried out in connection with the supply of some fire clays.

Manufacturing methods for these varied products are interesting but generally outside the scope of this book. Two details may perhaps be mentioned. In many plants, it is usual to store the clay or shale out-of-doors for an appreciable period before use. Broken up and exposed to the atmosphere in this way, the clay or shale will disintegrate and soften. This result, although often so unfortunate in civil engineering construction, is here an advantage, since the weathered material can be worked so easily. Color is often important in connection with structural clay products, especially in the case of hollow tiles to be used for partition walls. With uniform clays, this can be controlled during manufacture; but it is possible—and the practice is often followed—to introduce during manufacture iron oxide (and other substances) to ensure a uniform color of the desired shade.

Sources of Further Information

Refer to the end of the preceding section.

19.6. Building Stone.—"The style of a national architecture" wrote John Ruskin, "may evidently depend, in a great measure, upon the nature of the rocks of the country."[19.3] Although modern methods of construction have in many instances reduced building stone to a mere facing, this forthright statement yet remains a telling reminder of the close connection between Geology and the successful use of *building stone*, as specially cut and trimmed rock is generally termed. The granite of Egypt and the marble and limestone that typify so many European countries bear this out, as does also such a typically popular description as "The Limestone City" in relation to Kingston, Ontario, Canada. The relationship has provided one of the few points of contact between civil engineering and Geology that appear to have been generally and continuously recognized, at least to some degree, throughout the past century.

The study of the Geology of building stones may, indeed, have sometimes induced too great a regard for geological features to the neglect of other physical characteristics or even quarrying methods.

This was well shown at a very early date in connection with the stone used for the construction of the "new" Houses of Parliament in Westminster, England, about the year 1840. In the year 1838, a commission was appointed to select a suitable building stone for the work; William Smith and Thomas de la Beche, the first Director of the

FIG. 19.3.—Doorway of the Jermyn Street Geological Museum, London (for many years the headquarters of the Geological Survey of Great Britain) showing the excellent preservation of stonework, referred to in the text.

(Geological Survey of Great Britain photograph. Crown Copyright, reproduced by permission of The Controller of H. M. Stationery Office, London.)

Geological Survey of Great Britain, were members. After an exhaustive examination of many supplies of stone, a dolomitic limestone from Anston, near Mansfield, was selected as the most suitable. In the year 1861, just about twenty years later, another government commission had to be appointed to consider the "Decay of the Stone of the New

Palace of Westminster," so serious had the disintegration of the selected stone become. The fault was not in the geological selection but appears to have been due to the fact that at the quarry there was no inspection of stone as it was shipped. It has been stated that if only 15 per cent of the original stone had been rejected at the quarry (as showing signs of weathering or traversed by vents), trouble would have been obviated. Confirmation of this is given by the use of the same stone, but quarried and used under strict supervision, for the construction of the Jermyn Street Museum and Survey Headquarters in London within about one mile of the Parliament Buildings. After almost a century's exposure to the London atmosphere, the stone still presented a splendid appearance when the building was recently closed.*[19.4]

Geology, therefore, is only one of the factors on which the suitability of a building stone depends; but it is one of prime importance, since unless the geological nature and mineralogical content of a natural rock are satisfactory, other tests of its suitability as a building stone are useless. The following brief discussion of significant geological features applies generally to cut stone used for structural purposes and the stone shapes that are used for special road work, particularly in Europe. Ornamental stones and slate for roofing will not be considered, since their principal use is in architectural practice wherein long usage of stones from recognized quarries results in general satisfaction.

The adoption of natural stone as a building material for a civil engineering structure, in preference to an artificial material such as concrete, will usually be based on the following factors: availability of suitable rock, relative cost of cut stone delivered on the job, appearance of dressed stone, strength of stone, and durability of stone. The relation of Geology to each of these features will be considered.

Availability of Suitable Rock.—This is an obvious prerequisite for the use of cut stone as a structural material, and one that can be determined only by means of a geological survey. In open country, this may be a simple matter; but in drift-covered areas, the services of a trained geologist may be essential. It is of some importance to note that the economic necessity for using stone in preference to concrete usually increases with distance from settled country and so from areas in which building stone supplies have been accurately surveyed. Thus it is that civil engineers engaged on pioneer work

* It is, perhaps, of more than passing interest to note that the establishment of the British Museum of Practical Geology, which was for so long the headquarters of the British Geological Survey, was largely due to this investigation, the Jermyn Street building being provided to accommodate the many specimens of British building stones assembled by the Commission of Enquiry. It was opened in 1851, superseding a smaller building opened in 1841.

have frequently to make original investigations of sources of rock suitable for use in their work.

Relative Cost of Cut Stone.—Cut stone will not usually be used on civil engineering works, except as decoration, unless it proves to be an economical alternative to concrete. In isolated localities, this economy may be well defined; but in more normal practice, accurate calculations of cost may have to be made before a decision can be reached. Delivered cost of the stone will depend on ease of quarrying, extent of handling, length of haul, and cost of dressing (including wastage). Quarrying and dressing costs will be determined by the nature of the rock to be worked, massive formations of hard igneous rocks being obviously more expensive to work than blocky limestone which can be quarried without the aid of explosives and readily dressed. Geological investigations will assist, therefore, in cost determination, particularly in enabling an estimate of the essential homogeneity of the rock mass to be made.

Appearance of Dressed Stone.—The final appearance of dressed stone in a structure will depend on its texture, its mineralogical composition, and the weathering to which it is exposed. In civil engineering work, a beautiful appearance is not often the controlling factor in the selection of a building stone, although a satisfactory appearance will always be desirable for any stone that is to be exposed to public view. Uniformity of color and texture will usually be desirable; they can be checked up generally by visual examination of the rock mass before quarrying, coupled with an appreciation of its geological nature. The natural color of a rock is determined mainly by its mineral content. Thus, green may be due to the presence of chlorite or glauconite; and red and pink, to the presence of iron oxide. Normally, the color of a rock has to be accepted without question, but the mineralogical examination that is always necessary may sometimes reveal a condition of incipient discoloration. An interesting case of serious discoloration of grey granite, used for the reconstruction of Chelsea Bridge, London (the granite turning yellow quite rapidly), was investigated by Dr. B. H. Knight and reported by him in one of the few petrological papers to be found in the proceedings of civil engineering societies.[19.5] Dr. Knight's research showed that the staining occurred only in granite containing both white and dark mica and was due to the presence of a peculiarly unstable type of dark mica which can be detected microscopically. The staining can be artificially induced but not permanently obviated.

This investigation serves to demonstrate the valuable aid that the microscopic investigation of thin rock sections can give to civil engineers. There are, however, some injurious minerals that can be detected in a potential building stone with the unaided eye. Flint or chert (amorphous silica) is often found in the form of concretions in limestone; it is always injurious, and the use of cherty rock is to be avoided. Its presence affects the working of the stone and causes uneven weathering, sometimes leading to fracture of exposed stones.

Two-mica granite from higher spargo, Cornwall. Unstable uni-axial biotite mica at A, in the same section as stable bi-axial biotite mica B. × 25.

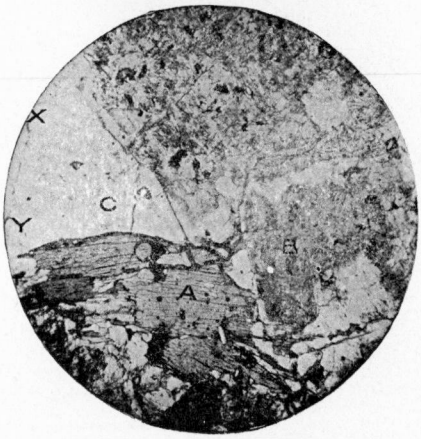

Two-mica granite from Creetown, Kirkcudbrightshire. Unstable uni-axial biotite mica at A, Breaking down and staining in decomposed feldspar B, and in fissured quartz C. The action is progressive, and will ultimately stain the whole surface of the stone. × 25.

FIG. 19.4.—Photomicrographs of two specimens of granite, obtained by Dr. B. H. Knight during his investigation on the rapid staining of certain British granites.

(*Reproduced by permission of Dr. B. H. Knight, and the Council of the Institution of Civil Engineers, London, from reference* 19.5.)

Mica may cause trouble if present in appreciable quantities in some types of limestone and sandstones, since, if segregated along bedding planes, it may lead to splitting of the rock. Pyrite, if present as anything more than fine grains uniformly spaced in the rock, is a potential source of trouble, as it weathers easily and may form sulphuric acid in the process. Fossil beds occasionally form planes of weakness in sedimentary rocks, but their presence is not generally objectionable.

Strength and Weight of Building Stones.—These physical properties are of obvious importance in connection with the use of stone in civil engineering work. They are dependent on the geological nature of the rock, as a glance at any table of weights and strengths of building stones will make clear. As one indication of the influence of geological nature on these properties may be noted the fact that fine-grained rocks are usually stronger than coarse-grained rocks. The strength of porous rocks (particularly sandstones) is dependent on the quantity of water present in them; tests should therefore always be carried out on both wet and dry samples. The variation in strength may amount to 50 per cent of the test results for the dry stone, so that ample factors of safety applied to the latter are advisable. In the case of sedimentary rocks, tests should always be made at right angles to the normal bedding plane of the rock. The weight of building stone is of importance mainly in structures such as dams in which stability is dependent on the weight of the structure. Close-grained igneous rocks will usually be found to be the heaviest of normally occurring formations, having thus a special advantage for this class of work.

Durability of Building Stones.—Durability is a prime requirement of a building stone, once it has been selected for use in civil engineering work. It is a property dependent on many factors and affected by varying conditions. It is also a property the testing of which is still a matter of debate, so that no recognized standards are available. The Geology of the rock from which the cut stone is obtained is again of paramount importance.

Expressed in simple terms, the durability of a stone is its capacity for retaining its original size, strength, and appearance throughout a long period of time while performing the work intended for it in the structure concerned. Reference is usually made to the stone exposed on the surface of structures; it is subject to the combined action of the atmosphere and of the extremes of temperature to which it is subjected. Essentially, therefore, durability is related directly to the normal weathering of rocks. The first and most obvious test of a building stone is therefore to be made by visiting the quarry from which it came and there examining the evidences of weathering on the naturally exposed surfaces of the untouched rock. If this were always possible, and were an invariable practice, some troubles with building stones would be avoided.

Conditions of exposure at a quarry may not be comparable with those to which the cut stone is to be subjected. Especially is this true if the stone is to be used in city areas where the atmosphere may be highly charged with soot and the waste gases of industry, which may tend to form weak acids in combination with atmospheric moisture. (The sulphuric acid content of soot layers resting on building ledges has been found to amount to as much as 5 per cent.) A second test, of a very general nature, is therefore to examine stone obtained from the same working in the same quarry which has been exposed to

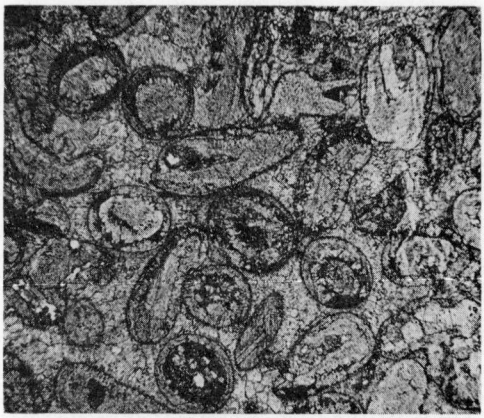

Oölitic Montreal limestone, parallel to bedding.

Queenston limestone (Ontario), parallel to bedding.

FIG. 19.5.—Photomicrographs of Canadian limestones. Magnification × 8½.

(Reproduced by permission of the Dept. of Mines and Resources, Canada, from Pub. No. 733, "Canadian Limestones for Building Purposes" by M. F. Goudge, Ottawa, 1933.)

the same type of atmospheric conditions as those contemplated, comparing the result of such observations with the length of exposure.

In the case of stone for architectural work, coming from established quarries, such observations can usefully be made. The possibility of making them will be the exception rather than the rule in civil engineering practice, thus necessitating a more fundamental study. A prime consideration is that natural building stone is porous and that its porosity will have an important bearing on the physical process of weathering. Driven rain can find its way into a porous stone, carrying with it salts in solution which may thus be deposited in the pore spaces; in time, they may cause trouble. Similarly, water trapped in the pore spaces and there frozen may lead to surface fracturing. The relative significance of these two processes is in doubt, much weathering being probably due to a combination of the two in addition to other causes. Determinations of porosity are a valuable guide, despite this, to the probable durability of a building stone. Similar physical tests include alternate drying and wetting, freezing and thawing, soaking in salt solutions and drying out, all three being in simulation of natural conditions to which the stone may be exposed.

Porosity is but one aspect of the internal texture of a rock, and is determined generally by its geological history and mineralogical nature. This can be studied with the unaided eye to a degree, but it can be fully investigated only by the examination of thin sections viewed through a petrological microscope. It is by this means that the rock-forming minerals are distinguished; their nature determines the chemical nature of the weathering that may take place at the rock surface. The petrological examination of thin rock sections is thus an important part of the testing of building stones; it probably represents the most important contact between petrological technique and civil engineering (see page 12). Rarely does a civil engineer have to attempt such work unaided; this will usually be the work of an expert. The engineer should, however, be able to appreciate expert work and its application to his own problems. A few leading points may therefore be suggested. Ready recognition can be made of coarse-grained rocks such as granite with its quartz, feldspar, and mica constituents; the existence of typically large crystals will immediately suggest that the rock will not be suitable for use in extremes of temperature in view of the differing coefficients of expansion of minerals. The finer grained rocks, basalt, for example, do not suffer from the same disadvantage; they include some of the more useful rock types for civil engineering work. Watch must be kept for differential weathering of rock minerals; all should preferably be fresh when the rock is quarried. Minerals that weather easily can readily be detected; they include pyrite, biotite (dark mica), feldspar, and nephelite. Their presence will often serve as a note of warning, usually suggesting the need of further investigation. The foliated metamorphic rocks cannot usually be considered as building stone because of the effect of foliation on weathering; slate is a peculiar exception.

The sedimentary rocks are widely used; and owing to their nature and the wide variation possible in their character, they present the most serious problems in connection with durability. Porosity will be one of the first things to be noted during a microscopical investigation; grain shape is also important, especially in the case of sandstones. Of even greater importance is the nature of the cementing material joining the sand grains; it may be siliceous, ferruginous, or calcareous. If of the first type, the rock should be sound, some siliceous sandstones being used even for acid pickling baths. Ferruginous cementing material is not satisfactory, as a rule; but calcareous cementing material is particularly dangerous. Although possibly good initially, it is very susceptible to the attack of even weak acids, as from the atmospheric moisture. Calcareous sandstones are known that contain 80 per cent of silica and yet are useless as building stones. By way of contrast, certain siliceous limestones containing 80 per cent of calcium carbonate are excellent stones, the silica content appearing to act as a protective coating to the main rock mineral. Extraneous materials constitute another danger in limestones; the minerals mentioned in connection with igneous rocks are still dangerous if present. In addition, clay is sometimes found in limestone and may prove to be dangerous because of its expansion when wet, thus leading to rapid disintegration of the stone.

These are a few of the phenomena that may be revealed by petrological study of a potential building stone. It is a testing method of great value; and although it may not be an infallible guide, it should be an invariable part of any building stone investigation.

Notes on the Use of Building Stone.—The large number of variable factors influencing the durability of building stone suggest that there may be restrictions to be observed in connection with its use that will assist in its preservation. Some of these may be noted. The advisability of always setting sedimentary building stones parallel to their bedding planes has been mentioned; it is most important. If dressed stone is to be used, this direction should be determined before the stone is dressed. In the case of porous stones, every possible precaution should be taken to facilitate drainage of the stone, particularly as some stones are more pervious the wetter they get. No ledges should be left unprotected on which water can collect, particularly if the exposure is in a city area.

The greatest care must be taken in the selection and use of the mortar used for jointing. Portland cement mortar should not be used with sandstones, and its use with some limestones may be inadvisable. Pitting of high-quality limestone (and sandstone) has been observed by Mr. S. H. Ellis in buildings in Liverpool, England. Although the exact reason for this is still not certain, the accompanying efflorescence suggests that portland cement backing to the stone is the cause.[19.6] All mortar for stone should be kept as low in lime and cement content as possible. Some authorities suggest that pointing should be left rough and not troweled off, in order to ensure that the joint is not sealed tight. Limestones and sandstones should never be used together in alternating positions.

The preservation of building stone is a matter of importance in architectural work; it rarely arises in civil engineering. The subject is debatable, but it seems now to be generally agreed that repeated washing with water is an efficient and practicable solution. The best solution of all, however, is to prevent initial decay by the selection of durable stone, a task in which it will be seen that Geology can be of great value.

Sources of Further Information

WILKINSON, G.: "Practical Geology and Ancient Architecture of Ireland," John Murray, London, 1845; an interesting old volume, mentioned to show the early recognition of this application of Geology.

BOWLES, OLIVER: "The Stone Industries," McGraw-Hill Book Company, Inc., New York, 1934.

ECKEL, E. C.: "Building Stones and Clays," John Wiley & Sons, Inc., New York, 1912.

HOWE, J. A.: "The Geology of Building Stones," Edward Arnold & Company. London, 1910.

KNIGHT, B. H.: "Road Aggregates, Their Uses and Testing," Edward Arnold & Company, London, 1935.

MARSH, J. E.: "Stone Decay and Its Prevention," Basil Blackwell & Mott, Ltd., Oxford, England, 1926.

MERRILL, G. P.: "Stones for Building and Decoration," John Wiley & Sons, Inc., New York, 3d ed., 1910.

SCHAFFER, R. J.: "Weathering of Building Stones," H. M. Stationery Office, London, 1936; contains a useful bibliography.

WARNES, A. R.: "Building Stones, Their Properties, Decay, and Preservation," Ernest Benn, Ltd., London, 1926, together with some of the references given at the end of Chap. XX and in Appendix D.

19.7. Limes and Plasters.—Although their importance in civil engineering work has declined with the steady advance in the use of concrete and concrete products, limes are still used in connection with building work and masonry construction and so are of some interest to the engineer. Lime is pure calcium oxide and is never found in nature; impure forms are prepared by calcining limestones, the calcium carbonate of which loses its carbon dioxide, yielding calcium oxide mixed with whatever impurities were present in the original limestone.

The distinguishing property of lime is the slaking that takes place when water is put into contact with it, calcium hydroxide resulting from the strong chemical reaction of the two. The slaking manifests itself by a general effervescence and disintegration of the usual lumps in which "quicklime" is obtained from a lime kiln. The "slaked lime" has the property of setting when exposed to the atmosphere, and for this reason it is used in building mortars.

The impurities in limestone may vary in quantity from a slight trace to a large percentage of the total volume of rock. They determine in large measure the nature of the lime that will be obtained when the limestone is calcined. The impurities cannot often be distinguished by eye, but they can be roughly determined by solution of the calcium carbonate in hydrochloric acid. When an engineer is faced with testing a limestone as a possible source of lime, his easiest course will usually be to calcine a small quantity of the natural rock and then examine the properties of the resulting lime. The more usual impurities are siliceous and clay materials; they have the effect of slowing up the slaking of the lime and developing its "hydraulic" property, *i.e.*, its ability to set when immersed in water. *Hydraulic limes*, as they are called, are commonly yellowish in color, in contrast to the white of pure quicklime; they set slowly and have little strength unless mixed with sand.

The impurities that cause these variations in the properties of limes naturally depend on the nature of the formation of the limestones from which the limes are made. They may therefore vary considerably, even between the top and bottom of a bed of rock. In the quarrying of limestone for lime making, care has therefore to be exercised with regard to the exact character of the rock being obtained. For example, it has been found that in the well-known Lias beds of England, variations in the quantity of aluminum silicates present from 8 to 64 per cent may occur, the corresponding quantity of calcium carbonate varying from 90 to 34 per cent.[19.7] Limestones that contain fossils may produce a lime with variable and uncertain slaking properties due to the peculiar calcining of some fossil material. Small particles that will not slake may occur, subsequently taking up water and tending to disintegrate the mass of material surrounding them.

Manufacture of lime is carried out in coal- or wood-heated kilns, calcining taking place at about 900°C., or somewhat less if steam is present in the kiln. Higher temperatures will result in clinkering or partial fusion of the rock if clay impurities are present. The resulting quicklime will be obtained in blocks corresponding to the original blocks of rock; it is usually slaked in special boxes before being used.

Various names such as *fat lime, rich lime,* and *poor lime* are used to denote different grades; the rate and nature of slaking are a usual criterion. These descriptive terms should be accepted and used with caution, as they are essentially local in significance. If the lime has an appreciable percentage of magnesium oxide in it, derived from the magnesium compounds present in the original rock, it will be designated as a *magnesian* or *dolomitic* lime.

The plasters that play so important a part in finishing processes in building construction are derived mainly from deposits of gypsum, the hydrous sulphate of calcium. This occurs as natural deposits,

FIG. 19.6.—Working face in gypsum quarry, Dingwall, Cape Breton Island, Nova Scotia, Canada.

(Reproduced by permission of National Gypsum (Canada), Ltd., Mr. J. W. McFarland, Vice President.)

usually as a solid rock mass, but in some places as unconsolidated sandy material. It is fairly widely distributed, one of the most notable deposits of high-grade gypsum being in the extreme north of Nova Scotia, on Cape Breton Island, Canada. It is often of a high degree of purity, having probably originated by the evaporation of oceanic waters and the consequent concentration of the dissolved salts. It is often found in association with common salt. Gypsum will frequently be found to change to anhydrite, the anhydrous calcium sulphate, some of which will generally be found in most gypsum deposits. Unlike gypsum, anhydrite is of no commercial value and so has to be carefully distinguished and discarded. (The trouble that the action of water on anhydrite can cause, changing it to gypsum with a corresponding increase in volume, has been mentioned in previous chapters.) When pure gypsum is heated to a temperature of between 250 and 400°F., it loses most of its combined water and gives an amorphous substance commercially known as *plaster of Paris*. When this is mixed with water, it takes up by chemical combination as much as was previously lost and sets as a hard and compact mass. Special plasters may be obtained by variation of the calcining process and by the admixture of other materials with plaster of Paris.

Sources of Further Information

ECKEL, E. C.: "Cements, Limes and Plasters," John Wiley & Sons, Inc., New York, 1922.

Geological survey publications will usually be found to deal with the local distribution of limestone and gypsum deposits suitable for commercial use.

19.8. Cements.—Because of the general availability of modern portland cement, prepared by expert manufacturers to rigid standards of quality, a civil engineer has but seldom to give close attention to its origin. So widely is cement used, however, and so far reaching has been its effect on modern construction that a brief note on the geological aspect of its preparation is warranted. Portland cement was first patented in 1824 by an Englishman, Joseph Aspdin, and so named because he obtained his limestone from the famous Portland quarries. The cement is an artificial product in that it is made by partially fusing together a specially prepared mixture of natural materials; the word *artificial* is used in distinction to the natural cements which are available and which are mentioned later. The usual mixture for cement manufacture consists of limestone (to give the necessary lime), clay or shale (to give the necessary silica and alumina), and a small quantity of iron oxide. The clinker obtained by heating this mixture is finely ground to give the gray amorphous powder familiar to all civil engineers. The limestones used can have a wide variation in hardness, texture, and chemical composition; but magnesia, free particles of silica, and sulphur are undesirable constituents. Clays used in cement manufacture are often impure, but they cannot contain gravel or any free solid particles. Relative proportions of the several materials can be determined by calculation when the nature of the constituents is known, this being one of the several phases of cement manufacture that are the work of experts. Portland cement is distinguished by its slow and even setting when mixed with water and its relatively high strength. Its exact chemical composition is complicated; but by means of modern methods of analysis, some knowledge of this is now being obtained.

In addition to standard portland cement, special cements are available for a variety of unusual purposes. Thus, white cement for decorative use is obtained by the use of clays that burn to a final white color. Aluminous cement, having an unusually high alumina content, is produced in France by fusing together a mixture of limestone, bauxite, and coke; it is unusually quick in setting and is said to be more resistant to the action of sea water than ordinary portland cement. Tests have shown clearly that it is more resistant to the action of acidic moorland waters.[19.8] In Canada, a special high-silica cement has been invented which is extremely resistant to the peculiar alkaline soil conditions found in some parts of the west of the Dominion.[19.9]

FIG. 19.7.—Quarry face at the Montreal east plant of the Canada Cement Co., Ltd., somewhat weathered as this part of the face has not recently been worked. The limestone here obtained gives a natural cement mixture.
(*Reproduced by permission of Canada Cement Co., Ltd., Montreal.*)

Although artificially prepared cements are now in general use, they represent an ingenious adaptation of natural materials to give an imitation of a natural product. If a limestone has a very high clay content and certain other impurities, it may give a natural cement when calcined at a higher temperature than that used for the production of lime. Although not common, some deposits of this kind are used—sometimes with small admixtures—for the manufacture of cement. Natural cements are usually found to set more rapidly than portland cement and to have a somewhat lower strength; they are naturally somewhat variable in composition. Natural cement is sometimes called *Roman cement;* it appears to have been discovered by a Mr. Parker of London, England, in 1796, who took out a patent in that year for the manufacture of what he called Roman cement from the septaria of clayey limestone found in the London clay formation of the island of Sheppey, England. It was found later that other deposits would yield similar natural cements.

Mention of the name Roman in connection with cement is a reminder of one of the most fascinating aspects of early Roman engineering—the discovery and use of *pozzuolana.* This is a volcanic ash found in great beds near Naples at Puteoli from which was derived the ancient name *pulvis Puteolanus.* Puteoli is now called *Pozzuoli,* from which the modern name has been derived. The Romans found that if this ash was mixed with lime instead of the sand ordinarily used, a watertight cement was obtained which would set under water. The use of this mixture has been found in the lining of the water channel in the Aqua Marcia, dating as early as 144 B.C. It was subsequently the principal constituent of the concrete so boldly used in later Roman structures. Modern Italian builders still use the same material; in other countries, similar deposits of volcanic ash are used to make pozzuolana cement. Smeaton used pozzuolana in the construction of the Eddystone Lighthouse (1757), one of the pioneer structures of modern engineering. The name is sometimes loosely used; a slag cement made up of a mixture of finely ground blast furnace slag and lime is sometimes thus described. It would seem that the name should rightly be restricted to cement containing natural volcanic ash.[19.10]

Source of Further Information

DAVIS, A. C.: "Portland Cement," Concrete Publications, Ltd., London, 1934.

19.9. Sands.—Sands constitute an important part of the unconsolidated materials described as *soils* and are therefore discussed in the next chapter. They are used as materials of construction not only as filling and for providing porous foundation blankets (as for roads) but also to a wide extent as a filtering medium and as a constituent of mortars and concrete. Their use in conjunction with clays in forming a natural highway wearing surface has been mentioned. In view of these important applications, some brief notes will therefore be here included additional to those that will be found on page 577.

First it must be pointed out that the term *sand* is used to describe naturally granular material of a certain grain *size* irrespective of the shape of the grains, their uniformity, and their mineral composition. The formation of sands during the process of rock weathering and the way in which they are transported before reaching their final position affect all three of these leading characteristics of sand. The mineralogical nature of sand is important, especially in view of the popular belief that all sands are composed of quartz particles. Although quartz is frequently a major constituent, pure silica sand is the exception rather than the rule. Depending on the nature of the rock from which it was formed and on the erosive action to which it has been subjected, many other minerals may be found as constituents. Mica will sometimes be present; feldspar is more common; even particles of shale occasionally occur. The presence of such materials in sand make a detailed examination of many sands imperative before their use as concrete aggregate, for example. Feldspar is not a thoroughly stable mineral, and it seems probable that its presence in some sands is responsible for "hair cracking" in some concretes. Shale particles may be equally injurious, as may also be the presence of any soluble salts.

Specifications for sand for concrete usually exclude all sands containing deleterious substances and restrict the clay and silt content of sand to a small amount, often 5 per cent or less. They usually exclude also sand that is not "clean," a term used to signify freedom from organic matter such as tannic acid derived from surface vegetation. Geology can here often assist by suggesting the mode of origin of sands, the probable extent of deposits, and the probable origin of organic matter; it thus presents a guide to the condemnation of either a whole deposit or just an upper layer.

Uniformity of grain size is of importance in connection with special sands used for standard concrete tests and for use as a filtering medium. For the latter purpose, sands should have a high silica content in addition to uniform grain size. The latter property is described by the use of *uniformity coefficients* and *effective size* statistics, definitions of which are to be found in engineering texts; it is determined by their origin and mineral content. Sands that have been subject to marine action are often fine grained and fairly uniform. River sand may have similar properties but is liable to be more variable. Glacial sands frequently are mixed with gravel and are not generally uniform. These several types of sand have fairly characteristic grain shapes none of which, incidentally, are rounded as a general rule. Rounded grains are generally an indication of a wind-deposited sand (see Fig. 20.23, page 579).

Sources of Further Information

MARR, J. E.: "Deposition of the Sedimentary Rocks," Cambridge University Press, Cambridge, England, 1929.

TWENHOFEL, W. H.: "Treatise on Sedimentation," Williams & Wilkins Company, Baltimore, Maryland, 2d ed., 1932.
Standard specifications such as those of the A. S. T. M.

19.10. Crushed Rock.—Crushed rock is used for a variety of purposes in civil engineering; it has the advantage over natural gravel of having angular particle shape, and its limiting size and grading can be controlled as desired. As railway ballast it is widely used; its corresponding use as road-forming material has already been discussed (see page 250). As a special type of fill material, especially for drainage work, it is most useful. It is very widely used as coarse aggregate for concrete making; as the resulting concrete is in effect an artificial conglomerate of the aggregate used, the geological significance of this application will be discussed in the next section.

A common practice on construction jobs is to utilize any rock that has to be excavated as a source, when crushed, of concrete aggregate and for other purposes. If the rock is suitable for concrete work, a qualification discussed later, this practice may lead to great economy in construction. It comes within the experience of many civil engineers, and it might therefore be expected that the operation of rock-crushing plants and their correlation with the nature of leading rock types would have received wide attention in civil engineering literature. This does not appear to be the case; the papers listed at the end of this section are only the more important of the few on the subject known to the author. That by Messrs. Miller and Sarjant, records in detail the results of an interesting study of various types of rock crushers and their performance with different rock types; the following notes are based upon its conclusions.

The effect of different rock types is well shown by a record of crusher jaws lasting only 10 days when crushing granite, whereas a pair of exactly similar jaws lasted 7 years when used for crushing limestone! This surprising difference is best explained by variations in the *toughness* and *hardness* of rocks. Hardness is generally measured by testing the abrasive quality of rock in a machine in which a sample is pressed on a revolving steel disk, and toughness is usually measured by impact tests on rock samples. The research already noted showed that rocks can be classified generally into groups having distinctly different abrasive qualities and that within these groups the factor that principally influenced the wear on crusher parts was the toughness of the rock. It was also found that the strength of rock specimens tested to failure in compression could be related to their behavior when being crushed into rock fragments. The shape of the fragments obtained from a crusher was found to be largely influenced by natural planes of weakness in the rock but, in general, particle shape could be influenced also by the nature and direction of the blow struck by the crusher mechanism, *i.e.*, by the type of crusher being used. Grain size was suggested as a factor of importance, coarse-grained rocks being found poorer in quality than corresponding fine-grained rocks. Geological age, in the case of sedimentary rocks, was also noted as a matter of significance.

The variable factors are so numerous that no very specific statements can be made with regard to the crushing qualities of individual rock types. The following general statements are a guide. *Granites* possess variable crushing qualities, although their behavior when crushed is related to their resistance to impact. *Porphyries* are, in general, very tough and hard. *Basalts* naturally show a wide variety of results, but crushing strength is found to be related to porosity, possibly produced by partial decomposition. *Quartzites* are moderately tough, but *gneisses* are much less so, because of their foliated texture. In many sedimentary rocks, the nature of the cementing material governs their performance during crushing.

These notes are an indication of the correlation of geological and physical properties of rock types with their performance when being crushed. A careful geological examination of samples of any rock to be crushed should therefore be made before any plans for rock crushing are completed. If these can be supplemented by physical tests on samples and also by petrological examinations, a good general idea of the probable behavior of the rock during crushing should be obtained. Special attention should always be paid to the natural planes of cleavage. The natural tendency to break along these planes can be offset to some degree by the use of a suitable crusher type. Variation in particle shape can have an appreciable effect on the workability of concrete and therefore on the necessary amount of cement to be used. A change to a correct type of crusher may sometimes result in a saving of one bag of concrete per cubic yard; the monetary saving that this can mean on a large construction job will readily be imagined.

Sources of Further Information

KNIGHT, B. H.: The Interpretation of Physical Tests of Road Making Stones, *Good Roads (London)*, November, 1933.

KNIGHT, B. H.: "Road Aggregates, Their Uses and Testing," Edward Arnold & Company, London, 1935.

LOVEGROVE, E. J.: "Attrition Tests on British Road Stone," H. M. Stationery Office, London, 1929.

MILLER, W. T. W., and R. J. SARJANT: The Evolution of Various Types of Rock Crushers for Stone and Ore, and the Characteristics of Rock as Affecting

Abrasion in Crushing Machinery, *Inst. Civil Eng. Min. Proc.* (*London*), **239**: 39 (1934).

19.11. Concrete.—Concrete is one of the three main materials of construction utilized in modern civil engineering work. It is, perhaps, that which should be most closely studied by civil engineers. The quality of a given piece of timber cannot be varied, and steel has also definite physical properties when supplied for use. The quality and strength of concrete, on the other hand, are under the control of the engineer in actual charge of construction. This provides at once an opportunity and a responsibility which are being met by continued study of and instruction in concrete technique.

The use of concrete in modern practice probably dates from about the year 1850; but as in the case of cement, this was a rediscovery rather than an innovation. The fame of the Romans as the master builders of olden days is due in no small measure to their use of concrete, *structura caementicia*, which they made from the pozzuolana cement already mentioned. This they mixed with broken fragments of stone and tile; they poured it into timber forms, the marks of which can still be seen in places; and they even used pumice as aggregate when lightweight concrete was needed. The great dome of the Pantheon and other vaulted roofs testify to their skill in the use of this early concrete, lacking only reinforcing steel to make it the counterpart of that used today.[19.10]

Despite the number of factors that determine the final character of concrete, the nature of the aggregate is always of prime importance. When it is recalled that in an average concrete, 70 per cent by volume, and possibly 80 per cent by weight, consists of the coarse aggregate used, this importance will be more readily appreciated. The attention usually devoted to the testing of coarse aggregate is not in accord with its importance in the finished product. Compared to the care exercised in the testing of cement, the checks made on the suitability of aggregates are sometimes insignificant. Although the old saying that the strength of a concrete is the strength of its aggregate cannot be upheld as generally accurate, it does give a vivid indication of the general importance of the aggregate in a concrete mix. It must be added that a good aggregate will not ensure good concrete, the possible variations in grading, mixing, placing, and curing (quite apart from climatic conditions during setting) being so great that they may outweigh by far the initial advantages presented by an unusually sound and strong aggregate. On the other hand, the use of a poor aggregate is one of the easiest ways of obtaining concrete that will be a source of constant trouble.

A satisfactory aggregate must meet four main requirements: It must be clean; it must be durable; it must be correctly graded in size; and it must be uniform both in grading and in quality. Grading can be controlled in the crushing and screening plant in the case of a crushed rock. Uniformity can be similarly controlled, by close inspection of the supplies of aggregate. Cleanliness can be obtained if

FIG. 19.8.—Vaulting of Roman concrete (*structura caementicia*), with brick facing, in Trajan's Market Hall, Rome.

(Reproduced by permission of the Society for the Promotion of Hellenic Studies, London.)

suitable measures are taken. And durability can be determined before an aggregate is selected for use. Both these properties are dependent on the geological history and composition of the rock to be used.

Cleanliness of coarse aggregate is similar to that required of fine aggregate, organic impurities being the most objectionable form of "dirt." These impurities are usually in the form of weak organic acids or their derivatives, which may seriously impair the setting of the cement. One of the earliest occasions on which the serious effects of organic matter on concrete was noted was during the construction of the Quinze and Kippewa Lake dams, in the upper reaches of the Ottawa River, Quebec, by the Department of Public Works, Canada, in 1910–1911. The organic matter could not be detected by eye, since it was found to have formed transparent coatings round the particles of the siliceous aggregate used. Eventually it was found that by washing the aggregate with a 3 per cent solution of commercial caustic soda, it could be cleaned. This was therefore done in a special plant. The concrete made with aggregate not so treated disintegrated rapidly.[19.11] A colorimetric test, using a similar solution of caustic soda, is now almost universally used as an indicator of the presence of organic matter in aggregate. It is of special interest to note that these organic acids will not be encountered in calcareous sands or in limestone or similar calcareous rocks, with the calcium carbonate of which the acids can react. This simple geological fact will often prove a useful guide in locating good aggregate. In "limestone districts," clean aggregates will usually be widespread; this is the case in Ontario, Canada, west of the Niagara escarpment, apart from a small area near Windsor.

Washing of aggregate with water is often necessary not only for removing organic impurities but also for washing out surface dirt and dust. It is also imperative with some types of crushed rock in order to remove the rock dust attached to the rock fragments as they leave the crushing plant, which may be a potential source of trouble. Minute particles of feldspar formed during the crushing of some granites have been found to affect the quality of concrete, and other injurious minerals present in rock may have similar serious effects. The detection of these injurious minerals is related to the study of the durability of a proposed aggregate in a way similar to the use of petrology in building stone investigations. Much of what was said with regard to the durability of building stones applies equally well to the durability of rock used as aggregate. It can therefore be investigated by physical tests on porosity, by freezing and thawing, and by soaking in salt solutions—always coordinated with petrological study of the rock under the microscope. Injurious minerals in building stones will be injurious minerals in aggregate and for the same reasons. Possibly the most serious defect is the presence of any material that will swell when in contact with water.

This last property is of unusual importance in connection with concrete, since many types of rock that would not be considered as building stones will often be used as concrete aggregate. The economic "temptation" offered by rock that has to be excavated on a construction project, as providing cheap concrete aggregate, has already been mentioned. It has probably been responsible for much of the poor-quality concrete to be seen today. In many cases, engineers may have been forced to use unsuitable rock in this way against their better judgment. Limestone with any clay content, or so-called *argillaceous limestone*, is a rock often encountered that looks sound and durable when freshly quarried. Argillaceous shales and slates are similarly undesirable; another photograph shows an extreme case of disintegration arising from the use of shale as aggregate, the rock being seen in place in the foreground. (Other types of shale, if relatively strong, can

FIG. 19.9.—Disintegration of concrete, due in part to the use of a local argillaceous limestone as aggregate.

be used safely, although the typically laminated structure is always a source of weakness.) Clay in any form is therefore to be avoided in concrete aggregate; petrological examination will assist in its detection.

Porosity of rock to be used as aggregate is just as important as in the case of building stones. In addition to the usual penetration of rain water into exposed aggregate, it must be remembered that when the concrete is placed in position, it has a high water content so that the aggregate will be in a saturated condition. This may lead to strange results, especially if the concrete is subjected to frost before the outer part has had time to set permanently. The freezing of water thus trapped in the aggregate may sometimes lead to the disruption of otherwise sound concrete. This is a reminder that the aggregate used is only one of the variables involved in the preparation of good concrete. Possibly for this reason it is a little difficult to specify exactly desirable characteristics for a high-grade concrete aggregate. The prohibition of such generally recognized materials as gypsum, anhydrite, and chert is frequently inserted. Some specifications also limit the clay content to a very small percentage if not entirely. Metamorphic rocks of doubtful value can also be excluded by reference to fragments having a foliated structure. There remain a number of possible potentially dangerous mineral constituents, but to list them

FIG. 19.10.—Disintegration of concrete pedestal, due in part to use of local shale as aggregate. The shale, *in situ*, can be seen in the foreground. Lower part of pedestal has been "patched," but disintegration has continued.

all is impossible in the usual specification. To overcome this difficulty, a general clause restricting the use of aggregates to those obtained from approved sources of supply may sometimes be possible without hardship to contractors; alternatively, the engineer may himself arrange for the supply of satisfactory aggregate on the job. By following these precautions, close check can be kept on this important constituent of concrete.

Finally, an interesting sidelight on the relation of Geology to concrete—fortunately not of general application—may be mentioned; it is described in the record of the tunneling work carried out by the British Army in the World War.[19.12] Captain W. B. R. King, who served as Geologist to the Chief Engineer in France, acting on information supplied by Dr. H. H. Thomas, was able to prove, by the relatively simple device of studying the concrete used for the construction of "pillboxes" on the Ypres and Vimy fronts, that the Germans were transporting war material through the neutral Dutch canal system! The aggregate was a peculiar form of basalt which is found only on the Rhine above Cologne at Niedermendig, a simple geological fact thus used to good effect.

Sources of Further Information

BOSWELL, P. G. H.: The Geology of Sands and Aggregates for Concrete Making, *The Quarry (London)*, **24**: 242 (1919).

HOLMES, A.: Geological and Physical Characteristics of Concrete Aggregates, *British Fire Prevention Committee Red Book* 256, H. M. Stationery Office, London, 1922.

KNIGHT, B. H.: "Road Aggregates, Their Uses and Testing," Edward Arnold & Company, London, 1935.

LOUGHLIN, G. F.: Quality of Different Kinds of Natural Rock for Concrete Aggregate, *Jour. Concrete Inst.*, **23**: 319 (1927).

VIENS, E.: Does Cement Protect Poor Quality Aggregates? Yes and No, *Journal Am. Concrete Inst.*, **5** (new ser.): 437 (1934).

Discussions of the matter are also to be found in the proceedings of such bodies as the A. S. T. M.

PERIODICALS AND TRADE JOURNALS

So much valuable information is available in the periodicals and trade journals concerned with materials of construction that this partial list is presented as a guide to wider reading on some of the matters discussed in this chapter:

Cement and Lime Manufacture (monthly), Concrete Publications, Ltd., London.
Cement, Lime and Gravel (monthly), Quarry Manager's Journal, Ltd., London.
Concrete and Constructional Engineering (monthly), Concrete Publications, Ltd., London.
Crushed Stone Journal, National Crushed Stone Association, Washington.
Publications of the National Sand and Gravel Association, Washington.
Publications of the National Lime Association, Washington.
Pit and Quarry (monthly), Complete Service Publishing Co., Chicago, Illinois.
Quarry Managers' Journal (monthly), Quarry Managers' Journal, Ltd., London.
Quarry and Roadmaking (monthly), Colliery Guardian Co., Ltd., London.
Rock Products (monthly), Tradepress Publishing Corp., Chicago, Illinois.
Sands, Clays and Minerals, A. L. Curtis, Chatteris, Cambridge, England.

CHAPTER XX

SOIL AND SOIL MECHANICS

> I had ample opportunity to witness the striking contrast between what we expected when digging into the earth or loading it, and what really happened. . . . I came to the United States and hoped to discover the philosopher's stone by accumulating and co-ordinating geological information. . . . It took me two years of strenuous work to discover that geological information must be supplemented by numerical data which can only be obtained by physical tests carried out in a laboratory.
>
> KARL VON TERZAGHI.[20.1]

The unconsolidated materials found in the earth's crust, generally described as soils, constitute so large a part of the actual surface of the earth that few civil engineering operations, apart from rock tunneling, can be carried out without an encounter with soil of some type. Foundations cannot always be carried to solid rock, the founding of structures on unconsolidated material being probably the most important part of foundation engineering. Troubles met with during the construction of civil engineering works, and after their completion, are frequently due to failure of unconsolidated strata. Despite their wide and important significance in all phases of civil engineering work, soils were not studied and investigated by civil engineers in any general way until relatively recent years. The contrast between this neglect and the past century of progress in the study and investigation of practically all other materials used by civil engineers is so marked as to be indeed a paradox. Nobody sells soils, however; they are generally a part of the natural order which the civil engineer has to accept as the basis of his operations. It may be that the absence of commercial incentive is linked in some way with this past neglect; the development of testing and research laboratories operated under public control, for public benefit, has probably had to be the necessary incentive to the surprising progress in scientific soil study during the last two decades.

Terminology has been another stumbling block. The word soil is popularly used to denote the loose material constituting the surface of the earth, in which plant life grows; in this sense, a layer no more than a few inches thick and containing humus is usually imagined. The word is used technically to describe this top layer in scientific agricultural studies. Surface material, generally containing organic matter, is of little consequence to the civil engineer. His use of the word soil applies to all unconsolidated material in the earth's crust, extending possibly to depths of several thousand feet below the surface. This wide use of the word is not strictly in accord with geological terminology, in which its use is sometimes restricted to products of rock weathering that have not been transported from their original position. It may be more accurate and more in accord with popular usage to describe this mantle of unconsolidated material, with the exception of the topsoil, as *subsoil*, and this compound word is often so used. It has not been generally adopted; Subsoil Mechanics is a term fortunately never heard. Soil will therefore be the word here used to denote generally the unconsolidated material of the earth's crust encountered in the ordinary course of civil engineering work.

It is of some value to consider briefly the wide range of interests associated with soils. The geologist is naturally concerned with all such unconsolidated material, many geologists devoting themselves particularly to its study. Agriculturalists are especially interested in the upper soils on which plant life depends; as agriculture has developed along scientific lines, so has the detailed study of soils from the agricultural viewpoint. The development of what used to be called the *useful earths, e.g.*, the many clays used in ceramics and manufacturing processes and the sands used in glassmaking and molding, has called for detailed study of the nature of particular types of soil. Special methods of investigation and testing have been evolved in all these connections, the chemist and the physicist particularly having been of great service. These allied soil studies are now linked with the special soil investigations developed in connection with engineering work. Particularly is this true of physical and mechanical testing methods; some of those now regularly used in engineering laboratories were first evolved in connection with other work. Several notable works on the relation of Geology to agriculture have been published, and these contain data of value in connection with engineering as well as agricultural work.

20.2. Historical Note.—In the earliest days of modern civil engineering, the characteristics of soils appear to have received some attention notably with reference to the pressure exerted on retaining walls. The theoretical work of Coulomb (1773) and of Rankine (1856) is fairly well known and appreciated. In France, remarkable pioneer work was performed by Alexander Collin as early as 1846 in connection with slides occurring during the construction of French canals. Not only did Collin investigate the curved sliding surface which he had noted in clay slips, but he experimented on the properties due to changes in moisture content.[20.2] Throughout the remainder of the nineteenth century, attention to the properties of soils appears to have been spasmodic. A paper by Sir Benjamin Baker (The Actual Lateral Pressure of Earthwork) presented to the Institution of Civil Engineers in London on April 5, 1881, is a notable exception; the importance of geological structure in all such studies was stressed in the ensuing discussion.[20.3]

Not, apparently, until after the turn of the century was anything more than isolated soil investigations carried out, although notable individual contributions were made in the United States by Dr. Milton Whitney and Dr. George E. Ladd. As Dr. Terzaghi has pointed out, three major investigations undertaken in the early part of the present century directed the attention of civil engineers to soil problems and may rightly be regarded as the start of modern soil studies. These were the studies made in Sweden of the landslides that had occurred on the State Railway system; the investigation of difficulties encountered during the construction of the Kiel Canal in Germany; and the intensive analysis made of the great landslides that interfered so seriously with the completion of the Panama Canal. These investigations did not stand alone. In England, a paper was read by Mr. A. L. Bell at the Institution of Civil Engineers on January 19, 1915 (The Lateral Pressure and Resistance of Clay, and the Supporting Power of Clay Foundations), which, together with the long discussion that it provoked, contained much suggestive material on the mechanics of soils. The necessity for obtaining undisturbed samples was mentioned by the author—a practice that has been developed in subsequent years.[20.4] The paper was the precursor of a notable group of British publications on earth pressure and soil properties. In the United States of America, Dr. C. M. Strahan had begun his investigations of natural-soil roads which led to presentation of his first paper in 1914; this may rightly be regarded as the start of modern stabilized soil-road practice (although some isolated papers had been published previously). In 1913, the American Society of Civil Engineers set up their first Committee on the codification of bearing values for soils, and the work of this and of subsequent similar committees has been of great value.

These few references indicate the awakening toward the potentialities of scientific soil study that featured the early years of the present century and led eventually to the general recognition of Soil Mechanics as an important scientific aid to many branches of civil engineering. The title was not used to denote an entirely new branch of scientific study but rather as a convenient name for the coordinated group of soil studies that were found to be necessary in the application of soil analyses and investigations to the several branches of civil engineering work. Prominently associated with this pioneer work has been Dr. Karl von Terzaghi, until recently Professor at the Technische Hochschule in Vienna and well known in North America by reason of the posts that he has there occupied. By his publications in *Engineering News-Record* and his work for the United States Bureau of Public Roads, at Massachusetts Institute of Technology, and at Harvard University, Dr. Terzaghi gave an impetus to scientific soil study in North America that has had most fruitful results, and his European publications and activities have been similarly effective for engineers of the Old World. Dr. A. Casagrande, of Harvard University, has also contributed much to Soil Mechanics progress, his work in organizing the First International Conference being particularly noteworthy.

Other pioneer names could be mentioned did space permit; suffice it to say that in the records of the First International Conference on Soil Mechanics and Foundation Engineering, held at Harvard University in June, 1936, there is assembled a collection of invaluable data from most of the leading research workers in the field, which is thus conveniently available for general use.[20.5] This Conference, with its truly international atmosphere and its record of solid achievement, appeared to set at rest any doubts that may still have existed as to world-wide interest in scientific soil study and the value of practical applications of laboratory soil tests. In consequence, it is now no more than usual practice to discuss modern concepts of soil behavior under the accepted title of Soil Mechanics.

20.3. Soil Mechanics.—Along what lines have modern soil studies advanced? Three broad divisions may be distinguished: Methods of soil sampling have been greatly improved; the mechanical analysis and corresponding classification of soils have been systematized; and gradually there is being built up a generally recognized group of laboratory tests, the results of which can be applied with confidence to practical design work. Soil Mechanics naturally includes developments other than these. The purpose of this note, however, is to indicate the general lines of present-day soil investigations in order that their relation with Geology may be the more readily appreciated, so that a brief discussion of these three main divisions will suffice.

Soil Sampling.—The fact was recognized from an early stage in the development of soil testing that, if laboratory tests were to be fully effective, the condition of the samples of soils to be tested should correspond as closely as possible to that of the soil in its natural location. The acknowledged effect of even slight variations of the moisture content of soil samples on test results was a leading factor in such considerations. Other possible variations were subsequently noted, and the great differences that were found between the behavior

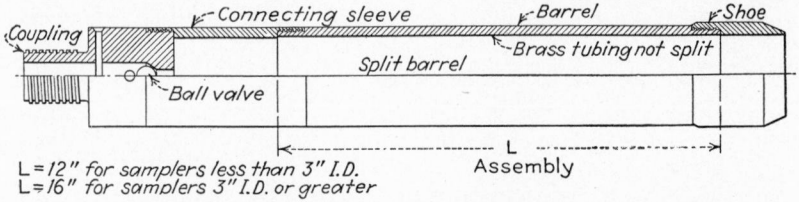

FIG. 20.1.—Diagrammatic drawing of a modern soil sampling tool—a model designed by Messrs. Moran and Proctor, of New York, and now known by their name.
(Reproduced by permission of Messrs. Sprague and Henwood Inc., Scranton, Pennsylvania.)

of certain clays in their natural and in a disturbed state confirmed the desirability of being able to obtain all clay samples as undisturbed as possible. The development of special devices for this purpose has been a notable advance in connection with soil-sampling practice. Details of the main types of sampling tools that have been evolved are to be found in publications dealing with boring operations. Essen-

FIG. 20.2.—Equipment used for the field sampling of soils by the Ohio Department of Highways.
(Reproduced from Bulletin No. 99, "Soil Mechanics as applied to Highway Engineering in Ohio" of the Engineering Expt. Station of Ohio State University, by permission of the Director, and the Ohio Dept. of Highways.)

tially they consist of sampling tubes (of a few inches diameter) which can be attached to the end of a pipe or rod by means of which they are lowered into a previously bored hole (cased if necessary) and forced into the undisturbed ground at the bottom. They are then withdrawn and disconnected either in part or in whole but in a way that ensures that the sample is left well protected, the exposed surfaces usually being coated with paraffin wax the moment the sample is brought to the surface.

Samples so obtained are still far from perfect, the friction caused by the sides of the sampler being still a disturbing element, but they represent a great improvement over the older kind of churned-up sample. The devices mentioned are for use in cohesive soils, but a method has been devised whereby, through the use of a special bituminous injection, relatively undisturbed samples of granular materials such as sand may be similarly obtained.[20.6] Improvements in equipment and technique continue to be made, not only in connection with

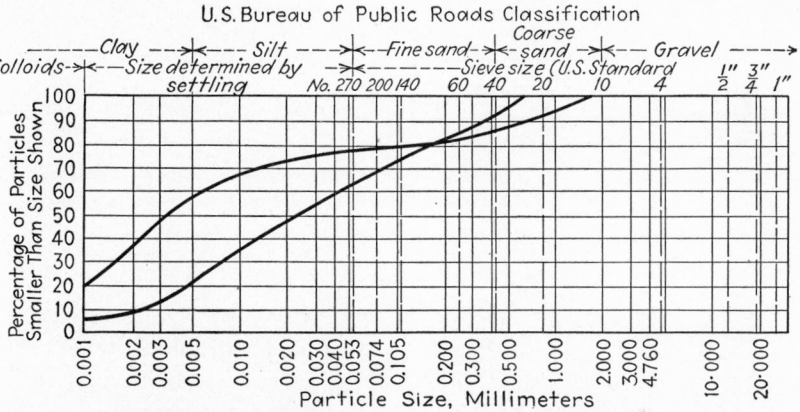

FIG. 20.3.—Particle-size accumulation curve, showing two analysis curves for types of clay.

the large-diameter samples obtained with these special samplers but also in connection with more portable boring and sampling outfits, which permit simple soil surveys to be made easily and economically. In the promotion of the wider use of simple test-boring outfits, the interest aroused by Soil Mechanics has been a most potent influence.

Soil Analysis and Classification.—Standard methods of mechanical soil analysis include the use of sieves and elutriators and the use of delicate hydrometers for particles too small to be investigated by these means. The latter method, following the lines suggested by Professor Bouyocos, is that usually followed in engineering laboratories. All methods result in data as to the proportions of grains of different sizes contained in the soil sample being tested. The results are usually plotted graphically, on semilogarithmic paper for convenience. The curves so obtained are familiar to all soil students; Fig. 20.3 is an example. On the basis of particle size, certain classifications of soil types have been evolved. Unfortunately, there is no recognized standard, but the one shown in Fig. 20.3 (the system adopted by the United States Bureau of Public Roads) is widely used and is typical of accepted classifications.

It will be noted that soils usually contain particles of several classifications, as these groups of soil solids are found mixed together to give the natural soil formation. Again, there are as yet no universally recognized combined proportions of soil solids to which specific soil type names can be applied, but the system in most general use is

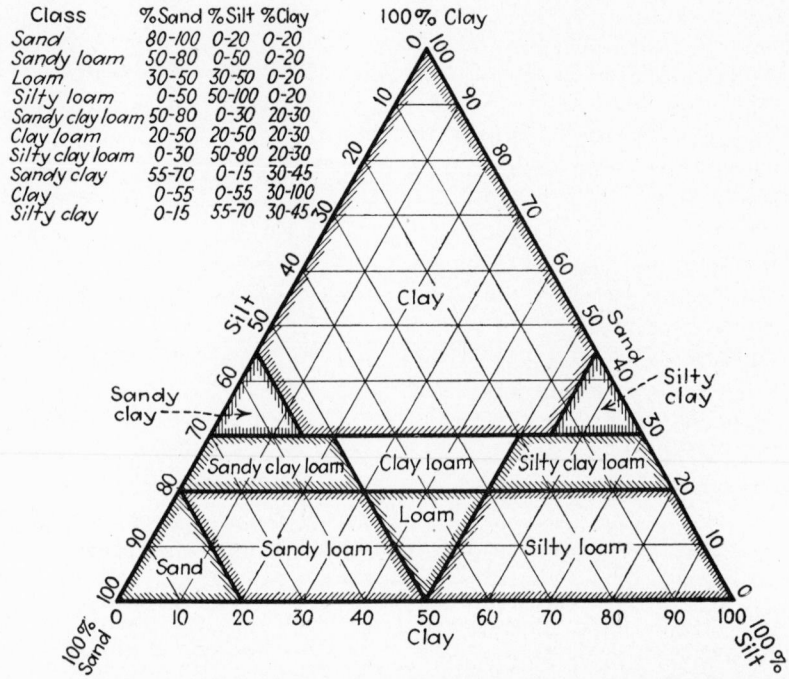

FIG. 20.4.—Trilinear soil-classification chart (for mechanical analysis).
(Reproduced by permission of the U. S. Bureau of Public Roads.)

that adopted by the United States Bureau of Public Roads, which is illustrated in Fig. 20.4. This useful chart provides a ready means of classifying a soil, with respect to its particle size, when once a mechanical analysis has been made.

These charts symbolize real achievements in Soil Mechanics; they do not, however, take account of the shape of particles or of their mineral nature and in this respect are incomplete. Mechanical analysis alone is not sufficient for accurate classification of all soils, although it does provide a most useful guide.

Laboratory Soil Tests.—So variable are soils in composition and behavior that their performance when under load or when subjected to unusual conditions such as forming the wearing surface of a road can be predicted with certainty only after tests have been made on samples of the soil to be used. Practice in the past almost exclusively involved field tests—trial and error in the case of soil road mixtures, and loading tests on small areas in connection with the bearing capacity of soils. Field tests still have their value, but it has been a leading part of the achievement of Soil Mechanics to have transferred the scene of soil testing to the laboratory. Certain critical limits for the moisture content of soils have been found to have considerable significance; standard means of evaluating these are now available. In specially designed apparatus, the shearing strength and consolidation of soil samples under load can be determined. Penetration tests are quite general; permeability tests have a certain value, and the number of other widely recognized useful tests is steadily increasing. All these tests, it will be remembered, are performed on samples of a natural soil formation—samples assumed to be typical. If the test results are to be applied with certainty, some assurance must be felt that the samples are typical.

Relation with Geology.—The fundamental importance to engineers of the geology of unconsolidated material has been demonstrated throughout this book. To the geologist, soils are rocks, so that their origin, composition, and present structural arrangement are just as important to him as those of the solid material known to the engineer as *bedrock*. Practically all that has been herein written, therefore, with regard to the application of geological methods and principles in civil engineering and all that has been suggested with regard to the necessity for close study of local geology in connection with civil engineering work apply equally well whether the local formations are solid rock or unconsolidated material.

Consideration of the foregoing brief notes on Soil Mechanics will show that the geology of soils may be, and often is, of more than ordinary importance to civil engineers. Sampling, for example, can be carried out effectively only if something is known of the probable uniformity and regularity of the soil strata penetrated during boring (unless innumerable samples are taken), and this can be the case only if the geology of the local soils has been studied. A study of local geology is therefore an imperative counterpart of all soil sampling operations. Soil classification is similarly related to Geology. It is possible to encounter a soil that, when analyzed mechanically, can be shown to conform to the particle-size requirements of a clay and yet contains no clay matter at all, having the characteristics of a granular material. This single case is sufficient to show how important the geological nature of a soil can be in relation to its correct classification, and many similar cases could be cited. Particle shape and the mineralogy of the soil grains are other geological factors of importance. Finally, with regard to soil tests, the results obtained can be applied to design work only if it is known that they are typical for all the soil to be encountered, and study of the local geology can most readily give assurance on this point. Some test results, especially those for clays, are notably affected by the geological history of the soil, particularly by the pressure to which it has been subjected. Even a general study of the local geology will assist in determining the main features of the soil deposit under study.

Thus the relation between Geology and Soil Mechanics may be traced out, a relation so intimate and so important that it is apt to be overlooked in the enthusiasm engendered by practical soil studies involving the application of clear-cut laboratory technique to previously indefinite problems. Yet it is a relation the neglect of which can invalidate much useful work in Soil Mechanics. Inevitably, it has to be a peculiarly *local* correlation in view of the innumerable varieties of natural soils. The elements of the geology of soils are common to all local applications, however, and so should be appreciated by civil engineers. The origin of soils should be clear, and the various processes by which soils may have come to occupy their present position should be generally familiar. As these elements are briefly outlined in the following notes, it will be seen that a study of soil transportation leads to considerations of quite general importance, especially with regard to glacial action, so that this contact between Geology and Soil Mechanics is of far wider significance than at first appears.

20.4. The Origin of Soils—Rock Weathering.—Examination of almost any rock surface that has been exposed to the atmosphere for an appreciable period will show that it has been noticeably affected by this exposure; in many cases, disintegration of the surface layer will be apparent. It is as a result of this complex action caused by atmospheric factors throughout recent geological time that soils and, similarly, the mechanically formed sedimentary rocks have been created. The process of change resulting from exposure of rocks to the influence of the atmosphere is known as *weathering*. Dr. G. P. Merrill has suggested that the word should be

> . . . applied only to those superficial changes in a rock mass brought about through atmospheric agencies, and resulting in a more or less complete disin-

tegration of the rock as a geological body . . . it does not include those deeper-seated changes . . . during which the rock mass as a whole maintains its individuality and geological identity.

The term *alteration* is used to describe these internal rock changes which are often of the nature of hydration and lead to the formation of new minerals in the rock mass.[20.7]

So variable and uncertain are atmospheric influences that the process of rock weathering is always most complex; it naturally varies

FIG. 20.5.—"Nature's Architecture," unusual scenery on the Makran coast, Iran, attributed to the erosion of jointed and stratified rocks by torrential rains, and the action of sand-laden winds.

(Photograph by The Times. Copyright, The Times, London; reproduced by special permission.)

in character at different localities, at different elevations above the sea, at different times of the year, and with different rock types. It is possible, however, to classify generally the several agencies responsible for weathering, although the relative significance of each will vary according to local circumstances. It is often difficult to separate weathering from erosion, the process by which the products of weathering are removed from their natural location, since the two are often simultaneous. For convenience, and in accord with the distinct identities of the two processes, they will be considered separately in the following notes.

The action of the atmosphere itself, *i.e.*, of the mixture of gases that constitute the air, is practically negligible as a cause of weathering; only when water (or moisture) is present are the constituents of the air significant. Movement of the atmosphere, in the form of

FIG. 20.6.—Exfoliating granite on lower Quarter Dome, Yosemite National Park, California.

(*Photograph by F. E. Matthes. Reproduced by permission of the Director, U. S. Geological Survey.*)

strong wind, is an important agency in arid parts of the world, in addition to being a leading means of erosion. In dry areas, where winds can pick up and carry along in suspension solid particles such as sand grains, the grinding, or "sandblasting," effect of this solid matter on exposed rock faces can be a very effective mechanical agent of rock weathering. Figure 20.5 shows an extreme example of the type of scenery resulting from this action, examples of which on a smaller scale will be generally familiar. Variations in the temperature of the atmosphere are another cause of weathering attributable to the atmosphere itself; if the changes occur suddenly and cover an appreciable range of temperature, the internal stresses set up in an exposed rock mass may be sufficient to cause flaking off of surface layers. *Exfoliation* is a name frequently applied to this phenomenon, which is not at all uncommon. Temperature changes are usually of more effect in the case of igneous rocks than in that of sedimentary rocks, especially with the coarser grained igneous rocks, the several minerals in which may have coefficients of expansion that are sensibly different from one another.

Temperature plays an important part in one of the mechanical processes of weathering caused by water—the effect of freezing water. Rain falling on rock surfaces will naturally fill up any exposed cavities, open cracks and joints in the rock mass, and tend to fill up the empty pore spaces of the rock. If freezing takes place, this entrapped water will exert a powerful disrupting force on the surrounding rock; the corresponding disintegration will often be considerable even in a short period of time. Running water, in addition to being an important factor in erosion, will also act as a weathering agent if it is carrying solid matter in suspension and as a bed load. The removal of solid material from a river bed is primarily a process of weathering although becoming immediately a feature of transportation. The corresponding action of moving ice, as in a glacier, is equally important over large areas, and the quantity of soil formed in this way has been appreciable.

Water plays another part in the general process of weathering, because of the fact that, in nature, it is practically never found in its pure state. The impurities usually found in water give to it certain chemical properties which result in the weathering processes which are among the most important of all. Oxidation is one chemical reaction that may take place when rocks are in contact with rain water; it is especially notable in rocks containing iron, as the brown staining of weathered surfaces often shows. The detailed reactions are complex; but hydrated iron oxides, carbonates, and sulphates are some of the products. All these reactions are accompanied by an increase in volume, with subsequent disintegration of the original rock mass. Finally, there is the action of rain water as a solvent to be considered, in some localities the most potent of all aspects of weathering. The action is usually thought of in connection with limestones, but many other rock types are also affected although not to the same degree. Although pure water is a poor solvent of the more common minerals, rain water containing an appreciable carbonic acid content will, in time, decompose nearly all the rock-forming minerals.

There are a few other weathering agencies, but all are relatively unimportant. Possibly the most interesting of these minor contributors to the general process are living creatures and plants. It is believed by some that the action of bacteria is of real significance in connection with rock weathering, but the matter is still awaiting full investigation. Ants and similar insects may assist the weathering process by carrying underground grains of soil, and so soil moisture, thus allowing the chemical action of water to penetrate into a rock mass. Finally, the effect that plant growth, and especially the spreading of tree roots, can have in splitting up solid rock masses and thus in allowing other weathering agencies access to the interior of the rock is a phenomenon frequently to be observed.

The possible complexity of the combined action of all or some of these agencies of weathering will now be clear. When the general process is considered in relation to the measure of geological time, then the immensity of this great part of the natural cycle will be apparent. Speculations as to rates of weathering are interesting but of little practical significance; what is of value to the engineer is the evidence of weathering that he encounters in his work and, in connection with soils, the variety of material resulting from weathering in the immediate geological past. Wind-cut topography may not be within the purview of every engineer, but evidence of the erosive power of running water will be familiar to all who work out-of-doors. Glacial action can be traced only in areas that have been ice covered in the past; in these regions, various typical topographical features may be noted (see page 573). The chemical action of water is usually to be seen by more detailed observation; the weathered appearance of most exposed rock faces in temperate regions is due to this agency. Observations of rock outcrops in the field will show how weathering conforms to features of rock masses such as joints and bedding planes, how the extent of weathering may vary with the nature of the rock, how weathering may result in the original rock form being retained until the decomposed rock is disturbed, and how in other cases the weathering is of such a nature that the decomposed or disrupted rock is automatically separated from the parent rock mass.

Observations of this nature show also how the nature and rate of weathering are directly dependent on the type of rock exposed. Thus, to take extreme cases, a limestone may be partially dissolved away leaving a cavity (such as many a famous cavern) and a residue of tough ferruginous clay, whereas a granite may weather only for a thin surface layer into "rotten rock," ultimately yielding sand and clay. Considering general rock types, many of the sedimentary rocks are themselves composed of weathered products (sandstone is a good example), so that it is to be expected that further weathering will be due merely to mechanical action. The other great group of sedimentary rocks, the limestones, are liable to be decomposed chemically. In the case of pure limestones, decomposition can be complete; with impure limestone, many records exist of a loss of weight of well over 60 per cent, the residue frequently being a tough clay.

The weathering of igneous rocks is not so simple; the precise process of weathering varies with the mineralogical content of the rock and also to some extent with the grain size. Fortunately, although the processes are complex, the recurrence of a few main mineral types in the more common igneous rocks enables a broad outline of the process of weathering to be obtained. Naturally, the harder minerals will offer more resistance to mechanical weathering agencies than some of the softer types. Similarly, the more stable the mineral in a chemical sense the greater will be its resistance to chemical change. In consequence, free quartz, being at once hard and chemically stable, is the most refractory of all common minerals to the action of weathering. Some silicates are but little affected, notably the harder types such as tourmaline and zircon. The weathering of mica depends on the form in which it occurs and the rock of which it is a constituent. Of the feldspars, the potash varieties are generally more refractory than the soda lime types.

TABLE R.—SOME LEADING ROCK-FORMING MINERALS, ARRANGED IN THE ORDER OF THEIR RESISTANCE TO WEATHERING, THE MAIN PRODUCTS OF ALTERATION BEING INDICATED

Quartz	Retains identity
Micas	Variable—clay minerals and hydrous silicates
Feldspars	Soluble matter and clay minerals
Hornblende, Augite	Hydrous silicates
Olivine	Serpentine and iron oxide

In Table R are listed representative minerals together with notes on the products resulting from their weathering. From an examination of even such a simplified statement as this, it will readily be appreciated that the chemical reactions causing the changes noted are involved and in many cases of long duration. Essentially, the combined effect on rock masses is a breaking down of the mineral contents into relatively stable products of decomposition. The weathering process does not stop at this point. Even the most stable minerals are in time affected by contact with rain water; and if this has leaked through an upper layer of humus, carrying with it a small percentage of organic acids, its effect as a weathering agent will be increased. By the prolonged action of this moisture on silicate minerals, certain secondary products of weathering are gradually formed and these range down to colloidal material and consist of what may generally be termed the *clay minerals.*

Study of clay minerals is still in its infancy; the technical difficulties of analysis of such fine-grained and intimately mixed materials have themselves been a deterrent to progress. These secondary minerals may be generally classed into three main groups—

1. Montmorillonite—beidellite minerals.
2. Halloysite—kaolinite minerals.
3. Hydrated ferric oxide and alumina.

Many of the peculiar physical properties of clay are now known to be due to the properties of these minerals, especially those relative to moisture content. They are chemically reactive (particularly groups 1 and 3), being acidic in nature, and they possess the important absorptive property of base exchange. This property may eventually prove to be of great importance in the treatment of clays responsible for difficulties in engineering construction. Research on these minerals is today in progress; notable work on them has already been done in England and in the United States by Professor H. E. Winterkorn of the University of Missouri, in the laboratories of the Illinois Geological Survey, etc. Dr. G. E. Ekblaw and Mr. Ralph E. Grim, responsible for the Illinois work, have already been able to state:[20.8]

> (1) . . . that the concept that all clays contain a mysterious universal clay substance, sometimes called clayite, to which they owe their properties, is fallacious; (2) that the concept that all clay materials are composed essentially of the mineral kaolinite is erroneous; (3) that an alternative concept that these materials are simply heterogeneous assemblages of almost any species of minerals existing in very small particles and that therefore their properties are entirely dependent upon the size-grade distribution of the component particles is also erroneous; and (4) that they contain little or no amorphous material.
>
> On the contrary, this work has shown that clays actually are composed of crystalline constituents throughout—even the colloidal fraction is composed of crystalline particles, the colloidal properties depending on smallness of particle size and also on the shape of the individual particles.

The potential value of the further investigation of the mineral contents of clays is unusually significant.

To summarize these notes: Soils are seen to be the aggregation of the products of rock weathering, the formation of which is due to the simultaneous operation over long periods of time of one or more of several weathering agencies, the general nature of which is known. The coarser particles found in natural soil formations, above about 0.053 mm. in diameter, consist generally of the more stable primary products of weathering—minerals relatively hard and chemically stable. The finer particles consist mainly of the secondary products of the weathering of silicate minerals, called clay minerals, the properties of which are still under active investigation.

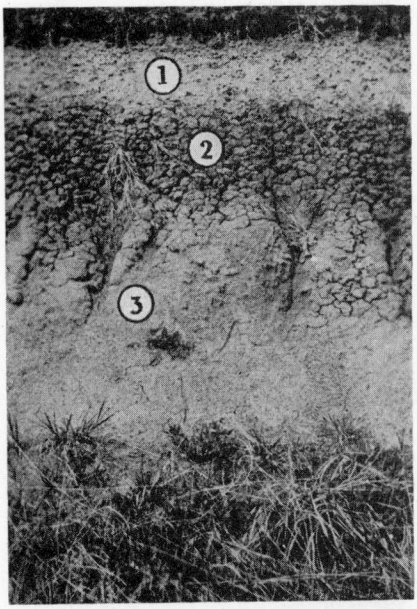

Fig. 20.7.—Section through residual soil, showing (1) light-textured upper soil (2) heavy, plastic, sticky clay; and (3) decomposed rock.

(*Reproduced by permission of the U. S. Bureau of Public Roads, from Public Roads, September,* 1931.)

20.5. Residual Soils.—The erosion of the products of rock weathering, which is so frequently associated with the weathering process, is not universal. Over many parts of the surface of the earth where the natural rock surface is approximately horizontal, weathering of the rock surface has proceeded without the soils so produced being moved from their original position. Soils of this nature are known as *residual soils.* The name is accurately descriptive, since the soils are actual residues of the original rock, all soluble materials having been leached out by the long continued seepage of groundwater. This chemical disintegration naturally becomes less active as depth increases, the alteration of the original rock gradually becoming less until finally the natural unaltered rock is reached. This gradual downward change from the soil to original rock is a distinctive feature of residual soils, and one by which they can readily be distinguished. Another characteristic is that the soils contain no minerals foreign to the locality; the entire mineral content is directly related to that of the underlying rock. Particles are sharp edged and not rounded; fragments of the original rock may sometimes be found, unaltered, in the disintegrated stratum.

In other respects, residual soils may vary. They may retain the structure and even appearance of the original rock until disturbed in some way; on the other hand, some residual clays show no evidence at all of their origin. Some residual soils are sandy in nature, but residual clays are more common than sands. Others possess peculiar characteristics which have led to their being given special local and sometimes regional names. Only one of these warrants mention in this discussion; this is *laterite*. The name is generally applied to soils of tropical countries, notably of India, which, because of the absence of organic material in the topsoil and consequent lack of humic acid to act as a leaching agent, have a high alumina and iron oxide content. Clays of this nature are soft when naturally moist but harden on exposure to the atmosphere. The name is sometimes applied to clays of more temperate countries such as the United States which have a high silica content, but its use in this way is largely governed by local practice.

Residual soils are to be found only in regions that have not been greatly disturbed during recent geological time. They are not generally found in areas that have been subject, for instance, to glaciation. Some of the more recent soil deposits which are, in a sense, residual soils are not subject to this restriction. They are the deposits of organic and inorganic material accumulated by vegetation; they usually occur in the presence of water, peat bogs and marshes being obvious examples. Although not widely distributed, they are frequently a source of trouble to civil engineers, especially during the construction of transportation routes. Their method of formation and variable organic content tend to make each occurrence a matter for individual local study; their moisture content may be as high as 80 per cent.

20.6. Transported Soils.—Erosion has been repeatedly mentioned as one of the great modifying influences at work today on the surface of the earth. It has been responsible already for the formation of most of the sedimentary rocks, covering three quarters of the earth's surface; and, similarly, it is today ceaselessly at work in removing the products of rock weathering and depositing them elsewhere. Thus it is forming soils some of which will in time become consolidated into the sedimentary rocks of the future.* Chapter XV included a discussion of present-day erosive processes and the corresponding

* It has been estimated that 78 million cubic miles of the earth's crust have been eroded in this way, the resulting sediments giving 80 per cent shale, 15 per cent sandstone, 5 per cent limestone, and the corresponding unconsolidated materials. (CLARKE, F. W., reported in *Nature*, p. 334, May 30, 1912.)

silting. This study must now be extended to cover the range of recent geological time during which the materials that are today classed as transported soils have been deposited in their present positions; much of Chap. XV may usefully be read in conjunction with the following notes.

Although gravity is usually thought of as the agency ultimately responsible for the movement of disintegrated rock by erosive processes, the action of the wind is an additional agency of consequence.

FIG. 20.8.—Deep cut through loess, on U. S. Highway No. 61, south of Vicksburg, Mississippi.

(Reproduced by permission of the Mississippi State Highway Dept., Mr. R. A. Harris, Chief Engineer.)

Deposits of soil formed by the wind are known as *aeolian* deposits. Sand dunes are an obvious and notable example of materials so deposited; the extent of some suggests their age. Their frequent existence near seacoasts shows that arid conditions are not imperative for the formation of aeolian deposits, although wind action is most potent in dry regions. Wind-deposited sand may show traces of stratification; its mineral contents will usually be fresh, since it has been formed by a mechanical weathering agency; and the grains will be well rounded, in some cases almost spherical, this being the most notable characteristic of aeolian deposits.

A special type of aeolian deposit, widely distributed throughout the world, is that known as *loess*. It is very fine grained, often of a yellow color, and exhibits no stratification but occurs in great massive beds which may be hundreds of feet thick. The material is largely composed of silt-sized particles which under microscopic examination can be seen to consist of sharp-edged fresh particles of minerals such as quartz, feldspar, calcite, and mica; it has also some clay content. It can fitly be described as an accumulation of wind-blown dust, its dustlike nature when in the air being evidenced in those areas in which it forms the surface of the ground—notably in China where ancient roads have cut themselves into the loess as deep trenches, where caves are excavated in it for human habitations, and where its characteristic color has given names such as the *Yellow River*. A somewhat similar deposit found in the southwestern United States is known as *adobe;* it is used for making bricks and other molded articles by sun drying.

The most obvious type of gravitational erosive action is that which leads to the accumulation of rock fragments at the foot of exposed faces of rock, generally classed as *talus*, and in extreme cases to the occurrence of avalanches. Here the force of gravity acts directly on the products of mechanical weathering. These accumulate until disturbed in some way, often by erosion of the toe of the talus slope by running water. In the case of extensive accumulations of talus, the mass of disintegrated rock may eventually reach a state of unstable equilibrium, and itself start to move. If the movement is sudden, an avalanche will result, but it may be exceedingly slow, to be noticed only by long-term observations. The term *rock glacier* is sometimes applied to such phenomena. Clearly, they are of importance in civil engineering work, since, if unrecognized and covered by a mantle of topsoil, they may be utilized as a foundation bed, possibly with disastrous results. Study of the local Geology in association with test borings and drilling should disclose the true nature of the material; a gradation from disintegrated rock near the surface to fragments of fresh rock at greater depths will be a main distinguishing feature, all minerals encountered being correlated with the adjacent natural rock formations.

Although aeolian deposits and talus are important types of deposits, the most important and most widely distributed types of transported soil are those which have been moved by water and by the slow movement of ancient glaciers. *Glacial deposits*, as the latter are known, are so important and so closely related to topography that they will be treated separately in the next section. *Aqueous deposits* are also important and constitute probably the largest group of transported soils. Aqueous deposits may properly be called *sediments*, as they have been formed by deposition from either standing or moving water in a manner entirely similar to that in which silting can be observed to take place at the present time. They are generally

FIG. 20.9.—Rock screes or talus on Wast Water (one of the English lakes), Cumberland; material derived from crags of Borrowdale Volcanic Series.

(Geological Survey of Great Britain photograph. Crown Copyright, reproduced by permission of The Controller of H. M. Stationery Office, London.)

grouped into two main divisions: *marine* sediments and *continental* sediments. The former were deposited in sea, and the latter in fresh, water but in a variety of ways: some by river flow, some as deposits in fresh water lakes, some as deltaic and detrital cone formations. All the processes of sedimentation can be observed today, in some form or other, so that although there are many aspects of sedimentation as yet imperfectly understood, it is possible to classify the main ways in which material is deposited from moving water.

Considering continental or terrestrial sediments and, in the first place, the headwaters of a running stream, transportation of disintegrated rock or eroded soil will be effected by the stream flow, since its bed gradient will be steep. As such a stream changes slope on entering a level area of country, its velocity will decrease, and much of its load will be deposited as an *alluvial cone*, or *alluvial fan*. In the past, much material has been deposited in this way, probably at all stages of the earth's history. Deposits of this nature are to be distinguished in the vicinity of most important mountain ranges, and they are often of surprising depth and extent. Much of the sedimentary material to be found in the coastal plain of the United States of America is thought to have been formed in this way. The deposits show stratification, as would be expected from the nature of their formation; and in view of the changes in the volume of flow that may take place in all steeply graded streams, adjacent strata may show a great variation in the nature of the sediment. Apart from this minor variation, all alluvial cone deposits show a major variation from the coarse sediment (possibly including boulders) at the apex to fine clays at the outer reaches. If the process of erosion continues without serious earth movement, the materials in an alluvial cone may be themselves eroded as the life history of the responsible stream develops.

The final stage of river development is also a notable factor in sedimentation. After a stream has reached maturity, any increase in the volume of sediment brought down from its higher reaches will result in a steady building up of the river bed, the river being then said to be *aggrading*. Several notable examples of this process are to be seen today. Corresponding *natural levees* are built up at the sides of the normal river channel, and in times of flood the river water will extend beyond these, covering a large area on either side (if in flat country) on which material is deposited with decrease in velocity; this area is known as the *flood*, or *alluvial, plain*. Material deposited in this way will again be stratified but all of fine-grained material. The deltas formed where such rivers debouch into the sea have already been described (page 400). Ancient deltaic deposits exhibit the same structural arrangement of beds and so are featured by regular stratification and uniformity of the grain size of the material that they contain. River terraces, or *alluvial terraces*, are topographical features often to be seen which frequently assist in the location of deposits of fine-grained sediments. They may be old flood plains below which the river has scoured out a deeper bed, possibly through interference with the headwaters of the river. Alternatively, they may be evidence of earth movement that has led to a major alteration in the regime of the stream.

In desert regions, rivers may be found flowing (often intermittently) into inland basins that are either dry or covered only with a shallow depth of water. The sedimentation that takes place in such regions, when the rivers in flood bring down large quantities of material with them, is obvious. If earth movement accompanies sedimentation, the depth of the deposits may become considerable, and this has happened in the past. The material will be stratified and, as none can escape,

FIG. 20.10.—A "model" alluvial cone; early stage of erosion of a soil bank escaping through channels in the snow, and accumulating on the shore ice of the Kaministikwia River, Ontario, Canada.

(*Photograph by H. B. R. Craig. Reproduced by permission of the Ontario Dept. of Mines, from 45th Annual Report, Part VII*, 1936.)

will be varied in nature, often including layers of common salt and gypsum. Rivers flowing into lakes will tend to form deltaic deposits similar to those at outfalls into the sea. Characteristics of the deposits will be similar to those of marine deltas.

Marine deposits are usually considered as having been formed in one of three regions: the shore zone; the shelf, or shallow water, zone; and the deep sea zone (extending beyond the edge of the continental shelf). It is helpful to recall that all marine sediments are derived from the material transported by rivers and that eroded from sea-coasts. Except in the shore zone, therefore, fine-grained sediments are to be expected. The material to be found in the shore zone, although often of considerable local present-day importance, is really of a temporary nature, since it varies with changes in shore line and is itself constantly being changed. It is predominantly coarse in nature, usually to be classed as sand and gravel; its observed irregularity is characteristic.

Deposits in the shelf, or shallow water, zone, often considered as extending to the 100-fathom line, are of great importance. The region is today one of ceaseless change, the movement of the sea gradually sorting out the material deposited on the sea bed, the finer material gradually working down into the deeper water. Interesting as are present-day deposits in this zone, the corresponding deposits of pre-

FIG. 20.11.—River terraces, junction of the Nicola River with the Thomson Valley, British Columbia.

(Reproduced by permission of the Geological Survey of Canada.)

vious eras are of greater importance, upward crustal movements involving parts of previous shallow water zones having converted many of these sediments to land formations. They cover surprisingly large areas and are often of considerable depth (due to repeated earth movement). They are usually flat and display stratification; they may show gradation from gravel to finer materials, but clays will sometimes be found close in to original shore lines. Although usually inorganic, many shallow water deposits contain lime-secreting organisms which have been responsible for the formation of limestones. Ripple marks are a frequent occurrence in deposits of this nature; and when they are of clay, "mud cracks" may still be seen, caused originally by the action of the sun on the clay when wet. Shallow water sediments sometimes display the tracks of land animals, and their fossil contents are often appreciable. The nature of the fossils will often enable distinction to be made between a marine shallow water sediment and one laid down in fresh water; the general characteristics of the two groups are markedly similar. There remain for consideration the deposits of the ocean depths. Characteristic sediments are known to be muds, organic oozes, and clays of varying color but remarkable uniformity in texture. Of great present-day interest, they are of little significance in the study of soils, since the existence of materials from the ocean depths now forming part of the surface of the earth has been traced in the case of a few islands only. Marine deposits when spoken of geologically are therefore generally to be regarded as shallow water sediments.

20.7. Glacial Action and Glacial Deposits.—Just over one hundred years ago, Louis Agassiz, a Swiss naturalist, first noted the existence of boulders so far removed from their natural bedrock that he decided they could have been transported only by glacial action. It was many years before his revolutionary idea was generally accepted; the persistence of the word *diluvial*, suggesting the great flood as the origin of deposits now known to be glacial, is an interesting reminder of this conservative attitude. Today, the importance of glacial action in the geological cycle is generally appreciated. The large erratic blocks so often preserved in public places when discovered in urban areas are significant indications of the power of this great natural process. Examples of such blocks are known that weigh well over 1,000 tons, and some—even of this magnitude—are found at great heights *above* their original positions as well as great distances away. Many readers will be familiar with boulders of this kind and so will appreciate the importance of glaciation as a factor in erosion. This was, at one time, probably the most important of all erosional processes, when one fifth of the land area of the world is estimated to have been covered by ice. It is this ice age, as it is often called, that is of such importance in connection with present-day soils and also with many civil engineering operations.

The existence of glaciers today in several parts of the world, notably in the European Alps and in North America, and the covering of Greenland and the Antarctic Continent with ice sheets have made these phenomena generally familiar. Without these present-day examples, the state of the world during glacial epochs would be difficult indeed to imagine. Even with these examples available, it is still far from easy to imagine practically the whole of Canada and a large part of the northern United States, most of northern Europe and Asia (including the British Isles), and much of South America covered with ice sheets varying in thickness up to several thousand feet. This is now known to have been the case during the recent ice age which occurred during the Pleistocene era. The proof of this has been a

FIG. 20.12.—Sketch map showing centers of radiation and area covered in North America during the Pleistocene ice age.
(*After Chamberlin and Salisbury.*)

fascinating accumulation of carefully observed facts linked to an outline of the effects of glaciation which are about to be described. It is known also that the ice receded at least three times, the interglacial periods being to some extent warmer than present-day climates, certainly in the Toronto district, Canada, where some of the most important interglacial deposits yet to be examined have been found.

Speculation as to why these incursions of ice took place is still a matter of conjecture. The engineer has to accept the accredited fact that glaciation did occur over a large part of the earth's surface and, recognizing this, to avail himself of all that a study of Glacial Geology can suggest as to the possible effect on his work of glacial features. Figures 20.12 and 20.13 show, in broad outline, the areas in North America and Europe that were covered with ice and in which, therefore, Glacial Geology is of practical importance.

FIG. 20.13.—Sketch map of Europe showing the extent and trend of the maximum glaciation (compare with Fig. 20.12).

(Based, by permission on a plate in "The Quaternary Ice Age" by W. B. Wright, 2d edition 1937, Macmillan and Company, Ltd., London, and The Macmillan Company of Canada, Ltd., Toronto.)

In considering glacial action, distinction should be made between the two main types of glacier, the *mountain*, or *valley*, *glacier*, slowly moving down the valley that it fills; and the *continental glacier*, or *ice sheet*, over-riding mountains and valleys and of considerable extent and thickness, that over Greenland being today over 5,000 ft. thick. Movement of the ice is similar in both cases, although rates of travel may vary considerably; some ice sheets remain stationary for long periods. Many interesting observations of the rate of travel of existing glaciers have been made.[20.9]

The action of glaciers in eroding and transporting material off the surface of the earth is well defined; the effects of the two types on topography is naturally different. As the mass of ice passes over the earth's surface, it may pick up loose and unconsolidated material directly, incorporating this into the lowest part of the ice mass. If some of this material is in the form of rock fragments and boulders and is pressed against solid bedrock as the ice passes over it, an abrasive action will result—the bedrock will be eroded, and more loose material (rock flour) will be available for picking up by the moving ice. Finally, it is known that thick ice sheets exercise a "plucking," or "quarrying," action in tearing large blocks of rock away from their natural position by adherence of the ice to the rock surface and that these blocks also are carried along in the lower part of the moving ice formation.

This erosive action has a definite effect on the topography of the land covered by the glacier. In the case of continental glaciers, it results in a general leveling of the surface, high points of land being naturally eroded to a greater extent than lower lying ground. The ice sheet will follow the contour of the ground that it covers and erode constantly throughout the whole course of its movement. Thus it is estimated that glaciation removed on the average probably as much as 100 ft. from the top of that great area of pre-Cambrian rock now known as the *Canadian Shield.* Much of this area is today bare rock, for weathering since the last glacial period has been insignificant. On exposed rock surfaces, the characteristic polishing and markings caused by glaciation are to be seen. These were probably effected by rocks embedded in the bottom of the ice sheet. The polishing is often remarkably smooth and uniform, broken only by the grooves and scratches (striae) caused by projecting rock points, markings also found on some rock fragments encountered in glacial deposits. The polishings and markings provide good clues to glaciation, should this be in doubt; by the study of these marks and by tracing the rocks found in glacial deposits back to their origins it has been possible to discover the positions of the centers from which radiated the ice movements indicated in Figs. 20.12 and 20.13.

The main erosive effect of mountain glaciers is to smooth out the valleys through which they pass, often deepening them considerably and almost always giving to them a characteristic U-shaped cross section in distinction to the V shape of a young valley being eroded by a running stream. A corresponding result is that valleys are straightened, the projection of hillsides between the valleys of side streams being removed. Frequently, if the side-stream valleys do not have glaciers in them, they will remain at their original elevation while the main valley is deepened. After the glacier has melted, this will give rise to *hanging valleys*, a topographical feature responsible for many waterfalls and so for potential water power development sites. The deepening of valleys by glaciation has led to many beautiful fiords and similar inlets from the sea when a general depression of the land surface has followed the glacial period. Another topographical feature caused by erosion is the *cirque*, or steep-sided rounded end of a valley that has been glaciated, differing considerably from the typical valley formation caused only by running water. Cirques

FIG. 20.14.—Glacial Valley, Flint Creek Mountains, near Anaconda, Montana, showing typical U shape.
(*Photograph by W. C. Alden. Reproduced by permission of the Director, U. S. Geological Survey.*)

result from the initial stages of glaciation; snow and ice collect in the head of the valley and gradually accumulate debris eroded from the adjacent valley slopes. *Roches moutonnées* are outcrops of rock that, having been passed over by a glacier, exhibit a striking and characteristic form, smoothed and rounded on the side opposed to the glacial movement and in a natural state on the opposite face.

The effect of glaciation on topography cannot be considered in relation to erosion alone; simultaneous deposition of the products of erosion has occurred and is frequently related to some of the erosive features already mentioned. Rock debris is carried in three ways by moving ice: at the bottom of the ice sheet; on its surface; and within the ice sheet, this material having worked its way in from the surface of the ground. While moving, ice deposits little of its load; but during the process of melting, the *glacial deposits* which are today of great importance were laid down. The melting was not confined to the end of the glacier, a process such as can be observed today, but must be thought of as being much more extensive owing to climatic changes of unusual degree.

Material deposited by glaciers is generally considered as being divided into two classes: glacial deposits proper; and glacio-fluvial deposits, laid down by water issuing from the glacier. Glacio-lacustrine and glacio-marine deposits are those laid down in lakes and the

FIG. 20.15.—*Roche moutonnée* of Leven Schist, with evidence of plucking, near Achriabhach, Glen Nevis, Inverness-shire, Scotland.

(Geological Survey of Great Britain photograph. Crown Copyright, reproduced by permission of The Controller of H. M. Stationery Office, London.)

sea bed adjacent to glaciers. Considering first deposits from the glacier itself, moraines will immediately come to mind, as they are often so striking a feature of modern glaciers. A *terminal moraine* is the deposit formed by wastage of the end of a glacier; the moving glacier may have pushed forward the material deposited as its front face melted away, or else the melting of the glacier was just equaled in speed by its forward movement. Terminal moraines are therefore found as ridges across valleys, often of considerable height and in plan usually convex down the valley. *Lateral moraines* and *medial moraines* are the deposits formed by the material carried on the surface of the glacier at the sides and in its center, often in ridges parallel to its movement and called by the same names. They are frequently of considerable height, especially lateral moraines, some of which are over a thousand feet high; in plan, they follow the direction of ice movement. Finally, the term *ground moraine* is applied to what is probably the most important of all types of glacial deposits—that deposited from the bottom of the glacier itself, consisting of material previously carried within the ice sheet and at its lower surface. In thickness, this type of deposit may vary from 500 ft. to nothing; it produces a fairly level surface, featured by gentle undulations. Occasionally, the surface of the ground moraine will be interrupted by small

FIG. 20.16.—The Crillon Glacier, Lituya Bay, Alaska. Note the surface debris (mainly from the tributary glacier on the left) and the alluvial fan of glacio-fluvial sand and gravel being built up on the right. As the glacier ends in an arm of the sea, there is no terminal moraine.

(*Photograph by Alaskan Aerial Survey, U. S. Navy. Reproduced by permission of the U. S. Forest Service, Washington D.C.*)

oval-shaped hills of unstratified clayey material, with their long axes in the direction of ice movement. They are called *drumlins* and are thought to be due to deposition through crevasses in the ice.

Melting of glaciers produces streams of water which, in the course of running away, carry with them some at least of the debris previously carried by the ice from which they flow. Restricted as to channel, and swift flowing at first, these streams gradually deposit their loads with decrease in velocity and so give rise to stratified deposits of the debris from the glacier. In a manner comparable to that causing alluvial fans at points of change in river flow, *valley trains* are formed by the outwash waters from mountain glaciers as they leave the cover of the ice and start their flow down the valley. In the case of continental glaciers, *outwash plains* are the corresponding formation, being naturally on a larger scale. Some glacial streams may flow under the ice sheet in tunnels, and as they carry debris with them, this will tend to be deposited in long winding ridges corresponding to the hidden streams. As the ice melts, these ridges will be exposed; they are called *eskers*, and some of them are several miles in length. *Kames* are the corresponding deposits sometimes formed at the mouths of such hidden glacial streams; they are roughly conical hills composed of poorly stratified glacial deposits. If a glacier had

FIG. 20.17.—Typical moraine topography, Ontario, Canada.
(*Photograph by Dr. A. P. Coleman.*)

its end located in the waters of a lake, depositional features already described would be somewhat modified, and a deltaic formation would result, *glacial deltas* being a recognized although infrequent type of glacio-fluvial deposit.

If it is recalled that the different types of glacial deposit are associated with the results of the erosive action of the corresponding glaciers, it will be appreciated that the modification of topography, made possible as a result of glaciation, is surprising. Many major modifications have been detected and studied. Thus it is now known that in North America Lake Algonquin, a glacial lake which occupied the present basins of lakes Huron, Michigan, and Superior, had outlets successively (or simultaneously) past the site of Chicago into the Mississippi, through the St. Clair River and Niagara, and through Georgian Bay. Similar examples from other regions can be found in

FIG. 20.18.—Excavating typical glacial drift (by jetting and sluicing), Northern Ontario, Canada.

(Photograph by R. F. Legget. Reproduced by permission of the Power Corporation of Canada, Mr. L. C. Jacobs, Construction Manager.)

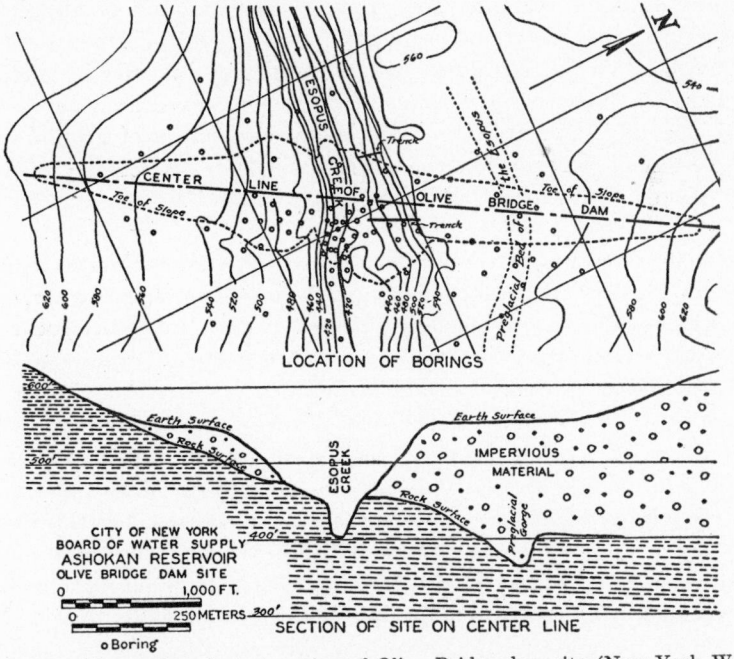

FIG. 20.19.—Plan and geological section of Olive Bridge dam site (New York Water Supply) showing buried river valley or preglacial gorge.

(Reproduced, by permission, from "New York City and Vicinity," ed. by Dr. C. P. Berkey, Guidebook 9 of XVI International Geological Congress.)

publications dealing with local glacial features. Of far more importance to the civil engineer are the minor changes that glaciation effected, and these are particularly related to the deposition of glacial deposits in valleys that have been subject to glacial erosion.

"Buried river valleys" have been frequently mentioned; their formation may now be appreciated as being due to glacial or subglacial erosion of a deep valley in solid rock, afterward partially filled in by a ground moraine or other type of deposit. After the ice has gone, the ordinary river flow will have to carve out a new channel in the surface of the drift, and it is entirely possible that the new course will have no relation to the buried valley. A typical example is shown in Fig. 20.19. Terminal moraines often form natural barriers across the ends of eroded rock valleys. If impervious, they may be the cause of lakes in the valleys; and, similarly, they may sometimes be relied upon to form a part of the impervious structure needed for the formation of a reservoir. In all such cases, owing to their unstratified and variable nature, detailed subsurface examination must be made to obviate all possibility of leakage. In addition to covering up buried valleys, glacial drift will hide from sight all other irregularities of the bedrock surface; in some regions, these may be frequent and of appreciable magnitude, and they may account for some of the strange results obtained when putting down test borings or well shafts in glaciated areas. The latter is not a frequent operation, since the nature of the drift does not, as a rule, promote the accumulation of groundwater. Glacio-fluvial deposits, on the other hand, and particularly outwash plains and valley trains are frequently good aquifers.

The detailed nature of glacial deposits must finally be considered. The material transported by glaciers varies from the largest type of boulder to clays, all gradations from one to the other being included; it is known generally as the *drift*. Professor W. O. Crosby found that the drift around the Boston area, United States, on the average and excluding all large stones, contained 25 per cent of gravel, 20 per cent of sand, 40 to 45 per cent of fine sand, and less than 12 per cent of clay. The material will either have been picked up by the glacier during its travel from the natural surface of the ground or else eroded by the glacial action and then transported. If this eroded material is that which is ground off the surface of bedrock, it will usually be of the nature of *rock flour*, finely ground fresh rock particles (its mechanical analysis placing it generally in the class of silt) with no plastic properties. As deposited directly from the glacier, all gradations of transported material will be thoroughly mixed together; the resulting unstratified, unsorted mixture of unconsolidated material is known as *till.* The term *boulder clay,* because of its usual boulder content, is often used to describe this unsorted deposit, but the name is not strictly correct, since frequently the till will be found to contain no clay at all. A large percentage of the rock fragments will be found to be angular in shape rather than rounded; many will exhibit scratches similar to the striae found on polished bedrock surfaces; and the rock types encountered in the till will vary, often bearing no relation at all to any of the local rocks near the deposit.

FIG. 20.20.—A deep section in boulder clay, at Moniack Burn, South Clunes, near Beauly, Inverness-shire, Scotland.

(Geological Survey of Great Britain photograph. Crown Copyright, reproduced by permission of The Controller of H. M. Stationery Office, London.)

This same material will constitute the load carried away by outwash streams from the glacier and thus forms the glacio-fluvial deposits already described. During transportation by the running water, and through the mechanical nature of its deposition, the material will be sorted in size—at least to some degree—the coarser material being deposited first, and the finest last. Glacio-fluvial deposits are consequently stratified to some extent and consist of materials approximating to the different soil groups. The exact nature of the deposits depends on the process of deposition. Thus kames, because of their irregular mode of formation and although stratified, will usually consist of materials that vary so much between strata as to make the deposit useless as a source of one particular type of soil such as sand or gravel. Eskers are not commonly encountered but are generally found to consist of coarser materials (sands and gravels) due to their formation from running water. Valley trains, outwash plains, and glacial deltas all provide fairly regular deposits of gravels, sands, silts, and possibly clays, the type of soil met with depending on the location selected for trial relative to the formation of the deposit.

It will be seen that glacial deposits will not generally exhibit uniformity as a leading feature. This necessitates the exercise of the

FIG. 20.21.—A kame, near Glen Ross, Ontario, Canada, showing structure of the deposit.

(Photograph by Dr. A. P. Coleman. Reproduced by permission of Dr. Coleman, and the Ontario Dept. of Mines; photograph reproduced in 45th Annual Report, Part VII, 1936.)

greatest possible care in their investigation before use. Attention must be devoted not only to the extent of the deposits but also, if it is planned to use the soils for any purpose, to their mineralogical nature. As will be realized from the nature of their formation, glacial deposits will often consist of disintegrated rock that has weathered in a mechanical sense only, all the constituent minerals being fresh. If, therefore, any of these minerals are liable to weather chemically on exposure to the atmosphere, discretion must be exercised in deciding on the use of the deposit concerned, say as concrete aggregate. Similarly with respect to the finer deposits, guard must be kept against mistaking rock flour for clay; but with modern methods of soil testing, this distinction is easy to make.

Finally, brief mention must be made of glacial boulders. Erratic blocks were mentioned in a general reference at the outset of this

TABLE S.—CONDENSED SUMMARY OF DRIFT DEPOSITS AND TOPOGRAPHY (FOR CONTINENTAL GLACIERS)[1]

Name	Topography	Material		Bedding	Confused with	Diagnostic features
		Kind	Sorting			
Terminal moraine, stony	Ridge parallels ice border, small knobs, kettles, irregular skyline, "short hills"	Till-sand gravel	None, very poor	None, sorted parts very irregular	Sand dunes, gullies, pitted outwash, drumlins	Irregular skyline trend
Terminal moraine, clayey	Ridge parallels ice border, smooth or gullied	Till	None	None	Ground moraine rock ridge	Trend
Terminal moraine, deltaic	Cones or mesalike hills, ridges between kettles	Sand, gravel, silt	Very poor	Top horizontal, inclined below	Outwash, esker, kame	Inclined beds, trend, level tops, relation to standing water
Roche mountonnée	Like drumlin, but ends reversed	Rock	.....	.....	Drumlin, rocdrumlin	Material form
Drumlin	Oval hill, long axis parallels ice movement	Till	None	None	Esker, rocdrumlin	Material form
Ground moraine	Smooth gentle slopes, except rock hills	Till	None	None	Terminal moraine, lake deposits, outwash	Material, rolling topography
Till plain	Flat	Clay, till	None	None	Outwash plain, lake plain	Material
Outwash level	Flats, terraces, valley bottoms	Gravel, sand	Good	Horizontal, small cross-bedding	Ground moraine, lake deposits	Material, sorting, topography
Outwash pitted	Kettles in flats or knob and sag with even skyline, terraces	Gravel, sand	Good	Horizontal, slumped near kettles, small cross-bedding	Terminal moraine, kames	Material sorting, even skyline mesas
Esker	Winding, interrupted ridge parallels glacial movement	Gravel, sand	Poor	Irregular, anticlinal cross section	Drumlin, bar	Trend, material, irregular top
Crevasse filling	Ridge, high parts, level top	Gravel, sand, silt	Good	Horizontal, inclined, or irregular	Esker	Top on same level as outwash or delta
Kame—see stony moraine	Knolls, kettles, irregular skyline	Gravel, sand	Poor	Irregular faulted	Pitted, outwash, terminal moraine	Material, sorting, bedding, irregular skyline
Beach	Shelf, steep slope or cliff above, slope below, local kettles	Gravel, sand, boulders	Excellent to poor	Dip down front of terrace	Accidental break in slope	Nearly level, best on headlands
Bar	Ridge, curving, level top, rare kettles	Gravel, sand	Excellent to poor	Anticlinal section	Esker	Excellent sorting, level top
Lake bed	Flat, gently rolling, some depressions	Clay, silt, sand	Excellent scattered stones	Horizontal	Ground moraine, outwash	Relation to topography clays, silts, valves

[1] Reproduced, by permission and with modifications, from F. T. Thwaites, "Outline of Glacial Geology," Edwards Bros., Inc., Ann Arbor, Michigan, 1937.

section; they typify the boulders that *may* be met with in glacial drift. Examples from construction practice have already been cited to show what trouble undetected boulders can cause. One of those encountered during the construction of the cutoff trench for the Silent Valley Dam for the Belfast Water Supply, Northern Ireland (page 312), was described as being "as big as a cottage." Although this

FIG. 20.22.—A boulder pavement (of glacial boulders), Sunnyside, Toronto, Ontario, Canada. The effect of such a pavement, if not penetrated, on test-boring results can be readily imagined.

(Photograph by Dr. A. P. Coleman. Reproduced by permission of Dr. Coleman, and the Ontario Dept. of Mines; photograph reproduced in 41st. Annual Report, Part VII, 1932.)

case is unusual, it illustrates vividly the dangerous possibilities of inadequate subsurface exploration in glacial drift.

Boulders of rock type identical with the local bedrock will be but rarely encountered, so that, if an engineer once recognizes the fact that he is working in drift-covered country, an examination of the first diamond-drill core obtained when "rock" is struck should tell him whether the drilling is actually in bedrock or in a boulder. It will almost certainly give him this valuable information if he appreciates the leading features of the main rock types; if he is in doubt, consultation with a geologist will be an obvious course to pursue. There are few other ways in which Geology can be of such direct assistance in the elucidation of an engineering problem.

20.8. Soil Characteristics.—In the study of soils, there are two extreme positions—that of the geologist who is interested primarily in the origin of soils and that of the civil engineer who has no direct interest in origins but is vitally interested in types of soils and the properties peculiar to each type. As is the case with so many extreme positions, they are tenable only in theory. In practice, the geologist has to differentiate among gravels, sands, and clays—at least in general terms—and the engineer, if he is to be able properly to appreciate soil properties, must give some consideration to the origin of the particular soils with which he has to deal. A common meeting ground is thus established for the geologist and engineer, each interested in soils, the unifying factor being the consideration that both have to give to soil characteristics. Already there has been intercourse on this common ground, and in the future the correlation of Soil Mechanics and Geology is bound to be further developed. Note may well be taken, therefore, of certain geological features affecting soil characteristics, and this will be done without venturing further into more detailed aspects of modern soil testing work.

Considering soils in general, the nature of their formation and deposition confirms all that was suggested previously with regard to soil sampling. Residual soils will vary in character from the surface downward, as also will many sediments. If the latter are stratified, this fact may assist sampling operations, as the respective strata are identified. River deposits, especially from the swifter flowing streams, are liable to be variable in content, as are also several types of glacial deposit. Marine deposits tend to be more uniform, as do flood plain sediments from mature stream flow. Whatever the origin of the soil, however, a general study of the local geology prior to the taking of soil samples will be of great value. When once the subsurface survey operations have been carried out, records will be plotted, accurately and to reasonable scales. The resulting cross section, indicating the different soil types encountered, is often referred to as the *soil profile.* This term is frequently met with in highway subgrade work in which soil characteristics are of great importance. A study of soil profiles, both shallow and deep, in conjunction with local geological data will often suggest locations where further samples may usefully be obtained. Especially is this the case with river and glacial deposits. Only when this study of the record has been made can the sampling be regarded as complete.

Special note must be made of the terminology presently employed for soil classification. Each of the main group terms used—*gravel, sand, silt,* and *clay*—is based on the *size* of particles composing the soil, generally without respect to their mineralogical nature. This is in striking contrast to the usual practice in geological terminology. Occasionally, compound names, such as *quartz sand* or *limestone gravel,* are used to describe materials more accurately, but such usage is still infrequent. The more usual compound names have relation to the physical properties of the soil, *e.g.*, *fine sand* and *tough clay*. It is probable that a more frequent use of geological compound names will develop as soil studies progress.

Terminology in the field, as in the recording of test-boring results, provides another problem in that the selection of the names used to describe the soils encountered is still a matter of individual judgment. Experienced boring operators can usually be relied upon to differenti-

FIG. 20.23.—A large pocket of glacial drift in a hollow in the Bunter Pebble Beds, near Hopwas, Staffordshire. At the left can be seen a bed of horizontal Bunter Sandstone, resting on a wedge of gravel with a high dip due to current (false) bedding.

(Geological Survey of Great Britain photograph. Crown Copyright, reproduced by permission of The Controller of H. M. Stationery Office, London.)

ate correctly between different soil mixtures, but they will be working to their own individual standard of comparison. Consequently, it is advisable, in all soil exploratory work, to correlate the soil names used by those responsible for taking field notes with the results of laboratory analysis and test, if a uniform and recognized standard is to be adhered to throughout the work.

In this respect, as in others, laboratory work on soils can be of assistance to the geologist as well as to the engineer. In the laboratory, for example, the shape of sand grains can be determined, and their mineralogical nature ascertained, this information often being of real assistance to purely geological studies of sedimentary origins.

In some cases, it can be obtained by geological study in the field without the aid of laboratory facilities.

Gravels.—Gravels are accumulations of unconsolidated rock fragments resulting from the natural disintegration of rock, being limited in size to 2 mm. diameter. From this minimum size, gravel may range up to the largest pieces of rock normally encountered. Names commonly given to larger fragments are *pebbles* (from 4 to 64 mm.), *cobbles* (64 to 256 mm.), and *boulders* (having a diameter greater than 256 mm., or about 10 in.); these are the sizes given by the Committee on Sedimentation of the Division of Geology and Geography of the National Research Council, United States.

Gravels are widely distributed but rarely found without some proportion of sand, and possibly silt, unless these finer constituents have been washed out subsequent to the deposition of the gravel. From the nature of this deposition, they are characteristic shallow water or river deposits (including the river flows from glaciers); they may also have been beach formed, and some gravels are found in kames. The shape of the individual rock fragments and their relative mineralogical freshness will depend on the history of the gravel, but variations from rounded gravel to angular fragments may be met with. Gravel that has been transported to such an extent that the individual fragments are rounded will be found to consist of the more resistant rock types, often a mixed collection; but angular gravels, being comparatively new, may consist of fragments of even so relatively soft a rock as limestone.

Sands.—Sand is the name applied to fine granular materials, derived from either the natural weathering or the artificial crushing of rocks, that range in size from 0.053 to 2.0 mm. in diameter. Various subdivisions are used by different authorities to denote the gradations of sand-particle sizes; the most usual of these is to fix a diameter of 0.42 mm. as the dividing line between fine and coarse sand. As it occurs naturally, a sandy soil mixture will frequently have some clay and silt mixed with the sand-sized grains. If the clay and silt content does not exceed 20 per cent, the mixture is called a sand.

The origin and therefore the occurrence of sands are similar to those of gravels, the two often being found together in one deposit. Beach sands, which may today be found far inland as a result of earth movements, are the most uniform type of deposit; river sand often contains relatively large amounts of gravel, silt, and clay. Glacial deposits may contain good sand beds, but they are apt to be uncertain. Residual sands are sometimes found. As sands are often derived from the continued disintegration of gravel, it follows that, in general, they will be composed of the harder and more stable minerals. The popular conception that all sands consist of quartz particles is, however, far from true. Although sands consisting of almost pure quartz do occur, most of them contain at least a small percentage of other minerals. Some consist primarily of minerals other than quartz, calcareous sands sometimes being encountered in limestone districts. As a rule, their mineral content will be found to be fairly stable, but this may not always be the case with glacial sands which may contain fresh minerals liable to weather on exposure.

The shape of sand particles may vary from completely rounded grains to angular fragments. The former are rare and are found generally in desert regions. The age of a sand deposit is no criterion of the shape of the grains but depends principally on the rock from which the sand was formed and its history during transportation. Thus, quartz grains from a granite rock will retain an irregular shape; from felsite they will be more truly crystalline; and from gneiss and schists, they may be flat scalelike particles. Long travel in water will tend to make the grains rounded, but it is a slow process, as shown by the fact that sand reaching the mouth of the Mississippi is composed of angular grains. As a rule, river sands are more angular than those found in lake and marine deposits.

Clays.—Soil particles smaller in diameter than 0.005 mm. and displaying in the mass the property of being plastic when wet are known as clay. The name is correspondingly applied to a soil mixture consisting of clay solids but that may contain also up to 70 per cent of sand and silt, although the clay and silt content combined must exceed 50 per cent. Such a mixture will still display the characteristic properties of a pure clay. The essential qualification of plasticity being shown by the soil mixture when wet is particularly to be noted; it is possible to grind up quartz so that it conforms to the size requirements of a clay, and yet it has no characteristic clayey properties. It may also be noted that clay solids include particles of colloidal size; the colloidal content of clays is of great importance although not yet fully investigated. Clays are now known to be crystalline in nature and not amorphous, as was previously supposed; particle shape is related to their peculiar physical properties.

As will be noted from a study of soil origins and deposition, clays may be formed by all the main processes associated with rock weathering. They may be residual or transported. Transported clays may be flood plain river deposits, of varying thickness and probably mixed with sand; lake deposits, now elevated as a result of land movement; estuarine or marine clays similarly raised from their original positions.

The last are a particularly important group and occur in many parts of the world and as large deposits. Finally, glacial clays (often called boulder clays) constitute a widespread series of deposits, although it is to be remembered that some "boulder clays" contain no clay at all. The name is correctly applied, however, to many glacial deposits,

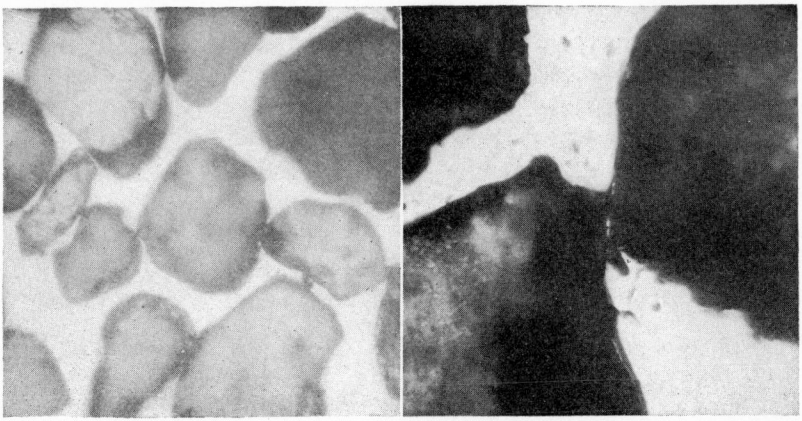

Beach sand. Crushed angular sand.

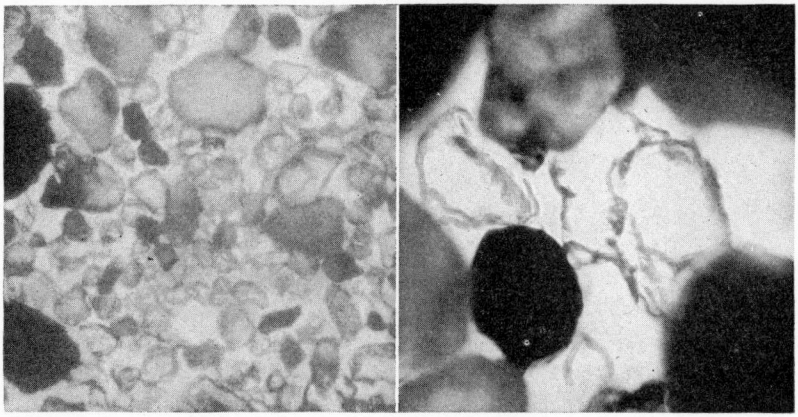

Glacial sand. Subangular river sand.

FIG. 20.24.—Photomicrographs of some typical forms of sand grain.

(*Reproduced by permission of the U. S. Bureau of Public Roads, from Public Roads, June,* 1931.)

although the mixture of sand, gravel, and boulders usually present in such deposits often renders them useless as a commercial source of clay.

Unlike other soil types, clays are susceptible to pressure, shales being the resulting rock type when they are subjected to large enough pressures for considerable periods. Even before passing into shale, clays may be considerably affected by pressure, becoming *consolidated* (or *preconsolidated*, as it is sometimes expressed). The necessary pressure may have been due to the presence of great depths of other deposits, subsequently eroded, or to that of an ice sheet possibly several thousand feet thick. Pressure of this magnitude cannot easily be duplicated in the ordinary laboratory; still less can the range of geological time be even remotely simulated in experimental work.

FIG. 20.25.—Varved clays at Wawbewana, Ontario. ('Varves' are laminae, usually two, representing an annual deposit of clay; they occur in glacial clays, in some lake deposits now forming, and possibly in other sediments.)
(*Reproduced by permission of the Geological Survey of Canada.*)

Thus it is that when once a sample of such consolidated clay has been thoroughly disturbed (by remolding), it will lose its original properties, and these cannot be regained. Although consolidated clays can be detected in a laboratory by means of suitable tests, it is clearly of great assistance if this can be foretold from study of the local geology—and this is usually possible.

20.9. The Future of Soil Studies.—To talk about the future is always hazardous; to write about it is almost foolhardy. So recent, however, has been the growth of scientific soil studies, and so great are the potentialities that they appear to present for future usefulness to the engineering profession, that a chapter such as this would be incomplete without at least a glance beyond present achievements. All that will be so observed is the correlation of Geology with soil studies, the future of the latter alone being a theme worthy of detailed review.

Of more than ordinary significance is the number of Soil Mechanics experts who have been led to their soil studies through their knowledge of Geology. Dr. Karl von Terzaghi is preeminent in this regard and also as a soil expert; the quotation at the head of this chapter explains his approach in his own words. In reference to problems in engineering related to soils, it has been said that "Geology is not enough"; and the attainments of Soil Mechanics go far to show how true this is. At the same time, it may usefully be remembered in this connection that "Soil Mechanics is not enough." Without a judicious application to practice only when correlated accurately with knowledge of the relevant local geology, much laboratory work will be wasted. To describe the correlation would be to go over old ground anew; suffice it to say that one future trend of Soil Mechanics work must be closer cooperation with local geological studies.

The word *local* may well be stressed. So variable are soil types and soil formations that, useful as generalizations may be, every soil problem is liable to present peculiar problems and so must be studied afresh. This implies that appreciation of the significance of soil studies in the future will become more widespread among engineers. Already there are signs of this awakening. As an indication may be mentioned the fact that already (in 1939) about 30 of the state and provincial highway departments in North America have set up laboratories especially for the study of soils. Comparisons between the use and testing of concrete and of soils cannot be pressed too far; but in that they are to some extent comparable, it may be suggested that the day may come when soil testing is as familiar a counterpart of civil engineering construction as is concrete testing today.

Mention of construction leads to thoughts of the use to which soils may be put by engineers. Normally, they represent a part of the earth's crust to be traversed by transportation routes or removed to make way for some engineering structures. In two major directions, soils are actually used as construction materials: in the building of earth dams and as providing natural surfacing material for secondary roads. Dam construction will always be relatively restricted in scope, but secondary road construction is probably the most widespread of all civil engineering operations. It is a field that is yet far from being completely serviced. And in the prosecution of the great amount of work that still lies ahead in road building, the scientific use of soil mixtures is bound to play an increasingly important part. This will mean not only an active continuation of soil research work but also—of equal importance—a great extension of local geological study in the attempt to discover local deposits of clays, sands, and gravels that may be used economically in this way.

Geology thus appears again as an aid to civil engineering. And as this particular application is developed in connection with the use of the most humble of all natural materials for the convenience of mankind, so may engineers come to appreciate the more widely that observation of Sir Benjamin Baker—himself a master engineer of the past—that

"Nature never deceives you, if you watch her long enough."

Suggestions for Further Reading

A fair number of books exist dealing with soils from various angles although usually in connection with agriculture. The following are representative:

KEEN, B. A.: "The Physical Properties of the Soil," Longmans, Green & Company, London, 1931.

LYON, T. L., and H. U. BUCKMAN: "Nature and Properties of Soils," The Macmillan Company, New York, 1929.

MERRILL, G. P.: "Rocks, Rock Weathering and Soils," The Macmillan Company, New York, rev. ed., 1911. Special mention must be made of this book, which remains an authoritative work.

ROBINSON, G. W.: "Soils, Their Origin Constitution and Classification," Thomas Murby & Co., London, 2d ed., 1936.

WOODWARD, H. B.: "The Geology of Soils and Sub-strata," Edward Arnold & Company, London, 1912.

ON AGRICULTURAL GEOLOGY

EMERSON, F. V.: "Agricultural Geology," John Wiley & Sons, Inc., New York, 2d ed., 1928.

RASTALL, R. H.: "Agricultural Geology," Cambridge University Press, Cambridge, England, 1916.

ON SEDIMENTARY DEPOSITS

MARR, J. E.: "Deposition of the Sedimentary Rocks," Cambridge University Press, Cambridge, England, 1929.

TWENHOFEL, W. H.: "Treatise on Sedimentation," Williams & Wilkins Company, Baltimore, Maryland, 2d ed., 1932.

ON GLACIAL DEPOSITS

THWAITES, F. T.: "Outline of Glacial Geology," Edwards Bros., Inc., Ann Arbor, Michigan, 1937; contains many references to relevant literature.

On Soil Mechanics

Hogentogler, C. A.: "Engineering Properties of Soil," McGraw-Hill Book Company, Inc., New York, 1937; extensive references listed to most of the other publications, in English, on Soil Mechanics.

Proceedings of the First International Conference on Soil Mechanics and Foundation Engineering, Harvard University, Cambridge, Massachusetts, 3 vols., 1936.

PART III
APPENDICES

APPENDIX A

A SHORT GLOSSARY OF GEOLOGICAL TERMS COMMONLY ENCOUNTERED

This glossary is merely an introduction to geological terminology. An attempt has been made to include in it the more common specialized words met with in general references to Geology, words such as may be encountered by civil engineers in the ordinary course of their work. Compound names such as *gabbro-syenite* have not generally been included, and no mention is made of the names of minerals —whether used to describe minerals only or, as in such cases as *serpentine*, rocks as well. Reference may be made to Table A on page 13 for particulars of the more common minerals. Letters such as (H) refer to the authorities cited, listed at the end of the glossary.

Accessory. A term applied to minerals occurring in small quantities in a rock, and whose presence or absence does not affect its diagnosis (H).

Acidic. A term applied to rocks in which silicic acid (silica) or quartz predominates (R).

Adobe. An impure calcareous clay, widely used in western United States for making semidried bricks (R).

Aeolian. A term applied to deposits whose constituents have been carried by, and laid down from, the wind (H).

Agglomerate. Contemporaneous pyroclastic rocks containing a predominance of rounded or subangular fragments greater than 32 mm. in diameter, lying in an ash or tuff matrix and usually localized in volcanic necks. The form of the fragments is in no way determined by the action of running water (N).

Alluvium. The deposits made by streams in their channels and over their flood plains and deltas. The materials are usually uncemented and are of many kinds and dimensions (N).

Andesite. A volcanic rock, generally porphyritic, composed essentially of plagioclase, together with one or more of the mafic minerals, biotite, hornblende, and pyroxene (H).

Aphanitic (Rock). Refers to rocks that are so fine grained that the individual minerals cannot be distinguished by the naked eye (R).

Arenaceous. Like or pertaining to sand. An example is arenaceous limestone or a sandy limestone (R).

Argillaceous. An adjective meaning clayey and applied to rocks containing clay (N).

Ash (Volcanic). Uncemented pyroclastic debris consisting of fragments mostly under 4 mm. in diameter (N).

Batholith. A huge intrusive body of igneous rock, without a known floor—in contrast to a laccolith, which rests on a floor. Like other intrusive bodies, batholiths become accessible to human observation only as a result of their exposure by erosion (L).

Basalt. A microlithic or porphyritic igneous rock of a lava flow or minor intrusion, having an aphanitic texture as a whole or in the groundmass and composed essentially of plagioclase and pyroxene, with or without interstitial glass (H).

Basic. A term applied to igneous rocks having a relatively low percentage of silica, the limit below which they are regarded as basic being about 52 per cent (H).

Bed (of Rock). A single layer, or stratum, of rock, which may be either an inch or a hundred feet or more in thickness (R).

Bedding Plane (of Rock). Refers to the plane of junction between different beds or layers of rock (R).

Bentonite. The plastic residue from the devitrification and attendant chemical alteration of glassy igneous material, usually volcanic tuff or ash. It swells greatly with the addition of water and forms a milky suspension (after N and R).

Boss. A transgressive intrusion of igneous rock like a stock or dome. The term is also applied to forms of less regular outline than the latter and is therefore of wider application (H).

Boulder. A detached rock mass, somewhat rounded or otherwise modified by abrasion in transport; 10 in. (256 mm.) has been suggested as a convenient lower limit for diameter (N).

Boulder Clay. A tenacious unstratified deposit of glacial origin consisting of a hard matrix (rock flour) packed with subangular stones of varied sizes (after H).

Breccia. A clastic rock made of coarse angular or subangular fragments of varied or uniform composition (and of either exogenetic or endogenetic origin) (H).

Calcareous. An adjective applied to rocks containing calcium carbonate (N).

Chalk. A fine-grained somewhat friable foraminiferal limestone of Cretaceous age occurring in Britain, northwestern Europe, and elsewhere (H).

Chert. A more or less pure siliceous rock, occurring as independent formations and also as nodules and irregular concretions in formations (generally calcareous) other than the chalk. The fracture is generally splintery rather than concoidal (H).

Clastic. A term applied to rocks composed of fragmental material derived from preexisting rocks or from the dispersed consolidation products of magmas (H).

Clay. A fine-grained unconsolidated material which has the characteristic property of being plastic when wet and which loses its plasticity and retains its shape upon drying or when heated (after R and N).

Claystone. Indurated clay. If possessed of bedding cleavage or fissility approximately parallel to bedding, it is also a *shale* (*cf. Sandstone*) (N).

Cleavage. (1) The property of minerals . . . whereby they can be readily separated along planes parallel to certain possible crystal faces.

(2) The property of rocks, such as slates, that have been subjected to orogenic pressure, whereby they can be split into thin sheets . . . (H).

Cobble. A rock fragment between 64 and 256 mm. in diameter, rounded or otherwise abraded in the course of aqueous, aeolian, or glacial transport (N).

Concretions. Nodular or irregular concentrations of certain authigenic constituents of sedimentary rocks and tuffs . . . (H).

Conglomerate. A cemented clastic rock containing rounded fragments corresponding in their grade sizes to gravel (H).

Crystal. A body, generally solid but not necessarily so, whose component atoms are arranged in definite space lattices, crystal faces being a commonly developed outward expression of the periodic arrangement of atoms (H).

Cumulose Deposits. Sedentary accumulations of carbonaceous matter with very little detrital sediment (H).

Current- or Cross-bedding. A structure of sedimentary rocks, generally arenaceous, in which the planes of deposition lie obliquely to the planes separating the larger units of stratification (H).

Detritus. Fragmental material, such as sand and mud, derived from older rocks by disintegration. The deposits produced by the accumulation of detritus constitute the detrital sediments (H).

Diabase. See under *Dolerite.*

Diorite. A phanerocrystalline igneous rock composed of plagioclase and mafic minerals such as hornblende, biotite, and augite, hornblende being especially characteristic (H).

Dip. The angle of inclination of the plane of stratification with the horizontal plane (R).

Dolerite. An igneous rock occurring as minor intrusions, consisting essentially of plagioclase and pyroxene, being distinguished from basalt by its coarser grain and the absence of glass. American authors use the term *diabase* in a sense synonymous with dolerite (after H).

Dolomite. A carbonate rock, consisting predominately of the mineral dolomite (H and R) (*cf.* dolomitic limestone and magnesian limestone, in which dolomite and calcite are present, the latter predominating).

Drift. A more general and looser term than *till* used to describe all the material handled by glacial action whether moved or not.

Dyke. An injected wall-like intrusion, cutting across the bedding or other parallel structures of the invaded formations and having a thickness narrow in proportion to its length (H).

Ejectamenta. A general term for pyroclastic materials ejected from a volcanic vent (H).

Endogenetic. A term applied to geological processes originating within the earth and to rocks that owe their origin to such processes (H).

Epigene. A general term for geological processes originating at or near the surface of the earth (H).

Epigenetic. A term now generally applied to ore deposits of later origin than the rocks among which they occur (H).

Erosion. The wearing away of portions of a rock by such natural agencies as stream or ice action (R).

Essential Minerals. The essential constituents of a rock are those minerals whose presence is necessary in it by definition (R).

Exfoliation. The "peeling" of a rock surface in sheets due to changes of temperature or to other causes (R).

Exogenetic. A term applied to geological processes originating at or near the surface of the earth and to rocks that owe their origin to such processes (H).

Faults. Abrupt breaks in the continuity of the beds or strata with the elevation or depression of beds on one side of the plane at the fault (R).

Felsite. An igneous rock, with or without phenocrysts, in which either the whole or the groundmass consists of cryptocrystalline aggregates of felsic minerals, quartz and orthoclase being those characteristically developed (H).

Felsitic. A term applies to the cryptocrystalline texture seen in the groundmass of quartz-felsites and similar rocks (H).

Ferruginous. An adjective applied to rocks containing iron.

Flint. A more or less pure siliceous rock composed mainly of granular chalcedony together with a small proportion of opaline silica and occurring in nodules, irregular concretions, layers, and veinlike masses. The fracture is concoidal, whereas in *chert* a splintery fracture is more usual (H).

Foliation. A structure represented most characteristically in schists, due to the parallel disposition of layers or lines of one or more of the conspicuous minerals of the rock, the parallelism not being a direct consequence of stratification (H).

Formation. A term applied stratigraphically to a set of strata possessing a common suite of lithological and/or faunal characteristics (H).

Gabbro. A phanerocrystalline rock consisting of labradorite, or bytownite, and augite (generally diallage) (H).

Gneiss. A foliated or banded phanerocrystalline rock (generally, but not necessarily, felspathic and of granitic or dioritic composition) in which granular minerals, or lenticles and bands in which they predominate, alternate with schistose minerals, or lenticles or bands in which they predominate. The foliation of gneiss is more "open," irregular, or discontinuous than that of schist (H).

Granite. A phanerocrystalline rock, consisting essentially of quartz and alkali-feldspars with any of the following—biotite, muscovite, and amphiboles and pyroxenes (H).

Granitic. A term applied to irregularly granular textures like that of a nonporphyritic granite (H).

Gravel. An unconsolidated accumulation of rounded rock fragments predominantly larger than sand grains. Depending on the coarseness of the fragments, the material may be called *pebble gravel, cobble gravel,* or *boulder gravel* (after N).

Grit. A coarse sandstone of angular grain. Variation from the grain-size range of from ½ to 1 mm. in diameter should be indicated as coarse grit, fine grit, etc. (N)

Groundwater. A term that should be applied only to the basal or bottom water in the zone of rock fracture below the earth's surface, in which interstitial water occurs, *i.e.*, in the zone of saturation (M).

Gumbo. A provincial term used to describe local varieties of sticky, highly plastic clay soil (after N).

Hardpan. A term to be avoided, in view of its wide and essentially popular local use for a wide range of materials. If it has to be used, the following definition should be taken as a guide: "An horizon of accumulation that has been thoroughly cemented to an indurated, rocklike layer that will not soften when wet. The term hardpan is not properly applied to hard clay layers that are not cemented or to those layers that may seem indurated when dry but which soften and lose their rocklike character when soaked in water. The true hardpan is cemented by materials that are not readily soluble and is a hard layer that definitely and permanently (in nature) limits downward movement of roots and water" (U. S. Bureau of Public Roads publications).

Holocrystalline. A term applied to igneous rocks completely made up of crystals (H).

Magma. A comprehensive term for the molten fluids generated within the earth from which igneous rocks are considered to have been derived by crystallization or other processes of consolidation (H).

Magnesian Limestone. See under *Dolomite.*

Marble. A general term for any calcareous or other rock of similar hardness that can be polished for decorative purposes (petrologically restricted to granular crystalline limestones) (H).

Megascopic. A general term, more appropriate than *macroscopic,* applied to observations made on minerals and rocks by means of the naked eye or pocket lens but not with a microscope (H).

Metamorphic Rocks. Rocks derived from pre-existing rocks by mineralogical, chemical, and structural alterations due to endogenic processes, the alteration having been sufficiently complete throughout the body of the rock to have produced a well-defined new type (H).

Mica-schist. A schist composed essentially of micas and quartz; the foliation is mainly due to the parallel disposition of the mica flakes (H).

Mud. A mixture of water with clay and/or silt together with materials of other dimensions (N).

Mudstone. A general term (including siltstone, claystone, and shale) to be used only when a more precise one is not possible (*cf. sandstone, limestone*) (N).

Natural Bed. The surface of a stone (cut rock) parallel to its stratification.

Oölite. A rock made up of spheroidal or ellipsoidal grains formed by the deposition of successive coats of calcium carbonate around a nucleus (H).

Ophitic. A term applied to microscopic rock texture to designate a mass of longish interlacing crystals, the spaces between which are filled with minerals of later crystallization (R).

Outcrops (of Rock). Those places where the underlying bedrock comes to the surface of the ground and is exposed to view (R).

Pebble. A rock fragment between 4 and 64 mm. in diameter, which has been rounded or otherwise abraded by the action of water, wind, or glacial ice (after N).

Pegmatite. A term applied to graphic granite and extended to coarse-grained modifications of granite characterized by irregular segregation of particular minerals rather than by interpenetration (H).

Phanerocrystalline. A term applied to igneous rocks in which all the crystals of the essential minerals can be distinguished individually by the naked eye; contrasted with *aphanitic* (H).

Phenocryst. A term applied to isolated crystals visible to the unaided eye and lying in the mass of a rock of igneous origin (R).

Phreatic. Pertaining to *groundwater* (M).

Plutonic. A general term applied to major intrusions and to the rocks of which they are composed, suggestive of the depths at which they were formed in contradistinction to most minor intrusions and all volcanic rocks (H).

Porphyrite. A term applied to hypabyssal rocks of marked porphyritic structure and andesitic composition. The phenocrysts are generally plagioclase and mafic minerals, and the groundmass is holocrystalline (after H).

Porphyry. A term first given to an altered variety of porphyrite on account of its purple color and afterward extended by common association to all rocks containing conspicuous phenocrysts in a fine-grained or aphanitic groundmass. The resulting texture is described as *porphyritic* (H).

Hypabyssal. A general term applied to minor intrusions, such as sills and dykes, and to the rocks of which they are made, to distinguish them from volcanic rocks and formations, on the one hand, and plutonic rocks and major intrusions such as batholiths, on the other (H).

Hypogene. A general term intended to include both plutonic and metamorphic classes of rocks, *i.e.*, for rocks formed within the earth (H).

Igneous (Rocks). Those which have been formed or crystallized from molten magmas, the source of which has been the earth's interior (R).

Inclusion. A general term for foreign bodies (gas, liquid, glass, or mineral) enclosed by minerals; also extended in its English usage to connote *enclosures* of rocks and minerals within igneous rocks, such as *cognate* and *accidental* inclusions (H).

Indurated. A term technically restricted to compact rocks that have been hardened by the action of heat (H). Occasionally the term is applied also to sediments hardened by impregnating solutions (R).

Intermediate (Rocks). Applies to rocks intermediate in silica content between the *basic* and the *acidic* groups (R).

Intrusive (Rocks). Igneous rocks that have cooled and solidified from magma under the cover of the rock masses of the outer shell of the earth; they therefore become accessible to view only after they have been exposed by erosion of the overlying rock (after L).

Joints (Rock). Rock fractures on which there has been no appreciable displacement parallel with the walls. They may sometimes be characterized, according to their relation with the rock bedding, as *strike joints* and *dip joints* (after L).

Kaolin. A white, or slightly stained, clay, due to the decomposition of a highly felspathic rock and therefore also containing a variable proportion of other constituents derived from the parent rock (H) (*cf. Kaolinisation*, the process by which a feldspar passes—or "weathers"—into kaolin) (after R).

Laccolith. A dome-shaped intrusion with both floor and roof concordant with the bedding planes of the invaded formations, the roof being arched upward as a result of the intrusion (H).

Laterite. A residual deposit, often concretionary, formed as a result of the decomposition of rocks by weathering and groundwaters and consisting essentially of hydrated oxides of aluminum and iron (H).

Lime Rock. A natural material, found in the southern part of the United States. It is composed essentially of calcium carbonates with varying percentages of silica. Lime rock hardens upon exposure to the elements, and some varieties provide excellent road metal (R).

Limestone. A general term for bedded rocks of exogenetic origin, consisting predominantly of calcium carbonate (H).

Lithology. The study of rocks as based on the megascopic observation of hand specimens. In its French usage, the term is synonymous with *Petrography* (H).

Loam. A detrital deposit containing nearly equal proportions of sand, silt, and clay, these terms referring to the grade size of the particles (H) (see Fig. 20.4).

Loess. A widespread deposit of silt or marl, a buff-colored, porous, but coherent deposit traversed by a network of generations of grass roots (H).

Macroscopic. See under *Megascopic.*

Mafic. A mnemonic term for the ferromagnesian and other nonfelsic minerals actually present in an igneous rock (H).

(H) HOLMES, A.: "The Nomenclature of Petrology," Thomes Murby & Co., London, 1920.

(L) LONGWELL, C. R., A. KNOPF, and R. F. FLINT: "Textbook of Geology, Part One: Physical Geology," John Wiley & Sons, Inc., New York, 1932.

(M) For a full discussion of groundwater terminology, see O. E. MEINZER: Outline of Groundwater Hydrology with Definitions, *U. S. Geol. Survey Water Supply Paper* 494, 1923, on which the three definitions given are based.

(N) Reports, from 1930 to 1938, of the Committee on Sedimentation of the Division of Geology and Geography of the National Research Council, Washington, in which will be found—in addition to much valuable information on sedimentation—studies of the nomenclature of various groups of sediments.

(R) RUNNER, D. G. A Glossary of Geological Terms for the Highway Engineer (rev.), *Roads and Streets*, Chicago, Illinois, May 1, 1937. See also BROWN, V. J., and D. G. RUNNER: "Engineering Terminology," Gillette Publishing Company, Chicago, Illinois, 1938.

(T) TWENHOFEL, W. H.: "A Treatise on Sedimentation," Williams & Wilkins Company, Baltimore, Maryland, 2d ed., 1932.

Glacial Geology.—For definitions of topographical features associated with glacial action, see Table S (p. 573).

APPENDIX B

GEOLOGICAL SURVEYS OF THE ENGLISH SPEAKING WORLD

Brief details of the geological surveys of North America and the British Empire are here presented as a convenience for those civil engineers who may wish to get into touch with their local survey, with regard to either geological publications or the progress of any of their work in which the survey might be interested. Great care has been taken in the preparation of this list, a copy of each note having been submitted to the director of each survey mentioned; but in view of the lapse of time between this check and publication, the accuracy of the information cannot be guaranteed. It is hoped that the particulars given will be used not only as suggestions of sources of information useful to engineers but also as enabling engineers readily to make contact with scientists to whom new excavation work can often give information of great interest and inestimable value. The geological departments of all universities will also be found to be appreciative of advice of this nature. For a review of world-wide geological survey work, see R. S. G. Stokes: The Geological Surveys and Societies of the World, anniversary presidential address, *Geol. Soc. South Africa Min. Proc. Trans.*, **38**: xxiii (1936).

Australia:

Commonwealth of Australia: Department of the Interior

Australia does not possess a federal geological survey comparable to those of Canada and the United States. The several sovereign states of the Commonwealth maintain their own geological surveys.

The Federal Government does maintain a small geological advisory staff with headquarters at Canberra, F.C.T. This consists of

Dr. W. G. Woolnough, Commonwealth Geological Adviser.
Dr. Arthur Wade, Chairman of the Commonwealth Technical Oil Advisory Committee.

With them is associated Dr. L. Keith Ward, Director of Mines of South Australia, as a part-time member of the Oil Advisory Committee.

Until recently, all three were engaged in advising the Government in relation to the expenditure of funds voted by Parliament to assist in the search for petroleum in Australia and New Guinea. Recently, the question of availability of supplies of iron ore within the Commonwealth has become acute. Dr. Woolnough has been instructed to devote the whole of his time to research on this subject.

Mr. E. A. Kodyen was recently appointed Commonwealth Inspector of Drilling to guide the operations of drilling for oil for which the Government has imported up-to-date plant from America. Miss Irene Crespin is Commonwealth Palaeontologist.

Northern Australia:

Aerial, Geological and Geophysical Survey of Northern Australia (Governments of the Commonwealth, Queensland and Western Australia)

Director: Sir Herbert Gepp
Headquarters: Canberra, F.C.T.

This Survey was organized in June, 1934, as a result of conferences between members of the participating governments, for the purpose of making a geological and geophysical reconnaissance of the mineral-bearing regions of North Australia, in order to advance the development of this part of the Commonwealth through its mineral industry. Through the cooperation of state governments, members of the staffs of state geological surveys are associated with the work, Mr. P. B. Nye (of Tasmania) being Executive Officer and Dr. W. G. Woolnough (Commonwealth Geological Adviser) Geological Consultant to the Executive Committee. Geophysical work is being done, and aerial photography carried out (by the Royal Australian Air Force) over a wide area. Four reports have been issued, accompanied by maps, and already some prospective mines have been located as a result of the survey work.

Australia: New South Wales:

Geological Survey of New South Wales

Government Geologist: L. F. Harper
Headquarters: Department of Mines, Sydney

The discovery of gold in New South Wales in 1851 apparently led to the appointment of a government geologist, but it was not until a Department of Mines was created in 1874 that a geological survey came into being. The Survey Staff now constitutes one branch of the Department of Mines. Associated with it are the Geological Survey Laboratory and a Mining and Geological Museum (both at Sydney) which are under the direction of the Government Geologist. The work of the Survey is largely related to mining, advice being supplied to the mining community, and surveys conducted of known ore reserves. Surveys are also made of Coal Fields and areas likely to contain subsurface waters. Geophysical surveys of suitable areas, mainly auriferous and stanniferous deep leads, are undertaken. The services of the Survey are in constant demand by other government departments, and advice on civil engineering matters, especially on the foundations of large dams, is often supplied. A large number of free assays of economic materials are made annually.

Publications.—Annual reports, from the year 1875, constitute the main body of publications. A Mineral Resources Series of papers is also issued, as well as bulletins dealing with specific materials. Memoirs of the Survey are classified into General, Palaeontological, and Ethnological issues. A series of records is also issued in addition to a large selection of maps. All publications are obtainable from the headquarters of the Survey, Sydney.

Australia: Queensland:

Queensland Geological Survey

Chief Government Geologist: L. C. Ball
Headquarters: Brisbane

The Geological Survey of Queensland was founded in 1879, its main functions being to advise the Department of Mines generally, to prepare reports and maps from field investigations, and to provide mineral and rock collections to educational establishments. Fieldwork consists of the inspection of individual mineral deposits and mines and of groups or fields, the carrying out of geological investigations and mapping in so far as directly economic work permits, and the undertaking of consultative work for other departments and organizations.

Publications.—Almost 300 publications, unclassified, have been issued by the Survey as well as an extensive series of maps. Copies are obtainable from the Librarian, Queensland Survey, Brisbane.

Australia: South Australia:

Geological Survey of South Australia

Director of Mines and Government Geologist: L. Keith Ward
Headquarters: Department of Mines, Flinders Street, Adelaide

The Survey is a part of the Department of Mines and is interlocked with other branches, the general and clerical work being performed by one staff for all branches. The geologists of the Department, who constitute the Survey staff, are engaged mainly on economic work concerned with the development of the mineral resources of the state and also water supply problems; underground water sources are of great importance to this state. In addition to the economic work, the attempt is made to gather information which is embodied in the successive editions of the geological map of South Australia, published at intervals by the Survey. The Department carries out well boring and diamond drilling work, all under geological supervision, and its geologists are freely consulted on engineering matters.

Publications.—In addition to the map mentioned, annual reports are published, and a series of bulletins (on general subjects) and mining records, all obtainable from the Department and from Harrison Weir, Government Printer, North Terrace, Adelaide.

Australia: Tasmania:

Geological Survey of Tasmania

Government Geologist: P. B. Nye
Headquarters: Department of Mines, Hobart

The Geological Survey of Tasmania was founded in 1907, as a division of the State Department of Mines. It carries out investigations into economic geology, with incidental study of areal and stratigraphical problems and investigations into mineral deposits and groundwater resources.

Publications.—These include about 40 general bulletins, several special reports and records, and some papers on mineral resources and underground water supplies; copies are obtainable from Survey Headquarters.

Australia: Victoria:

Geological Survey of Victoria

Director: W. Baragwanath
Headquarters: Department of Mines, Treasury Gardens, Melbourne

The Geological Survey of Victoria was founded in 1851, under A. R. C. Selwyn, shortly after gold had been discovered in the state. For the first 17 years, reconnaissance and detail surveys received attention. Then for a short time, the survey work ceased, and Selwyn received an appointment in Canada. On resumption of work early in the seventies, the Survey devoted attention to areas of economic importance, chiefly gold fields, but areas of black and brown coal were also dealt with. In 1903, under J. W. Gregory, a reorganization of the Survey took place, with the objective of a more even balance of geological and mining surveys, and this program has been followed, according to financial support, to the present time. The Survey operates its own boring branch and laboratories and has also a geological museum.

Publications.—Eleven *Geological Progress Reports* were published between 1874 and 1899; in the two years following, nine monthly *Progress Reports* appeared. From 1902 onward, regular bulletins, memoirs, and records have been issued. In addition, an extensive series of geological maps has been published. A *Mining and Geological Journal,* issued twice yearly since 1937, has taken the place of the records. Copies are obtainable from Survey Headquarters and the Government Printer, Melbourne.

Australia: Western Australia:

Geological Survey of Western Australia

Government Geologist: F. G. Forman
Headquarters: Beaufort Street, Perth

Following individual geological investigations carried out mainly by five geologists between the years 1847 and 1895, three of them working officially as government geologists, the Geological Survey of Western Australia was created in 1896 for the purpose of investigating the geological structure, mineral resources, mining industries, and underground water supplies of the state. The Survey has its own laboratory and museum. An excellent review of the work of this Survey will be found in a paper by Mr. A. Gibb Maitland, sometime Government Geologist, read before the West Australian Natural History Society on May 10, 1910.

Publications.—Annual reports, bulletins (about 100 having already been issued on all branches of the Survey's work), and geological maps are published by the Survey; copies are obtainable from Survey Headquarters.

British Colonial Possessions:

British Colonial Possessions (exclusive of those mentioned later)

The following parts of the British Empire have no geological surveys. Geological and relevant information can generally be obtained from the official mentioned.

Bahamas: Surveyor General.
Barbados: Director of Public Works.

Bermuda: Director of Public Works.
British Guiana: Secretary, Department of Lands and Mines.
British Honduras: Surveyor General.
Ceylon: Surveyor General.
Cyprus: Director of Land Registration and Surveys.
Falkland Islands: Director of Public Works.
Fiji: Commissioner of Works.
Gambia: Land Officer and Surveyor.
Gibraltar: Director of Public Works.
Hong Kong: Director of Public Works.
Jamaica: Surveyor General.
Turks Island: Superintendent of Works.

Leeward Islands:

Antigua: Superintendent of Public Works, British West Indies.
St. Christopher and Nevis: Superintendent of Public Works, St. Christopher and Nevis, British West Indies.
Dominica: Colonial Engineer and Surveyor General, Dominica, British West Indies.
Montserrat: Superintendent of Works and Roads, Montserrat, British West Indies.

Malta: Superintendent of Public Works.
Mauritius: Director of Public Works and Surveys, Port Louis.
St. Helena: Superintendent of Public Works.
Seychelles: Superintendent of Public Works and Surveys.
Somaliland: Director of Agriculture and Geology, Berberra.
Trinidad: Surveyor General, Lands and Forests, Port-of-Spain.

Western Pacific:

Gilbert and Ellice Islands: Superintendent of Works, Public Works Department.
British Solomon Islands: Commissioner of Lands and Crown Surveyor.

Windward Islands:

Grenada: Superintendent of Works.
St. Lucia: Superintendent of Works.
St. Vincent: Superintendent of Works.

Canada:

The Geological Survey of Canada

Headquarters: National Museum, Ottawa

The Geological Survey of Canada was founded in 1842 and is thus only seven years younger than that of Great Britain. It is also exceeded in age by four of the state surveys of America. In view of the undeveloped state of Canada a century ago, this early start makes the Canadian Survey one of the most notable of all geological surveys. Its early history is largely the history of Sir William Logan, whose pioneer work is well known; he resigned in 1869, two years after the foundation of the Dominion of Canada as it is known today. The Survey's work was not interrupted by Confederation, but it was widened considerably in succeeding years. In 1880, its headquarters were transferred from Montreal to Ottawa, where they remain today. The Survey is now a division of the federal Department of Mines and Resources. Its work has included much exploration and investigation of mineral deposits and water supply problems in addition to the study of Canadian geology and the preparation of geological maps.

Publications.—The Survey issues preliminary reports, economic series memoirs, maps, etc. Inquiries should be made to the Chief, Bureau of Geology and Topography, Department of Mines and Resources, Ottawa.

Canada: Provinces:

The Canadian Provinces

As natural resources are now generally under the control of the respective provinces that make up the Dominion of Canada, it is to be expected that the provinces will be interested in geological work at least in so far as it is related to mining. This is generally the case, and the departments and chief officials are as follows:

Nova Scotia: Department of Public Works and Mines, A. E. Cameron, Deputy Minister, Halifax.

New Brunswick: Department of Lands and Mines, W. J. Wright, Provincial Geologist, Fredericton.

Quebec: Department of Mines and Fisheries, A. O. Dufresne, Director, Bureau of Mines, Quebec City.

Ontario: Department of Mines, H. F. Rickaby, Deputy Minister, Toronto.

Manitoba: Department of Mines and Natural Resources, G. E. Cole, Director of Mines, Winnipeg.

Saskatchewan: Department of Natural Resources, E. Swain, Superintendent of Mines, Regina.

Alberta: Department of Lands and Mines, J. W. Hamilton, Superintendent of Mining Lands, Edmonton.

British Columbia: Department of Mines, J. F. Walker, Deputy Minister, Victoria.

The provinces of Quebec and Ontario have each published many notable geological papers, etc., those of Ontario (many complete with maps) being in the form of sectional annual reports of the Department of Mines, issued yearly since 1891.

Federated Malay States:

Geological Survey Department, Federated Malay States

Director: E. S. Willbourn
Headquarters: Batu Gajah

The Geological Survey Department began to make a geological survey of Malaya when Mr. J. B. Scrivenor was appointed Geologist to the Federated Malay States Government in September, 1903. When he retired, in 1931, a preliminary general survey of the greater part of the Malay Peninsula had been completed, including the Federated Malay States, the unfederated States, and the Straits Settlements. For many years, Malaya has been the largest tin producer in the world. As a result of mapping done by the Geological Survey, a very important extension of one tin field was located in Selangor. A record of all the prospecting done by miners is kept by the Geological Survey Department. This is proving of great assistance to the Government and to the mining community. More than two thirds of the area of Malaya is still under jungle in which very little detailed geological mapping has been done. An increase in the staff of the department has enabled an examination of some of this virgin country to be begun.

Advice is given by the Department to the Government and to the public not only on mining matters but in reply to requests for information about road location, dam and bridge foundations, water supply, landslides, etc. A chemical laboratory is maintained to conduct assay work.

Publications.—In addition to a certain number of maps and annual reports, the Department has issued about 20 special reports on different geological matters. All publications are obtainable from the Director at Batu Gajah.

Gold Coast:

Geological Survey of Gold Coast Colony

Director: N. R. Junner
Headquarters: Prestea
London Office: 75 Cornwall Gardens, South Kensington

The Department was founded in 1913, and from its inception it has been closely associated with mineral developments. It is engaged in geological mapping and investigation of the mineral resources of the Gold Coast, and in the preliminary investigation of water supplies and construction of waterworks in the Northern Territories. Important deposits of manganese ore, diamonds, and bauxite have been discovered by the Survey. Detailed surveys and reports are made on the chief mineral-bearing areas. In addition, assistance and advice are given to mining companies, prospectors, government departments, and local inhabitants on questions of water power, water supply, local geology, the determination of rocks and minerals, and the investigation of mineral resources.

Publications.—These center around *Annual Reports* issued from 1913 onward. *Bulletins* and *Memoirs* have also been issued on special subjects, and in the *Sessional Papers* much useful information is contained. All publications are issued by the Government Printing Office, Box 124, Accra, Gold Coast Colony, and by the Crown Agents for the Colonies, Millbank, London, S.W. 1.

Great Britain:

Geological Survey of Great Britain

Director: E. B. Bailey
Headquarters: Geological Survey and Museum, Exhibition Road, South Kensington, London, S.W.7

As early as the year 1814, a geologist was appointed to assist in the work of the Trigonometrical Survey of Scotland; in 1832, a similar appointment was made for Ireland. Three years later, the idea of forming a geological survey of Great Britain was suggested by the Ordnance Survey; and on the recommendation of the distinguished geologists then consulted, Henry Thomas de la Beche was appointed in 1835 as the first Director at the age of thirty-nine. He was the first of a distinguished company, later directors including Murchison, Ramsay, and Geikie. Throughout its long history, the Survey has undergone several reorganizations, but its functions today remain but little different from those originally conceived, as expressed by Lyell in 1838: " . . . as calculated to promote geological science . . . but also as a work of great potential utility, bearing on agriculture, mining, road-making, the formation of canals and railroads, and other branches of natural industry." Since 1919, it has been under the Department of Scientific Research and is thus now closely linked with other official research and scientific bodies, such as the Building Research Station and the National Physical Laboratory. A most interesting account of the Survey is to be found in its Centenary volume, "The First Hundred Years of the Geological Survey of Great Britain," by Sir John Flett (a recent Director), published in 1937 by H. M. Stationery Office, London.

Publications.—These are naturally varied and extensive, including *Memoirs* of different types, *Annual Reports*, and (in recent years) *Summaries of Progress*, in addition to geological maps and sections covering the whole of Great Britain to various scales. A full list of publications is available and is to be obtained, as are also the publications themselves, from H. M. Stationery Office, York House, Kingsway, London, W.C.2.

India:

Geological Survey of India

Director: A. M. Heron
Headquarters: Geological Museum, Library, and Office, Calcutta

The Geological Survey of India is one of the oldest in the world. A geologist was appointed to the staff of the Trigonometrical Survey of India in 1818, but the modern Survey may be said to have started in 1845. The Survey today has a permanent staff of 26 officers and maintains a well-equipped laboratory and a scientific library of over 80,000 volumes. The preparation of a complete geological map of India is naturally the ultimate goal of the Survey, in the course of the field-work for which many mineral investigations are made. The Survey is freely consulted by the Central Government and all local governments and Indian states on mineral and engineering questions. The Director is consulted on mineral policy; and in these and many other ways, the Survey provides a vital service to the mining, metallurgical, and chemical industries of India as well as to Indian engineers. Interesting reviews of the work of the Survey are to be found in the presidential addresses to the Mining, Geological and Metallurgical Institute of India, by Dr. C. S. Fox in 1936 and Dr. A. M. Heron in 1938.

Publications.—These consist of *Memoirs* which are devoted to complete surveys dealing with either a particular region or a mineral industry; *Records*, consisting of shorter papers on general geological subjects; *Palaeontologia Indica*, giving results of paleontological work by the Survey; and an extensive series of maps. Copies are to be obtained from the Survey Office and from the office of the High Commissioner for India, Aldwych, London, England.

India: Hyderabad:

Hyderabad Geological Survey Department

Director: K. Mirza
Headquarters: Hyderabad, Deccan

This Geological Survey was organized in 1932, and the staff now consists of the Director and four assistant superintendents. A journal (quarterly or half yearly) is published, and a series of *Bulletins;* copies of publications are obtainable on application to the Survey Headquarters.

India: Mysore:

Mysore Geological Department

Director: B. Rama Rao
Headquarters: Bangalore, Mysore

The Department has been in existence from 1894; and in addition to conducting geological survey work, it assists in the investigation and regulation of the extensive mining operations carried out in this Indian state. Publications, in English, include *Records* (issued since 1894), *Memoirs*, *Bulletins*, and *Maps*. Copies are obtainable from the Superintendent, Government Press, Bangalore, Mysore, and from the Director.

Kenya:

Mining and Geological Department of the Colony and Protectorate of Kenya

Government Geologist: C. S. Hitchen
Headquarters: Nairobi

Formed in 1903, this Department has already issued a number of *Reports* on local geology with special reference to economic features. Copies may be obtained from the Government Printer at Nariobi, Kenya, and from the Crown Agents for the Colonies, Millbank, London, S.W.1.

Newfoundland:

Geological Survey of Newfoundland

Government Geologist: A. K. Snelgrove
Associate Government Geologist: C. K. Howse
Headquarters: St. John's

On May 1, 1866, Sir William Logan, Director of the Geological Survey of Canada, transmitted to the Government of Newfoundland a report on the Geology of Newfoundland by a member of his staff, Mr. Alexander Murray. Thereafter Mr. Murray and Mr. J. P. Howley served officially with the Newfoundland Government, their collected reports forming two notable geological survey volumes (issued in 1881 and 1918). From 1926 to 1929, Dr. H. A. Baker was Government Geologist. Geological survey work was interrupted until the Commission of Government took office in 1934, when a Geological Section of the Department of Natural Resources was organized. In 1937, the title *Geological Survey of Newfoundland* was resumed, and the Survey was made responsible to the Commissioner for Public Utilities. The present Survey works in close collaboration with Princeton University, Princeton, New Jersey, an outstanding example of international cooperation. The functions of the Survey are to map and study mineralized areas; to administer the mining laws and generally assist in the mining development of the country; to publish reports on fieldwork on the geology and mineral resources of

Newfoundland; and, through the Surveys Division of the Department of Natural Resources, to inaugurate topographical surveys.

Publications.—The reconstituted Survey has issued more than a dozen *Bulletins* of which five are Princeton University Contributions to the Geology of Newfoundland. *Bulletin* 5 is a bibliography covering the years 1818–1936. A number of *Information Circulars* have also been issued. Copies of all these publications are to be obtained from the offices of the Geological Survey at St. John's.

New Zealand:

The New Zealand Geological Survey

Director: J. Henderson
Headquarters: 156, The Terrace, Wellington

Official geological survey work was carried out in the individual provinces of New Zealand as early as 1858 and in succeeding years, but the appointment of James Hector as Director of the Geological Survey of New Zealand in 1865 marks the formation of the Survey as it is known today. Reorganized in 1904, the Survey today carries out field and laboratory investigations and cooperates with government departments, engineers, and others interested in geology and its applications. It operates as a branch of the Department of Scientific and Industrial Research. An interesting review of the history and work of the Survey is to be found in a paper entitled The Organization and Functions of a State Geological Survey, by P. G. Morgan, to be found in the *New Zealand Journal of Science and Technology*, **2**: 289 (1919).

Publications.—These include *Annual Reports* from 1907 onward; about 40 general and economic bulletins, *Palaeontological Bulletins*, and maps. Copies are to be obtained from the Government Printer, Wellington.

Nigeria:

Geological Survey of Nigeria

Director: R. C. Wilson
Headquarters: Kaduna Junction

The Survey personnel includes a headquarters staff; five geologists engaged on fieldwork; and a water supply section of four engineers, six foremen, and two drillers who are engaged in drilling and shaft sinking for water supply for towns, villages, quarantine stations, along trade routes, and along the Nigerian Railway. The Survey is engaged on the investigation of the economic minerals of Nigeria and in the assembly and distribution of information concerning them. Geologists are also engaged in prospecting for underground water supplies in the more arid areas, and work in connection with soil surveys, erosion, river control, and afforestation is carried out in collaboration with other departments. A laboratory, library, and museum are maintained.

Publications.—These consist of *Annual Reports* (1929 and onward), a number of *Bulletins*, a series of *Occasional Papers*, a map series of the Nigerian tin fields and the Nigerian gold fields, and a provisional geological map of Nigeria. Copies may be obtained from the Headquarters of the Survey and from the Crown Agents for the Colonies, Millbank, London, S.W.1.

Northern Rhodesia:

Northern Rhodesia has no geological survey, but a considerable amount of geological surveying has been carried out by the Anglo-American Corporation under Dr. Bancroft.

Nyasaland:

Geological Survey of Nyasaland

Director: F. Dixey
Headquarters: Geological Survey Department, Zomba

As with other African geological surveys, that of the Nyasaland Protectorate was an outcome of the World War and was founded in 1918. One general purpose of the institution of such surveys was to initiate the search for minerals of special value economically. The staff of the Nyasaland Survey has always been small; and in 1931, as a retrenchment measure, the Survey suspended its normal duties and concentrated on the development of water supplies. Despite these handicaps, it has progressed toward completing an initial rough geological survey of the Protectorate. Much assistance has been given with water supply schemes, and advice on road materials, bridge foundations, etc. Naturally, considerable attention has been devoted to mineral resources.

Publications.—Over 50 publications have been issued, including *Annual Reports*, some general maps, *Bulletins*, and other papers. Special mention may be made of the series of *Water Supply Papers*, and the Director is the author of "A Practical Handbook of Water Supply," mentioned at the end of Chap. XVI. Copies of all are obtainable from the Survey Headquarters and from the Government Printer, both at Zomba.

South Africa:

The Geological Survey of the Union of South Africa:
(Geolosiese Opname van die Unie van Suid Afrika)

Director: S. H. Haughton
Headquarters: Office of the Geological Survey, Pretoria

The Geological Survey of the Union of South Africa was founded in 1910. Prior to that date, geological work had been carried out by the Geological Commission of the Cape of Good Hope from 1896 until amalgamation in 1910 with the Geological Survey of the Transvaal, established in 1903, the two combining to form the Union Survey. In addition, work had been done by the State Geologist of the South African Republic from 1897 to 1900 and by the Geological Survey of Natal and Zululand up to the year 1908. Functions of the present Survey, which is a division of the Department of Mines, include geological mapping, the investigation of mineral resources, the study of underground water problems, the examination of rocks and minerals for the public, the maintenance of a geological museum, and the investigation of the value of geophysical prospecting methods as applied to South African conditions.

Publications.—Those of the older bodies, up to the year 1910, consisted generally of maps and annual reports. Publications since 1910 include many more maps, some *Annual Reports*, *Bulletins*, and a number of general and palaeontological memoirs. All publications are obtainable from the Government Printer, Pretoria, but not from the Survey.

Southern Rhodesia:

Geological Survey of Southern Rhodesia

Director: B. Lightfoot
Headquarters: Salisbury

The Geological Survey was formed originally in 1910 by the British South Africa Company which then owned the mineral rights. At present, the staff consists of the Director, six geologists, one chemist, one mineralogist, four draftsmen, one librarian, and one clerk. The work done is predominantly economic, including detailed surveys of gold mining areas and the mapping of all underground workings. Mineral analyses and water investigations are undertaken at Headquarters.

Publications.—These include 34 *Bulletins*, 30 short *Reports*, and geological maps and are obtainable from the Geological Survey Office, P.O. Box 366, Salisbury.

Tanganyika:

Geological Division, Department of Lands and Mines, Tanganyika Territory

Chief Geologist: F. B. Wade
Headquarters: Dodoma

In 1926, a Geological Survey Department started work in the Territory under the directorship of Dr. E. O. Teale and functioned as a government department until 1935, when it was amalgamated with the Mines, Survey and Lands Departments, becoming a division of the new organization under the guidance of F. B. Wade as Chief Geologist.

Although no radical changes have been made in the aims of the new Division, its staff has been nearly doubled in the last few years. Its organization is designed to deal as far as possible with all aspects of Economic Geology. These are concerned chiefly with mineral investigations and water supply, embracing detailed surveys of known mineral fields and the exploration of unknown areas, followed by more detailed survey if needed. Deposits of gold, coal, copper, salt, soda, diamonds, corundum, nickel, mica, limestone, and tin have been investigated. Engineering questions considered by the organization have included several construction problems, soil erosion, railway location, road and building material, natural fertilizers, and reservoir sites. Special work in connection with water supply has been concerned mainly with boring for water, both choice of site and actual boring being done by the Division. A chemical, petrological, and metallurgical laboratory, a repair shop for boring equipment, and a well-equipped museum are maintained.

Publications.—In addition to *Annual Reports*, geological maps and reports are issued periodically. Short *Papers* are printed at an early date after the completion of special field examinations. *Bulletins* giving more complete accounts of investigations are published at longer intervals. Copies of all are obtainable from the Chief Geologist, Dodoma.

Uganda:

Geological Survey Department of Uganda

Director: E. J. Wayland
Headquarters: Entebbe

This Survey was first mooted in 1918, when E. J. Wayland was offered the post of Geological Adviser. He reached Uganda in January, 1919; in a few years, three more geologists joined the staff. Scientific personnel now includes six geologists, a chemist and petrologist, a laboratory assistant, a topographer, and a field drilling staff. Advice is continually being given with regard to specific engineering problems as well as on such matters as building materials, lime, road materials, soil erosion, and particularly water supply. A laboratory is maintained at headquarters.

Publications.—*Annual Reports* have been issued since 1920, and these contain much useful information. In addition, a number of *Memoirs*, *Occasional Papers*, and *Bulletins* have been published. All may be obtained on application to the Director, at Entebbe.

United States:

Geological Survey

United States Department of the Interior

Director: W. C. Mendenhall
Headquarters: Washington, D.C.

The Geological Survey of the United States came into being on July 1, 1879, in response to an Act of Congress passed on March 3 of that year. Prior to that date, considerable official geological survey work had been done by federal organizations, and even more by the states. Among the earlier federal surveys, that under Hayden, which began as early as 1853; King (1867) along the Fortieth Parallel; Wheeler (1869); and Powell were outstanding. Today, the Survey is at least as extensive in its scope, operations, and personnel as any other in the world. Its publications are known to geologists of all countries. It operates as a bureau of the Department of the Interior and has five field branches—(1) Geologic, (2) Topographic, (3) Water Resources, (4) Conservation, (5) Alaskan—and a central administrative organization. The main functions of the Survey as defined by its basic law include the study of the geologic structure, mineral resources, and products of the national domain and the classification of the public land. Its actual operations include many aspects of economic geology, topographic mapping, and systematic studies of water resources, as the names of its main divisions make clear. It cooperates with the various state geological surveys, both formally and informally, and with other units of the government service. *Bulletin* 227 (1904) of the Survey describes its origin, development, organization, and operations as at the date of publication. A much more recent sketch of the organization and its work appears in *The Scientific Monthly*, **XXVI** (February, 1933).

Publications.—These are extensive, numbering over 2,000 volumes of reports, thousands of geologic maps, and millions of topographic maps. The regularly published index constitutes a pamphlet of over 250 pages. The publications include *Annual Reports*, *Bulletins*, *Water Supply Papers*, *Professional Papers*, maps, *Monographs*, *Folios*, *Special Publications*, and press notices. Copies may be obtained on application to the Superintendent of Documents, Government Printing Office, Washington, D.C.

United States, States of the Union:

Geological Surveys of the States of the Union

As will be seen from following the list, a majority of the states of the Union maintain their own geological surveys, the functions of which are usually related to areal and economic geological problems. It is impossible to do more than list these surveys in this place, in view of their number. Many of them far surpass in size and achievement some of the more isolated surveys already mentioned in detail, but this uneven attention is felt to be desirable, as the state surveys are generally so well known, whereas those in more distant parts of the world are but little heard of in normal work. Tribute must be paid to the high standard of state geological survey publications and to their frequent use far outside state boundaries.

State geological survey work may be said to have begun with surveys made in North Carolina in 1824. The first real state survey was that of Massachusetts, started in 1830 by Hitchcock. The history of the development and checkered careers of the many subsequent state surveys is long and complicated. Those interested may refer to *Bulletin* 465 of the United States Geological Survey, by C. W. Hayes, entitled The State Geological Surveys of the United States. Apparently, only four states have never had a geological survey at some time in their history.

State	Name of survey	Director	Headquarters	Publications from
Alabama	Geological Survey of Alabama	Dr. Walter B. Jones, State Geologist	University	Headquarters
Arizona	Arizona Bureau of Mines	G. M. Butler, Director and State Geologist	Tuscon	Headquarters
Arkansas	Arkansas Geological Survey	George C. Branner, State Geologist	443–447 State Capitol, Little Rock	Headquarters
California	Division of Mines of State Department of Natural Resources	Walter W. Bradley, State Mineralogist, Chief of Division	Ferry Building, San Francisco	Headquarters
Colorado	Geological Survey Board	John T. Joyce, Secretary, State Bureau of Mines	State Bureau of Mines State Museum Building, Denver	Prof. P. G. Worcester, Department of Geology, University of Colorado
Connecticut	Connecticut Geological and Natural History Survey	Dr. W. E. Britton, Superintendent	Box 1106, New Haven	State Librarian. Hartford
Florida	Florida Geological Survey	Herman Gunter, Geologist	Tallahassee	Headquarters
Georgia	Division of Mines, Mining and Geology, Georgia Department of Natural Resources	Capt. Garland Peyton	Room 425, State Capitol, Atlanta	Headquarters
Idaho	Idaho Bureau of Mines and Geology	A. W. Fahrenwald	Moscow	Headquarters
Illinois*	State Geological Survey	M. M. Leighton, Chief	Urbana	Headquarters
Indiana	Department of Conservation, Division of Geology	Ralph E. Esarey, State Geologist	401 State Library Building, Indianapolis	Headquarters
Iowa	Iowa Geological Survey	Dr. A. C. Trowbridge	Iowa Geological Survey, Iowa City	Headquarters

State	Name of survey	Director	Headquarters	Publications from
Kansas	State Geological Survey	Raymond C. Moore, State Geologist	University of Kansas, Lawrence	Headquarters
Kentucky	State Department of Mines and Minerals, Geological Division	D. J. Jones, State Geologist	Box 680, Lexington	Headquarters
Louisiana	Louisiana Geological Survey of the Department of Conservation	Cyril K. Moresi, State Geologist	Department of Conservation, Civil Court House Building, New Orleans	Headquarters
Maryland	Maryland Geological Survey	E. B. Mathews, State Geologist	Maryland Geological Survey, Baltimore	Headquarters
Michigan	Geological Survey Division, Department of Conservation	Richard A. Smith, State Geologist and Chief	State Office Building, Lansing	C. A. Paquin, Chief of Division of Education, Department of Conservation, Lansing
Minnesota	Minnesota Geological Survey	W. H. Emmons	University of Minnesota, Minneapolis	University of Minnesota Press, Minneapolis
Mississippi	Mississippi State Geological Survey	Dr. William Clifford Morse	University	Headquarters
Missouri	Missouri Geological Survey and Water Resources	H. A. Buehler, State Geologist	Rolla	Headquarters
Montana	State Bureau of Mines and Geology	Francis A. Thomson	Montana School of Mines, Butte	Headquarters
Nebraska	The Nebraska Geological Survey Department Conservation and Survey Division	Dr. G. E. Condra, State Geologist and Director of the Conservation and Survey Division	Nebraska Geological Survey, University of Nebraska, Lincoln	Headquarters
Nevada	Nevada State Bureau of Mines	J. A. Fulton	Nevada State Bureau of Mines, Reno	Headquarters

State	Name of survey	Director	Headquarters	Publications from
New Jersey	Division of Geology and Topography, State Department of Conservation and Development	Meredith E. Johnson, State Geologist	Department of Conservation and Development, State House Annex, Trenton	Headquarters
New Mexico	State Bureau of Mines and Mineral Resources of the New Mexico School of Mines	E. H. Wells	Socorro	Headquarters
New York	Section of Geology, New York State Museum	Dr. D. H. Newland, State Geologist	State Education Building, Albany	Headquarters
North Carolina	Division of Mineral Resources, Department of Conservation and Development	H. J. Bryson, State Geologist	Raleigh	Headquarters
North Dakota	North Dakota Geological Survey	Frank C. Foley, State Geologist	University of North Dakota, University Station, Grand Forks	Headquarters
Ohio	Geological Survey of Ohio	Wilber Stout, State Geologist	Geological Survey of Ohio, Orton Hall, Room 106, Ohio State University, Columbus	Headquarters
Oklahoma	Oklahoma Geological Survey	Robert H. Dott	Oklahoma Geological Survey, Norman	Headquarters
Pennsylvania	Topographic and Geologic Survey	George H. Ashley, State Geologist	Harrisburg	Headquarters
South Carolina	South Carolina Geological Survey	Stephen Taber, State Geologist	University of South Carolina, Columbia	Headquarters
South Dakota	State Geological Survey	Dr. E. P. Rothrock, State Geologist	Vermillion	Headquarters

State	Name of survey	Director	Headquarters	Publications from
Tennessee	Tennessee Division of Geology	Walter F. Pond, State Geologist	Nashville	Headquarters
Texas	Bureau of Economic Geology of the University of Texas	E. H. Sellards	Bureau of Economic Geology, University of Texas, Austin	Headquarters
Vermont	Geological Survey of Vermont	Elbridge C. Jacobs, State Geologist	University of Vermont, Burlington	Headquarters
Virginia	Virginia Geological Survey	Dr. Arthur Bevan, State Geologist	Brooks Museum, University of Virginia, Box 1428, University Station, Charlottesville	Headquarters
Washington	Division of Geology, Department of Conservation and Development	Harold E. Culver, Supervisor	Pullman	Department of Conservation and Development, Olympia
West Virginia	West Virginia Geological and Economic Survey	Paul H. Price, State Geologist	Morgantown	Headquarters
Wisconsin	Wisconsin Geological and Natural History Survey	E. F. Bean, State Geologist	Science Hall, Madison	Headquarters
Wyoming	Geological Survey of Wyoming	S. H. Knight, State Geologist	University of Wyoming, Laramie	Headquarters

* A special note must be added with regard to the Illinois Geological Survey, since one of its divisions is devoted to Areal and Engineering Geology. So far as the author is aware, this is the only special division of a geological survey devoted to the applications of Geology to Civil Engineering. The Division was formed in 1927, with Dr. George E. Ekblaw as its head, and arose from the work done in previous years by Dr. M. M. Leighton, first as a member of the Survey staff and later as Chief of the Survey, a position that he holds today. Many of the topics featured in this book have called for the attention of the Engineering Division. Several publications, to be found noted in the Survey's official lists, reflect this attention to the applications of Geology to Civil Engineering.

APPENDIX C

GEOLOGICAL SOCIETIES AND PERIODICALS

As may be noted from the references cited throughout this book, information regarding the application of Geology to civil engineering is usually to be found in the records of engineering societies and associated periodicals and independent journals. It is not surprising, however, that special features of some of these applications are of such interest geologically that they are commented upon in geological circles. There is, of course, no special organization devoted to this one branch of Economic Geology; but as there is a Society of Economic Geologists, in whose publication civil engineering matters are occasionally discussed, details of its work may be of interest and of use.

The Society of Economic Geologists.—The Society, which has its headquarters in the United States of America, is accurately described by its name. It holds regular meetings, usually in North America, but its proceedings include discussions of problems in Economic Geology (and in the application of Geology to civil engineering) in all parts of the world. The Society has no separate journal of its own but, by arrangement with the *Journal of Economic Geology*, uses this as its official publication and includes therein its contributions. It has regional vice-presidents representing North and South America, Europe, Asia, Africa, and Australasia, respectively. The Secretary is Mr. W. D. Johnson, Jr., of the United States Geological Survey, Washington.

Economic Geology.—This is an independent international journal having the largest world circulation of any journal relating to Economic Geology. It also serves as the *Bulletin of the Society of Economic Geologists* whose contributions constitute one fifth to one quarter of its pages. It is published eight times a year and includes papers relating to Geology and civil engineering contributed from all parts of the world. Its Editor is Dr. Alan M. Bateman of Yale University, New Haven, Connecticut, and its Business Manager is Mr. W. S. Bayley, Urbana, Illinois.

Annotated Bibliography of Economic Geology.—This journal, published twice a year, was started in 1928 under the National Research Council of the United States. It is now sponsored by the Society of Economic Geologists and published by the Economic Geology Publishing Company. It is financed by a group of geological societies, the Economic Geology Publishing Co., private companies and individuals. Each volume contains some 2,500 abstracts of papers (and index) relating to the subject of Economic Geology and published throughout the world. Each includes several hundred abstracts of paper relating to engineering, many of which deal with Geology and civil engineering. The business manager is Mr. W. S. Bayley, Urbana, Illinois.

In the study of the geology of locations in which they have to carry out their work, civil engineers may sometimes need to refer to geological publications other than those of their local geological survey. These will usually be the official proceedings of geological societies or geological journals. Because some civil engineers may be drawn, through its contacts with their own work, to a study of pure Geology, it may be useful to state briefly the main particulars of the geological societies and periodicals of the English speaking world, in addition to those already mentioned.

Societies

The Geological Society of London (Burlington House, London, W.C.1, England).—The Society is one of the oldest in existence devoted to one branch of science, having been founded on November 13, 1807. It publishes brief *Proceedings* and its well-known *Quarterly Journal,* now approaching its hundredth volume.

The Geological Society of America (419 West 117th Street, New York City).—Founded in 1888, the Society is the North American counterpart of the London Society. It publishes a monthly *Bulletin, Annual Proceedings, Memoirs,* and *Special Papers.*

International Geological Congress.—The congress meets every three years in different countries. Its publications contain articles relating to Geology and civil engineering.

There exist also many specialized and local geological societies such as the American Association of Petroleum Geologists and the Liverpool and Manchester Geological Societies in England, the existence of which will usually be known to engineers practicing in the respective districts or fields of work. Detailed mention may be made of the Geologists' Association of London (General Secretary, Mr. A. G. Bell, 34 Sherard Road, Eltham, London, S.E.9), an essentially popular body which exists to "foster the progress and diffusion of the science of geology." It holds meetings in London and valuable field excursions not only in all parts of Great Britain but also on the continent of Europe. It publishes *Proceedings;* many engineers are members. No equivalent appears to exist in North America.

Periodicals

In addition to the regular publications of the various geological societies and of journals that deal with special branches of pure and applied Geology in the mining and petroleum fields, the following periodicals are of value to geologists and cover the whole science as their sphere of interest.

The *Geological Magazine:* a monthly journal, published in London by Dulau and Co., 29 Dover Street, London, W.1; it incorporates *The Geologist,* which was founded in 1864.

The American Journal of Science: a monthly journal, founded in 1818 by Benjamin Silliman, published at Yale University, New Haven, Connecticut.

The Journal of Geology: A semiquarterly journal, published by the University of Chicago Press, Chicago, Illinois.

Finally, it may be noted that valuable publications of geological interest, including some relating to applications in civil engineering, are to be found in the records of the great general scientific societies such as the Royal Society of Great Britain, the Royal Society of Canada, the American Association for the Advancement of Science, and the British association of the same nature. Scientific journals such as *Nature* and *Science* occasionally contain articles of relevant interest, and the proceedings of the various Mining Engineering institutions include some papers that touch upon Geology as applied in civil engineering (as references listed in Appendix D will indicate).

APPENDIX D

REFERENCES CITED IN THE TEXT

Chapter I

1.1. Quoted in HUNTER, W. H.: "Rivers and Estuaries," Longmans, Green & Company, London, 1913, as an extract from "The Conservation and Improvement of Tidal Rivers."

1.2. This brief note is based generally upon information given in an interesting volume by Sir A. Geikie, "Founders of Geology," Johns Hopkins Press, Baltimore, Maryland, 1901.

1.3. Mentioned by Prof. Boyd Dawkins in his James Forrest Lecture of 1898, On the Relation of Geology to Civil Engineering, *Inst. Civil Eng. Min. Proc.*, (*London*), **134:** 275 (1898).

Chapter II

2.1. For an introduction to this quotation, and so to the book from which it came, the author is indebted to the late Dr. G. P. Merrill who used it in his book "Rocks, Rock Weathering and Soils," The Macmillan Company, New York, rev. ed., 1911.

Chapter III

3.1. Quoted in LONGWELL, C. R., A. KNOPF, and R. F. FLINT: "A Textbook of Geology," p. 8, Part I, Physical Geology, John Wiley & Sons, Inc., New York, 1932.

3.2. KEITH, SIR A.: An African Garden of Eden, *John o' London's Weekly*, **35:** 890 (1934).

Chapter IV

4.1. Quoted by T. Sheppard in "William Smith, His Maps and Memoirs," p. 92, A. Brown & Sons, Ltd., Hull, England, 1920.

Chapter V

5.1. See pp. 254–255 of 1.3.

5.2. MCCALLUM, R. T.: The Opening-out of Cofton Tunnel, London, Midland, and Scottish Railway, *Inst. Civil Eng. Min. Proc.*, (*London*) **231:** 161 (1931).

5.3. GRUNER, H. E.: Hydro Electric Power Development on the Rhine, *Inst. Civil Eng.* Special Lecture (*London*) (1935).

5.4. Contractors Win Suit to Recover Losses Due to Engineer's Misrepresentation, *Eng. News-Record* (*New York*) **93:** 21 (1924).

5.5. After this chapter had been written, the author came across the reference to this matter in Engineering and Contracting Procedure for Foundations, *A.S.C.E. Manual* 8, 1934, in which a similar classification is suggested.

5.6. WALKER, T. A.: "The Severn Tunnel; Its construction and Difficulties, 1872–1887," R. Bentley and Son, London, 1888.
5.7. CARPMAEL, R.: Cementation in the Severn Tunnel, *Inst. Civil Eng. Min. Proc.*, (*London*) **234**: 277 (1933).

Chapter VI

6.1. Statement made during the discussion of 12.4.
6.2. JACOBY, H. S., and R. P. DAVIS: "Foundations of Bridges and Buildings," p. 585, McGraw-Hill Book Company, Inc., New York, 2d ed., 1925.
6.3. HUSBAND, J.: The San Francisco-Oakland Bay Bridge, *Structural Eng.* (*London*), **14** (new ser.): 170 (1936).
6.4. For a general description of these works, see W. T. HALCROW: The Lochaber Water Power, *Inst. Civil Eng. Min. Proc.*, (*London*) **231**: 31; costs in a personal communication from Mr. Halcrow.
6.5. DOW, A. L.: Foundation Exploration in Deep Water, *Eng. News-Record* (*New York*), **119**: 635, (1937).
6.6. SIMPSON, W.: "Foundations: Methods and Appliances," p. 117, Constable & Company, Ltd., London, 1928.
6.6. See illustrated articles in *Construction Methods* (*New York*), p. 36, April, 1932; p. 38, July, 1932. *See also* Redesign and Construction of Prettyboy Dam, *Eng. News-Record* (*New York*), **111**: 63 (1933).
6.8. JOHNSTON, H. R.: Underground Exploration with Calyx Drills, *Eng. News-Record* (*New York*), **120**: 436 (1938).
6.9. U. S. Standard Form 23, Art. 4, as rev. September 9, 1935.
6.10. Bigger than Boulder, *Eng. News-Record* (*New York*), **120**: 647 (1938).
6.11. SUTHERLAND, R. A.: French Build High Dam in Narrow Limestone Canyon, *Eng. News-Record* (*New York*), **115**: 706 (1935).

Chapter VII

7.1. JONES, O. T.: Geophysics, James Forrest Lecture, 1935, *Inst. Civil Eng. Min. Proc.* (*London*), **240**: 689 (1936).
7.2. SHAW, H.: "Applied Geophysics," p. 11, *Sci. Mus. Pub.* (*London*), H. M. Stationery Office, 1936; quotations made by permission of the Controller.
7.3. See p. 699 of 7.1.
7.4. See p. 24 of 7.2.
7.5. See p. 43 of 7.2.
7.6. For names of other geophysical prospecting companies, see note at end of chapter.
7.7. DOLMAGE, V.: Geology and Geophysics in Engineering, *The Canadian Eng.* (*Toronto*), **67**: 19 (1931). See also *Eng. News-Record* (*New York*), **105**: 364 (1930).
7.8. CROSBY, I. B., and S. F. KELLY: Electrical Subsoil Exploration and the Civil Engineer, *Eng. News-Record* (*New York*), **102**: 270 (1929).
7.9. LEONARDON, E. G.: Electrical Exploration Applied to Geological Problems in Civil Engineering, *Geophysical Prospecting, A.I.M.E. Trans.*, **97**: 99 (1932).
7.10. LUGEON, M., and C. SCHLUMBERGER: The Electrical Study of Dam Foundations, *Water and Water Eng.* (*London*), **35**: 609 (1933) (rep. from *Mining Mag.*).

7.11. McLintock, W. F. P., and J. Phemister: A Gravitational Survey over the Buried Kelvin Valley at Drumry, near Glasgow, *Royal Soc. Edin. Trans.*, **56**: 141 (1931).

7.12. Shepherd, E. R.: Subsurface Exploration by Earth Resistivity and Seismic Methods, *Public Roads* (*Washington*), **16**, 57 (1935).

7.13. Wilcox, S. W.: Prospecting for Road Materials by Geophysics, *Eng. News-Record* (*New York*), **114**: 271 (1935).

7.14. Bruckshaw, J. M., and F. Dixey: Ground Water Investigation by Geophysical Methods, *Water and Water Eng.* (*London*), **36**: 261 and 368 (1934) (repr. from *Mining Mag.*).

7.15. Cox, A. H., D. A. B. Davies, and T. G. Williams: "Report on Certain Trials of Geo-electric Methods in S. Wales," p. 33, H. M. Stationery Office, London, 1935. Quoted by permission of the controller.

7.16. Geophysical Prospecting for Water, editorial in *South African Min. Eng. Jour.* (*Johannesburg*) **47**: 44 (1936).

7.17. Williams, N.: 260 Geophysical Crews Active in Field Exploration Work, *Oil and Gas Jour.*, **35**: 25 (1936).

7.18. Ayvazoglou, W.: "Geophysical Abstracts," *U. S. Geol. Survey Bull.* 895*D*, latest issue 1938.

Chapter VIII

8.1. For some of these historical notes the author is indebted to C. Prelini: "Tunnelling," D. Van Nostrand Company, Inc., New York, 6th ed., 1912.

8.2. Information from the late R. S. Cole. See also A. J. Dean: The Lake Copais, Boeotia, Greece: Its Drainage and Development, *Jour. Inst. Civil Eng.* (*London*), **5**: 287 (1937).

8.3. Fox, F.: The Simplon Tunnel, *Inst. Civil Eng. Min. Proc.* (*London*), **168**: 61 (1907).

8.4. The Lötschberg-Simplon Railway and its Construction, *The Engineer* (*London*), **112**: 633 (1911).

8.5. The Las Raices Tunnel, Chile, *Civil Eng.* (*London*), **28**: 335 (1933).

8.6. Berkey, C. P., and J. F. Sanborn: Engineering Geology of the Catskill Water Supply, *Am. Soc. Civil Eng. Trans.* (*New York*), **86**: 1 (1923). See also C. P. Berkey,: Geology of the New York [Catskill] Aqueduct, *N. Y. State Museum Bull.* 146, 1911. See also J. T. Kemp: Geological Problems Presented by the Catskill Aqueduct of the City of New York, *Jour Canadian Min. Inst.*, **14**: 472 (1911).

8.7. Tunnel Looped under a Fault, *Eng. News-Record* (*New York*), **119**: 220 (1937).

8.8. Haskins, G.: The Construction, Testing and Strengthening of a Pressure Tunnel for the Water Supply of Sydney, N.S.W., *Inst. Civil Eng. Min. Proc.* (*London*), **234**: 25 (1932).

8.9. Schoklitsch, I.: "Hydraulic Structures," trans. by S. Shulits and L. G. Straub, A. S. M. E., and J. R. Freeman Trust, 2 vols., New York, 1937.

8.10. Information in personal communication from Mr. J. W. Pfau, Chief Engineer, New York Central Railroad Co.

8.11. Information in personal communication from Le Chef du Service de la Voie et des Batiments du Sud-Est, S. N. C. F., Paris, France.

8.12. Keays, R. H.: Construction Methods on the Moffatt Tunnel, *Am. Soc. Civil Eng. Trans.* (*New York*), **92**: 69 (1928).

8.13. ANDERSON, D.: The Construction of the Mersey Tunnel, *Jour. Inst. Civil Eng.* (*London*), **2**: 473 (1936). BOSWELL, P. G. H.: The Geology of the New Mersey Tunnel, *Liverpool Geol. Soc. Proc.*, **17**: 160 (1937).
8.14. See 1.3, pp. 255–6.
8.15. KEAYS, R. R.: Inflows Block Construction of Aqueduct Tunnel for Athens, Greece, *Eng. News-Record* (*New York*), **106**: 978 (1931).
8.16. See 5.7.
8.17. PEACH, B. N.: The Locaber Water Power Scheme and Its Geological Aspect, *Water and Water Eng.* (*London*), **32**: 71 (1930) (repr. from *Inst. Min. Eng. Trans.*). See also 6.4.
8.18. GRIEVE, W. G., and B. NEWMAN: "Tunnellers," p. 123, H. Jenkins, London, 1936.
8.19. Hydraulic Mining on Twin Cities Sewer Tunnels, *Eng. News-Record* (*New York*), **115**: 627 (1935).

Chapter IX

9.1. *Inst. Civil Eng. Min. Proc.* (*London*), **1**: 61 (1841).
9.2. See 5.2.
9.3. A New Interoceanic Railway in Central America, *Eng. News-Record* (*New York*), **88**: 474 (1927).
9.4. BURR, E.: Remove Subaqueous Ledge above Rapid Transit Tunnel, *Eng. News-Record* (*New York*), **89**: 1021 (1922).
9.5. GWYTHER, R. D.: Improvements to St. Helier's Harbour, Jersey, *Inst. Civil Eng. Min. Proc.* (*London*), **238**: 100 (1934).
9.6. Tough Shale Bored and Sawed for Fort Peck Spillway Gate, *Eng. News-Record* (*New York*), **116**: 37 (1936).
9.7. Wheeler Dam Construction Enters Final Year, *Eng. News-Record* (*New York*), **115**: 259 (1935).
9.8. See 8.18, p. 157.
9.9. LEGGET, R. F.: The Wellpoint System; Application to an Excavation in Waterlogged Ground in Canada, *Civil Eng.* (*London*), **31**, 229 (1936).
9.10. Information in personal communication from Mr. Thomas F. Moore, The Moretrench Corporation, Rockaway, New Jersey.
9.11. MCHAFFIE, M. G.: Southampton Docks Extension, *Jour. Inst. Civil Eng.* (*London*), **9**: 184 (1938).
9.12. THORESEN, S. A.: Shield-driven Tunnels near Completion under the Schelde at Antwerp, *Eng. News-Record* (*New York*), **110**: 827 (1933).
9.13. JAPP, SIR HENRY: New Methods of Excavating Quicksands and Unstable Ground, *The Engineer* (*London*), **156**: 422 (1933).
9.14. GUTMANN, I.: The Moscow Subway; C. N. PINCO: Highlights of Moscow Subways, *Eng. News-Record* (*New York*), **116**: 515–525 (1936).
9.15. Cheap Levees by Hydraulic Fill at Lake Okeechobee, *Eng. News-Record* (*New York*), **115**: 81 (1935).
9.16. ANDERSON, A. P.: Some Studies of Drilling and Blasting in Highway Grading, *Public Roads* (*Washington*), **12**: 293 (1932).
9.17. JENKIN, C. F.: The Pressure on Retaining Walls, *Inst. Civil Eng. Min. Proc.* (*London*), **234**: 103 (1933).
9.18. TERZAGHI, Karl: Record Earth-Pressure Testing Machine, *Eng. News-Record* (*New York*), **109**: 365 (1932). For test results, see *Eng. News-Record* (*New York*), **112**: 136, etc.

Chapter X

10.1. In a personal communication from Prof. P. G. H. Boswell, attributed to Prof. W. W. Watts, London.

10.2. Information received from Mr. F. W. Furkert, New Zealand, and also from F. W. Furkert: Remedial Measures on the Arapuni Hydro-electric Scheme of Power Development on the Waikato River, New Zealand, *Inst. Civil Eng. Min. Proc.* (*London*), **240**: 411 (1935).

10.3. See 3.1, p. 329.

10.4. Heck, N. H.: Earthquake History of the United States, *Special Pub.* 149, *U. S. Coast and Geodetic Survey*, 1928, in which original sources of information are listed.

10.5. Furkert, F. W.: The Effect of Earthquakes on Engineering Structures, *Inst. Civil Eng. Min. Proc.* (*London*), **236**: 344 (1933).

10.6. See 3.1, p. 296.

10.7. Land Subsidence of 15 In. Follows Earthquake in Utah, *Eng. News-Record* (*New York*), **114**: 322 (1935).

10.8. Meinzer, O. E.: Compressibility and Elasticity of Artesian Aquifers, *Econ. Geology* (*New Haven, Connecticut*), **23**: 263 (1928).

10.9. Stohsnet, E. E.: Santa Clara Valley Subsidence Has Now Reached 5 Feet, *Eng. News-Record* (*New York*), **118**: 479 (1937).

10.10. Collins, J. J.: New Type Seawall Built for Subsiding Lake Shore in Venezuela, *Eng. News-Record* (*New York*), **114**: 405 (1935).

10.11. See *The Engineer* (*London*), **51**: 123 (1881).

10.12. Ladd, G. E.: Landslides, Subsidences and Rock-falls, *A.R.E.A. Proc.* (*Chicago, Illinois*), **36**: 1092 (1935).

10.13. Atwood, W. W.: Relation of Landslides and Glacial Deposits to Reservoir Sites, *U. S. Geol. Survey Bull.* 685, 1918.

10.14. Emerson, F. E.: 180 Ft. Dam Formed by Landslide in Gros Ventre Canyon, *Eng. News-Record* (*New York*), **95**: 467 (1925). See also W. C. Alden: Landslide and Flood at Gros Ventre, Wyoming, *A.I.M.E. Trans.* (*New York*), **76**: 347 (1927).

10.15. Ries, H., and T. L. Watson: "Engineering Geology," p. 366, John Wiley & Sons, Inc., New York, 5th ed., 1936.

10.16. Ladd, G. E.: Bank Slide in Deep Cut Caused by Drought, *Eng. News-Record* (*New York*), **112**: 324 (1934).

10.17. Gordon, G.: Arch Dam of Ice Stops Slide, *Eng. News-Record* (*New York*), **118**: 211 (1937).

10.18. Fox, F.: "Sixty Three Years of Engineering," p. 47, John Murray, London, 1924.

10.19. Nolan, T. R.: Slips and Washouts on the Hill Section of the Assam-Bengal Railway, *Inst. Civil Eng. Min. Proc.* (*London*), **218**: 2 (1924).

10.20. Allen, G. H.: Arresting a Slide on an Indiana Highway, *Civil Eng.* (*New York*), **6**: 384 (1936).

10.21. Hill, R. A.: Clay Stratum Dried out to Prevent Landslips, *Civil Eng.* (*New York*), **4**: 403 (1934). And from information in a personal communication from Mr. R. A. Hill.

10.22. Collin, A.: "Recherches expérimentales sur les glissements spontanés des terrains argilleux," Paris, 1846. See also Appendix of 11.14.

10.23. For a review of this subject, see D. W. Taylor: Stability of Earth Slopes, *Jour. Boston Soc. Civil Eng.*, **24**: 197 (1937).

10.24. Report of the Commission appointed to investigate Turtle Mountain, Frank, Alberta, *Canadian Geol. Survey Mem.* **27** (1912).

10.25. CARPMAEL, R.: Noteworthy Engineering Work on the Vriog Cliffs near Barmouth, *G. W. Ry. Mag.* (*London*), p. 118, March, 1938. Information also from an unpublished paper by Mr. H. H. Reynolds, kindly loaned by Mr. Carpmael.

Chapter XI

11.1. Quoted in GREGORY, J. W.: "The Story of the Road," Alexander MacLehose & Co., London, 1931.

11.2. BENNETT, F. J.: The Influence of Geology on Early Settlements and Roads, *Geol. Assoc. Proc.* (*London*), **10**: 372, 1888.

11.3. MCADAM, J. L.: "Remarks on the Present System of Road Making," Longman, Hurst, Rees, Orme, and Brown, London, 7th ed., 1823.

11.4. BOSWELL, P. G. H.: Some Geological Characteristics of Road Stones, paper read to *Institute of Quarry Managers* (*London*), April, 1921.

11.5. BURTON, V. R.: Fill Settlement in Peat Marshes, *Public Roads* (*Washington*), **7**: 233 (1927).

11.6. Heat Treatment of Soils as Base for Road Construction, *Commonwealth Eng.* (*Australia*), **21**: 399 (1934).

11.7. ALLEN, H., and A. W. JOHNSON: Adding Water to Subgrade Levels Up Pavement, *Eng. News-Record* (*New York*), **113**: 464 (1934).

11.8. SMITH, N. L.: Notable Dual Road Completed, *Eng. News-Record* (*New York*), **121**: 45 (1938).

11.9. Accident Investigation 2222, Bureau of Safety, Interstate Commerce Commission (in Summary 74).

11.10. MORUM, S. W. F.: The Treatment of Mud-runs in Bolivia, *Jour. Inst. Civil Eng.* (*London*), **1**: 426 (1935).

11.11. Historical notes from various sources, including A. P. GEST: "Engineering (Our Debt to Greece and Rome)," Longmans, Green & Company, New York, 1930.

11.12. READE, T. M.: The Advantage to the Engineer of a Study of Geology, *Liverpool Eng. Soc. Trans.*, **10**: 36 (1888).

11.13. RINGERS, J. A.: Construction of the New Ijmuiden Lock, *Eng. News-Record* (*New York*), **104**: 769 (1930).

11.14. Report of the Committee of the National Academy of Sciences on Panama Canal Slides; *Nat. Acad. Sci.* (*Washington*) *Mem.* **18** (1924).

11.15. KIRKPATRICK, SIR C. R. S.: The Development of Harbour and Dock Engineering, Vernon Harcourt Lecture, *Inst. Civil Eng.* (*London*), 1926.

11.16. SCHLUMBERGER, C., and P. J.-M. RENAUD: Étude géophysique sous marine exécutée dans le Port d'Alger, *Annales des Ponts et Chaussées*, **4** (1935).

11.17. BUCKTON, E. J.: The Construction of Haifa Harbour, *Inst. Civil Eng. Min. Proc.* (*London*), **239**: 544 (1936).

11.18. STANTON W. F., and A. G. LE CLERCQ: The Improvement of the Port of Valparaiso and Extension of the Breakwater, *Inst. Civil Eng. Min. Proc.* (*London*), **233**: 199 (1931).

11.19. Channel Tunnel Committee Report, Economic Advisory Council (*London*), Cmd. 3513, H. M. Stationery Office, 1930.

11.20. ELLSON, G.: Dover Train-ferry Dock, *Jour. Inst. Civil Eng.* (*London*), **7**: 223 (1937).

Chapter XII

12.1. See "British Bridges," p. 173, Public Works, Roads and Transport Congress, London, 1933.

12.2. See pp. 175–176, 12.1.

12.3. WADDELL, J. A. L.: "Bridge Engineering," **2**: 1544, John Wiley & Sons, Inc., New York, 1916.

12.4. ALLEN, P.: The George's River Bridge, New South Wales, *Inst. Civil Eng. Min. Proc.* (*London*), **232**: 183 (1932).

12.5. MORAN, D.: Sampling and Soil Tests for Bay Bridge, San Francisco, *Eng. News-Record* (*New York*), **111**: 404 (1933). See also other articles in the same journal on other aspects of this bridge, *e.g.*, C. H. PURCELL, C. E. ANDREW, and G. B. WOODRUFF: Deep Open Caissons for Bay Bridge, **113**: 227 (1934).

12.6. In a personal communication from Mr. W. M. Scott of Winnipeg.

12.7. MILLARD, L. W.: The Mortimer E. Cooley Bridge, *Civil Eng.* (*New York*), **7**: 617 (1937).

12.8. BUCKTON, E. J., and H. J. FEREDAY: The Demolition of Waterloo Bridge, *Jour. Inst. Civil Eng.* (*London*), **3**: 472 (1936). See also F. H. MACINTOSH: Waterloo Bridge, *Building* (*London*), April, 1932.

12.9. MACKENZIE, C. J.: The Broadway Bridge, Saskatoon, *Eng. Jour.* (*Montreal, Canada*), **17**: 3 (1934).

12.10. Bridging the St. Lawrence at the Île d'Orléans, *Eng. News-Record* (*New York*), **112**: 356 (1934).

12.11. Road Foundation Problems, *Civil Eng.* (*London*), **32**: 442 (1937).

12.12. Mentioned in BROOKE-BRADLEY, H. E.: Bridge Foundations, *Struct. Eng.* (*London*), **12**: (new ser.): 103 (1934), from older records.

12.13. PILKEY, O. H.: Arresting Abutment Shifting on a Bascule Bridge, *Eng. News-Record* (*New York*), **108**: 725 (1932).

12.14. WENTHOLT, DR. L. R.: The Construction of Two New Canals for Inland Navigation in the Netherlands, Special Lecture, *Inst. Civil Eng.* (*London*), 1935.

12.15. From a paper by Mr. W. T. Everall, Description of Several Unusual Structures adopted in Bridge Construction (No. 160), kindly supplied by the Chief Engineer, N.W. Rly., India.

12.16. Described in The Strengthening of an L.M.S. Railway Bridge, *British Steel Piling Co., Ltd., Bull.* 4-1935; geological data kindly supplied by Messrs. W. H. Gyles and D. C. Bean.

12.17. See pp. 186 and 190, 12.1.

12.18. DEAN, A. W. H.: Construction of a Submergible Road-bridge over the Nerbudda River near Jubbulpore, Central Provinces, India, *Inst. Civil Eng. Min. Proc.* (*London*), **239**: 178 (1936).

12.19. WOODRUFF, G. B.: An Overturned 19,000 Ton Caisson Successfully Salvaged, *Eng. News-Record* (*New York*), **106**: 275 (1931).

12.20. BOULTON, G. O.: Construction of the Grey Street Bridge, Brisbane, Australia, *Civil Eng.* (*London*), **26**: 55 (1931).

12.21. Failure to Clean Bottom Results in Defective Pier Bases, *Eng. News-Record* (*New York*), **95**: 221 (1925).

12.22. Quoted on p. 167 of A. P. GEST: "Engineering," 11.11, as from Vitruvius, Book V, c.12, Harbours and Breakwaters.

12.23. Contractors Win River Battle, *Eng. News-Record* (*New York*), **119**: 13 (1937).
12.24. Sinking Open Cofferdams through Glacial Drift, *Eng. News-Record* (*New York*), **114**: 1 (1935).
12.25. ALEXANDER, F. W.: Maintenance of Substructure of the Lethbridge Viaduct, *Eng. Jour.* (*Montreal, Canada*), **17**: 523 (1934).
12.26. What about Your Foundations? *Ry. Eng. and Maintenance*, **31**: 572, 1935.
12.27. Two Railroad Bridge Failures Laid to Inadequate Inspection, *Eng. News-Record* (*New York*), **111**: 687 (1933).

Chapter XIII

13.1. Mentioned in *Civil Eng.* (*New York*), **3**: 209 (1933), as appearing on a tablet on St. John's River Bridge, Jacksonville, Florida.
13.2. For further historical notes, see PRELINI, C.: Some Dams of the Ancients, *Eng. News-Record* (*New York*), **87**: 556 (1921).
13.3. Letter from L. J. MENSCH in *Eng. News-Record* (*New York*), **100**: 674 (1928).
13.4. LAPWORTH, H.: The Geology of Dam Trenches, *Assoc. Water Engs. Trans.* (*London*), **16**: 25 (1911).
13.5. JUTSIN, J. D.: The Design of Earth Dams, *Am. Soc. Civil Eng. Trans.* (*New York*), **87**: 1 (1924) (list on pp. 133–134).
13.6. See 13.4, p. 23. See also T. U. TAYLOR: The Austin Dam, *U. S. Geol. Survey Water Supply Paper* 40 (1900).
13.7. RANSOME, F. L.: Geology of the St. Francis Dam Site, *Econ. Geology* (*New Haven, Connecticut*) **23**: 553 (1928).
13.8. COYNE, A.: The Construction of Large Modern Dams, *Structural Eng.* (*London*), **15** (new ser.): 70 (1937).
13.9. MCILDOWIE, G.: The Construction of the Silent Valley Reservoir, Belfast Water Supply, *Inst. Civil Eng. Min. Proc.* (*London*), **239**: 465 (1936).
13.10. GANNETT, F.: Unmapped Eroded Limestone Complicates Corewall Job, *Eng. News-Record* (*New York*), **115**: 328 (1935); Caverns under Dam Corewall Set a Nice Repair Problem, **116**: 492 (1936).
13.11. DEACON, G. F.: The Vyrnwy Works for the Water-supply of Liverpool, *Inst. Civil Eng. Min. Proc.* (*London*), **126**: 24 (1896).
13.12. BERKEY, C. P.: Responsibilities of the Geologist in Engineering Projects, *A. I. M. E. Tech. Pub.* 215, p. 4, 1929.
13.13. See 6.11.
13.14. See 6.7.
13.15. Gorge Excavation Confirms Geological Assumptions, *Eng. News-Record* (*New York*), **111**: 761 (1933). See also C. P. BERKEY: Geology of Boulder and Norris Dam Sites, *Civil Eng.* (*New York*), **5**: 24 (1935).
13.16. TWENHOFEL, W. H.: "A Treatise on Sedimentation," p. 17, Williams & Wilkins Company, Baltimore, Maryland, 2d ed., 1932.
13.17. FITZMAURICE, M.: The Nile Reservoir, Assuan, *Inst. Civil Eng. Min. Proc.* (*London*), **152**: 71 (1903).
13.18. REEVES, F. and C. P. ROSS: A Geologic Study of the Madden Dam Project, Alhajuela, Canal Zone, *U. S. Geol. Survey Bull.* 821, 1931. See also F. H. KELLOG: Clay Grouting at Madden Reservoir, *Eng. News-Record* (*New York*), **109**: 395 (1932).

13.19. Morris, S. B., and C. E. Pearce: Concrete Gravity Dam for Faulted Mountain Area, *Eng. News-Record* (*New York*), **113**: 823 (1934).

13.20. Earthquake-proof Dam in Chile, *Eng. News-Record* (*New York*), **107**: 725 (1931).

13.21. Three Dams on San Andreas Fault Have Resisted Earthquakes, *Eng. News-Record* (*New York*), **109**: 218 (1932). See also N. A. Eckart: Development of San Francisco's Water Supply to Care for Emergencies, *Seismological Soc. America Bull.* 3, **27**: 1937.

13.22. Foundation Procedure at Owyhee Dam, *Eng. News-Record* (*New York*), **106**: 178 (1931).

13.23. Unique Cutoff Construction and Arched Foundation Features of Rodriguez Dam, *Eng. News-Record* (*New York*), **105**: 600 (1930). See also C. P. Williams: Foundation Treatment at Rodriguez Dam, *Am. Soc. Civil Eng. Trans.* (*New York*), **99**: 295 (1935).

13.24. For discussion, see p. 58, 13.12. See also G. W. Christians: Asphalt Grouting under Hales Bar Dam, *Eng. News-Record* (*New York*), **96**: 798 (1926).

13.25. Wilbur, E. M.: Grouting Checks Leakage in Earthfill Dam, *Eng. News-Record* (*New York*), **115**: 499 (1935).

13.26. Barnes, A. A.: Cementation of Strata below Reservoir Embankments, *Inst. Water Engs. Trans.* (*London*), **32**: 42 (1927).

13.27. Rands, H. A.: Grouted Cutoff for the Estacada Dam, *Am. Soc. Civil Eng. Trans.* (*New York*), **78**: 447 (1915).

13.28. Fox, J. R.: Precementation of a Reservoir Trench, *Inst. Wat. Engs. Trans.* (*London*), **32**: 69 (1927).

13.29. Taylor, M.: Cementation of Abitibi Dam Foundations, *Canadian Eng.* (*Toronto*), **66**: 5 (1934); and information from Mr. R. L. Hearn, Dominion Construction Corp., Toronto.

13.30. Thousands of Holes Grouted under Norris Dam, *Eng. News-Record* (*New York*), **115**: 699 (1935).

13.31. Extensive Rock Grouting at Boulder Dam, *Eng. News-Record* (*New York*), **114**: 795 (1935). See also V. L. Minear: The Art of Pressure Grouting, *Reclamation Era* (*Washington*), **27**: 56 (1937).

13.32. San Gabriel River Flood Control, *Eng. News-Record* (*New York*), **114**: 113 (1935) (bibliography on p. 116).

Chapter XIV

14.1. Smith, William: On Retaining Water in the Rocks for Summer Use, *Phil. Mag.* (*London*) (new ser.), **1**: 415 (1827).

14.2. Swales, J. K.: The Broomhead Reservoir, *Water and Water Eng.* (*London*), **36**: 565 (1934).

14.3. Porosity of Reservoir Prevents Water Storage, *Eng. News-Record* (*New York*), **96**: 561 (1926).

14.4. Bryan, K.: Geology of Reservoir and Dam Sites, *U. S. Geol. Survey Water Supply Paper* 597 (a), p. 7, 1928.

14.5. See, for further details, Dole, R. B.: Use of Fluorescein in the Study of Underground Water, *U. S. Geol. Survey Water Supply Paper* 160, 1906.

14.6. A Curious Water Wheel, *The Engineer* (*London*), **142**: 590 (1926).

14.7. Strange, W. L.: Reservoirs with High Earthern Dams in Western India, *Inst. Civil Eng. Min. Proc.* (*London*), **132**: 137 (1898).

14.8. WADE, L. A. B.: Concrete and Masonry Dam Construction in New South Wales, *Inst. Civil Eng. Min. Proc.* (*London*), **178**: 1 (1909).
14.9. THOMSON, D. H.: Water and Water Power, *Liverpool Eng. Soc. Trans.*, **44**: 105 (1923).

Chapter XV

15.1. HERSCHEL, C.: "The Two Books on the Water Supply of the City of Rome of Sextus Julius Frontinus," p. 21, Dana Estes and Co., Boston, Massachusetts, 1899.
15.2. JOHNSON, D. W.: "Shore Processes and Shoreline Development," John Wiley & Sons, Inc., New York, 1919.
15.3. Beach Erosion Studies by Federal Board, *Eng. News-Record* (*New York*), **111**: 281 (1933).
15.4. ASH, W. C., and O. B. RATTENBURY: Vizagapatam Harbour, *Jour. Inst. Civil Eng.* (*London*), **1**: 235 (1935).
15.5. HOWARD, E. A.: Permeable Groins of Concrete Check Beach Erosion, *Eng. News-Record* (*New York*), **114**: 594 (1935).
15.6. See 13.17.
15.7. PIERCE, R. C.: Measurement of Silt-laden Streams, *U. S. Geol. Survey Water Supply Paper* 400, p. 41, 1916.
15.8. FORTIER, S., and H. F. BLANEY: Silt in the Colorado River and Its Relation to Irrigation, *U. S. Dept. Agr. Tech. Bull.* 67, 1928.
15.9. See 15.8. See also C. S. HOWARD: Suspended Matter in the Colorado River 1925–1928, *U. S. Geol. Survey Water Supply Paper* 636, p. 15, 1929.
15.10. See p. 374 of discussion of B. M. HELLSTROM: The Perak River Hydroelectric Power Scheme, *Inst. Civil Eng. Min. Proc.* (*London*), **239**: 253 (1936).
15.11. VETTER, C. P.: Why Desilting Works for the All-American Canal, *Eng. News-Record* (*New York*), **118**: 321 (1937).
15.12. KENNEDY, R. G.: "Hydraulic Diagrams for Channels in Earth," 2d ed., Edinburgh, 1907.
15.13. STEVENS, J. C.: The Silt Problem, *Am. Soc. Civil Eng. Trans.* (*New York*), **101**: 207 (1936).
15.14. Report of discussion at American Society of Civil Engineers meeting, *Eng. News-Record* (*New York*), **114**: 131 (1935).
15.15. SONDEREGGER, A. L.: Modifying the Physiographical Balance by Conservation Measures, *Am. Soc. Civil Eng. Trans.* (*New York*), **100**: 284 (1935).
15.16. BUTCHER, A. D. D., and J. D. ATKINSON: The Causes and Prevention of Bed Erosion, with Special Reference to the Protection of Structures Controlling Rivers and Canals, *Inst. Civil Eng. Min. Proc.* (*London*), **235**: 175 (1934).
15.17. See LANE, E. W.: Retrogression of Levels in Riverbeds below Dams, *Eng. News-Record* (*New York*), **112**: 836 (1934). See also original report, marked For Official Use Only, dated December 16, 1929, by the Islam Enquiry Committee into the failure of the Islam Weir, S.V.P., September 19, 1929, Lahore, India, 1930.
15.18. ELLIOT, D. O.: "The Improvement of the Lower Mississippi River for Flood Control and Navigation," U. S. Waterways Experiment Station, Vicksburg, Mississippi, 3 vols., 1932. See also numerous articles in *Eng. News-Record* and *Civil Eng.* (*New York*) in recent years, *e.g.*, G. R. CLEM-

ENS: Straightening the Father of Waters, *Eng. News-Record* (*New York*), **116, 269** (1936).

15.19. See p. 63, 3.1.

15.20. See p. 65, 3.1.

15.21. See p. 62, 3.1.

15.22. WHEELER, W. H.: "Tidal Rivers," p. 165, Longmans, Green & Company, London, 1893.

15.23. See p. 145, 15.22.

15.24. See p. 365, 15.22. Also C. PEEL: The Mersey Estuary, paper published by Manchester and District Association of Institution of Civil Engineers (England), 1926.

15.25. See p. 191, 15.22. Also from information from personal communication with Mr. Billings Wilson, Assistant General Manager, The Port of New York Authority.

15.26. See p. 142, 15.22.

15.27. See "The Addresses and Papers of James B. Eads," ed. by E. McHenry, Slawson and Co., St. Louis, Missouri, 1884. See also papers by E. L. Corthell in *Am. Soc. Civil Eng. Trans.* (New York).

15.28. JOHNSTON, W. A.: Sedimentation of the Fraser River Delta, *Geol. Soc. Canada Mem.* 125 (No. 107 Geol. Ser.), 1921.

15.29. GIBSON, A. H.: Tidal and River Models, Vernon Harcourt Lecture, *Jour. Inst. Civil Eng.* (*London*), **3**: 699 (1936).

15.30. Data from U. S. Soil Conservation Service. See recent articles, *e.g.*, H. O. HILL: Soil Conservation in the Texas Blacklands, *Civil Eng.* (*New York*), **8**: 109 (1938).

Chapter XVI

16.1. Quoted in GEST, 11.11, from *Nat. History*, **XXXVI**, 15.24, 101–125.

16.2. VAN ESVELD, W. H. C.: De Watervoorziening in die Oudheid, *Water en Gas*, p. 37, March 7, 1930.

16.3. See, *e.g.*: WIEGAND, T., and H. SCHEADER: "Priene," G. Reimer, Berlin, 1904.

16.4. See 15.1, p. 19.

16.5. LAMBARDE: "Perambulation of Kent," p. 152, London, 1826, quoted in *Meteorological Mag.* (*London*), February, 1928.

16.6. BOWMAN, J., German Waterworks Practice, *Water and Water Eng.* (*London*), **38**: 665 (1936).

16.7. See discussion of 16.9, p. 530.

16.8. The New Filter Gallery at Des Moines, *Eng. Record* (*New York*), **65**: 468 (1912): and information from Mr. Dale L. Maffit, General Manager, Des Moines Water Works.

16.9. GOURLAY, H. J. F.: The Water-supply of Kano, Northern Nigeria, *Inst. Civil Eng. Min. Proc.* (*London*), **237**: 454 (1935).

16.10. ARMSTRONG, R. W.: The Delaware River Water Supply of the City of New York, *Civil Eng* (*New York*), **8**: 581 (1938).

16.11. See *Eng. News-Record* (*New York*), **113**: 129, Hetch Hetchy Water Supply No. (1934); also publications of the city of San Francisco.

16.12. HINDS, J.: Colorado River Water for California, *Civil Eng.* (*New York*), **7**: 573 (1937). See also **5**: 2 (1935).

16.13. From information in a personal communication from Mr. W. H. Pearce, City and Water Engineer, Gibraltar; see also C. D. Jamieson, The Municipal Water Supply of Gibraltar, *Journal of Royal Army Medical Corps (British)* August, 1935.

16.14. PARKER, P. à. M: "The Control of Water," p. 207, George Routledge & Sons, Ltd., London, 2d ed., 1925.

16.15. EAST, L. R., Water Supply Problems in Australia, *Inst. Civil Eng. Selected Eng. Paper* 141 (*London*), 1933.

16.16. See bulletins of the Geological Survey of South Australia, *e.g.*, R. LOCKHART JACK: Some Developments in Shallow Water Areas in the North East of South Australia, No. 11, 1925.

16.17. HALCROW, W. T., G. B. BROOK, and R. PRESTON: The Corrosive Attack of Moorland Water on Concrete, *Inst. Water Eng. Trans.* (*London*), **33**: 187 (1929).

16.18. See 13.9.

16.19. WOODWARD, H. B.: "The Geology of Water Supply," Edward Arnold & Company, London, 1910.

Chapter XVII

17.1*a*. *Pittsburgh:* Discussion of FORSTER BROWN, E. O.: Underground Waters in the Kent Coal Field and Their Incidence in Mining Development, *Inst. Civil Eng. Min. Proc.* (*London*), **215**: 27 (1923).

b. *India:* MEINZER, O. E.: The History and Development of Ground-water Hydrology, *Jour. Washington Acad. Sci.*, **24**: 6 (1934).

c. *Germany:* See 16.6.

d. *United States:* See *b*.

e. *London:* SMITH, B.: Geological Aspects of Underground Water Supplies, Royal Society of Arts Cantor Lecture, *Water and Water Eng.* (*London*), **38**: 223 (1936).

17.2. See 17.1*b*.

17.3. JACK, R. LOCKHART: The Geology of the County of Jervois . . . with Special Reference to Underground Water Supplies, *Geol. Survey South Australia Bull.* 3, p. 29, 1914.

17.4. See 17.1*e*, p. 224.

17.5. See 8.3.

17.6. GREGORY, J. W.: The Origin and Distribution of Underground Waters, *Water and Water Eng.* (*London*), **29**: 321 (1927).

17.7. SCOTT, F. W.: Pietermaritzburg-Riet Spruit Deviation, *Inst. Civil Eng. Min. Proc.* (*London*), **207**: 229 (1921).

17.8. See discussion of 17.1*a*, p. 77.

17.9. Report of discussion at British Association meeting in Aberdeen, *Water and Water Eng.* (*London*), **36**: 623 (1934).

17.10. See 17.1*e*, p. 226.

17.11. See 17.1*e*, p. 225.

17.12. Hot Well Water Cooled for Use in City Distribution System, *Eng. News-Record* (*New York*), **118**: 130 (1937).

17.13. Corrosiveness of Aqueduct Water Reduced by Gas Drainage, *Eng. News-Record* (*New York*), **117**: 790 (1936). See also Drains Remove CO_2 from Tunnel, **120**: 885 (1938).

17.14. DIXEY, F.: "A Practical Handbook of Water Supply," p. 183, Thomas Murby & Co., London, 1931.

17.15. MEINZER, O. E.: Plants as Indicators of Ground Water, *U. S. Geol. Survey Water Supply Paper* 577, 1927.

17.16. ELLIS, A. J.: The Diving Rod; A History of Water Witching, *U. S. Geol. Survey Water Supply Paper* 416, 1934 (repr.).

17.17. Divination by Rods, *Water and Water Eng.* (*London*), **34**: 147 (1932).

17.18. GREGORY, J. W.: Water Divining, *Water and Water Eng.* (*London*), **29**: 453 (1927).

17.19. See, *e.g.*, WALTERS, R. C. S.: The Hydro-geology of the Chalk of England, *Inst. Water Engs. Trans.* (*London*), **34**: 79 (1929).

17.20. MEINZER, O. E., and N. D. STEARNS: A Study of Groundwater in the Pomperaug Basin, Connecticut with Special Reference to Intake and Discharge, *U. S. Geol. Survey Water Supply Paper* 597(b), 1928.

17.21. MEINZER, O. E.: Outlines of Methods for Estimating Ground-water Supplies, *U. S. Geol. Survey Water Supply Paper* 638(c), 1932.

17.22. MCDANIELS, A. B.: Groundwater Cutoff Wall Provides New Water Supply, *Eng. News-Record* (*New York*), **113**: 757 (1934).

17.23. *Karez.* THOMPSON, R. O. C.: Capturing Water in the Desert, *Eng. News-Record* (*New York*), **120**: 327 (1938). *Keghriz:* L'alimentation en eau des régions désertiques de la Perse et du Caucase, *L'Eau* (*Paris*), p. 86, 1933.

17.24. CUMMING, G. A.: Underground Water Circulation, *Water and Water Eng.* (*London*), **38**: 319 (1936).

17.25. JACK, R. LOCKHART: Geological Structure and Other Factors in Relation to Underground Water Supply in Portions of South Australia, *Geol. Survey South Australia Bull.* 14, 1930.

17.26. See 17.3, p. 28; 16.15, p. 10.

17.27. BROWN, J. S.: A Study of Coastal Ground Water, *U. S. Geol. Survey Water Supply Paper* 537, 1925.

17.28. Reports of the Interstate Conferences on Artesian Water: No. 1: Sydney, 1912: No. 2: Brisbane, 1914; No. 3: Adelaide, 1921; No. 4: Perth, 1924; No. 5: Sydney, 1928, issued by various state government printers; copies kindly supplied by Dr. L. Keith Ward, Director of Mines, South Australia.

17.29. SUTER, R.: Engineering Report on the Water Supplies of Long Island, *New York State Cons. Dept. Div. Water Power and Control Bull. G.W.*-2, 1937.

17.30. Underground Water Storage in California's South Coastal Basin, *Eng. News-Record* (*New York*), **115**: 733 (1935).

17.31. Water Scheme for Nassau, Bahamas, *Civil Eng.* (*London*), **29**: 177 (1934). See also J. O. RIDDEL: Excluding Salt Water from Inland Wells, *Civil Eng.* (*New York*), **3**: 383, 1933.

17.32. Public Water Supply from Colliery, *Water and Water Eng.* (*London*), **38**: 294 (1936).

17.33. See 17.1*e*, p. 417. See also J. L. FRANCIS and J. C. ALLAN, Driving a Mines Drainage Tunnel in North Wales, *Inst. Min. and Met., Trans.* (*London*), **41**: 236 (1932).

17.34. MARR, J. C.: Drainage by means of Pumping from Wells in Salt River Valley, Arizona, *U. S. Dept. Agr. Dept. Bull.* 1456, 1926.

17.35. LANE, D. A.: Artificial Storing of Groundwater by Spreading, *Jour. Am. Water Works Assoc.*, **28**: 1240 (1936).

17.36. 1. *Australia:* See 17.28, *Report of 5th Conference*, p. 9, Government Printer, Sydney, 1929.

2. *South Dakota:* MEINZER, O. E.: Progress in the Control of Artesian Water Supplies, *Eng. News-Record* (*New York*), **113**: 167 (1934).

3. *Paris:* LEMOINE, P., R. HUMERY, and R. SOYER: L'Appauvrissement de la nappe des sables vertes de la région parisienne, *Comptes rendus* (Paris), p. 1870, May 23, 1934.
4. *Chicago:* BURDICK, C. B.: Ground Water as a Source of Supply, *Eng. News-Record* (*New York*), **105**: 398 (1930).
5. *London:* The Falling Water Level of the Chalk under London, *Water and Water Eng.* (*London*), **35**: 440 (1933).

17.37. See 17.36b. *See also:* MCCLURE, T. M.: Plugging Old Artesian Wells to Stop Underground Water Loss, *Eng. News-Record* (*New York*), **116**: 425 (1936).

17.38. The Geology and Underground Water of Rangoon, *Water and Water Eng.* (*London*), **35**: 539 (1933).

Chapter XVIII

18.1. SAWTELL, H. E.: Foundations of the Boston, Mass., Parcel Post Building, *Jour. Boston Soc. Civil Eng.*, **22**: 29 (1935).

18.2. PRENTIS, E. A., and L. WHITE: "Underpinning," frontispiece, Columbia University Press, New York, rev. ed., 1931.

18.3. NOETZLI, F. A.: Multiple Arch Retaining Wall Damaged by Slip, *Eng. News-Record* (*New York*), **98**: 146 (1927); also Cahuenga Pass Arch Retaining Wall Continues Slipping, **99**: 681 (1927).

18.4. For a fuller discussion, see, *e.g.* K. VON TERZAGHI: The Science of Foundations—Its Present and Future, *Am. Soc. Civil Eng. Trans.* (*New York*), **93**: 270 (1929).

18.5. *First Internat. Conference on Soil Mechanics and Foundation Eng. Proc.*, 3 vols., *Harvard University* (*Cambridge, Massachusetts*) **3**: 150 (1936) (abstract of complete report).

18.6. Earths and Foundations, *Special Comm. Progress Rept., Am. Soc. Civil Eng. Proc.* (*New York*), **59**: 777 (1933).

18.7. MOHR, H. A.: Exploration of Soil Conditions and Sampling Operations, *Harvard University* (*Cambridge, Massachusetts*) *Graduate School Eng. Bull.* 208 (Soil Mechanics Ser. 4), June, 1937.

18.8. Undiscovered Substratum of Peat Complicates Foundation Job, *Eng. News-Record* (*New York*), **91**: 192 (1923).

18.9. OGDEN, P. H.: Adjustable Water Tanks at Heanor, *Civil Eng.* (*London*), **33**: 131 (1938).

18.10. See 18.5, **3**: 150.

18.11. KNIGHT, W. J.: Open Pier Foundations of St. Louis Telephone Building, *Eng. News-Record* (*New York*), **99**: 536 (1927).

18.12. Consolidating the Foundations of the Leaning Tower of Pisa, *Civil Eng.* (*London*), **31**: 289 (1936) (from *Schweizerische Bauzeitung*, **107**: 272).

18.13. STANTON, SIR T. E.: Engineering Research, James Forrest Lecture, *Inst. Civil Eng. Min. Proc.* (*London*), **232**: 400 (1923).

18.14. Modern Engineering to Save Mediaeval Tower, *Eng. News-Record* (*New York*), **91**: 505 (1923).

18.15. SNOW, B. F.: Tracing Loss of Groundwater, *Eng. News-Record* (*New York*), **117**: 1 (1936).

18.16. Artificial Groundwater for Wood Piles, *Eng. News-Record* (*New York*), **107**: 70 (1931).

18.17. See 10.18, p. 196.

NAME INDEX

(This index includes only names mentioned in the text; names of authors of books and articles included in lists at the ends of the chapters and in the appendices are not listed here.)

(*Q.*) indicates the inclusion of a quotation.

A

Agassiz, L., 4, 561
Agricola, G., 449
Alexander the Great, 266
Aristotle, 25 (*Q.*)
Aspdin, J., 526
Atkinson, A. A., 364
Atkinson, J. D., 393
Atwood, W. G., 226
Austen, Godwen, 9

B

Bailey, E. B., 179
Baker, Sir B., 324, 540, 582 (*Q.*)
Banks, Sir J., 451
Beech, T. de la, 515
Bell, A. L., 540
Berkey, C. P., 67, 147 (*Q.*), 154, 317, (*Q.*), 322, 496
Berry, C. S., 88
Bertrand, M., 265
Binger, W. D., 154
Boswell, P. G. H., 67, 170, 250
Bouguer, P., 121
Boulton, W. S., 435
Bouyoucos, G. J., 543
Brown, J., 179
Brown, W. L. Lowe, 85 (*Q.*), 275
Bruckshaw, J. M., 135
Bryan, K., 358
Buckle, H. T., 304 (*Q.*)
Butcher, A. D. D., 393

C

Calver, E. K., 3 (*Q.*)
Carlyle, T., 473 (*Q.*)
Carpenter, E. E., 87
Carpmael, R., 79
Casagrande, A., 541
Castelli, Abbot, 402
Cavendish, H., 122
Caxton, W., 450
Clarke, F. W., 554*n.*
Clarkson, C. H., 88
Coales, F. G., 88
Collin, A., 241, 539
Coode, Sir J., 402
Coulomb, C., 207, 539
Creager, W. P., 167
Crosby, I. B., 496
Crosby, W. O., 147, 570
Cuevas, J., 503

D

Daft, 125
Darwin, C., 5
Davis, R. P., 89
Dawkins Boyd, 61 (*Q.*), 173
Deacon, G. F., 316, 335 (*Q.*)
Dean, A. C., 88, 492
De Castro, 117
Department of Public Works, Canada, 407, 534
Dixey, F., 135, 446
De Saussure, 4
Dolmage, V., 129

E

Eadbald, King, 416
Eads, J. B., 405
Eanswide, St., 416
Eckel, E. C., 68, 71
Ekblaw, G. E., 552 (*Q.*)

Ellis, S. H., 523
Eötvos, Baron R. von, 122
Ethelred, King, 270
Everall, W. T., 287

F

Farey, J., 451 (*Q.*)
Farnham, F. C., 133
Ferguson, H. B., 396
Fitzmaurice, Sir M., 114 (*Q.*)
Forman, R., 270 (*Q.*)
Fougerolles, Mons., 267
Fox, Sir F., 170, 373
Francois, A., 340
Frontinus, S. J., 369 (*Q.*), 415 (*Q.*)
Furkert, F. W., 217

G

Gaillard, D. D., 264 (*Q.*)
Geikie, Sir A., 324
Gibson, A. H., 407
Giesecke, F. E., 490
Gish, O. H., 127
Glenn, L. C., 334
Gould, R. G., 154
Grant, U. S., 396
Greathead, J., 340
Greenley, H., 43 (*Q.*)
Grim, R. E., 552 (*Q.*)
Gruner, H. E., 70 (*Q.*)

H

Halley, E., 430
Hammurabi, King, 474 (*Q.*)
Hatton, T. C., 491
Hawkesley, T., 316, 336
Hayes, Dr., 265
Heim, A. von St. G., 224
Herzberg, B., 460
Hiskia, King, 415
Hooke, R., 3
Hornibrook, M. R., Ltd., 293
Howe, Dr., 265
Hutton, J., 4

I

Irvine, L. R. H., 254

J

Jacoby, H. S., 89
Jenkin, C. F., 207
Joosten, H., 201
Justin, J. D., 167, 306

K

Kemp, J. F., 147
Kendall, P. F., 436
Kennedy, R. G., 390 (*Q.*)
King, W. B. R., 536
Knight, B. H., 517

L

Ladd, G. E., 224, 540
Lapworth, H., 306, 434 (*Q.*)
Latham, Baldwin, 450
Lehmann, J. G., 4
Leland, J., 281
Le Mesurer, Reverend, 452
Lister, M., 41 (*Q.*)
Logan, Sir. W. E., 4
Lovebone, Vicar, 281
Lugeon, M., 131 (*Q.*)
Lyell, Sir C., 5

M

McAdam, J. L., 249 (*Q.*)
Macaulay, Lord, 115 (*Q.*)
MacDonald, A. B., 265
Macdonald, A. E., 497
McIldowie, G., 509
Maclure, W., 4
Mallet, R., 118
Maskelyne, N., 121
Meinesz, V., 122
Meinzer, O. E., 223, 429, 448
Menes, King, 305
Merrill, G. P., 546 (*Q.*)
Merriman, T., 147
Miller, W. T. W., 530
Mitchell, J., 123
Mohr, H. A., 95, 481
Morrison, Fynes, 271 (*Q.*)
Morton, G. H., 169
Moses, 448
Municipal Engineers' Society of New York, 497

N

Necho, King, 260
Nero, Emperor, 260
Newton, I., 121
Nicol, W., 5

O

Olaf, King, 270

P

Page, J. C., 87
Palissy, B., 430
Parker, 528
Parker, P. à M., 424 (*Q.*)
Pearce, W. H., 424
Perrault, P., 430
Pittman, E. J., 364
Playfair, J., 4
Pliny, 414 (*Q.*), 462
Polo, Marco, 448
Powell, W. H., 87
Prentis, E. A., 492
Ptolemy II, King, 260

Q

Quiller-Couch, Sir A., 247 (*Q.*)

R

Rankine, J. M., 207, 539
Ransome, F. L., 308 (*Q.*), 333
Raymond, R. W., 450 (*Q.*)
Reade, T. M., 170, 261 (*Q.*)
Rehboch, T., 393
Reid, H. F., 241
Rennie, J., 490
Reynolds, O., 407
Ritchie, H. C., 484
Robinson, T., 68 (*Q.*)
Rodio, G., 485
Rooney, J., 127
Ruskin, J., 514 (*Q.*)

S

Saavedra, A. de, 263
Sanborn, J. F., 148
Sarjant, R. J., 530
Schaffer, R., 436
Schappler, R. C., 133
Schlumberger, C., 127, 131 (*Q.*)
Schoklitsch, I., 160 (*Q.*)
Scott, Sir W., 11 (*Q.*)
Sheppard, T., 452
Simpson, W., 102
Singewald, J. T., 105
Six Companies Inc., 347
Smeaton, J., 528
Smith, B., 431 (*Q.*)
Smith, H. A., 12
Smith, J. Waldo, 147
Smith, W. Sooy, 477
Smith, William, 5, 36, 81, 353 (*Q.*), 430, 451, 515
Solomon, King, 415
Spear, W. E., 147
Stansfield, J., 495
Steno, N., 3
Stephenson, G., 256
Stephenson, R., 173 (*Q.*), 430
Stevens, J. C., 390
Stevens, J. F., 265
Strahan, C. M., 252, 540
Strange, W. L., 361
Sweeney, J., 99

T

Taft, William Howard, 265
Telford, T., 249, 490
Terzaghi, K. von, 208, 224, 481 (*Q.*) 484, 538 (*Q.*), 540, 541, 581
Thales of Miletus, 462
Thomas, H. H., 536
Thomson, D. Halton, 367
Tschebotareff, G. P., 499
Twain, Mark, 211

U

U.S. Beach Erosion Board, 376
U.S. Bureau of P. Roads, 133, 252, 544
U.S. Bureau of Reclamation, 347, 389
U.S. Soil Conservation Service, 410

V

Vermuyden, C., 371
Vernon Harcourt, L. F., 407

Viele, 492
Vitruvius M. Pollio, 294 (*Q.*), 448
Vogt, F., 46, 448

W

Walker, T. A., 79
Walters, R. G. S., 450
Werner, A. G., 4
Westergaard, H. M., 327
White, L., 148, 492
Whitney, M., 540
Williams, 125
Williams, H., 43 (*Q.*)
Winterkorn, H. E., 552
Wood, S. M., 382
Worcester, J. R., 496
Wyllie, J., 442*n*.

X

Xerxes, King, 260

Z

Zurcher, P., 265

SUBJECT INDEX

A

Abutments, special bridge, 280
Accident, due to gas in clay, 491
Adits for water supply, 455
Adobe, 556
Airplane, use of, in geological surveys, 47
Africa, artesian well at Thebes, 460
  Kano water supply, 419
  geophysics and groundwater, 136
  littoral drift, Durban, 377
  seepage into tunnel, 433
  springs at Kivu, 433
  underground dams, 452
    (*See also* Algeria; Egypt)
Aggregate, concrete, 532
  for concrete roads, 251
Agriculture and soils, 539
Air conditioning, groundwater for, 438
Airport construction, 269
Airports, Toronto, 269
Algeria, Cheurfas Dam, 310
  groundwater movement, 434
  Habra Reservoir, 390
  geophysical survey, 266
Alluvial cone, 558
Alluvial plain, 558
Alluvial terrace, 558
Alteration of rocks, 547
Aluminous cement, 527
  resistance of, to water action, 426
  for tunnel lining, 163
Analysis, soil, 543
Anchoring loose rock, 244
Andesite, 23
Anhydrite, 163, 525
Anticline, 30
Aqueduct, Colorado River, 422
Architecture and Building Stone, 514
Arsenic in Groundwater, 442*n*.
Artesian Basin, Great Australian, 462
  Great Dakota, 461
Artesian basin, Murray River, 463
    intake, 431
Artesian water, under dock, 200
Artesian wells, Berlin, Germany, 461
  Boronga, Australia, 461
  leakage from, 471
  Ottershaw, England, 461
  St. Augustine, Fla., 461
  St. Louis, Mo., 461
  sealing leaky, 472
Asphalt grouting, 334
Atmosphere, action of, 548
Auger borings, 97
Australia, artesian basins, 463
  artesian well, Boronga, 461
  baked clay roads, 254
  bridge at Brisbane, 293
  bridge, George's River, 273
  cataract reservoir, 364
  divining tests, 449
  groundwater movement, 434
  Murray River Basin, 431
  pressure tunnel failure, 158
  soak wells, 456
  soil erosion, 409
  springs in mine, 433
  tray wells, 456
  unusual water supplies, 425

B

Bahamas, water supply, 466
Baked clay roads, 254
Ballast of baked clay, 512
Bank protection, 397
Barachois, 376
Bars in estuaries, Aberdeen, Scotland, 402
    Mersey, England, 403
Basalt, 23
Beach formation, 373
Beer, water for, 426
Belgium, Albert Canal, 200
  Scheldt Tunnel shafts, 200

Bibliography, geology and civil engineering, 81
Blasting for fill settlement, 257
Bolivia, mud runs on railway, 259
Boston, proposed building code, 501
Botany and soil control, 411
Boulder clay, 20, 77
Boulders, and bridge piers, 298
  and dam foundations, 312
  glacial, 572
  and sheet-pile walls, 209
Bournes, 446
Braces between bridge abutments, 286
Brazil, underground dams, 452
Breakwater, of scuttled ships, 379
  Valparaiso, Chile, 266
Bricks, clay for, 510
  color of, 512
  Staffordshire blue, 426, 512
Bridge pier design, 280
  lack of preliminary data, 273
Bridges:
  Anacostia River, 302
  bascule, Toledo, Ohio, 286
  Broadway, Sask., 283
  Burford, England, 284
  Chelsea, England, 517
  Grosvenor, England, 285
  historical note, 270
  Île d'Orléans, Que., 283
  Iskander, F.M.S., 88
  Kohala, India, 287
  London, England, 270, 282
  L.M.S., Railway, England, 287
  over Meuse-Waal Canal, Holland, 286
  Mid-Hudson, New York, 293
  Mississippi River, La., 294
  Mortimer E. Cooley, Michigan, 280
  near Hargis, N.M., 302
  Nerbudda, India, 291
  San Francisco-Oakland Bay, cost of borings, 90
    pier design, 283
    soil sampling, 99, 275
  Suisan Bay, California, 294
  Vauxhall, England, 290
  Waterloo, England, 282
  over Welland Canal, Ont., 286
  Westminster, England, 290
Building foundation, design, 475
  well points and excavation, 198
Building foundations, Austin, Tex., 490
  Battersea, England, 88, 492
  Boston, Mass., 477
  Detroit, Mich., 502
  Heanor, England, 484
  Lochaber, Scotland, 90
  London, England, 476, 486, 490
  Mexico City, Mexico, 503
  Milwaukee, Wis., 489
  Montreal, Que., 476
  New York, N.Y., 198, 476
  Philadelphia, Pa., 481
  Pisa, Italy, 485
  St. Louis, Mo., 485
  Toronto, Ont., 197
Building Regulations, 499
Building settlement, Boston, Mass., 487
  Montreal, Que., 490
  Philadelphia, Pa., 481
  Strasbourg Cathedral, 487
  Winnipeg, Man., 498
Building stone, appearance of, 517
  availability of, 516
  cost of, 517
  discoloration of, 517
  durability of, 519
  for Parliament buildings, 515
  strength of, 519
  use of, 522
  weight of, 519
Bulbs of pressure, 480
Buried valleys, 570
  and bridge foundations, 273
  and dam foundations, 312
  River Kelvin, Scotland, 133
  and road construction, 254
  and test borings, 94
Burma, groundwater legislation, 472

C

Caissons for bridge piers, 292
Calyx drill holes, at Grand Coulee Dam, 105
  at Norris Dam, 105
  at Prettyboy Dam, 105, 320
  at Watts Bar Dam, 106
Canadian Shield, 564

Canals:
Albert, Belgium, 200
All American, 388
historical note, 260
Kiel, Germany, 540
leakage from, 261
Manchester ship, England, 393
Panama, 262, 540
Welland, Ont., 262
Canary Islands, adits for water supply, 455
Carbon dioxide, from clay, 491
in groundwater, 440
in tunnels, 440
Catchment area, River Alwen, Wales, 367
River Avon, Scotland, 367
River Exe, England, 367
River Pomperaug, Conn., 451
Caves, in limestone, 193, 194
Cellular cofferdams, 296
Cement, 526
aluminous, 163, 426, 527
Cementation, François process, 340
Cephalonia, unusual waterwheel, 359
Chalk, leakage through, 268
porosity of, 435
"Swallow holes" in, 284
Chalk caves under Arras, 193
Chat Moss, crossing of, 256
Check dams, 260
Chemical composition of rocks, 15
Chemical consolidation, and building foundations, 485
for dam foundations, 339
Joosten process in tunnels, 171
of waterlogged ground, 201
Chert, 21
Chesil Bank, England, 402
Chicago well method, 477
Chile, Las Raices Tunnel, 144
rock-fill dams, 328
China, Hwang Ho River, 399
loess deposits, 556
Shanghai geology, 499
Cirques, 565
City geological surveys, Boston, Mass., 496
New York, N.Y., 496
New Orleans, La., 497
Philadelphia, Pa., 499
City geological surveys, Shanghai, China, 499
Winnipeg, Man., 497
Clay, baked for roads, 254
for bricks, 509
burned as riprap, 512
composition of, 552
consolidated, 580
depressions in London, 492
gas from, 491
lifting church, 490
porosity of, 435
protection of, 195
puddle, 508
sealing above tunnel, 176
for tile, 513
tunneling in, 181
water supply from, 437
weathering of, 514
Cleaning rock surfaces, for bridge piers, 294
for dam foundations, 350
Coal mining, under dam, 364
by dredging, 370
geophysical survey and, 135
Cofferdams, for Cape Cod Canal bridges, 298
at Grand Coulee Dam, 296
Concrete, detection of, 536
Concrete roads, 251
Cone of depression, 456
Conference, Soil Mechanics, 541
Conglomerate, 21
drainage of cut in, 190
Contract plans, 74
Cooling water supply, 439
Core drilling, 102
Corrasion, 383
Cosmogony, 6
Costs, of bridge substructures, 272
of building stone, 517
of geophysical surveys, 131
of preliminary work, 86
of San Francisco Bay bridge, preliminary work, 90
of test borings, 90, 96, 171
of test pits, 101
of tunnels, 86, 87
Crushed rock, 530
Crystallography, 12
Cycle, geological, 26

D

Dam construction, anchors to bedrock, 312
effects of, 309
foundation grouting, 340
historical note, 305
uplift pressure, 335
Dams, Abitibi Canyon, Ont., 342
Assouan, Egypt, 324, 385
Bonneville, Ore., 362
Boulder, Colo., 87, 321, 346, 388
Calderwood, Tenn., 395
Cheurfas, Algeria, 310
Colorado, Texas, 306
Conowingo, Md., 395
Estacada, Ore., 340
Fifteen Mile Falls, N. H., 131, 208
Fort Peck, Mont., 191
Grand Coulee, Wash., 105, 232, 296
Hales Bar, Tenn., 334
Huguenot, N. Y., 339
Kinder, England, 340
Kippewa Lake, Que., 534
Le Sautet, France, 114, 319
Madden, Panama Canal Zone, 326, 338
Morris, Cal., 327, 329
Norris, Tenn., 105, 343
Ontelaunee, Pa., 314
O'Shaughnessy, Cal., 422
Owyhee, Ore., 330
Parker, Cal., 423
Prettyboy, Md., 105, 320
Puentes, Spain, 305
Quinze, Que., 534
Rodriguez, Mexico, 331
Sarrans, France, 131
St. Francis, Cal., 308
San Andreas, Cal., 329
San Gabriel, Cal., 348
Scout Dyke, England, 341
Sennar, Egypt, 325, 393
Shasta, Cal., 113
Silent Valley, Northern Ireland, 312, 509
underground, 452
Upper Crystal Springs, Cal., 329
Vyrnwy, Wales, 315
Watts Bar, Tenn., 106
Wheeler, Tenn., 192
Deltas, Cairo, Roda guage record, 370
depth of, 405
growth of, 400
Depletion of groundwater, Australia, 471
Chicago, Ill., 471
Dakota, 471
London, England, 471
Long Island, N. Y., 464
Paris, France, 471
Southern California, 466
Desilting at Imperial Dam, 388
Deterioration, of concrete, 426, 534
of timber, 487
Detrital cones, 466
Dew ponds, 424
Diamond drilling, 102
Diorite, 22
Dip, 25
Discoloration of granite, 517
Disconformity, 34
Diving for pier inspection, 301
Divining, water, 448
Docks, Dover train ferry, England, 268
Southampton, England, excavation, 200
fossils from, 38
Dolerite, 23
Dolomitic limestone, 21
Domes over "swallow holes," 284
Dowsers, British Society of, 449
Drainage, of airports, 269
near Bath, England, 431
from colliery for supply, 467
of dam foundations, 336
of Lake Copais, Greece, 140
Lake Fucino Tunnel, 140
of London tube tunnel, 469
of mines, Wales, 468
of roadbeds, 255
of rock cuttings, 190
Dredging, of bars, 402
for coal, 370
Lake Okeechobee, Fla., 204
Mississippi cutoffs, 396
Drift, 570
Drift maps, 47
Drilling and blasting, 206
Drumlins, 567
Dry-weather flow, 367
Dumb wells, 469

Dunes, movement of sand, 411
Durability of building stone, 519
Dust storms, 409
Dyes for tracing leakage, 359
Dykes, 15

E

Earth, dimensions of, 11
Earth dams, Ewden Valley, England, 355
  Kinder, England, 340
  Scout Dyke, England, 341
Earth dam construction, across active fault, 329
  material for, 506
  sealing with grout, 339
Earth excavation, 76
Earth movement, at Arapuni, New Zealand, 213
  at Lake Maracaibo, Venezuela, 223
  at New Madrid, Mo., 26
  in Santa Clara Valley, 223
  in Utah, 223
Earth movements, due to tides, 219
  across earth dams, 329
  and earthquakes, 217
  and shoreline types, 373
  of sedimentary rocks, 16
Earth pressure, on bridge abutments, 285
  on retaining walls, 207
Earthquakes, Assam, India, 220
  and bridge design, 290
  and dam design, 327
  Eastern Canada, 219
  effect on dams at faults, 328
  San Francisco, 217
  Tokyo, Japan, 217
Earthquake intensities, 220
Education, in geology for engineers, 63
Egypt, ancient tunnels, 140
  Assuan Dam, 324, 385
  Cairo, Roda gauge record, 370
  granite of, 514
  Nile delta, 370
  Sennar Dam, erosion at, 393
    foundations, 325
  silt in Nile valley, 387
Electrical coring, 128
Electrical geophysical methods, 125
Engineering societies and geological surveys, 495
English channel tunnel, proposed, 267
Erosion, of coal, 370
  coastal, Lake Michigan, 381
  differential, river, 383
  due to river flow, 382
  glacial, 564
  below lock sluices, 393
  at Lyme Regis, England, 372
  at Niagara Falls, 385
  of rocky coasts, 372
  below Sennar Dam, Egypt, 393
  soil, 409
Eskers, 568
Estuaries, navigation in, 400
Everest, Mount, 11, 16
Excavation, earth and rock, 76
  of rock, 188
  of soil, 194
Exfoliation of rock, 549
Experience, practical, 66
Extrusive rocks, 14

F

Failure, of arch bridge, Manitoba, 278
  of bridge at Cornwall, Ont., 273
  of Colorado Dam, Texas, 306
  of dams, studies of, 306
  of Gleno Dam, Italy, 351
  of Islam Weir, India, 393
  of Puentes Dam, Spain, 305
  of St. Francis Dam, Cal., 308
  of two bridges, United States, 302
Faults, arched under dam, 333
  construction joint over, 330
  crossed by aqueduct, 423
  dam across inactive, 329
  dams across active, 329
  in dam foundations, 325
  across Grand Canyon, 33
  in open excavation, 185
Felsite, 23
Field equipment, for geological mapping, 44
  for rock study, 19
Fill settlement by blasting, 257
Flint, 21
"Floating dams," 348
Floating rock breakers, 191

Folding, 30
effect on groundwater, 443
Fossils (*see* Palaeontology), 55
France, aluminous cement, 527
Caranca tunnel, 163
caves under Arras, 193
Col de Braus tunnel, 163
Le Sautet Dam, 114, 319
moving sand dunes, 411
Rhone delta, 400
sand in River Garonne, 404
Sarrans Dam, 131
well at Artois, 460
well at Grenelle, 434
Freezing, for slide prevention, 232
for tunneling, 143
for waterlogged excavation, 200
Fumoroles sufforoli, 433

G

Gabbro, 22
Gases in groundwater, 439
Geological maps, 47
Geological succession, 34
Geologists and civil engineering, 68
Geology, economic, 8
glacial, 7
name, 4
physical, 6
popular misconceptions, 9
stratigraphical, 6
subdivisions, 6
Geophysical surveys, Algiers Harbour, 266
Bridge River Tunnel, B.C., 128
Fifteen Mile Falls, N.H., 131
Minnesota Highway Department, 133
River Kelvin, Scotland, 133
St. Lawrence River, dam sites, 131
Sarrans Dam, France, 131
United States Bureau of Public Roads, 133
Geophysics, electrical methods, 125
and geology, 117
gravitational methods, 121
magnetic methods, 117
name, 115
seismic methods, 118
Germany, Berlin water supply, 418
Breslau water supply, 418
Dusseldorf water supply, 419
Frankfurt water supply, 418
groundwater supplies, 428
Hamburg water supply, 418
Kiel Canal, 540
Leipzig water supply, 434
Munich water supply, 446
Rhine water-power plants, 70
Ruhr water supply, 418
salt water studies, 460
springs at Karlsbad, 433
underground dams, 452
water divining, 449
Gibraltar, proposed tunnel, 130
water supply, 423
Glacial deposits, 566
deltas, 568
lakes, 568
Glaciated rock surfaces, 351
Glaciers, continental, 563
mountain (or valley), 563
rock, 556
Gneiss, 19, 21
Granite, 22
discoloration of, 517
Gravel, 577
water supply from, 437
Gravitational geophysical methods, 121
Great Lakes, flow through, 17
tilting of, 222
Greece, Boyati Tunnel, 173
Corinth Canal, 260
drainage of Lake Copais, 140
Mount Athos Canal, 260
unusual waterwheel, 359
Grit, 21
Groundwater, and building foundations, 486
causes fatal accident, 491
conservation, 470
in excavation, 196
and geophysics, 134
historical note, 429
in tunnel work, 167
lowering for construction, 172
movement, 434
quality of, 438
replenishment, 469
and sea water, 459

Groundwater, survey in Boston, 489
surveys, 447
temperature of, 438
Groundwater supply, Long Island, N.Y., 463
New Providence, Bahamas, 466
Southern California, 465
Grouting, asphalt in limestone, 334
at Boulder Dam, 347
cofferdam leaks, 297
of dam foundations, 337
high pressure for dam, 342
historical note, 340
at Le Sautet Dam, 319
limestone with clay, 338
limestone under dam, 314
with loam and cement, 339
prescription for mix, 298
program for dam, 344
in tunnels, 175
Groynes, design of, 380
permeable, 381
Gumbo, 199
Gypsum, in beer, 427
and landslides, 228
for plaster, 525

H

Hanging valleys, 565
Harbors, historical note, 265
Liverpool, England, 403
New York, N.Y., 404
Ostia, silting up, 266
Toronto, Ont., 375
Tyre, silting up, 265
Vizagapatam, 378
Hardness of water, 440
Hardness test for minerals, 20
"Hardpan," 76
Hawaii, adits for water supply, 455
sealing leaky wells, 472
Holland, bridge abutments, 286
lock construction, 261
salt water studies, 460
Hot springs, 433
Hydrogen bisulphide in water, 439
Hypabyssal rocks, 15

I

Ice age, 562
Igneous rocks, 13
Igneous rocks, as building stone, 521
crushing of, 531
water supply from, 438
weathering of, 551
India, earth dams, 506
groundwater irrigation, 428
irrigation practice, 390
Islam Weir failure, 393
Karez for water supply, 455
Kohala Bridge, 287
littoral drift, Madras, 377
Nerbudda Bridge, 291
silting up of leaks, 361
slips on Assam Railway, 235
Vizagapatam Harbor, 378
wire sausages, 399
Inspection and maintenance, 79
of bridge piers, 300
Intakes for water supply, 418
Intrusive rocks, 14
Iran, Keghriz for water supply, 455
Ireland, canal leakage in, 261
Iron in groundwater, 441
Irrigation, area in India, 428
Arizona, 468
assisted by pumping, 468
and silt in rivers, 388
Italy, delta of River Po, 400
fumoroles sufforoli, 433
Leaning Tower of Pisa, 485
water divining, 449

J

Jacking of foundations, 484
Jamaica, cockpit country, 224
Jersey, rock excavation, 191
Jetting for fill consolidation, 257
Jointing, 29
and water supply, 436

K

Kames, 568
Karez, of India, 455
Katavothras, 140
Keghriz, of Iran, 455

L

Lake Temiskaming, 34
Land reclamation, 371

Landslide, Assam Bengal Railway, 235
at Bonneville Dam site, 362
California, 241
Grand Coulee Dam, 232
Gros Ventre, Wyo., 227, 506
Indiana highway, 238
leakage through old, 359
Mount Grenier, 211
on L.N.W. Railway, England, 234
Panama Canal, 264
River Indus, India, 213
Rossberg, 211
Landslide types, 225
Landslides, affect bridge design, 287
due to drought, 229
preventive work, 231
recognition of, 227
studies of, 540
Laterite, 554
Leakage, detection of, 359
from artesian wells, 471
through dams, 352
from reservoirs, 355
Legislation and groundwater, 472
Les Eboulements, Canada, 232
Levees, natural, 558
on Mississippi, 396
Lighthouse, Beachy Head, 268
Eddystone, 528
Lime, 524
Limestone, 21
as building stone, 522
for cement, 526
for concrete, 534
and dam foundations, 314
flour, compact, 296
flow into, for mill, 360
grouted with clay, 338
leakage in canal through, 261
leakage through, at dam, 334
for lime, 524
porosity of, 436
and reservoirs, 356
water supply from, 437
weathering of, 550
"Limestone City," 514
Linings of tunnels, 155
Littoral drift, 376
at Durban, S. Africa, 377
at Madras, India, 377
Littoral drift, silting up of Ostia Harbor, 266
of Tyre Harbor, 265
at Vizagapatam, India, 378
and wharf construction, 379
Location of roads, 248
Lock construction, Holland, 261
Loess, 556

M

Macadam roads, 249
Magdalen Islands, 409
Magma, 14
Magnetic geophysical methods, 117
Malaya, silt buries town, 387
Malta, early tunnels, 140
Manganese in water, 441
Mapping, geological, extent of, 43
Maps, geological, 8, 41
interpretation of, 53
preparation of, 49
of William Smith, 5
topographical, 44
old, of cities, 492
Viele, of New York, 492
Marble, 18
Marine deposits, 559
Marshes, fill across, 256
"Megger" in geophysical work, 127
Metamorphic rocks, 14, 18
Methane, from artesian water, 463
in excavation, 198
in groundwater, 439
in tunnels, 422
Mexico, Mexico City geology, 503
Rodriguez Dam, 331
Microscope and building stone, 517
Mineralogy, 7, 12
Minerals and building stone, 521
in concrete aggregate, 536
in sand, 529
Mining, use of geology in, 9
Model experiments, for erosion below dams, 393
for estuaries, 407
for retaining walls, 209
Models of test borings, 110
Moraines, 566
Mortar for building stone, 522

Mud runs in Bolivia, 259
Museum, British Geological, 516

N

Natural cement, 527
New Zealand, earth movement at Arapuni, 213
  ripples observed on ground, 219
Niagara Falls, 385
Nile, silt deposits of, 370
Northern Ireland, Silent Valley Reservoir, 312, 426, 509

O

Observation wells for groundwater, 490
Obsidian, 23
Outwash plains, 568
Overbreak, 164
  elimination of, 193
  in Lochaber Tunnel, 167

P

Palaeontology, 6, 36
Peat, bogs, 554
  effect on water, 426
  removal from reservoirs, 426
Perched water table, 456
Periscope for drill holes, 346
Permeability tests for rock, 335
Perviousness, 435
Petra, rock caves, 193
Petrography, 7
Petroleum for tracing leakage, 360
Petrology, 7, 11
Photography, value of, 45, 56
Phreatophytes, 448
Piles, for bridge anchorage, 283
  in building foundations, 481
  composite, 489
  deterioration of timber, 487
  gravel, 485
  in hard clay, 209
  steel sheet, walls, 209
Pitting of building stone, 523
Plants, indicating groundwater, 448
  for slide prevention, 237
  for soil control, 411
Plaster of Paris, 526
Plutonic water, 432
Porosity, 435
Porphyries, 23
Portland cement, 526
Potholes, at Assuan Dam, Egypt, 385
  at Boulder Dam, 323
Pre-cementation of foundations, 341
Protection, of banks, 397
  of tracks, 258
Pumice, 23
Pumping, and groundwater depletion, 471
  hot air, to dry clay, 241
  interference in wells, 456
  from Mersey Tunnel, 171
  near St. Paul's, London, 490
  from Pittsburgh mines, 428
  sandstone, from tunnel, 181
  wells, for foundations, 485
Pumps, submersible, 200
Puzzuolana, 528

Q

Quality of water, 425, 438
Quarry, study of building stone in, 519
Quarrying, building stone, 514
  for Haifa Harbor, 266
  practice, 205
  for San Gabriel Dam, 349
Quartzite, 18, 21
Quicklime, 524
Quicksand, 199, 484

R

Rain water, salts in, 459
Railroads, Assam Bengal, landslides on, 235
  Bolivia, mud runs on, 259
  C.N.R., Vaudeuil Bridge, 282
    pier inspection, 301
  C.P.R., Lethbridge Viaduct, 300
    tracks on talus slope, 227
  drainage of cuts, 190
  G.W.R., protection tunnel, 245
  landslide in cut, 229
  L.M.S.R., opening out tunnel, 185
  L.N.W.R., slope of cuts, 234
  maintenance, 258
  N.Y.C.R., tunnel construction, 161

Stabilized soil roads, 251
Staffa, Isle of, 30
Staffordshire blue bricks, 426, 512
Steel sheet piling, for bank protection, 398
  for bridge piers, 289
  and retaining walls, 209
Stoneware, 513
Stratification, 17, 27
Stratigraphy, 6
Strike, 25
Strength of rock, 477
Structural clay tile, 512
Submergible bridges, 291
Subsidence, 223
  due to coal working, 289, 484
  Waterloo bridge piers, 282
    (*See also* Building settlement; Earth movement)
"Swallow holes," 284
Sweden, glaciation of, 17
  Gottenburg water supply, 418
  landslide investigations, 241, 540
Syncline, 30

T

Talus, 556
Tar, for grouting, 176
  macadam, 250
Temperature, of artesian water, 463
  of groundwater, 438
  in Lake Shore Mine, 160
  in Lochaber Tunnel, 160
  in Simplon Tunnel, 160
  underground, 159
T.V.A., geological reconnaisance, 71
  geological staff, 68
Test boring records, Boston, Mass., 496
  New Orleans, La., 497
  New York, N.Y., 496
  Shanghai, China, 499
Test borings, for building foundations, 492
  and building regulations, 500
  costs, 96
  and geology, 93
  with hand tools, 543
  for Hudson River crossing, 150
  importance of, 483
  inclined holes, 150
Test borings, lack of, for bridges, 273
  for San Francisco Bay bridge, 275
  for tidal power scheme, 99
  use of, 111
Test drilling, 102
  core recovery, 102
  large diameter holes, 103
Test pits, 100
Tidal power scheme, Passamaquoddy, Me., 99
Till, 571
Tilting of bridge piers, 283
Timber, deterioration of piles, 487
Track maintenance, 258
Training walls in estuaries, 404
  Fraser River, B.C., 405
  Mersey, England, 403
  Mississippi delta, 405
  New York Harbor, N.Y., 404
Transportation by water, 386
Tunnel construction, 159
  carbon dioxide in, 440
  chemical consolidation in, 201
Tunnel linings, 154
  troubles with Simplon, 155
Tunnels, Boyati, Greece, 173
  Breakneck, N.Y., 161
  Bridge River, B.C., 86, 128
  Chicago, Ill., in clay, 181
  Cofton, opening out of, 185
  Cripple Creek, in open, 141
  Egyptian, 140
  English channel, proposed, 267
  Euphrates, ancient, 140
  exploratory, 319
  First Narrows, Vancouver, 87
  Halkyn drainage, Wales, 468
  Hetch-Hetchy, Cal., 422
  Kilsby, England, 172
  Lake Fucino, drainage, 140
  Las Raices, Chile, 144
  Lochaber, Scotland, 179
  London tube railway, 62, 142, 202
  Lötschberg, 144
  Maltese, ancient, 140
  Mersey, England, 169
  Moscow subway, U.S.S.R., 202
  Moffatt, seepage into, 169
  New York, Ward's Island, 151
    Catskill, 147
    City, Nos. 1 and 2, 151

Stabilized soil roads, 251
Staffa, Isle of, 30
Staffordshire blue bricks, 426, 512
Steel sheet piling, for bank protection, 398
  for bridge piers, 289
  and retaining walls, 209
Stoneware, 513
Stratification, 17, 27
Stratigraphy, 6
Strike, 25
Strength of rock, 477
Structural clay tile, 512
Submergible bridges, 291
Subsidence, 223
  due to coal working, 289, 484
  Waterloo bridge piers, 282
    (*See also* Building settlement; Earth movement)
"Swallow holes," 284
Sweden, glaciation of, 17
  Gottenburg water supply, 418
  landslide investigations, 241, 540
Syncline, 30

T

Talus, 556
Tar, for grouting, 176
  macadam, 250
Temperature, of artesian water, 463
  of groundwater, 438
  in Lake Shore Mine, 160
  in Lochaber Tunnel, 160
  in Simplon Tunnel, 160
  underground, 159
T.V.A., geological reconnaisance, 71
  geological staff, 68
Test boring records, Boston, Mass., 496
  New Orleans, La., 497
  New York, N.Y., 496
  Shanghai, China, 499
Test borings, for building foundations, 492
  and building regulations, 500
  costs, 96
  and geology, 93
  with hand tools, 543
  for Hudson River crossing, 150
  importance of, 483
  inclined holes, 150
Test borings, lack of, for bridges, 273
  for San Francisco Bay bridge, 275
  for tidal power scheme, 99
  use of, 111
Test drilling, 102
  core recovery, 102
  large diameter holes, 103
Test pits, 100
Tidal power scheme, Passamaquoddy, Me., 99
Till, 571
Tilting of bridge piers, 283
Timber, deterioration of piles, 487
Track maintenance, 258
Training walls in estuaries, 404
  Fraser River, B.C., 405
  Mersey, England, 403
  Mississippi delta, 405
  New York Harbor, N.Y., 404
Transportation by water, 386
Tunnel construction, 159
  carbon dioxide in, 440
  chemical consolidation in, 201
Tunnel linings, 154
  troubles with Simplon, 155
Tunnels, Boyati, Greece, 173
  Breakneck, N.Y., 161
  Bridge River, B.C., 86, 128
  Chicago, Ill., in clay, 181
  Cofton, opening out of, 185
  Cripple Creek, in open, 141
  Egyptian, 140
  English channel, proposed, 267
  Euphrates, ancient, 140
  exploratory, 319
  First Narrows, Vancouver, 87
  Halkyn drainage, Wales, 468
  Hetch-Hetchy, Cal., 422
  Kilsby, England, 172
  Lake Fucino, drainage, 140
  Las Raices, Chile, 144
  Lochaber, Scotland, 179
  London tube railway, 62, 142, 202
  Lötschberg, 144
  Maltese, ancient, 140
  Mersey, England, 169
  Moscow subway, U.S.S.R., 202
  Moffatt, seepage into, 169
  New York, Ward's Island, 151
    Catskill, 147
    City, Nos. 1 and 2, 151

Sand movement, in arid regions, 411
at Durban, South Africa, 377
at Madras, India, 377
in River Garonne, France, 404
at Rockaway Point, N.Y., 376
Sand-lime bricks, 509
Sandstone, 21
as building stone, 522
porosity of, 435
water supply from, 437
"Sandwich" water bearing strata, 442
Scale of earthquake intensities, 220
Schist, 18, 22
Scour, around bridge piers, 290, 302
below spillways, 351, 395
Sealing leaky wells, 472
Sedimentary rocks, 14, 16
Sediments, 557
at Valparaiso, Chile, 266
Seepage through drift, 433
Seismic areas, 219
Seismic geophysical methods, 118
Seismographs, 191
Seismology, 7
Serpentine, 21
and landslides, 228
Settlement, of bridge abutments, 289
of bridge piers, 282
of buildings, 481
of viaduct pier, 300
of water tanks, 484
Sewer construction, 77, 199
Shaft excavation, by freezing, 200
Shale, 20
as aggregate, 535
excavated with saws, 192
protection of, 163, 192
water supply from, 437
Shore protection, 372
at Sheridan Park, Wis., 381
Shoreline types, 373
Toronto Island, 375
Shot drilling, 102
Sills, 15
Silt, and Boulder Dam, 388
burying town, 387
in Colorado River, 387
effect on dams, 391
measurement of, 387
in Mississippi River, 387
nature of, 387
Silt, removal for irrigation, 388
in reservoirs, 390
in River Nile, Egypt, 387
in San Juan River, 386
Silting up of leaks, 361
"Sinkholes," 224
near Norwich, England, 224
Skin friction, 279
Slate, 21
Slopes, of excavation, 195
of rock surfaces, 478
Soil action under foundations, 480
Soil, aeolian, 555
analysis, 543
erosion, 409
residual, 553
sampling, 541
stabilized roads, 251
terminology, 576
testing, 545
transported, 554
types, 577
use of word, 78, 538
Soil auger, 97
Soil Mechanics, and building foundations, 475
and building settlement, 483
conference on, 541
and dams on permeable foundations, 348
and dredging, 204
and earth dams, 507
and embankments, 204
future of, 581
and geology, 545
and slope stability, 241
and stabilized roads, 252
Soundness of rock, 326
Spain, Puentes Dam, 305
Specific yield, 451
Specifications, 74
Spreading groundwater, 469
Springs, 445
at Bath, England, 433
dam foundations, 351
intermittent, 446
at Karlsbad, Germany, 433
at Kivu, Africa, 433
at Munich, Germany, 446
in mine, Australia, 433
in Persian Gulf, 446

Reclamation of land, 371
Recording exploratory work, 107
Regulations, building, 499
Relocation of routes, 362
Reservoirs, abandoned, 354
  Boston, Mass., 426
  Cataract, Australia, 364
  Cedar River, Wash., 358
  Ewden Valley, England, 355
  Farmer's Union, Colorado, 226
  Habra, Algeria, 390
  Hondo, N.M., 354
  Jerome, Idaho, 354
  La Grange, Cal., 390
  Lake Macmillan, 390
  leakage from, 355
  Malad, Idaho, 356
  Much Kundi, India, 361
  New Lake Austin, Texas, 390
  Silent Valley, Northern Ireland, 426
  Yorkshire, England, 361
Retaining walls, and earth pressure, 207
  at Fifteen Mile Falls, 208
  movement of, Los Angeles, 479
  and soil study, 539
Rhabdomancy, 449
"Rhodesian Man," skull of, 36
Rhyolite, 23
Riprap of burned clay, 512
Rivers, Amazon, South America, 383
  Avon, England, 383
  Bulkely, B.C., 383
  Colorado, 383
  Derwent, England, 426
  Fraser, B.C., 405
  Ganges, India, 383
  Hwang Ho, China, 399
  Mersey, England, 403
  Mississippi, 396
  Missouri, 396
  Niagara, 385
  Severn, England, 400
River-training work, 260
  on Mississippi, 396
Road, baked clay, Australia, 254
  location, 248
  slab raised by well points, 255
  stabilized soil, 251
  surface material, 249
Roches mountonnées, 565
Rock, alteration of, 547
  crushing of, 531
  electrical resistance of, 135
  flour, 570
  glaciers, 556
  pressure, 462
  talus, 556
  use of word, 12, 78
  weathering, 546
Rock excavation, 76, 188
  with coal cutting machines, 191
  in East River, N.Y., 190
  of Flood Rock, N.Y., 190
  at Fort Peck Dam, 191
  in quarries, 205
  at St. Helier, Jersey, 191
  at Sunderland, England, 191
  at Wheeler Dam, 192
  with wire saws, 320
Rock falls, at Frank, Alta., 242
  protecting tracks from, 258
  at Vriog Cliffs, Wales, 245
Rock-fill dams, in Chile, 328
  San Gabriel, Cal., 348
Roman cement, 528
  concrete, 532
  road building, 249
  shore works, 371
  tunnels, 140
  use of wells, 430
  water supply, 415
Route location, 248
Runoff, 365
  abnormal, 368

S

Salinity of water, limits, 458
Salt, in rain water, 459
  in wells, 458
  water balance, 459
Sampling, practice, 101
  of soil, 541
  undisturbed, 540, 542
    at San Francisco Bridge, 99, 275
Sand, 528, 577
  water supply from, 437
  wind deposited, 555
"Sand Island" construction method, 293

Mud runs in Bolivia, 259
Museum, British Geological, 516

N

Natural cement, 527
New Zealand, earth movement at Arapuni, 213
  ripples observed on ground, 219
Niagara Falls, 385
Nile, silt deposits of, 370
Northern Ireland, Silent Valley Reservoir, 312, 426, 509

O

Observation wells for groundwater, 490
Obsidian, 23
Outwash plains, 568
Overbreak, 164
  elimination of, 193
  in Lochaber Tunnel, 167

P

Palaeontology, 6, 36
Peat, bogs, 554
  effect on water, 426
  removal from reservoirs, 426
Perched water table, 456
Periscope for drill holes, 346
Permeability tests for rock, 335
Perviousness, 435
Petra, rock caves, 193
Petrography, 7
Petroleum for tracing leakage, 360
Petrology, 7, 11
Photography, value of, 45, 56
Phreatophytes, 448
Piles, for bridge anchorage, 283
  in building foundations, 481
  composite, 489
  deterioration of timber, 487
  gravel, 485
  in hard clay, 209
  steel sheet, walls, 209
Pitting of building stone, 523
Plants, indicating groundwater, 448
  for slide prevention, 237
  for soil control, 411
Plaster of Paris, 526
Plutonic water, 432
Porosity, 435
Porphyries, 23
Portland cement, 526
Potholes, at Assuan Dam, Egypt, 385
  at Boulder Dam, 323
Pre-cementation of foundations, 341
Protection, of banks, 397
  of tracks, 258
Pumice, 23
Pumping, and groundwater depletion, 471
  hot air, to dry clay, 241
  interference in wells, 456
  from Mersey Tunnel, 171
  near St. Paul's, London, 490
  from Pittsburgh mines, 428
  sandstone, from tunnel, 181
  wells, for foundations, 485
Pumps, submersible, 200
Puzzuolana, 528

Q

Quality of water, 425, 438
Quarry, study of building stone in, 519
Quarrying, building stone, 514
  for Haifa Harbor, 266
  practice, 205
  for San Gabriel Dam, 349
Quartzite, 18, 21
Quicklime, 524
Quicksand, 199, 484

R

Rain water, salts in, 459
Railroads, Assam Bengal, landslides on, 235
  Bolivia, mud runs on, 259
  C.N.R., Vaudeuil Bridge, 282
    pier inspection, 301
  C.P.R., Lethbridge Viaduct, 300
    tracks on talus slope, 227
  drainage of cuts, 190
  G.W.R., protection tunnel, 245
  landslide in cut, 229
  L.M.S.R., opening out tunnel, 185
  L.N.W.R., slope of cuts, 234
  maintenance, 258
  N.Y.C.R., tunnel construction, 161

Tunnels, New York, Delaware, 422
East River Gas, 145
Hudson River, 145
subway, 62, 141
Pietermaritzburg Railway, 433
Ponlipio Hills, Italy, 140
pressure, 156
failure of, Australia, 158
River Scheldt, Belgium, 200
St. Paul, Minn., 181
Severn, England, clay sealing of, 169
grouting of, 176
old record, 79
Simplon, construction troubles, 144
hot water in, 433
lining troubles, 155
plutonic water in, 168
rock cover, 143
Soda Hill, Cal., 440
Winnipeg, Man., in clay, 181

U

Unconformity, 34
Underground dams, 452
Underpinning building, 481
U.S.S.R., geophysical surveys, 135
Moscow subway, 202
Uplift beneath dams, 335

V

Valley trains, 567
Venezuela, ground subsidence, 223
Ventilation, for slide prevention, 241
Vinegar, use of in tunnels, 140, 181
Volcanic deposits, New Zealand, 214
Volcanic eruption, Mount Katmai, 27
Vulcanology, 7

W

Wash borings, 98
Water, for beer, 426
connate, 433
divining, 448
fossil, 433
hardness of, 426, 440
internal, 432
plutonic, 432
quality of, 425, 438
underground movement, 434
Water, vadose, 431
witching, 448
Water power schemes, Arapuni, New Zealand, 213
Kinlochleven, Scotland, 143
Lochaber, Scotland, power house location, 90
tunnel selection, 143
tunnel construction, 179
Rhine, Germany, 70
Water supplies, from dew ponds, 424
historical note, 415
from impounding reservoirs, 421
by infiltration, 418
from iron-clad catchments, 425
from rock catchments, 425
Water supply, Accrington, England, 467
Berlin, Germany, 418
Birmingham, England, 421
Breslau, Germany, 418
Colorado River Aqueduct, 422
Des Moines, Iowa, 419
Dusseldorf, Germany, 419
Ewden Valley, England, 421
Frankfurt, Germany, 418
Germany, statistics, 428
Gibraltar, 423
Gottenburg, Sweden, 418
Hamburg, Germany, 418
Harrisburg, Va., 454
Kano, Northern Nigeria, 419
Liverpool, England, 421
Leipzig, Germany, 434
London, England, 443
statistics, 428
Long Island, N.Y., 463
Los Angeles, Cal., 421
Manchester, England, 421
Munich, Germany, 446
Nassau, Bahamas, 466
New York, Catskill, 147, 421
Delaware, 422
Richmond, England, 444
Riverside, Cal., 439
Ruhr, Germany, 418
San Francisco, Cal., 422
Scarborough, England, 431
Southern California, 465
Texas, from dunes, 412
Toronto, Ont., 418
United States, statistics, 428

Water table, 433
perched, 456
Water-tank foundations, 484
Wave action, 371
Weathering of rocks, 546
extent of, 324
Well points, assisting cofferdams, 299
for raising road slab, 256
for water supply, 418
for waterlogged ground, 196
Wells, artesian, 460
contamination of, 458
drilled too deep, 442
dumb, 469
Wells, Grenelle, France, 434
interference at Liverpool, 456
for London water supply, 428
observation, for groundwater, 490
St. Andrews, Scotland, 455
soak, 456
subartesian, 460
tray, 456
Wire sausage protection, 399
Wire saws for rock excavation, 320

Z

Zeolites, 441